Building Next-Generation Converged Networks

Theory and Practice

OTHER TELECOMMUNICATIONS BOOKS FROM AUERBACH

AUERBACH PUBLICATIONS
www.auerbach-publications.com
To Order Call: 1-800-272-7737 • Fax: 1-800-374-3401
E-mail: orders@crcpress.com

Building Next-Generation Converged Networks

Theory and Practice

Edited by
Al-Sakib Khan Pathan
Muhammad Mostafa Monowar
Zubair Md. Fadlullah

CRC Press
Taylor & Francis Group
Boca Raton London New York

CRC Press is an imprint of the
Taylor & Francis Group, an **Informa** business

CRC Press
Taylor & Francis Group
6000 Broken Sound Parkway NW, Suite 300
Boca Raton, FL 33487-2742

© 2013 by Taylor & Francis Group, LLC
CRC Press is an imprint of Taylor & Francis Group, an Informa business

No claim to original U.S. Government works

Printed on acid-free paper
Version Date: 20121203

International Standard Book Number: 978-1-4665-0761-6 (Hardback)

Visit the Taylor & Francis Web site at
http://www.taylorandfrancis.com

and the CRC Press Web site at
http://www.crcpress.com

All the seekers of knowledge and truth alongside my family.

—**Al-Sakib Khan Pathan**

My parents to whom I owe a lifetime.

—**Muhammad Mostafa Monowar**

My loving family. Their consistent support is an endless bounty
from the Almighty, to Whom I am ever grateful.

—**Zubair Md. Fadlullah**

Contents

PART III   NETWORK MANAGEMENT AND TRAFFIC ENGINEERING

PART IV   INFORMATION INFRASTRUCTURE AND CLOUD COMPUTING

PART V   WIRELESS NETWORKING

Preface

The telecommunications industry has seen a rapid boost within the last decade. New realities and visions of functionalities in various telecommunications networks have brought forward the concept of next-generation networks (NGNs). The competitions among operators for supporting various services, lowering of the cost of having mobile and cellular phones and smartphones, increasing demand for general mobility, explosion of digital traffic, and advent of convergence network technologies added more dynamism in the idea of NGNs. In fact, facilitating convergence of networks and convergence of various types of services is a significant objective of NGNs.

Although there is a considerable amount of research efforts underway to define the boundary and standards of NGNs, a proper boundary is yet to be finalized. NGNs are used to label the architectural evolutions in telecommunications and access networks. The term is also used to depict the shift to higher network speeds using broadband, the migration from the public switched telephone network (PSTN) to an Internet Protocol (IP)–based network, and a greater integration of services on a single network, and often is representative of a vision and a market concept. The NGN is also defined as a "broadband managed IP network." The IP address is sometimes used as the NGN is built around the IP. From a more technical point of view, the NGN is defined by the International Telecommunication Union (ITU) as a "packet based network able to provide services including telecommunication services and able to make use of multiple broadband, QoS-enabled transport technologies and in which service related functions are independent from underlying transport-related technologies." NGNs offer access by users to different service providers and support "generalized mobility which will allow consistent and ubiquitous provision of services to users" (ITU-T Recommendation Y.2001, approved in December 2004).

The Objective of This Book and Its Structure

The main goal of this book is to compile various works that contribute to the development of next-generation networks and technologies. We understand that the future is still looking hazy as blending of different technologies takes the definition of NGNs toward different directions. However, considering the gravity of the dynamism involved in this technology theme, we have divided the entire book into five major parts. The first part deals with the works on multimedia streaming in the networks in the future. The chapters include some basic information for general readers as well as in-depth information for the experts in the relevant areas. As we are now moving toward 4G and 5G or "So-and-So"G networks, multimedia streaming will play a very distinctive role in the future network settings. Not only high speed of multimedia traffic is needed but also high definition and resolution will be expected by the users. Hence, the first part addresses the critical aspects associated with these issues. In the second part, we have placed some chapters dealing with safety and security issues in networking. Also, basic Internet and cyber-security are considered that will also be relevant in any future network. In Part III, network management and traffic engineering issues are touched upon. This part may require some expertise or background knowledge as some mathematical modeling–based works are included.

In Part IV, we integrate the concept of cloud computing with general information infrastructure. It is expected that in the NGN, the information flow and pattern of information exchange will be different than those being currently employed. Hence, this part can give the readers some knowledge about the past achievements, present conditions, and future expectations in the information infrastructure-related areas. Finally, Part V contains some chapters dealing with various aspects of wireless networking. As many networks have now got wireless versions instead of the fixed wired connections, wireless networking will be an integral part of NGNs. Hence, this part could give the readers some flavor of wireless networking technologies without going into too much depth but keeping it relevant to NGN technologies.

What to Expect from the Book?

The book is mainly written for graduate researchers, students, regular industry researchers, university academics, and general networking readers. There is a combination of "easy-to-follow" chapters as well as chapters requiring some prior knowledge or expertise. Hence, the book can be a good reference item for the MS or PhD level students for gaining basic and in-depth knowledge on various issues of NGN development.

What Not to Expect from the Book?

The book is not written in a textbook style. Hence, the presented information is based often on the latest and most up-to-date research findings. It could be used for postgraduate level classroom teaching, but as the research fields demand, something *latest* today may not remain *latest* tomorrow. So, the basic standardized information presented in the book can be used with certainty, but the research findings or results may have some uncertainty factor involved with them.

MATLAB® is a registered trademark of The MathWorks, Inc. For product information, please contact:

The MathWorks, Inc.
3 Apple Hill Drive
Natick, MA 01760-2098 USA
Tel: 508 647 7000
Fax: 508-647-7001
E-mail: info@mathworks.com
Web: www.mathworks.com

Acknowledgments

First of all, as always, we are very thankful to the Almighty for giving us courage, strength, time, and physical fitness to complete this work. We would like to heartily thank all the authors who contributed to this book for its successful completion. A total of 56 authors from 16 different countries around the globe contributed to this book, without whose active support and brainstorming, the work would not have taken this current shape. The authors' cooperation in various cases, their timely responses, and adhering to the given guidelines for manuscript preparation are really praiseworthy. We hope that our work will be beneficial not only for the contributing authors for their careers but also for the wide variety audience related to the research topics and issues addressed in this book.

Dr. Al-Sakib Khan Pathan
International Islamic University Malaysia, Malaysia

Dr. Muhammad Mostafa Monowar
King Abdulaziz University, Saudi Arabia

Dr. Zubair Md. Fadlullah
Tohoku University, Japan

Editors

Al-Sakib Khan Pathan received a PhD degree in computer engineering from Kyung Hee University, South Korea, in 2009 and a BSc degree in computer science and information technology from the Islamic University of Technology (IUT), Bangladesh, in 2003. He is currently an assistant professor at the Computer Science Department in the International Islamic University Malaysia (IIUM), Malaysia. Until June 2010, he served as an assistant professor at the Computer Science and Engineering Department in BRAC University, Bangladesh. Prior to holding this position, he worked as a researcher at Networking Lab, Kyung Hee University, South Korea, until August 2009. His research interests include wireless sensor networks, network security, and e-services technologies. He is a recipient of several awards/best paper awards and has several publications in these areas. He has served as a chair, organizing committee member, and technical program committee member in numerous international conferences/workshops like HPCS, ICA3PP, IWCMC, VTC, HPCC, IDCS, etc. He is currently serving as the editor-in-chief of the *International Journal on Internet and Distributed Computing Systems* (IJIDCS), an area editor of *International Journal of Communication Networks and Information Security* (IJCNIS), editor of *International Journal of Computational Science and Engineering* (IJCSE), Inderscience, associate editor of IASTED/ACTA Press *International Journal of Computers and Applications* (IJCA) and Communications and Computer Security (CCS), guest editor of some special issues of top-ranked journals, and editor/author of six books. He also serves as a referee of some renowned journals. He is a member of the Institute of Electrical and Electronics Engineers (IEEE), USA; IEEE ComSoc Bangladesh Chapter; and several other international organizations. His email address is sakib.pathan@gmail.com.

Muhammad Mostafa Monowar is currently working as an assistant professor at the Department of Information Technology in King Abdulaziz University, Kingdom of Saudi Arabia. He is also an associate professor (on leave) at the Department of Computer Science and Engineering in the University of Chittagong, Bangladesh. He received a PhD degree in computer engineering from Kyung Hee University, South Korea, in 2011 and a BSc degree in computer science and information technology from the Islamic University of Technology (IUT), Bangladesh, in 2003. His research interests include wireless networks, especially ad hoc, sensor, and mesh networks, including routing protocols, MAC mechanisms, IP and transport layer issues, cross-layer design, and QoS provisioning. He has served as a program committee member in several international conferences/workshops like IADIS, DNC, IDCS, etc. He is currently serving as an associate editor of the *International Journal on Internet and Distributed Computing Systems* (IJIDCS) and a guest editor of some special issues of IJCSE. His email address is hemal.cu@gmail.com.

Zubair Md. Fadlullah is an assistant professor at the Graduate School of Information Sciences (GSIS), Tohoku University, Japan. He also served as a computer science faculty member at the prestigious international Islamic University of Technology (IUT) in Bangladesh. He is a member of the IEEE and ComSoc. He is also a member of the Japanese team involved with the prestigious A3 Foresight Project supported by Japan Society for the Promotion of Science (JSPS), NSFC of China, and NRF of Korea, which comprises prominent researchers in the field of networking and communications from the mentioned countries. Dr. Fadlullah holds a PhD in applied information sciences, which he obtained in March 2011 from Tohoku University. He has a noteworthy contribution toward research community through his technical papers in scholarly journals, magazines, and international conferences in various areas of networking and communications. Dr. Fadlullah has been serving as a technical committee member for several IEEE GC, ICC, PIMRC, WCNC, and WCSP conferences for a number of years. He is an associate editor of the *International Journal on Internet and Distributed Computing Systems* (IJIDCS) and a co-editor of the Special Issue (SI) on Wireless Networks Intrusion in *Journal of Computer and System Sciences* (Elsevier). He was a co-chair of the invited session on Smart Grid in WCSP'11. Furthermore, he has also been actively engaged in helping editorial members of prestigious IEEE transactions (including TVT, TPDS, TSG) to manage and delegate reviews in an efficient manner. His research interests are in the areas of smart grid, network security, intrusion detection, game theory, and quality of security service provisioning mechanisms. Dr. Fadlullah was a recipient of the prestigious Deans and Presidents awards from Tohoku University in March 2011 for his outstanding research contributions. His email address is zubair@it.ecei.tohoku.ac.jp.

Contributors

Abdelgadir Tageldin Abdelgadir
International Islamic University Malaysia
Kuala Lumpur, Malaysia

Md. Abdul Hamid
Hankuk University of Foreign Studies
Yongin-si, South Korea

M. Abdullah-Al-Wadud
Hankuk University of Foreign Studies
Yongin-si, South Korea

José M. Alcaraz Calero
Departamento de Informática,
    Universidad de Valencia
Valencia, Spain

Georgios Baltoglou
KTH
Stockholm, Sweden

Jorge Bernal Bernabé
University of Murcia
Murcia, Spain

Bhed Bahadur Bista
Iwate Prefectural University
Takizawa, Japan

Luca Caviglione
National Research Council of Italy (CNR)
Genova, Italy

Cristiano Cervellera
National Research Council of Italy (CNR)
Genova, Italy

Vigyan "Vigs" Chandra
Eastern Kentucky University
Richmond, Kentucky

Dimitris E. Charilas
National Technical University of Athens
Athens, Greece

Periklis Chatzimisios
CSSN Research Lab, Alexander TEI of
    Thessaloniki
Thessaloniki, Greece

Qiang Duan
The Pennsylvania State University Abington
    College
Abington, Pennsylvania

Zubair Md. Fadlullah
Tohoku University
Sendai, Japan

Zahid Farid
Universiti Sains Malaysia
Pulau Penang, Malaysia

Andreas P. Fatouros
National Technical University of Athens
Athens, Greece

Mostafa M. Fouda
Tohoku University
Sendai, Japan

and

Benha University
Giza, Egypt

Ioannis C. Fousekis
National Technical University of Athens
Athens, Greece

Félix J. García Clemente
University of Murcia
Murcia, Spain

M.S. Gaur
Malaviya National Institute of Technology
Jaipur, India

Soumya K. Ghosh
Indian Institute of Technology
Kharagpur, India

Antonio F. Gómez Skarmeta
University of Murcia
Murcia, Spain

Sumit Goswami
Indian Institute of Technology
Kharagpur, India

Mohammad Ghulam Rahman
Universiti Sains Malaysia
Pulau Penang, Malaysia

R. Hernandez-Aquino
ITESM-Monterrey
Orizaba, Mexico

Changcheng Huang
Carleton University
Ottawa, Ontario, Canada

Jesús D. Jiménez Re
University of Murcia
Murcia, Spain

Eirini Karapistoli
CONTA Lab, UoM
Thessaloniki, Greece

Nei Kato
Tohoku University
Sendai, Japan

Diallo Abdoulaye Kindy
CustomWare
Kuala Lumpur, Malaysia

Chhagan Lal
Malaviya National Institute of Technology
Jaipur, India

V. Laxmi
Malaviya National Institute of Technology
Jaipur, India

Soumya Maity
Indian Institute of Technology
Kharagpur, India

Roberto Marcialis
National Research Council of Italy (CNR)
Genova, Italy

Juan M. Marín Pérez
University of Murcia
Murcia, Spain

Gregorio Martínez Pérez
University of Murcia
Murcia, Spain

Qurban A. Memon
UAE University
Al Ain, United Arab Emirates

Sudip Misra
Indian Institute of Technology
Kharagpur, India

Sepideh Nikmanzar
Sahand University of Technology Tabriz
Tabriz, Iran

Athanasios D. Panagopoulos
National Technical University of Athens
Athens, Greece

Al-Sakib Khan Pathan
International Islamic University Malaysia
Kuala Lumpur, Malaysia

Akbar Ghaffarpour Rahbar
Sahand University of Technology Tabriz
Tabriz, Iran

Julio Ramírez-Pacheco
University of Caribe
Cancún, Mexico

Danda B. Rawat
Eastern Kentucky University
Richmond, Kentucky

Maria Salama
British University in Egypt
Cairo, Egypt

Kashif Saleem
King Saud University
Riyadh, Saudi Arabia

Ahmed Shawish
Ain Shams University
Cairo, Egypt

Anand Srinivasan
EION Inc.
Ottawa, Ontario, Canada

Chaynika Taneja
Directorate of Management Info. Systems and
    Tech., DRDO Hqrs
New Delhi, India

Sabu M. Thampi
Indian Institute of Information Technology
    and Management
Kerala, India

Homero Toral-Cruz
University of Quintana Roo
Chetumal, Mexico

Deni Torres-Román
CINVESTAV-IPN Unidad Guadalajara
Zapopán, Mexico

Cesar Vargas-Rosales
ITESM-Monterrey
Monterrey, Mexico

Pablo Velarde-Alvarado
Autonomous University of Nayarit
Tepic, Nayarit, Mexico

Gongjun Yan
Indiana University
Kokomo, Indiana

Mohammad Zulhasnine
Carleton University
Ottawa, Ontario, Canada

MULTIMEDIA STREAMING I

Chapter 1

Request-Based Multicasting in Video-on-Demand Systems

Sepideh Nikmanzar and Akbar Ghaffarpour Rahbar

Contents

1.1 Introduction

Video-on-demand (VoD) is one of the fastest growing media services nowadays. VoD allows end users to access a library of on-demand movie titles stored in a video server, download them, and watch requested video contents using their set-top boxes (STBs) [1]. In addition to having immediate access to a video title, VoD provides videocassette recorder (VCR) operations, including pause, rewind, fast-forward, jump-forward, and jump-backward operations on video streams [2].

To provide commercial successes on VoD, not only does video traffic need to be transmitted over high-speed networks but also this transmission must be cheap for end users. The key issues of VoD delivery schemes are bandwidth requirements, storage requirements, and startup latency for users who want to watch a video. To tackle this challenge, researchers have investigated several video delivery techniques to improve the performance of VoD systems [3].

3

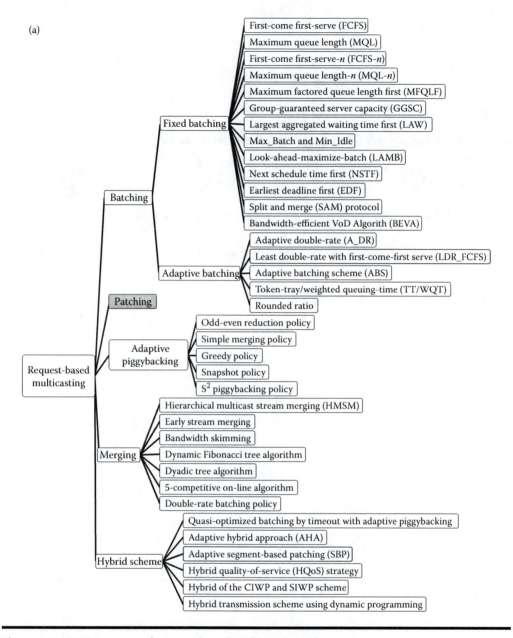

(a)

Figure 1.1   (a) Taxonomy of request-based multicasting schemes.

Video delivery techniques can be classified into the following: (1) server-initiation broadcasting or "server-push" or "reactive" [4–7], (2) caching-based delivery schemes [8–12], and (3) request-based multicasting or "client-pull" or "proactive" [4,7]. Server-initiation broadcasting techniques use a fixed number of channels to periodically broadcast video objects to a group of users. In caching-based delivery schemes, a regional cache server (proxy) delivers initial portion of a video to a client, whereas a central video server needs to transmit only the remaining portions of a video. A request-based multicasting scheme delivers video objects upon receiving a client request.

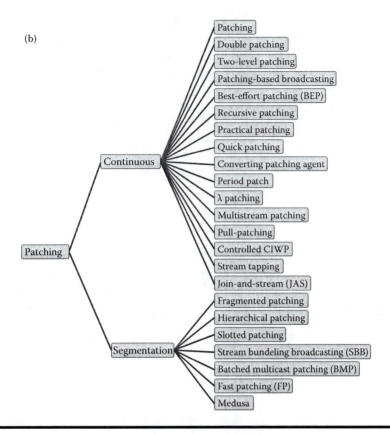

Figure 1.1 **(Continued) (b) Taxonomy of patching schemes.**

Therefore, it reduces the bandwidth requirements of VoD services, compared with the server-initiation broadcasting schemes. However, the access times experienced by users may be high. Various schemes proposed for request-based category try to improve the efficiency and scalability of VoD systems. In this chapter, our objective is to provide a comprehensive review of different approaches proposed for request-based video delivery as [13] batching, patching, piggybacking, stream merging, and hybrid techniques.

In batching schemes [14], requests for the same video within a short time duration are grouped together and served by a single multicast stream. The patching or dynamic multicast [15] approaches exploit client-side buffer space to share multicast transmissions, instead of sending a unique stream to each user. Piggybacking approaches merge users on different channels by increasing the playback rate of later streams and decreasing the playback rate of early streams. The fourth approach, which is called stream merging [16], enables a client to receive a stream and simultaneously buffer another portion of the data from a different stream, thus enabling clients to share future streams with another client. The mentioned approaches can also be combined to form hybrid techniques to improve VoD efficiency.

A complete survey on multicast VoD services has been provided in [7]. This study performed a limited review on client-initiated multicast schemes until 2002, where its focus was on issues and problems associated with multicast VoD. Major scheduling policies for VoD streams have been reviewed in [17], which presents an overview on a few scheduling policies based on broadcasting, batching, caching, piggybacking, and look-ahead scheduling. Finally, a survey on user behavior in VoD systems and bandwidth-saving streaming schemes has been presented in [18]. This study also provides a limited overview on multicast streaming approaches, such as batching, patching, and stream merging. Due to its importance, we provide a comprehensive study on request-based video delivery multicasting schemes in this chapter. Figure 1.1a and b shows our classification on request-based multicasting schemes to be detailed in the following sections.

The remainder of this chapter is organized as follows: In Section 1.2, we describe various batching policies. In Section 1.3, we review different types of proposed patching approaches. Sections 1.4 and 1.5 present the common features and different types of piggybacking and stream merging, respectively. In Section 1.6, we review some hybrid techniques. Finally, Section 1.7 summarizes this chapter and provides future directions for research in VoD delivering.

1.2 Batching Solutions

In batching solutions, requests received by multiple users for the same video within a certain amount of time duration can be batched together and serviced by using a single multicast stream. The batching method was originally proposed by Anderson [19] in 1993. This multicasting technique is referred to as *scheduled multicast* because the server selects the next batch for multicasting according to some dynamic scheduling policy [20]. Batching of video requests reduces input/output (I/O) demands and improves throughput. They are different in the criteria used to select the next batch to service. We explore various scheduling policies in the following.

1.2.1 Fixed Batching

In the following batching schemes, the batching time/interval is fixed. The selection of batching time is performed before running the algorithm. The batch time is often set according to the users' waiting tolerance.

1) *First-come first-serve (FCFS)* [14]. The requests for all video objects join a single queue called *request queue*. When the server capacity for delivering a stream becomes free, FCFS selects the request with the longest waiting time from the request queue. The same multicast stream is assigned to all requests that ask for the same video. The FCFS is a fair policy since it treats every video independent of its popularity [14].

2) *Maximum queue length (MQL)* [14]. Under the MQL policy, each video object has a dedicated queue. A request for each video joins its corresponding queue. The MQL selects the batch with the longest queue length for multicasting. The MQL is unfair for cold movies since it prefers scheduling of popular videos [14].

3) *FCFS-n* [14]. FCFS-*n* reserves a fraction of the server channels for the *n* hottest movies. A new stream for delivering hot videos is initiated at regular intervals called *batching interval* [14]. The remaining cold movies are scheduled by the FCFS policy. The FCFS-*n* is slightly unfair for unpopular videos, but it guarantees maximum waiting time for popular videos. The reserved stream can be used to multicast cold movies if there is no request for hot movies [14].

4) *MQL-n* [14]. Reservation of a fraction of the server channels is performed like the FCFS-*n* policy, but the remaining cold movies are served according to the MQL policy. Since the MQL policy prefers popular videos by default, the MQL-*n* policy will not be interesting as well [14].

5) *Maximum factored queue length first (MFQLF)* [21]. The MFQLF policy considers both the waiting time of requests and the popularity of videos by scheduling the video with the maximum factored queue (MFQ) length [21]. The factored queue length is derived from an analytical model that optimizes the average waiting time of a user if there are no defections [21], where defection means that a user leaves the system without being serviced because its waiting time has exceeded from the maximum user waiting time tolerance. The factored queue length for video i is computed by $\sqrt{q_i \times \Delta t_i}$, where Δt_i is the interval since the last time that video i was scheduled and q_i represents the length of the queue corresponding to video i [21].

6) *Group-guaranteed server capacity (GGSC)* [22]. The GGSC family of scheduling algorithms preallocate the server channels to groups of objects. Video objects are grouped together according to their playback lengths. The GGSC algorithm schedules the requests within a group by the FCFS policy, which is referred to the GGSC$_w$-FCFS algorithm. Let C_m be the number of channels allocated to object m. An optimal value for C_m to minimize the average client waiting time is given by [22]

$$C_m = \frac{\sqrt{f_m L_m}}{\sum\limits_{i=1}^{M} \sqrt{f_i L_i}} \cdot C, m = 1,\dots,M$$

where C is the total number of available channels, L_m is the playback duration of object m, and f_m is the fraction of customers that request object m from all M objects stored in the video server. After allocating channel capacity to a group of objects, the channels become available at every cycle. The length of the cycle is equal to the playback length of objects in the group. A complete comparison between GGSC$_w$-FCFS against MFQ, FCFS, and FCFS-n policies shows that GGSC$_w$-FCFS is the most favorable among other batching policies [22].

7) *Largest aggregated waiting time first (LAW)* [20]: LAW schedules multicast streams by considering the arrival time of the awaiting requests [20]. In this policy, the *aggregated sum* S_i for each video is calculated by

$$S_i = t \times m - (a_{i1} + a_{i2} + \cdots + a_{im})$$

where t is the current time, m is the total number of waiting requests, and a_{ij} denotes the arrival time of the jth request for video i [20]. When a stream becomes free, LAW schedules the video with the largest values of S_i. Simulation results illustrate that LAW is fairer by about 65% than MFQ [20].

8) *Max_Batch and Min_Idle* [23]. These two heuristic batching schemes suggest effective batching by considering the next stream completion time and users' wait tolerance [23]. Based on video request frequencies, all videos are classified into sets of hot or cold videos and maintained into two video queues, which are denoted by H and C queues. The requests waiting in H that have reached or exceeded the *batch threshold* are denoted by L_H [23]. The *batch threshold* can be set to the size of the users waiting tolerance or a value less than the users' waiting tolerance [23].

Only one set H of queues are used by the Max_Batch scheme. When a stream becomes available or a request reaches its *batch threshold*, a decision is made to select a video from subset L_H with MQL for maximizing the influence of batching [23]. This strategy is referred to as *Max_Batch MQL* (or BMQ). Another strategy is called *Max_Batch with minimal loss* (or BML). The BML calculates the expected number of defections that occur until the transmission of the next stream becomes complete. The BML minimizes the number of losses (i.e., defections) in the system by choosing a video from the L_H queue with the highest expected defection [23].

In the Min_Idle scheme, if there is a request waiting in L_H, a video in L_H is selected based on the MQL (called the IMQ scheme) or the maximum expected loss (called the IML scheme). If L_H is empty, a cold video in set C will be chosen using FCFS. To avoid starvation of cold videos by the hot videos, the Min_Idle scheme conveys the long-waiting cold video queues into set H [23]. Hence, the Min_Idle scheme is an efficient batching scheme for both hot and cold videos.

9) *Look-ahead-maximize-batch (LAMB)* [24]. If the number of available server channels is more than the number of queues, LAMB assigns a channel to the queue whose head-of-the-line user is going to waive the service because of exceeding its waiting time tolerance. Otherwise, LAMB solves a maximization problem to decide whether to allocate a free channel to this queue or not [24]. The maximization problem is expressed as a 0–1 integer programming [24]. Simulations demonstrate that LAMB increases the number of admitted users in the batching interval, but it is an unfair policy for cold movies.

10) *Next schedule time first (NSTF)* [25]. The process of scheduling in NSTF is carried out based on *schedule times* assigned to each new request, where schedule time is equivalent to the "closest unassigned completion time" of an ongoing stream [25]. A queue with the closest scheduling time is chosen for multicasting under NSTF. For future coming requests that request video v, NSTF assigns the same schedule time allocated to previous waiting requests. The simulation results show that NSTF can guarantee very precise schedule times for waiting queues [25].

11) *Earliest deadline first (EDF)* [26]. The EDF scheduler is related to deadline-driven schedulers for multicasting of chunks, where a chunk is a small segment of video a few seconds in length. The deadline time corresponds to every chunk, that is, when it must be delivered before that specific time. The EDF scheduler transmits a chunk with the earliest deadline [26]. To reduce transmission of the same chunk for multiple users at different times, the EDF scheduler delays sending chunks and groups chunks to meet the deadline of user requests in the form of multicast groups. Two variants of schedulers based on the EDF policy are available [26].

   – *Work-conserving scheduler with limited multicast groups (EDF-Ls)* [26]. Here, the server schedules chunks in order of their deadline using a limited number of multicast groups [26]. The EDF-L drops some chunks if there are no sufficient available multicast groups for transmitting "unique" chunks.

   – *Non work-conserving deadline-driven dynamic scheduler (EDF-D)* [26]. Because of the nature of non work conserving, EDF-D delays transmitting the chunk until its deadline using adapting the number of multicast groups. Simulation results obviously indicate that the server bandwidth is declined by 65% and 58%, compared to unicast and cycle multicast, respectively [26].

12) *Split and merge (SAM) protocol* [2]. The SAM protocol provides interactive VoD by employing batching [14] to reduce video delivery costs. A number of requests are batched together and served by a multicast stream called *service stream* during normal playback [2]. When a user initiates VCR-like interaction (including stop, pause, resume, fast-forward, rewind, jump-forward, and jump-backward), the SAM protocol splits off the user from the original service stream

and temporarily dedicates a new stream (called *interaction stream*) to perform the requested interaction function [2]. After terminating an interaction, the user will be merged back to the nearest existing service stream. The SAM protocol uses a *synchronization* (*synch*) *buffer*, which is a circular buffer, located in an access node or the video server to synchronize two streams.

13) *Bandwidth-efficient VoD algorithm (BEVA)* [27]. In order to optimize utilization of all channels in the central server, all available channels are assigned to the first request (with the fastest transfer rate). For the second request, BEVA assigns one channel for transmission, whereas the rest of the bandwidth is still assigned to the first request. When the third request arrives, it is also assigned only one channel from the channels allocated to the first request. This procedure is described in Algorithm 1.1 obtained from [27]. After completing the first request, all free channels can be assigned to future coming requests.

ALGORITHM 1.1   BEVA [27]

When a new request arrives at time t_1:
If (there are no other requests currently being served) **then**
    Assign all the channels to this request
Else
    Identify request R_1 with more than one channel assigned
    If (number of channels used by $R_1 > 1$) **then**
        Free one channel from the identified request R_1
        Assign the channel to the newly arrived request
    Else
        Refuse the request
    End If
End If

1.2.2 Adaptive Batching

In adaptive batching, the batching time is adjusted dynamically. Adaptive batching schemes try to improve the performance of fixed batching schemes by means of altering the batching interval. The proposed techniques based on adaptive batching are given as follows:

1) *Adaptive double-rate (A_DR) batching scheme* [28]. The A_DR algorithm dynamically adapts the optimal batching time in agreement with the average arrival rate of current time. The A_DR algorithm improves the double-rate (DR) batching policy [29], so that the algorithm itself can find the optimal batching time.

   Notice that the DR batching policy initiates a multicast stream at intervals of W [29]. If a new customer arrives after beginning a multicast stream, a unicast stream is immediately initiated, so that the customers experience a very small delay. Then, the transmission rate of the unicast stream is doubled until the customer can merge into the multicast stream. The duration of doubling transmission rate r_D is given by

$$r_D = t_a \bmod W$$

where t_a is the new customer's arrival rate and *mod* is the modulus operator [29].

ALGORITHM 1.2   ADAPTIVE BATCHING ALGORITHM [28]

t_{mr} = –2× (batching time)
When a new request is accepted:
  1. Update average arrival rate κ
  2. Map κ to batching time w according to Table 1.1
  3. $r_D = t_a - t_{mr}$
  4. **If** ($r_D > w$)
  5. Initiate new multicast stream
  6. $t_{mr} = t_a$
  7. Else
  8. **If** (r_D > receiver buffer size)
  9. Initiate a new multicast stream
  10. $t_{mr} = t_a$
  11. Else
  12. Initiate a dedicated stream with double transmission rate
  13. **End**
  14. **End**

The A_DR batching algorithm is shown in Algorithm 1.2 obtained from [28]. New multicast streams are initiated at time t_{mr} = –2× (batching time). The algorithm determines the optimal batching time w by mapping the updated arrival rate by using Table 1.1 [28]. In line 4 of Algorithm 1.2, the algorithm decides whether a new multicast stream should be initiated or a dedicated stream with double transmission rate should be opened. If the receiver

Table 1.1   Mapping Table for a 2-h Movie

Arrival Rate (arrivals per second) ≤	Optimal Batching Time (minutes)
0.004	15
0.007	13
0.008	12
0.01	11
0.02	9
0.04	7
0.08	6
0.15	4
0.32	3
0.5	2

Source: Poon, W.-F. et al., *IEEE Transactions on Broadcasting*, 47, 66–70, 2001.

buffer does not have efficient space to hold r_D seconds of video data, the algorithm starts a new multicast stream. The simulation results show that the adaptive scheme outperforms the static scheme in terms of bandwidth requirements [28].

2) *Least DR with first-come-first-serve (LDR_FCFS)* [30]. The LDR_FCFS policy takes advantage of DR duration [29] similar to the A-DR batching scheme. Under light traffic load, LDR_FCFS behaves like FCFS, which carries out the scheduling based on the arrival time of requests [30]. Under heavy traffic load, a video with the minimum DR duration has the highest priority for scheduling [30]. Results illustrate that the LDR_FCFS policy outperforms other batching schemes, like FCFS, in terms of fairness and bandwidth requirements [30].

3) *Adaptive batching scheme (ABS)* [31]. ABS adapts the batching time by the updated arrival rates of users. Contrary to A_DR [28], which assumes that the arrival rate of requests is based on Poisson distribution, ABS follows the hyperexponential distribution pattern [32]. Table 1.2 provides the optimal batching interval *W* for different arrival rates under different lengths of videos shown with *L*.

The ABS algorithm, which makes the decision for initiating a new multicast stream or a dedicated stream with double transmission rate, is similar to the A_DR batching algorithm. Numerical results show that the number of multicast streams used in ABS is less than the fixed batching schemes [31].

4) *Token-tray/weighted queuing-time (TT/WQT)* [33]. TT/WQT is an adaptive batching policy that combines the token-tray (TT) allocation scheme with weighted queuing-time (WQT) [33].

Table 1.2 Optimal Batching Interval *W* for Different Arrival Rates in ABS

L = 150 min		*L = 120 min*	
Arrival Rate (arrivals per minute) ≤	*Optimal Batching Interval (minutes)*	*Arrival Rate (arrivals per minute)* ≤	*Optimal Batching Interval (minutes)*
0.02	18	0.02	17
0.03	14	0.03	16
0.04	12	0.04	15
0.05	11	0.05	14
0.06	19	0.06	13
0.07	9	0.07	12
0.1	8	0.08	11
0.2	6	0.09	10
0.3	5	0.1	9
0.4	4	0.2	8
0.5	3	0.3	7
0.7	2	0.5	6

Source: Jain, M. et al., *Journal of ICT,* 1–12, 2006.

The TT allocation scheme applies N tokens corresponding to each server channel, where N is the total number of channels in the system. A free channel has a token held in a leaky token bucket, as shown in Figure 1.2 [33]. The token bucket releases one token into the token tray every $\Delta\tau$ minutes [33]. Once a channel is assigned to a movie, the token of that channel is removed from the tray. When transmission of a movie finishes, a channel will be released and its token is dropped into the leaky bucket.

A dedicated queue i with weight W_i is considered in each video object. The TT/WQT policy defines a weighted function for calculating weight W_i of queue i by considering pay-per-view (PPV) of the jth request in the queue (which is denoted by PPV_j) and the waiting time for the jth request (which is denoted by WT_j) as follows:

$$W_i = \sum_{j=1}^{q_i} \text{PPV}_j \times (WT_j)^\alpha$$

where q_i is the length of queue i and α is a parameter that shows an important degree of the user's queuing time [33]. The video with the maximum weight has the highest priority to be transmitted on an idle channel. The TT/WQT offers low defection rate when the arrival rate varies [33].

5) *Rounded ratio* [34]. The "rounded ratio" algorithm determines the length of batch interval b_i' assigned to queue i according to the rounded ratio [34]. Suppose that there are M queues and N servers in the system. Parameter $0 < p_i < 1$ denotes the probability that queue i is chosen by a new arriving request to join $p_i \geq \cdots \geq p_M$. The rounded ratio is calculated by $\left\lceil \sqrt{p_i / p_M} \right\rceil$ [34]. The batch intervals are set ecursively, as shown in Algorithm 1.3 [34]. Video transmission is completed at interval $d = T/N$, where T is the interval in which the video starts, and N denotes the number of servers in the system. Analytical results show that the expected service latency of the "rounded ratio" is approximately twice the optimal value [34].

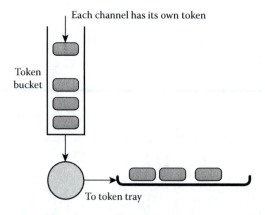

Figure 1.2   TT channel-allocation scheme. (From Wen, W. et al., *Computer Communications*, 25, 890–904, 2002.)

ALGORITHM 1.3   ROUNDED RATIO ALGORITHM [34]

1. **For** $i = 1$ to M

2. Set l_i as $\left\lceil \sqrt{\dfrac{p_i}{p_M}} \right\rceil$

3. $b'_M = d \displaystyle\sum_{i=1}^{M} l_i$

4. **For** $j \geq 1$
5. Let I_j be the interval

$$[d\,(\,j\,(b'_M + 1) + 1),\, d\,(\,j + 1)(b'_M + 1)]$$

6. Schedule queue M at time $d\,(\,j + 1)(b'_M + 1)$
7. **For** $i = M - 1$ down to 1
8. Schedule queue i for l_i times uniformly along interval I_j.

In this section, we have reviewed the proposed techniques for the fixed batching and adaptive batching categories. In the following, we will focus on the patching solutions, adaptive piggybacking, stream merging, and hybrid schemes. Batching scheduling is applied by the video server for selecting the next request at patching schemes [15]. The following section describes a large number of patching schemes that are much favorable in VoD systems.

1.3 Patching Solutions

Patching is a dynamic multicast scheme that tries to join a new client to the latest ongoing multicast stream. The basic idea of patching is to cache subsequent data from an ongoing multicast in the client's local buffer while playing the leading portion of video from the beginning. When the playback of the leading portion is ended, the client continues to playback the remainder of the video already buffered. An important objective of patching is to improve the efficiency of multicasting and to provide service request with no delay.

1.3.1 Continuous Patching

The basic idea of continuous patching is to transmit the video using a continuous stream on different channels without any partitioning a video into segments. There are different techniques proposed for this category:

1) *Patching* [15]. This patching scheme is called standard patching from now on. The server bandwidth is used to transmit either a *regular stream*, which multicasts the entire video, or a *patching stream* in order to multicast the beginning portion of a video [15]. If there is no regular multicast stream for the selected video in progress, the server activates a regular stream on a free channel. Otherwise, it transmits a patching stream on the free channel according

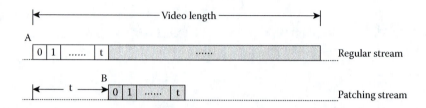

Figure 1.3 Original patching. (Cai, Y. et al., *Multimedia Tools and Applications*, 32, 115–136, 2007.)

to the status of the latest regular channel. The standard patching technique is illustrated in Figure 1.3 [35], where client A has arrived at time 0 and a regular stream has been initiated for client A. Client B arrives at time instant *t*. A patching stream with length *t* is initiated for client B. At the same time, client B caches ongoing regular stream in its local buffer. After playback of the patching stream, client B fetches video data from its buffer and plays back the ongoing movie.

The video server uses either *greedy patching* or *grace patching (GP)* to determine the portion of video data that should be multicast on the free channel [15]. In greedy patching, a new regular multicast is started only if the previous regular stream for the same video terminates streaming of the video [15]. GP will schedule a new regular multicast if the client buffer has enough space to buffer the missing portion of the video [15]. The time period after starting a regular multicast during which a patching stream should be initiated is referred to as the *patching window* [36]. Greedy patching considers the playback duration of the video as the patching window. On the other hand, the patching window in GP is determined by the client buffer size. Studies in [36,37] determine the optimal size for the patching window according to the request rate of the video. Patching is able to eliminate server latency and to offer better server throughput compared to batching schemes [15] and is able to achieve up to two times improvement than piggybacking [36].

2) *Double patching* [35] (*transition patching* [38]). In "double patching" or "transition patching," logical channels are used for transmitting *regular* streams, *long patching* streams (also referred to as *transition* stream in [38]), and *short patching* streams. A long patching stream can be shared by requests similar to a regular stream. A client caches video data not only from a regular stream but also from a long patching stream. The minimal distance between any two sequential regular streams is signified to *multicast window w_m* [35]. Double patching divides the *multicast window* into several *patching windows w_p*. The size of patching window w_p is the minimal interval between two consecutive long patching streams. The stream initiation pattern in "double patching" is shown in Figure 1.4 [35]. The first request is served by a regular stream. After a regular stream, the "double patching" scheduler initiates short patching streams for the requests arriving within the next w_p time units. These sequential short patching streams arrange a *patching group*. A long patching stream is initiated for the next first request that arrives after w_p time units. Notice that the long patching stream delivers an additional $2 \times w_p$ time units of data besides the missing portion of the video. The next patching group is repeated for the request arriving within the next w_p duration. If the arriving time of a new request is out of the multicast window of the latest regular stream, then a new regular stream will be initiated. Performance studies show that the double

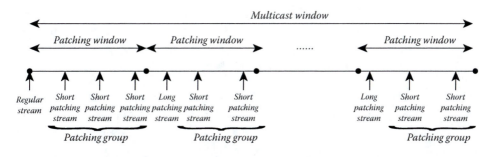

Figure 1.4 Pattern of stream initiation in double patching. (Cai, Y. et al., *Multimedia Tools and Applications*, 32, 115–136, 2007.)

   patching can double the performance of the standard patching in terms of server bandwidth requirement [35].

3) *Two-level patching* [39]. The "two-level patching" reduces the channel redundancy of the standard patching scheme [15] by introducing a two-level patching channel. Zero-level channel, one-level patching channel, and two-level patching channel are equivalent to the regular channel, long patching channel, and short patching channel in "double patching" [35], respectively. The entire duration of the video is divided into *periods*, and each *period* consists of subperiods called *time window*. The length of one-level patching is $(t_n + w_n)$, where w_n is the length of the nth time window and t_n is the difference of the arrival time between the first request in the nth time window and the first request in the corresponding *period* [39]. All requests in a *period* share the zero-level channel with others, and all requests in a time window share the one-level patching channel. Any other request in a time-window initiates two-level patching channels. The number of time-windows in a period (T) is computed from a recursive equation presented in [39]. Notice that the "two-level patching" requires low bandwidth at high request rates [39].

4) *Patching-based broadcasting* [40]. This technique improves the performance of the "two-level patching" scheme. It is proved that the average video transmitted by a two-level channel in a time-window is one-third of the time-window size [39]. This scheme derives the time-window size and optimal value of T by assuming that at least one request initiates a time-window [40].

5) *Best-effort patching (BEP)* [41]. In order to support admission control and continuous VCR interactivity for VoD service, BEP has been proposed. The BEP consists of three phases as follows [41]:
   – Admission: BEP services new requests by "transition patching" or "GP" in phase admission.
   – Interaction: The interaction phase is executed when a client operates VCR-like interactions. If the interaction time exceeds the capacity of the client's buffer, a patching channel will be initiated to transmit the desired playback point [41].
   – Merging: After completing the interaction, the client's video stream must be merged into the current regular multicast stream. The BEP proposes a dynamic merging algorithm [41] to perform merging interactions and regular streams.

   Theoretical analyses indicate that BEP performs better than the SAM protocol in terms of reducing bandwidth requirement for popular videos [41].

6) *Recursive patching* [42]. The basic idea of recursive patching is to cache multiple levels of transition streams (*T*-stream) recursively by a client station in the same way as in the transition patching [38]. The recursive patching proposes a new notation called *k-phase recursive patching* (*k*P-RP), which includes one regular stream and a new patching stream plus (*k*-2) transition streams [42]. Using this notation, 2P-RP and 3R-RP are equivalent to standard patching and transition patching, respectively. A new client will be served with a new patching stream, and at the same time, the client caches video data from the transition stream on level *k*-2 from the T_{k-2}-stream. In phase 2, the client will release the patching stream and start caching data from the T_{k-3}-stream. At the same time, the client continues caching data from the T_{k-2}-stream. In phase *k*-1, the client caches video data from the T_1-stream and the regular stream. Finally, in phase *k*, the client releases the T_1-stream to continue playback using the regular stream. Simulation results show that "recursive patching" can decline startup latency about 60%–80% compared to "transition patching" [42].

7) *Practical patching* [43]. "Recursive patching" sends the transition stream on level *i* (i.e., T_i-stream) that has additional data of 2 × w_i time units extra to *skew the T_i-stream to the latest T_{i-1}-stream for future requests* [43]. Note that w_i is the transition window length for level *i*. Practical patching eliminates the unnecessary data transmitted in *T*-streams. Every *T*-stream is initially scheduled with no additional data with the amount of *skew of the current request to the latest T_{i-1}-stream* [43]. When a new request arrives at the server during w_i time units of T_i-stream, practical patching dynamically expands the latest T_i-stream, T_{i-1}-stream,…,T_1-stream by appending additional data by 2 × *skew of the current request to T_i-stream*. Practical patching decreases both defection rate and service latency in comparison to the recursive patching [43].

8) *Quick patching* [44]. This patching is designed for overlay multicasting a wireless environment. Its goal is to deliver the main video stream using a scheme similar to asynchronous multicast [45] and to improve client viewing quality under lossy wireless links by retransmitting additional patch streams. Figure 1.5 illustrates the quick patching as follows [44]:
   a. After joining a multicast group, a new client receives the main video stream and buffers it until the cached data exceeds threshold *s*.
   b. When the data buffered exceed threshold *s*, playback will be started.

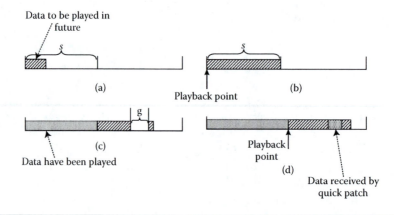

Figure 1.5   Illustration in quick patching. (Dai, H. and Chan, E. Quick patching: an overlay multicast scheme for supporting video on demand in wireless networks, *Multimedia Tools and Applications*, 36, 221–242 © 2008 IEEE.)

c. When a long error burst happens, a gap will be created between received packets. If traffic loss in some portion of the video is larger than g, a patch request is sent to the media server. The client will only request the same patch stream for at most n times (the "patch limit").

d. The server schedules a patch stream to send the lost portion of the data with a bigger rate than the playback rate. The old data will be overwritten by new data when filling the buffer.

Simulation experiments show that quick patching stabilizes client viewing quality in spite of wide fluctuations under error rate [44].

9) *Converting patching agent* [46]. This technique is a variation of greedy patching, but it converts the latest patching multicast into a regular multicast for a particular new request [46]. If the new request is out of the patching window of the latest regular multicast and also within the patching window of the latest patching multicast, then changing patching multicast into a regular multicast is carried out in order to transmit the whole video. Algorithm 1.4 shows the converting patching agent algorithm used in video servers [46]. Simulation results approve that the proposed technique can achieve better performance than greedy patching and GP in terms of defection rate and average service latency [46].

ALGORITHM 1.4   CONVERTING PATCHING AGENT ALGORITHM [46]

t: Current time of scheduling a new request
t_r: Start time of the latest regular multicast of video v
t_p: Start time of the latest patching multicast of video v on patching channel P
Patching Window: Size of patching window, that is, the size of client buffer
L: Playback duration of video v, that is, video length
D: Portion of video data that should be multicast on a new channel
$v[t_d]$: Video data of video v during the period from the beginning to time t_d

1. **If** $((t - t_r) \leq Patching\ Window)$ **then** $D = v[t - t_r]$
2. Else if $((t - t_p) \leq Patching\ Window$ and $(t_p - t_r) > Patching\ Window)$ **then**
   Convert the latest patching multicast into a regular multicast to make sure that patching channel P is used to transmit the entire video.
   $t_r = t_p$
   Set id of the latest regular multicast *as patching channel P*
   $D = v[t - t_r]$
3. **Else** $D = v[L - \text{Min}(Patching\ Window, L - (t - t_r))]$

---

10) *Period Patch* [47]. Period Patch presents a PERIOD delivering rule based on patching idea, where the interval between main streams (equivalent to regular stream) should be the integral times of a fixed period [47]. In Period Patch, multicast streams are created regularly, contrary to the standard patching scheme in which main streams are made randomly. The PERIOD delivering rule is according to the following equation:

$$MT = ST + n \times T_{buf}, \quad \text{where } n = 0, \pm1, \pm2, \pm3, \ldots$$

where MT is the movie time, ST denotes the system time when a regular stream is created, and T_{buf} is the size of the client buffer [47]. Simulation results show that Period Patch uses only 50% of streams of the standard patching scheme under the same simulation conditions [47].

11) λ *Patching* [48]. "λ patching" is based on the patching method [15] that allows a video server to adjust retransmission times for multicast streams. The optimal temporal distance between two regular streams (Δ_M) for one video depends on the length of video L and the current popularity of the video, which is expressed by $1/\lambda$ as follows [48].

$$\Delta_M = \sqrt{2 \times L/\lambda}.$$

The server dynamically recalculates Δ_M for every change in the request rate [48]. After that, it restarts regular streams in intervals with length Δ_M.

12) *Multistream patching* [48]. "Multistream patching" assumes that each client is able to receive $n + 2$ concurrent streams simultaneously. It adds n additional multicast patch streams between starts of two regular streams and plays them for length Δ_M. Figure 1.6 depicts multistream patching at $n = 1$. It initiates a multicast patch stream in the middle of two regular streams in every interval $[t_n, t_n + \Delta_M/2]$ [48]. This scheme decreases the average length of unicast patch streams [48].

13) *Pull-patching* [49]. Pull-patching utilizes *adaptive segmented HTTP streaming* [50] in standard patching [15]. HTTP streaming is a unicasting technique based on the idea of downloading 2-second video segments from the Internet. HTTP streaming is used for transferring patching streams and repairing packet loss or damages during the delivery of the multicast stream [49]. Popular video content is multicast at several bit rates from dedicated multicast servers. This prototype has been tested in a laboratory environment and has achieved good efficiency [49].

14) *Controlled client initiated with prefetching (CIWP)* [37]. The idea of controlled CIWP, which is also known as threshold-based multicasting [51], is similar to patching schemes. However,

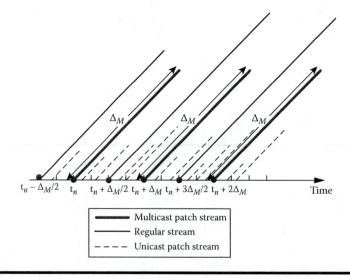

Figure 1.6 Stream setup example in multistream patching. (From Griwodz, C. et al., *ACM SIGMETRICS Performance Evaluation Review*, 27, 20–26, 2000.)

an optimal threshold T is applied by controlled CIWP to adjust the frequency at which a complete stream of video is started. The optimal threshold is given by

$$T = \left(\sqrt{2L\lambda + 1} - 1\right)/\lambda,$$

where L is the length of video and λ denotes the request rate for the video object [37]. The CIWP also uses a scheduler that determines the channels to satisfy a request immediately after receiving from the user [37].

15) *Stream tapping* [52]. Similar to patching schemes, the stream tapping uses an original stream for delivering the entire video or a *full tap stream* for providing β minutes of the start of the original stream for the same video. In addition, stream tapping uses a *partial tap stream* for providing data in any situation. The stream decision process is shown in Figure 1.7 [52] with the following descriptions:

a. If no original stream is activated for the same video, then the algorithm assigns an original stream to the request.

b. If an original stream is initiated within the last β minutes, then a full tap stream will start.

c. If an original stream is initiated over β minutes in the past, then the algorithm makes a decision either to assign a partial tap stream or an original stream. For each original stream in addition to all of the subsequent tap streams, the algorithm records the minimum average service time (AST) without the request m_{wo}. It also estimates the exact service time m_w for streams that should be assigned to the request. If the new AST is

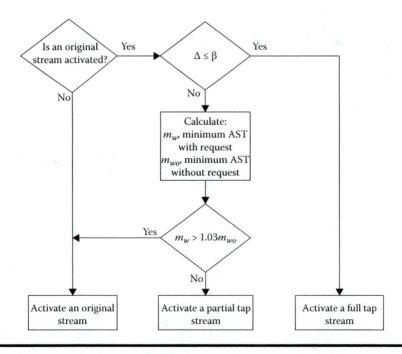

Figure 1.7 Decision process in stream tapping. (Carter, S. W. and Long, D. D. E., *Computer Networks*, 31, 111–123, 1999.)

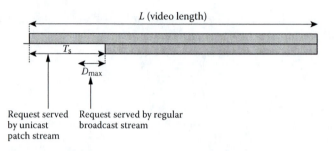

Figure 1.8   Operation of JAS for a movie. (Gary Chan, S. H. and Ivan Yeung, S. H. Client buffering techniques for scalable video broadcasting over broadband networks with low user delay, *IEEE Transactions on Broadcasting*, 48, 19–26 © 2002 IEEE.)

3% higher than the minimum AST (i.e., $m_w > 1.03 \times m_{wo}$), then an original stream is activated for the request [52]. Otherwise, it will assign a partial tap stream to the request.

Stream tapping decreases 20% of the bandwidth required by the traditional VoD scheme for popular videos [52]. If STB can tap data from any stream, not just from the original stream, it is called *extra tapping* [52]. If the server can use additional streams to quickly load more data, this option is called *stream stacking* [52].

16) *Join-and-stream (JAS)* [53]. JAS broadcasts/multicasts a video periodically at regular offset points (referred to as T_s) and uses a unicast stream to recover the missing portion of a video [53]. The JAS scheme is similar to the standard patching [15] scheme in terms of unicasting short streams. If the difference between the arrival time of a request and the starting time of the next multicast stream is less than D_{max} (where $D_{max} < T_s$), it joins the next regular multicast stream, as shown in Figure 1.8 [53]. Otherwise, the client is served by a unicast patch stream and buffers the nearest ongoing multicast stream. JAS is efficient for movies with intermediate request rates [53].

1.3.2 Segmentation Patching

In the following, we describe another category of patching schemes called segmentation patching that divides a video object into segments and delivers the segments via multicast streams or patching streams. The following section reviews the patching techniques that fit into the segmentation patching category.

1) *Fragmented patching* [54]. "Fragmented patching" provides client mobility in broadband wireless environments. A shared flow is equivalent to the regular stream that multicasts video content to multiple clients. A patch flow is a kind of unicast patching stream, but it is broken into segments. Due to extreme overhead of unicasting in mobile environments, patch flows are sent via broadcasting. If λ denotes the request arrival rate, each segment has the size of $1/\lambda$. The server calculates the average request rate and subsequently adjusts the segment length using the calculated value [54]. The server determines which segments have already been sent to previous clients and which segments must be newly sent [54]. Then, it records the segment numbers and time to be sent in a scheduling table. The clients receive the segments according to the scheduling table [54].

2) *Hierarchical patching* [55]. Hierarchical patching does not use periodically ongoing regular multicast streams similar to the GP, double patching, and fragmented patching. All downloads are always multicast in the form of patches. Let pt_i, j_i denote the j_ith patch created at time t_i for the ith client, where $0 \leq j \leq K_i$ and K_i is the number of patches that are newly created for this client [55]. The patch $pt_i, j_i = [a, b]$ created at time t_i that will be multicast at time $t_i + a$ is defined by

$$p_{t_i, j_i}(t) = \begin{cases} [a,b) & \text{if } t < t_i + a, \\ [t - t_i, b) & \text{if } t_i + a \leq t < t_i + b, \\ \phi & \text{otherwise.} \end{cases}$$

where interval $[a, b)$ denotes the part of a video between play time points a and b, where $0 \leq a < b \leq L$ [55]. The purpose of this scheme is to find a set of patches pt_k, j_k that cover the whole video for the requests. Table 1.3 shows an example of how the hierarchical patching schedules patches for a set of arrivals. Clients arriving at time t_i immediately start receiving patches from column 1 and can reuse the patches shown in column 2. As one can see, a patch may have any size. On average, 22 video channels must be used at the client side to download concurrently [55].

3) *Video data delivery using slotted patching* [56]. "Slotted patching" uses multicast channels and patching channels like other patching schemes, but it divides a video into time slots with uniform length T [56]. If there is at least one request in a time slot, "slotted patching" dispatches a patching channel in that time slot. The length of the patching channel for a request received in the rth time slot will be r time slots [56]. A multicast channel is transmitted after a fixed number of time slots determined by D/T, where D is the size of an interarrival time between two adjacent multicast channels [56].

The latency time for a user changes from zero to the length of a time slot, depending on the time instant when a request is accepted [56].

4) *Stream-bundling broadcasting (SBB)* [53]. Stream-bundling means bundling the server streams into multicast channels with bandwidth in an increasing order. The operation of this scheme is shown in Figure 1.9. [53]. Transmitting the main multicast stream is exactly the same as the way used in JAS (i.e., staggered manner). Slot interval T_s is divided into minislots of length D_{max}. The initial portion of the video is periodically multicast in each minislot, where the first minislot uses b bandwidth, the second one uses $2b$ bandwidth, and so on. A

Table 1.3 Arrival of Clients at Times 0, 2, 3, and 4.5 in Hierarchical Patching

Arriving Time	Starting Patch	Reuse Patch
$t_1 = 0$	$p_{0,1} = [0, L]$	–
$t_2 = 2$	$p_{2,1} = [0,2)$	$p_{0,1}(2) = [2,L)$
$t_3 = 3$	$p_{3,1} = [0,1), p_{3,2} = [2,3)$	$p_{0,1}(3) = [3,L), p_{2,1}(3) = [1,2)$
$t_4 = 4.5$	$p_{4.5,1} = [0,2), p_{4.5,2} = [3,4.5)$	$p_{0,1}(4.5) = [4.5,L), p_{3,2}(4.5) = [2,3)$

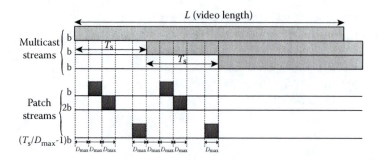

Figure 1.9 Operation of the SBB scheme. (Gary Chan, S. H. and Ivan Yeung, S. H. Client buffering techniques for scalable video broadcasting over broadband networks with low user delay, *IEEE Transactions on Broadcasting*, 48, 19–26 © 2002 IEEE.)

request arriving at time between $(k - 1) D_{max}$ and $k \times D_{max}$ ($1 \leq k \leq T_s/D_{max}$) from the start of the corresponding slot uses a patch channel with $k \times b$ bandwidth for a duration of D_{max} minutes; therefore, the request receives the first $k \times D_{max}$ minutes of the video segment [53]. Simulation results show that the bandwidth requirement of a system using both JAS and SBB can be reduced by a significant margin [53].

5) *Batched multicast patching (BMP)* [57]. Broadcasting the main stream and dividing time slot T_s into minislot of D_{max}, the BMP performs exactly like the stream-bundling scheme. As shown in Figure 1.10, if there is a request between time $(k - 1) D_{max}$ and $k \times D_{max}$ from the beginning of the corresponding slot, it will be served with a multicast patch stream with bandwidth b for duration of $k \times D_{max}$. Otherwise, no patch stream is transmitted at the end of that minislot [57]. The BMP scheme requires less server bandwidth in proportion to JAS and SBB schemes under both moderate and high arrival rates [57].

6) *Fast patching (FP)* [58]: In the JAS scheme, the patching stream is transmitted through unicasting channels, whereas FP broadcasts patching streams. FP implements modified segment allocation of fast broadcasting (FB) [59] at patching channels. FB divides a video into $2^N - 1$ uniform size segments and repeatedly broadcasts them through N channels. FB divides the T_s part of the video into 2^{N_p} equal size segments $S_1, S_2, S_3, ..., S_{2N_p}$, where N_p is

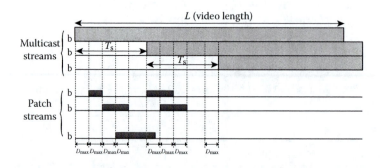

Figure 1.10 Operation of the BMP scheme. (Azad, S. A., Murshed, M. and Dooley, L. S., A novel batched multicast patching scheme for video broadcasting with low user delay. In *Proceedings of the 3rd IEEE International Symposium on Signal Processing and Information Technology*, 339–342, December 2003 © 2003 IEEE.)

Table 1.4   Segment Allocation in FP Scheme

Patching Channels	Transmitted Segments	Segments Repetition (In a Regular Offset Point T_s)
C_1	S_1	2^{N_p} times
C_2	$S_2 \sim S_3$	2^{N_p-1} times
C_3	$S_4 \sim S_7$	2^{N_p-2} times
⋮	⋮	⋮
C_{N_p}	$S_{2^{N_p-1}} \sim S_{2^{N_p}-}$	2 times

Source: Song, E. et al., Fast patching scheme for Video-on-Demand service. In *Proceedings of the Asia-Pacific Conference on Communications, Bangkok*, pp. 523–526, October 2007.

the number of patching channels. Allocating a segment at each patching channel is carried out according to the allocation rule presented in Table 1.4 [58]. FP broadcasts the entire video data periodically at regular intervals T_s similar to the JAS, SBB, and BMP schemes. In simulation, the waiting time of the FP scheme comes down exponentially with increasing patching channels [58].

7) *Medusa* [60]. Medusa is a stream scheduling scheme proposed for minimizing the server bandwidth for parallel video servers used in homogeneous fiber-to-the-building (FTTB) client network architecture. The scheme divides a video object into segments with uniform length T and transmits it in time interval length T. Medusa suggests using $T = \left\lceil \dfrac{L}{2b_c} \right\rceil$ for time interval, where b_c is the client bandwidth capacity in unit of stream. The parallel video server schedules a complete multicast stream for the first arrival request i. The later request j (where $i < j \le i + \lceil L/T \rceil - 1$) must be grouped into logical request group G_i. Parallel video server searches the *stream information list* maintained for group G_i to find out the segments that will be transmitted on a live patching multicast stream. It shares these segments from the patching stream among multiple clients; otherwise, it schedules a new patching multicast stream to transmit the missing segment. The client must buffer the later $\lceil L/T \rceil - (j - i)$ video segments from the ongoing multicast stream. The performance of the Medusa scheme significantly outperforms the batching schemes and the stream-merging schemes [60].

1.4 Adaptive Piggybacking Solutions

Adaptive piggybacking [61] tunes the display rates of subsequent ongoing streams for the same video until the streams can be merged into a single stream and form an entire group. This scheme takes the advantage of various display rate altering techniques [61]. For example, the play-out rate can be changed to a faster rate (or a slower rate) by replicating (or removing) frames. Another solution is to store replicated data with different display rates in the video server.

The goal of adaptive piggybacking policies is to evaluate all possible display rates for each request such that the expected stream demand is minimized [61]. Each policy makes a decision when one of the following events happens: (1) arrival of a new stream, (2) merging of two streams, (3) termination of display of an object, and (4) crossing the border of a catch-up window [61]. Note that catch-up window W_p is the maximum possible distance between two streams such that merging is beneficial. The following policies take into account slow speed (S_{min}), normal speed (S_n), and fast speed (S_{max}). In the following, we describe various piggybacking policies.

a. *Odd–even reduction policy* [61]. The main idea is to couple sequential arrivals for merging. If there is still a stream moving at the display speed of S_{min} (or S_{max}) in the catch-up window, the display speed of a new request is set to S_{max} (or S_{min}). Then, two streams with display rates S_{min} and S_{max} merge into a single stream. The algorithm for this technique is shown in Algorithm 1.5 [61].

b. *Simple merging policy* [61]. The main idea is to merge sequential streams to merging groups. One stream initiates a group, and all streams that arrive later to the system will merge into the group while the initiated stream is in the catch-up window. Aside from the catch-up window, this policy defines the maximum merging window W_p^m, which indicates the ultimate position where two streams can merge. When a new stream arrives at the system, the display speed of the new stream is set to S_{min}. While the first stream is within the catch-up window, other stream's display speeds are set to S_{max}. All newly arrived streams merge with the first stream within W_p^m. The algorithm for the simple merging policy is illustrated in Algorithm 1.6 [61].

ALGORITHM 1.5   ODD–EVEN REDUCTION POLICY [61]

Case arrival of stream i:
    If ((no stream, in front, is within W_p) or
        (Stream immediately in front is moving at S_{max}))
        $S_i = S_{min}$
    else
        $S_i = S_{max}$
    End if

Case merge of streams i and j:
    drop stream i
    $S_j = S_n$

Case window crossing (by stream i):
    If ($S_i = S_{min}$) and (no stream behind moving at S_{max} in W_p)
        $S_i = S_n$
    else
        S_i is unchanged
    End if

End

ALGORITHM 1.6   SIMPLE MERGING POLICY [61]

Case arrival of stream i:
  If no stream within W_p is moving at S_{min}
    $S_i = S_{min}$
  else
    $S_i = S_{max}$
  End if

Case merge of streams i and j
  drop stream i
  $S_j = S_{min}$

Case window crossing W_p^m
  $S_i = S_n$

End

---

c. *Greedy policy* [61]. The main idea is to merge streams as many times as possible during the playback of a video [61]. The greedy policy defines the "current" catch-up window (W_c) calculated relative to the current position in video playback [61]. For each new arrival, the speed adjustment is executed as in odd–even reduction policy. After a merging occurs, a new catch-up window W_c relative to the current position is computed. If no request exists within the current catch-up window, the request's speed is adjusted to S_n. Otherwise, if there is leading request with a display speed of S_n, then the speed of the leading request changes to S_{min}, and the speed of the request at the current position is set to S_{max} [61].

d. *Snapshot policy* [62]. The main idea is based on capturing snapshots from the positions of the stream at fixed intervals. Merging streams in snapshot policy occur in a *snapshot interval* given by $I = W/S_{max}$, where W is the optimal window size calculated for the generalized simple merging policy [62]. The speed of the first stream arriving within a snapshot interval is set to S_{min}. The speed of all other arriving streams within the same snapshot interval will be set to S_{max}. At the end of an interval, the snapshot policy applies the dynamic programming algorithm defined in [62] in order to adjust the speeds of all remaining streams that were admitted in that interval. The snapshot policy outperforms the simple merging policy and greedy policy in simulations [62]. The algorithm for the snapshot policy is displayed in Algorithm 1.7 [62].

e. *S^2 piggybacking policy* [63]. The S^2 policy has been proposed to minimize the number of displayed frames from the streams during the last *modified maximal merging window* [63]. The maximal merging window includes $\left\lfloor W_m \middle/ W \right\rfloor$ optimal windows, where W_m is the maximum catch-up window size and W is the optimal window size calculated in snapshot policy. The S^2 policy performs two levels of optimizations using the snapshot policy. It first applies the snapshot algorithm over the streams during the past snapshot intervals. At a later time, it again implements the snapshot algorithm on the streams obtained from the first stage at points with a variable (not fixed) interval [63]. Discrete-event simulations indicate that the

ALGORITHM 1.7   SNAPSHOT POLICY [62]

Compute snapshot interval I
Start interval counter

Case arrival of stream i:
   If first stream is within interval
      $S_i = S_{min}$;
   else
      $S_i = S_{max}$;
   End if

Case end of interval counter
   Solve dynamic programming problem on remaining new streams
   Reset interval counter

Case merge of stream i and j
   If within initial interval
      follow simple merging rules
   Else
      follow dynamic programming rules
   End if

S^2 piggybacking policy outperforms up to 8% the snapshot policy in high interarrival times [63].

1.5 Stream Merging Solutions

Stream merging is a technique for on-demand streaming using multicasting and client buffering. Clients have the ability to receive data from two multicast streams simultaneously. Each channel multicasts the same media object at different times. The simplest form of stream merging is patching [15] in which a secondary stream that delivers the prefix of the primary stream is allowed to merge with a primary stream.

The hierarchical multicast stream merging (HMSM) technique [16] is the original stream merging model. The HMSM takes advantage of patching/stream tapping and piggybacking, as well as dynamic skyscraper [64]. Merging of clients that request the same video object is performed repetitively into larger groups [16]. An example of the HMSM technique is demonstrated in Figure 1.11 [16] for a set of request arrivals. The crisscross lines in the figure represent one unit of transmitted stream in unit of time. Clients A, B, C, and D request a typical video at times 0, 2, 3, and 4, respectively. A new multicast stream is assigned to each new client in order to provide immediate service. A full stream of 10 units long is initiated for client A. Client A receives data only from stream A. Client B receives parts 1 and 2 from stream B and parts 3 and 4 from stream A. Client B merges with client A at time 4. Client D listens to stream C and merges with client C at time 5. When C and D merge, both C and D listen to stream A. Finally, C and D merge

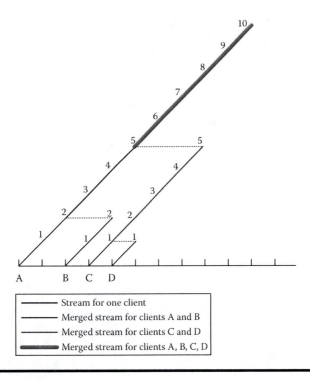

Figure 1.11 Example of HMSM. (Eager, D., Vernon, M., and Zahorjan, J., Minimizing bandwidth requirements for on-demand data delivery. *IEEE Transactions on Knowledge and Data Engineering,* **13, 742–757 © 2001 IEEE.)**

with stream A at time 8. Another alternative choice can be made if clients C and D separately merge with stream A. This variant structure requires higher server bandwidth for satisfying clients.

Online stream merging algorithms determine when and in what order to incorporate new arrivals in such a way that the total server bandwidth is minimized. An online algorithm operates without the knowledge about future requests. This is in contrast to an offline algorithm where client arrivals are acquainted ahead of time [13].

The following is a partial list of papers that introduce stream-merging algorithms:

1) Early stream merging [65] attempts to merge clients with the closest earlier stream still in the system.
2) Bandwidth-skimming policies [66] use a small fraction of the client reception bandwidth to merge transition streams using near-optimal hierarchical stream merging policies.
3) The dynamic Fibonacci tree algorithm [13] uses the Fibonacci tree structure to control how new arrivals should be merged with existing streams. The work in [13] also examines the natural greedy algorithms that incorporate a new arrival into the merge tree in order to minimize the merge cost of the resulted merge tree.
4) The dyadic tree algorithm [67] uses recursive interval partitioning to determine client receiving planes.
5) Five-competitive online algorithms [68] (with bandwidth usage not exceeding five times that of the optimal offline scheduling algorithm) can be used to schedule any request sequence using the embeddable binary merge trees on a grid.

6) DR batching policy [29] tries to merge requests to a single stream by means of doubling transmission rates and buffering. DR doubles the transmission rate, so that a new arriving customer is able to catch up with an existing multicast stream. Because of the buffering of early frames in client station, DR can be considered as a stream merging policy.

A comparative study on several proposed stream merging algorithms (such as early stream merging, dynamic Fibonacci tree algorithm, and dyadic tree algorithm) can be readily found in [69].

1.6 Hybrid Solutions

The hybrid of aforementioned techniques such as batching, patching, piggybacking, and even broadcasting [4] makes a new class of schemes that can provide the best performance. In the following, we review a number of hybrid schemes:

1) *Quasi-optimized batching by timeout with adaptive piggybacking* [70]. This method extends the algorithm of batching [14–21], where a timer is set for each arrival request. When the timer expires, a logical channel is allocated to satisfy waiting requests, as shown in Figure 1.12 [70]. The maximum waiting time in batching by timeout is equal to $(1/\lambda_k)$, where λ_k is the arrival rate of video k [70].

   The "quasi-optimized batching by timeout with adaptive piggybacking" combines batching by timeout with adaptive piggybacking by deriving the optimal catch-up window size [70]. This combination reduces the number of required channels by 25%, compared to the batching by replay [61] in simulations [70]. The work in [71] also investigates integrating piggybacking policies with the LAMB policy. It has been shown that a combination of both LAMB and odd–even piggybacking policy can admit 20% more than the number of users when a server uses LAMB only [71].

2) *Adaptive hybrid approach (AHA)* [72]. AHA uses the *Skyscraper broadcasting* (SB) scheme [73] for broadcasting the most popular videos and uses the LAW batching technique for multicasting less popular videos. Requests are maintained in a waiting queue corresponding to each video. Based on the estimation of the average interarrival times of the requests, average multicast interval, and request frequency, video queues are classified as either SB or LAW [72]. The LAW scheduler selects the video with the largest aggregated waiting time S_i to be served. At the same time, the SB scheduler reserves K channels and initiates SB [72].

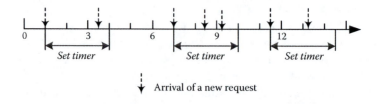

Figure 1.12   Batching by timeout (timeout = 3 min). (Kim, H. J. and Zhu, Y. *Allocation problem in VoD system using both batching and adaptive piggybacking.* **IEEE Transactions on Consumer Electronics, 44, 969–976, © 1998 IEEE.)**

3) *Adaptive segment-based patching (SBP)* [74]. Adaptive SBP is developed based on the batched-patching approach [74]. The batched-patching approach joins the standard patching [15] and the idea of batching. In batched patching, requests for the same video are delayed for a time slot with length d in order to be served with a patching stream.

SBP combines FB [59] and batched patching, which uses m regular channels to transmit a video. The following algorithm is executed in a video server for transmitting a video under the SBP scheme [74]:

a. Divide a video into N segments of equal size $S_1, S_2, ..., S_N$. The length of each segment is $d = L/(2^k - 1)$, where L is the video length. This algorithm selects appropriate k to keep a low delay.

b. Broadcast data segments $S_1, S_2, ..., S_{2^{k-m+1}-1}$ periodically in the first regular channel. Broadcast data segments $S_{2^{k-m+i}}, S_{2^{k-m+i}+1}, ..., S_{2^{k-m+i+1}-1}$ in the ith regular channel for $1 \leq i < m - 1$.

c. Create a patching stream to transmit missing segments for those clients who request the same video within a time slot.

Simulation results demonstrate that the SBP scheme requires less total bandwidth than the FB and original patching [74].

4) *Hybrid quality-of-service (HQoS) strategy* [75]. HQoS strategy joins batching with recursive patching. This scheme broadcasts the most popular videos with recursive patching and batching. Average popular videos are multicast along with recursive patching but without batching. Finally, the least popular videos are only unicast. The HQoS scheme assigns separate channels for broadcasting, multicasting, and unicasting and uses the remaining channels for recursive patching. The experimental results show that the HQoS strategy performs 35%–40% reduction in terms of the blocking ratio [75].

5) *Hybrid of the CIWP and SIWP scheme* [37]. The combination of CIWP and SIWP techniques uses controlled CIWP for scheduling cold videos and greedy disk-conserving broadcast [76] (a server-initiated broadcasting technique) for scheduling hot videos. A video classification algorithm divides videos into hot and cold according to the expected number of channels required for delivering video under CIWP or GSB. This hybrid scheme can enhance the overall performance of the VoD system [37].

6) *Hybrid transmission scheme using dynamic programming* [77]. This scheme uses FB for broadcasting the most popular videos and GP for multicasting the least popular videos. The number of channels allocated for periodic broadcasting and multicasting is determined by an optimization problem [77] in such a way that the user waiting time is minimized. Analytical results show that the hybrid scheme outperforms FB in terms of user waiting time [77].

1.7 Summary and Future Research Directions

For successful development of VoD, the VoD service providers need to provide low-cost services for a huge number of users. The cost of video delivery is the most important challenge for service providers. Request-based multicasting schemes introduce low-cost video transmission in a VoD system. The main purpose of this survey has been to provide a comprehensive study for all those techniques that were designed to support request-based multicasting. In particular, we have studied batching, patching, piggybacking, stream merging, and hybrid schemes. The schemes were classified according to the map presented in Figure 1.1.

Table 1.5 provides a qualitatively comparative summary for all the surveyed schemes. Multicasting techniques can significantly reduce the server and network bandwidth requirement. Batching is a

Table 1.5 Comparison among Request-Based Multicasting Schemes

	Category	References	Maximum Startup Latency	Bandwidth Requirements	Additional System Requirements	Support VCR Operations
Batching	Fixed batching	[2], [14], [20–27]	High	Single I/O stream	Not required	Yes (e.g., LAMB and SAM)
	Adaptive batching	[28], [30–31], [33–34]	Maximum batching interval	Single I/O stream (less than fixed batching)	Not required	N/A
Patching	Continuous	[15], [35], [37–44], [46–49], [52–53]	Low	Single multicast stream + multiple patching stream (depending on the scheme)	Client buffer	Yes (e.g., BEP)
	Segmentation	[53–58], [60]	Maximum segment size	Single multicast stream + multiple patching stream (depending on scheme)	Client buffer	N/A
Piggybacking		[61–63]	Low	One or two streams with different display rates	Hardware/additional copy	N/A
Stream merging		[13], [16], [29], [65–68]	Low	Multiple stream close to the minimum amount	Client buffer	Yes
Hybrid strategy		[37], [70–72], [74–75], [77]	Moderate	Moderate	Depends on the scheme	Yes (e.g., HQoS)

simple scheme that does not require additional buffer in client stations. It reduces bandwidth requirement since it services all batched requests with a single stream. On the other hand, it introduces startup latency for later request.

Patching, stream merging, and piggybacking can provide immediate service to each client due to the fact that all of these techniques apply the patching stream to send the initial portion of the requested video. The bandwidth requirement of patching is not optimal. Hence, the hierarchical stream merging uses heuristic policy to find the optimal merging plan, and therefore, the bandwidth usage of this technique is near the minimum required server bandwidth. However, clients must be able to buffer data from multiple streams. Because of the overload merging process, implementation of stream merging is complex. Piggybacking also needs either special hardware to adjust the display rate in real time or an additional storage to maintain multiple copies of a video with different display rates at the central server. Although request-based multicasting has some restrictions on supporting interactive playback control, a number of studies have been proposed to support VCR-like interactions such as LAMB, SAM protocol, BEP, and HQoS.

Which of the reviewed techniques fits best depends on the network infrastructure, location of the video server and its distance to viewers, and the required metric such as bandwidth consumption and startup latency. Many of these schemes are not practical in real networks, and they need to be evaluated with real parameters. The SAM provides the capability of sharing the same video stream, and therefore, it declines the costs of the VoD system. SAM seems to be a good scheme for VoD service providers. The S2 piggybacking policy constructs the best optimized merging tree, and therefore, transferred frames are minimized in proportion to other piggybacking policies such as the snapshot policy. This means a reduction in bandwidth consumption.

The following approaches could be the research directions in the future:

- Work on different schemes to support variable bit rate video delivery and also the continuous VCR-like control functions.
- Investigation on VoD delivery schemes in broadband networks, IP-based networks, and peer-to-peer structures.

References

1. Kevin C. Almeroth and Mostafa H. Ammar, On the performance of a multicast delivery video-on-demand service with discontinuous VCR actions. In *Proceedings of the IEEE International Conference on Communications (ICC)*, Seattle, Washington, USA, vol. 3, pp. 1631–1635, June 1995.
2. Wanjiun Liao and Victor O.K. Li, The split and merge protocol for interactive video-on-demand, *IEEE Multimedia*, vol. 4, pp. 51–62, October–December 1997.
3. Ramaprabhu Janakiraman, Marcel Waldvogel, and Lihao Xu, Fuzzycast: Efficient video-on-demand over multicast. In *Proceedings of the Twenty-First Annual Joint Conference of the IEEE Computer and Communications Societies (INFOCOM 2002)*, vol. 2, pp. 920–929, 2002.
4. Ailan Hu, Video-on-demand broadcasting protocols: A comprehensive study, In *Proceedings of the 20th Annual Joint Conference of the IEEE Computer and Communications Societies (INFOCOM 2001)*, Anchorage, AK, USA, vol. 1, pp. 508–517, 2001.
5. Steven W. Carter, Darrell D. E. Long, and Jehan-Francois Paris, *Video-on-Demand Broadcasting Protocols*. Academic Press, San Diego, 2000.
6. Tiko Kameda and Richard Sun, *Survey on VOD Broadcasting Schemes*. School of Computing Science, Simon Fraser University, 2006.
7. Huadong Ma and Kang G. Shin, Multicast video-on-demand services, *ACM SIGCOMM Computer Communication Review*, vol. 32, pp. 31–43, 2002.

8. Lixin Gao, Zhi-Li Zhang, and Don Towsley, Proxy-assisted techniques for delivering continuous multimedia streams, *IEEE/ACM Transactions on Networking (TON)*, vol. 11, pp. 884–894, December 2003.

9. Bing Wang, Subhabrata Sen, Micah Adler, and Don Towsley, Optimal proxy cache allocation for efficient streaming media distribution. In *Proceedings of the Twenty-First Annual Joint Conference of the IEEE Computer and Communications Societies (INFOCOM 2002)*, pp. 1726–1735, 2002.

10. Sridhar Ramesh, Injong Rhee, and Katherine Guo, Multicast with cache (mcache): An adaptive zero-delay video-on-demand service, *Circuits and Systems for Video Technology, IEEE Transactions on*, vol. 11, pp. 440–456, March 2001.

11. Kien A. Hua, Due A. Tran, and Roy Villafane, Caching multicast protocol for on-demand video delivery. In *Proceedings of S&T/SPIE Conference on Multimedia Computing and Networking (MMCN)*, vol. 3969, pp. 2–13, 2000.

12. Bing Wang, Subhabrata Sen, Micah Adler, and Don Towsley, Proxy-based distribution of streaming video over unicast/multicast connections. In *Proceedings of the 21st Annual Joint Conference of the IEEE Computer and Communications Societies (INFOCOM 2002)*, June 2002.

13. Amotz Bar-Noy and Richard E. Ladner, Competitive on-line stream merging algorithms for media-on-demand. In *Proceedings of the 12th Annual ACM-SIAM Symposium on Discrete algorithms (SODA)*, Philadelphia, PA, USA, pp. 364–373, 2001.

14. Asit Dan, Dinkar Sitaram, and Perwez Shahabuddin, Scheduling policies for an on-demand video server with batching. In *Proceedings of the Second ACM International Conference on Multimedia*, New York, 1994.

15. Kien A. Hua, Ying Cai, and Simon Sheu, Patching: A multicast technique for true video-on-demand services. In *Proceedings of the Sixth ACM International Conference on Multimedia*, New York, pp. 191–200, 1998.

16. Derek Eager, Mary Vernon, and John Zahorjan, Minimizing bandwidth requirements for on-demand data delivery. *IEEE Transactions on Knowledge and Data Engineering*, vol. 13, pp. 742–757, 2001.

17. Debasish Ghose and Hyoung Joong Kim, Scheduling video streams in video-on-demand systems: A survey. *Multimedia Tools and Applications*, Springer, The Netherlands, vol. 11, pp. 167–195, 2000.

18. Joonho Choi, Abu (Sayeem) Reaz, and Biswanath Mukherjee, A survey of user behavior in VoD service and bandwidth-saving multicast streaming schemes. *Communications Surveys and Tutorials, IEEE*, pp. 1–14, 2010.

19. David P. Anderson, Metascheduling for continuous media. *ACM Transactions on Computer Systems (TOCS)*, vol. 11, pp. 226–252, 1993.

20. Kien A. Hua, Jung Hwan Oh, and Khanh Vu, An adaptive hybrid technique for video multicast. In *Proceedings of the Seventh International Conference on Computer Communications and Networks*, Lafayette, LA, USA, pp. 227–234, October 1998.

21. Charu C. Aggarwal, Joel L. Wolf, and Philip S. Yu, On optimal batching policies for video-on-demand storage servers. In *Proceedings of International Conference on Multimedia Computing and Systems*, Hiroshima, Japan, pp. 253–258, June 1996.

22. Athanassios K. Tsiolis and Mary K. Vernon, Group-guaranteed channel capacity in multimedia storage servers. *ACM SIGMETRICS Performance Evaluation Review*, vol. 25, pp. 285–297, 1997.

23. Hadas Shachnai and Philip S. Yu, Exploring wait tolerance in effective batching for video-on-demand scheduling. *Multimedia Systems*, Springer, The Netherlands, vol. 6, pp. 382–394, 1998.

24. Nelson Luis Saldanha Da Fonseca and Roberto de Almeida Façanha, The look-ahead-maximize-batch batching policy. In *Proceedings of Global Telecommunications Conference (GLOBECOM)*, vol. 1a, pp. 354–358, December 1999.

25. Nabil J. Sarhan and Chita R. Das, A new class of scheduling policies for providing time of service guarantees in Video-On-Demand servers. *Management of Multimedia Networks and Services*, Springer, The Netherlands, vol. 3271, pp. 199–236, 2004.

26. Vaneet Aggarwal, Robert Caldebank, Vijay Gopalakrishnan, Rittwik Jana, K. K. Ramakrishnan, and Fang Yu, The effectiveness of intelligent scheduling for multicast video-on-demand. In *Proceedings of the 17th ACM International Conference on Multimedia*, New York, pp. 421–430, 2009.

27. Santosh Kulkarni, Bandwidth efficient video-on-demand algorithm (BEVA). In *Proceedings of the 10th International Conference on Telecommunications (ICT 2003)*, vol. 2, pp. 1335–1342, 2003.

28. Wing-Fai Poon, Kwok-Tung Lo, and Jian Feng, Adaptive batching scheme for multicast video-on-demand systems. *IEEE Transactions on Broadcasting*, vol. 47, pp. 66–70, 2001.
29. Wing-Fai Poon, Kwok-Tung Lo, and Jian Feng, Batching policy for video-on-demand in multicast environment. *Electronics Letters*, vol. 36, pp. 1329–1330, 2000.
30. Wing-Fai Poon, Kwok-Tung Lo, and Jian Feng, Scheduling policy for multicast video-on-demand system. *Electronics Letters*, vol. 37, pp. 138–140, 2001.
31. Madhu Jain, Vidushi Sharma, and Kriti Priya, Adaptive batching scheme for multicast near video-on-demand (NVOD) system. *Journal of ICT*, pp. 1–12, 2006.
32. Sarat Pothuri, David W. Petr, and Sohel Khan, Characterizing and modeling network traffic variability. In *Proceeding of the IEEE International Conference on Communications (ICC 2002)*, vol. 4, pp. 2405–2409, 2002.
33. Wushao Wen, Shueng-Han Gary Chan, and Biswanath Mukherjee, Token-tray/weighted queuing-time (TT/WQT): An adaptive batching policy for near video-on-demand system. *Computer Communications*, Elsevier, vol. 25, pp. 890–904, 2002.
34. Hadas Shachnai and Philip S. Yu, On analytic modeling of multimedia batching schemes. *Performance Evaluation*, Elsevier, vol. 33, pp. 201–213, 1998.
35. Ying Cai, Wallapak Tavanapong, and Kien A. Hua, A double patching technique for efficient bandwidth sharing in video-on-demand systems. *Multimedia Tools and Applications*, Springer, The Netherlands, vol. 32, pp. 115–136, 2007.
36. Ying Cai, Kien A. Hua, and Khanh Vu, Optimizing patching performance. In *Proceedings of the IS&T/SPIE Conference on Multimedia Computing and Networking (MMCN '99)*, pp. 204–215, 1999.
37. Lixin Gao and Don Towsley, Supplying instantaneous video-on-demand services using controlled multicast. In *Proceedings on the IEEE International Conference on Multimedia Computing and Systems*, Florence, pp. 117–121, July 1999.
38. Ying Cai and Kien A. Hua, An efficient bandwidth-sharing technique for true video on demand systems. In *Proceedings of the Seventh ACM international on Multimedia*, New York, 1999.
39. Dongliang Guan and Songyu Yu, A two-level patching scheme for video-on-demand delivery. *IEEE Transactions on Broadcasting*, vol. 50, pp. 11–15, 2004.
40. Satish Chand, Bijendra Kumar, and Hari Om, Patching-based broadcasting scheme for video services. *Computer Communications*, Elsevier, vol. 31, pp. 1970–1978, 2008.
41. Huadong Ma, G. Kang Shin, and Weibiao Wu, Best-effort patching for multicast true VoD service. *Multimedia Tools and Applications*, Springer, The Netherlands, vol. 26, pp. 101–122, 2005.
42. Ying Wai Wong and Jack Yui-Bun Lee, Recursive Patching: An efficient technique for multicast video streaming. In *Proceedings of the Fifth International Conference on Enterprise Information Systems (ICEIS)*, 2003.
43. Sook-Jeong Ha, Sun-Jin Oh, and Ihn-Han Bae, Practical patching for efficient bandwidth sharing in VOD systems. In *Proceedings of the Third International Conference on Natural Computation (ICNC)*, Haikou, pp. 351–355, August 2007.
44. Han Dai and Edward Chan, Quick patching: An overlay multicast scheme for supporting video on demand in wireless networks. *Multimedia Tools and Applications*, IEEE, vol. 36, pp. 221–242, 2008.
45. Yi Cui, Baochun Li, and Klara Nahrstedt, oStream: Asynchronous streaming multicast in application-layer overlay networks. *IEEE Journal on Selected Areas in Communications*, vol. 22, pp. 91–106, 2004.
46. Sook-Jeong Ha and Ihn-Han Bae, Design and evaluation of a converting patching agent for VOD services. *Agent and Multi-Agent Systems: Technologies and Applications*, Springer, The Netherlands, pp. 704–710, 2007.
47. Zhe Xiang, Yuzhuo Zhong, and Shi-Qiang Yang, Period Patch: An efficient stream schedule for video on demand. In *Proceedings of SPIE*, Boston, November 2000.
48. Carsten Griwodz, Michael Liepert, Michael Zink, and Ralf Steinmetz, Tune to lambda patching. *ACM SIGMETRICS Performance Evaluation Review*, vol. 27, pp. 20–26, 2000.
49. Espen Jacobsen, Carsten Griwodz, and Pål Halvorsen, Pull-patching: A combination of multicast and adaptive segmented HTTP streaming. In *Proceedings of the International Conference on Multimedia*, New York, pp. 799–802, 2010.

50. Dag Johansen, Håvard Johansen, Tjalve Aarflot, Joseph Hurley, Åge Kvalnes, Cathal Gurrin, Sorin Zav, Bjørn Olstad, Erik Aaberg, and Tore Endestad, DAVVI: A prototype for the next generation multimedia entertainment platform. In *Proceedings of the 17th ACM International Conference on Multimedia*, New York, pp. 989–990, 2009.

51. Lixin Gao and Don Towsley, Threshold-based multicast for continuous media delivery. *IEEE Transactions on Multimedia*, vol. 3, pp. 405–414, 2001.

52. Steven W. Carter and Darrell D. E. Long, Improving bandwidth efficiency of video-on-demand servers. *Computer Networks*, Elsevier, vol. 31, pp. 111–123, 1999.

53. Shueng-Han Gary Chan and S. H. Ivan Yeung, *Client buffering techniques for scalable video broadcasting over broadband networks with low user delay.* IEEE Transactions on Broadcasting, vol. 48, pp. 19–26, 2002.

54. Katsuhiko Sato, Michiaki Katsumoto, and Tetsuya Miki, *Fragmented patching: new VOD technique that supports client mobility.* In *Proceedings of the 19th International Conference on Advanced Information Networking and Applications*, pp. 527–532, March 2005.

55. Helmut Hlavacs and Shelley Buchinger, Hierarchical video patching with optimal server bandwidth. *ACM Transactions on Multimedia Computing, Communications, and Applications (TOMCCAP)*, vol. 4, p. 8, 2008.

56. Satish Chand, Bijendra Kumar, and Hari Om, Video data delivery using slotted patching. *Journal of Network and Computer Applications*, Elsevier, vol. 32, pp. 660–665, 2009.

57. Salahuddin A. Azad, Mohammad Murshed, and Laurence S. Dooley, A novel batched multicast patching scheme for video broadcasting with low user delay. In *Proceedings of the 3rd IEEE International Symposium on Signal Processing and Information Technology*, pp. 339–342, December 2003.

58. Eundon Song, Hongik Kim, and Sungkwon Park, Fast patching scheme for video-on-demand service. In *Proceedings of the Asia-Pacific Conference on Communications, Bangkok*, pp. 523–526, October 2007.

59. Li-Shen Juhn and Li-Ming Tseng, Fast data broadcasting and receiving scheme for popular video service. *IEEE Transactions on Broadcasting*, vol. 44, pp. 100–105, 1998.

60. Hai Jin, Dafu Deng, and Liping Pang, Medusa: A novel stream-scheduling scheme for parallel video servers. *EURASIP Journal on Applied Signal Processing*, vol. 2004, pp. 317–329, 2004.

61. Leana Golubchik, John C. S. Lui, and Richard R. Muntz, Adaptive piggybacking: A novel technique for data sharing in video-on-demand storage servers. *Multimedia Systems*, Springer, The Netherlands, vol. 4, pp. 140–155, 1996.

62. Charu Aggarwal, Joel Wolf, and Philip S. Yu, On optimal piggyback merging policies for video-on-demand systems. *ACM*, vol. 24, pp. 200–209, 1996.

63. Roberto De A. Façanha, Nelson L. S. Da Fonseca, and Pedro J. De Rezende, The S2 piggybacking policy. *Multimedia Tools and Applications*, Springer, The Netherlands, vol. 8, pp. 371–383, 1999.

64. Derek L. Eager and Mary K. Vernon, Dynamic skyscraper broadcasts for video-on-demand. *Advances in Multimedia Information Systems*, Springer, The Netherlands, vol. 1508, pp. 18–32, 1998.

65. Derek Eager, Mary Vernon, and John Zahorjan, Optimal and efficient merging schedules for video-on-demand servers. In *Proceedings of the Seventh ACM International Conference on Multimedia*, New York, 1999.

66. Derek L. Eager, Mary K. Vernon, and John Zahorjan, Bandwidth skimming: A technique for cost-effective video-on-demand. In *Proceedings of the ACM/SPIE Multimedia Computing and Networking*, San Jose, CA, USA, 2000.

67. Edward G. Coffman, Jr., Predrag Jelenkovic, and Petar Momcilovic, The dyadic stream merging algorithm. *Journal of Algorithms*, vol. 43, pp. 120–137, 2002.

68. Wun-Tat Chan, Tak-Wah Lam, Hing-Fung Ting, and Prudence W. H. Wong, On-line stream merging in a general setting. *Theoretical Computer Science*, Elsevier, vol. 296, pp. 27–46, 2003.

69. Amotz Bar-Noy, Justin Goshi, Richard E. Ladner, and Kenneth Tam, Comparison of stream merging algorithms for media-on-demand. *Multimedia Systems*, Springer, The Netherlands, vol. 9, pp. 411–423, 2004.

70. Hyoung Joong Kim and Yu Zhu, Channel allocation problem in VoD system using both batching and adaptive piggybacking. *IEEE Transactions on Consumer Electronics*, vol. 44, pp. 969–976, 1998.

71. Nelson L. S. Fonseca and Roberto A. Facanha, Integrating batching and piggybacking in video servers. In *Proceedings of the IEEE Global Telecommunications Conference*, San Francisco, CA, pp. 1334–1338, vol. 3, 2000.
72. Kien A. Hua, Jung Hwan Oh, and Khanh Vu, An adaptive video multicast scheme for varying workloads. *Multimedia Systems*, Springer, The Netherlands, vol. 8, pp. 258–269, 2002.
73. Kien A. Hua and Simon Sheu, Skyscraper broadcasting: A new broadcasting scheme for metropolitan video-on-demand systems. In *Proceedings of the ACM SIGCOMM on Applications, Technologies, Architectures, and Protocols for Computer Communication*, New York, 1997.
74. Yunqiang Liu, Songyu Yu, and Jun Zhou, Adaptive segment-based patching scheme for video streaming delivery system. *Computer Communications*, Elsevier, vol. 29, pp. 1889–1895, 2006.
75. D. N. Sujatha, K. Girish, Rajuk Venugopal, and Lalit Mohan Patnaik, An integrated quality-of-service model for video-on-demand application. *IAENG International Journal of Computer Science*, vol. 34, pp. 1–9, 2007.
76. Lixin Gao, Jim Kurose, and Don Towsley, Efficient schemes for broadcasting popular videos. *Multimedia Systems*, Springer, The Netherlands, vol. 8, pp. 284–294, 2002.
77. Salahuddin A. Azad and Manzur Murshed, An efficient transmission scheme for minimizing user waiting time in video-on-demand systems. *Communications Letters, IEEE*, vol. 11, pp. 285–287, 2007.

Chapter 2

P2P Video Streaming

Sabu M. Thampi

Contents

2.1 Introduction

Video has been an important medium for communications and entertainment for many decades. Movie is a form of entertainment that enacts a story by screening a series of images, giving the delusion of continuous movement. The trick was already known in second-century China but still raised curiosity up to the end of the nineteenth century. The invention of the motion-picture camera around 1888 allowed individual component images to be captured and stored on a single reel. For the first time, this has made possible the process of recording scenes in an automatic manner. In addition to that, a hasty transformation occurred with the development of a motion-picture projector to enlarge these moving picture shows onto a screen for an entire audience. Television broadcasting, after its invention in 1928, has attracted billions of people from different parts of the world to watch both live events and recorded videos simultaneously through their television sets. People moved from newspaper and radio to the more immersive experience of television as their primary source of entertainment and as a way to receive important information and news about the world [1]. For most of the twentieth century, the only ways to watch television were through over-the-air broadcasts and cable signals.

A third boost in the popularity of moving pictures came at the end of the twentieth century with the invention of the Internet and the World Wide Web (WWW). Web browsing and file transfer are the dominant services provided by the Internet. However, these kinds of service providing information about text, pictures, and document exchange no longer satisfy the demand of clients. Following the success of conventional radio and television broadcasting, research has been carried out into ways of delivering live media over the Internet to a personal computer. As a result, people have experimented with transmitting various multimedia data such as sound and video over the Internet. All multimedia contents were distributed no differently than any other ordinary files, such as text files and executable files. They were all transmitted as "files" using file-downloading protocols, such as ftp and http. The full file transfer, in download mode, can often suffer unacceptably long transfer times, which depend on the size of the media file and the bandwidth of the transport channel. For example, if downloaded from http://www.mp3.com, an MPEG Audio Stream Layer III (MP3) audio file encoded at 128 kb/s and of 5-min duration will occupy 4.8 MB of the user's hard disk. Using a 28.8k dial-up modem, it would take roughly 40 min to download the whole file [2]. As a result, an audio file might take more real time to download than the length of the audio being played. Video files, which carry much more information than audio files, entailed even longer download times [3]. Furthermore, there was no way for the users to "peek" into the content to see if it is the video they would like to watch. This was often inconvenient for users due to a long waiting time and a large amount of wasted resources when the content of the video turned out to be something in which they were not interested [4].

Internet evolves and operates basically without a central coordination, the lack of which was and is vitally important to its rapid escalation and evolution. However, the lack of management, in turn, makes it very difficult to guarantee proper performance and to deal systematically with performance issues. Meanwhile, the available network bandwidth and server capacity continue to be besieged by the mounting Internet utilization and the accelerating escalation of bandwidth-demanding content. As a result, Internet service quality perceived by customers is largely unpredictable and inadequate [5]. The current Internet is inherently a packet-switched network that was not designed to handle continuous time-based traffic such as audio and video. The Internet only provides best-effort services and has no guarantee on the quality of service (QoS) for multimedia data transmission [6].

Recent advances in digital technologies such as high-speed networking, media compression technologies, and fast computer processing power have made it feasible to provide real-time multimedia services over the Internet. Real-time multimedia, as the name implies, has timing constraints. For example, audio and video data must be played out continuously. If the data do not arrive in time, the play-out process will pause, which is annoying to human ears and eyes. Real-time transport of live video or stored video is the predominant part of real-time multimedia. *Streaming* is an enabling technology for providing multimedia data delivery among clients in various multimedia applications on the Internet. With this technology, the client can play back the media content without waiting for the entire media file to arrive. Thus, streaming allows real-time transmission of multimedia over the net. Internet streaming media changed the Web as we knew it, that is, changed it from a static text- and graphics-based medium to a multimedia experience populated by sound and moving pictures [7]. Web sites such as *YouTube* provide media content to millions of viewers. American National Standard for Telecommunications defines *streaming* as "a technique for transferring data (usually over the Internet) in a continuous flow to allow large multimedia files to be viewed before the entire file has been downloaded to a client's computer" [8]. The basic idea of video streaming is to split the video into parts, transmit these parts in succession, and enable the receiver to decode and play back the video as these parts are received, without having to wait for the entire video to be delivered. Thus, streaming enables near instantaneous playback of multimedia contents in spite of their sizes. Streaming media utilizes a very old concept called *buffering* to make feasible the playback of multimedia content as it is being downloaded. A buffer clasps a pool of content sufficiently large to stabilize the bumps in playback that may be caused by transitory server slowdown or network overcrowding.

Streaming diminishes the storage space and permits users to stop receiving the stream, if not interesting or satisfactory, before the entire file is downloaded. Streaming allows live and pre-coded content to be distributed. Live streaming captures audio/video signals from input devices (e.g., microphone and video camera), encodes the signals using compression algorithms (e.g., MP3 and MPEG-4), and distributes them in real time. Typical application of live streaming includes surveillance, broadcasting of special events, and distribution of information, that has the prime importance in real-time delivery. In live streaming, the server side has control over the selection of distribution content and the timing of their streaming. User involvement is typically limited to joining and leaving the running streaming sessions. Prerecorded or stored streaming distributes preencoded video files stored at a media server. Sample applications include multimedia archival retrievals, news clip viewing, and distance learning through which students attend classes online by viewing prerecorded lectures [4].

With the rise of broadband Internet connections, end users became able to receive video of acceptable quality on their home computers. Broadband has achieved mass-market penetration in several countries. According to the world's leading information technology research and advisory company, Gartner, worldwide consumer broadband connections will grow from 323 million connections in 2007 to 580 million in 2013. This ensures that a large number of consumers will have sufficient bandwidth to receive streaming video and audio in the near future. Now, streaming media is poised to become the de facto global media broadcasting and distribution standard, incorporating all other media, including television, radio, and film. According to an industry study [9], there are more than 60 million people listening to or watching streaming media each month, 58 U.S. TV stations performing live webcasting, 34 offering on-demand streaming media programs, and 69 international TV webcasters. The study also found that 6000 h of new streaming programming is created each week. The market for streaming content has grown substantially in Europe. For instance, the BBC, which reaches an audience of over 1 million a month, estimates that its streaming audience size is growing by 100% every 4 months. One of the leading French streaming sites, CanalWeb, boasts over 450,000 unique viewers per month, with video content watched for an average of 12 min. In the U.K., RealNetworks estimates that 500,000 users downloaded its player from the Big Brother Web site (www.bigbrother2000 .com). Big Brother U.K. reports it was serving at least 6000 simultaneous streams and 1.5 million streams per day. Market research firm NetValue reports that the average viewing time for these streams was 25 min. RealPlayer users are an increasingly international group, totaling over 48 million regular users, with approximately one-third of downloads/registrations now originating outside North America [9].

2.2 Architecture for Video Streaming

Figure 2.1 shows the architecture for video streaming, and it is divided into six areas as follows: *media compression, application-layer QoS control, media distribution services, streaming servers, media synchronization at the receiver side,* and *streaming media protocols.*

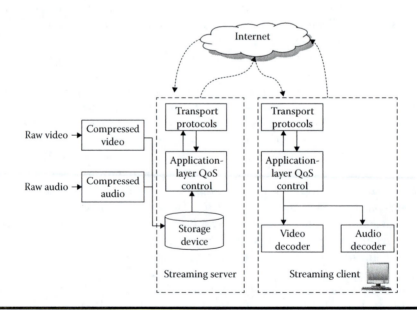

Figure 2.1   Video streaming architecture. (From Wu, D. et al., *IEEE Transactions on Circuits and Systems for Video Technology,* **11(3), pp. 282–300, 2001.)**

2.2.1 Media Compression

The large volume of raw multimedia data imposes a stringent bandwidth requirement on the network. Hence, for achieving better transmission efficiency, compression is widely employed. While video needs superior bandwidth requirements (56 kb/s–15 Mb/s) than audio (8–128 kb/s) and loss of audio is more infuriating to human than video, audio is given higher priority for transmission in a multimedia streaming system. For this reason, only video will be used for alteration to meet the QoS requirements [6]. In Figure 2.1, raw video and audio data are precompressed by video compression and audio compression algorithms and then saved in storage devices. Video compression is accomplished by utilizing the resemblances or redundancies that subsist in a normal video signal. Video compression reduces the irrelevance in the video signal by only coding video features that are perceptually important [10]. Video compression follows a standard for multimedia contents that encodes the content with a specific play rate. There are two major groups that define video encoders: the International Telecommunications Union (ITU) and the International Standards Organization (ISO). The ITU-T group (Telecommunication Standardization Sector of the ITU) defines the H.26x video formats, whereas the ISO group defines the formats that have materialized from committees of the Moving Pictures Experts Group: MPEG-x. The MPEG-4 standard is commonly designed for streaming media and compact disc distribution, video conversion, and broadcast television. MPEG-4 includes numerous features of MPEG-1, MPEG-2, and other associated standards. H.264 is also known as MPEG-4 part 10 or advanced video coding. Big Internet players such as Google/YouTube or Apple Tunes are founded on this standard.

2.2.2 Application-Layer QoS Control

Upon the client's request, a *streaming server* retrieves compressed video/audio data from storage devices, and then, the *application-layer QoS control* module adapts the video/audio bit streams according to the network status and QoS requirements. The application-layer QoS control involves *congestion control* and *error control*, which are implemented at the application layer. The former is used to determine the transmission rate of media streams based on the estimated network bandwidth, whereas the latter aims at matching the rate of precompressed media bit streams to the target rate constraint by using filtering [11].

Typically, for streaming video, congestion control takes the form of *rate control*. Rate control attempts to minimize the possibility of network congestion by matching the rate of the video stream to the available network bandwidth. Based on the place where rate control is taken in the system, rate control can be categorized into three types: *source-based, receiver-based, and hybrid-based*. With source-based rate control, only the sender (server) is responsible for adapting the transmission rate. In contrast, the receiving rate of the streams is regulated by the client in the receiver-based method. Hybrid-based rate control employs the aforementioned schemes at the same time, that is, both the server and the client are needed to participate in the rate control. Typically, the source-based scheme is used in either unicast or multicast environment, whereas the receiver-based method is deployed in multicast only [6].

The function of error control is to improve video presentation quality in the presence of packet loss. Error control mechanisms include *forward error correction (FEC), retransmission, error-resilient encoding, and error concealment*. With the FEC scheme, the received packets at the receiver end are FEC decoded and unpacked, and the resulting bit stream is then input to the video decoder to reconstruct the original video. *Error-resilient encoding* is executed by the source to enhance robustness of compressed video before packet loss actually happens. Even when an image sample or a block of samples are missing due to transmission errors, the decoder can try to estimate them based on surrounding received samples by making use of inherent correlation among spatially and temporally adjacent samples; such techniques are known as *error concealment* techniques [12].

2.2.3 Media Distribution Services

After the adaptation by the *application-layer QoS control* module, the *transport protocols* packetize the compressed bit streams and send the video/audio packets to the Internet. Packets may be dropped or experience excessive delay inside the Internet due to congestion. In addition to the application-layer support, adequate *network* support is necessary to reduce transport delays and packet losses. The network support involves *network filtering, application-level multicast*, and *content replication (caching)*. Network filtering maximizes video quality during network congestion. The filter at the video server can adapt the rate of video streams according to the network congestion status. The *application-level multicast* provides a multicast service on top of the Internet. These protocols do not modify the network infrastructure; instead, they employ multicast forwarding functionality solely at end hosts. *Content replication improves scalability of the media delivery system.*

2.2.4 Streaming Servers

Streaming servers play an important role in providing streaming services. To offer superiority streaming services, streaming servers are required to process multimedia data in real time, support VCR-like functions, and retrieve media components in a synchronous fashion. A streaming server generally waits for a Real-Time Streaming Protocol (RTSP) request from the viewers. When it gets a request, the server looks in the appropriate folder for a hinted media of the requested name. If the requested media is in the folder, the server streams it to the viewer using Real-Time Transport Protocol (RTP) streams.

2.2.5 Media Synchronization at the Receiver Side

With media synchronization mechanisms, the application at the receiver side can present various media streams in the same way as they were originally captured. An example of media synchronization is synchronizing the movements of a speaker's lips with the sound of his speech.

2.2.6 Protocols for Streaming Media

Streaming protocols provide means to the client and the server for services negotiation, data transmission, and network addressing. According to the functionalities, the protocols directly related to Internet streaming video can be classified as *network-layer protocol, transport protocol*, and *session control protocol*.

The network-layer protocol provides basic network service support, such as network addressing. The Internet Protocol (IP) serves as the *network-layer protocol* for Internet video streaming. *Transport protocol* provides end-to-end network transport functions for streaming applications. Transport protocols include User Datagram Protocol (UDP), Transmission Control Protocol (TCP), RTP, and Real-Time Control Protocol (RTCP). RTP and RTCP are upper layer transport protocols implemented on top of UDP/TCP. UDP and TCP support such functions as multiplexing, error control, congestion control, and flow control. RTP is a data transfer protocol. RTCP provides QoS feedback to the participants of an RTP session. *Session control protocol* defines the messages and procedures to control the delivery of the multimedia data during an established session. RTSP and the Session Initiation Protocol (SIP) are such session control protocols. RTSP is a protocol used in streaming media systems, which allows a client to remotely control a streaming media server, issuing VCR-like commands. It also allows time-based access to files on a server. SIP is a session protocol that can create and terminate sessions with one or more participants. It is mainly designed for interactive multimedia application, such as Internet phone and video conferencing [6].

2.3 Existing Streaming Networks

There are three important means in which a streaming service may be offered over the Internet. The first approach employs caching and replication for Web-based distribution of a small amount of streaming media. For large-scale service, streaming content is distributed through a content delivery network (CDN), which perks up the scalability of Web-based content sharing. The second method is to use a network specifically designed for the distribution of streaming content. A number of networks have been proposed that are specialized in on-demand delivery of video streams. These networks are called *on-demand multimedia streaming networks*. The third option, that is, live streaming systems, allows clients to simultaneously watch a number of television stations through the broadband Internet connectivity available at their homes.

2.3.1 Web-Based Distribution

Web-based distribution is the most frequently used technique to serve small streaming content. As the Internet has become a vital part of daily life, hundreds of millions of users currently connect to the Internet. Due to client/server-based computing models, Web-based content distribution architecture suffers from server overloading when a large number of user requests arrive. Hence, appropriate schemes are required to manage the server loads effectively. *Content caching* and *replication techniques* direct the workload away from possibly overloaded origin Web servers to deal with Web performance and scalability from the client side and the server side, respectively [13]. CDNs are another approach being widely employed to perk up Internet service quality.

Caching stores a copy of data close to the data consumer to allow faster data access than if the content had to be retrieved from the origin server. For prerecorded content, a streaming media-caching server can fetch and store the entire contents for a user. When other users request the similar content, the cache can deliver the stream directly out of its local storage. Web caching lessens the access latency, saves the central processing unit (CPU) cycle of a Web server, and reduces network bandwidth usage. However, it is not usually considered an excellent solution for streaming video content as caching of a video stream requires a very large buffer space [14]. *Replication* creates and maintains distributed copies of content under the control of content providers. This is obliging because client requests can then be sent to the adjacent and least loaded server. Several Web sites replicate their content at multiple servers, with the intention of reducing the load on the originating server. Replication also provides server redundancy in case of server and network failures. On the other hand, due to the unique nature of the WWW, its massive user community, document multiplicity, and access pattern replication seem to be not able to stand up fully to all of its conceptual promises with respect to latency and bandwidth reduction [15].

A *CDN* replicates content from the origin server to cache servers (also called replica servers) spread across the globe. Content requests are directed to the cache server closest to the user, and that server delivers the requested content. As a result, users get greater speed and higher quality. There are two general approaches for building CDNs: *overlay* and *network*.

In the *overlay approach*, application-specific servers and caches at several places in the network handle the distribution of specific content types, such as streaming media. Most of the commercial CDN providers such as Akamai and Limelight Networks follow the overlay approach for CDN organization. The core network (CN) components such as routers and switches play no active role in content delivery. The Akamai system has more than 12,000 servers in more than 1000 networks. In the *network approach*, the network components including routers and switches are equipped with code for identifying specific application types and for forwarding the requests based on predefined policies. Examples of this approach include devices that redirect content requests to local caches or switch traffic to specific servers, optimized to serve specific content types [16]. Aside from increased server capacity and resiliency, a CDN gives controlled load balancing and enhanced content accessibility. Operating servers in various locations creates several technical challenges, including how to direct user requests to suitable servers, how to manage failures, how to monitor and control the servers, and how to update software across the system [17,18]. The amount of load the network can manage is preset by the overall CDN capacity. Special events and programs frequently produce additional demands than what the network can handle in a short period of time, and CDN will not be able to bear those excess demands. At those locations where demands become high, a suitable mechanism that permits dynamic addition and removal of replica servers is required. Accordingly, in order for a CDN to be really successful, a large number of replication servers must be set up throughout the Internet. Such an arrangement may not be possible by small organizations [4].

2.3.2 On-Demand Multimedia Streaming

Video-on-demand (VoD), also known as *on-demand video streaming*, is a great way of viewing films and television programs. VoD service enables immediate distribution of video streams to users, from the beginning of the content, regardless of the time at which the service request arrives in relation to other on-going streaming sessions. Typically, these video files are stored in a set of central video servers and distributed through high-speed communication networks to geographically dispersed clients. Upon receiving a client's service request, a server delivers the video to the client as an isochronous video stream. VoD has become an extremely popular service on the Internet. For example, YouTube, a video-sharing service that streams its videos to users on demand, has more than 20 million views a day with a total viewing time of over 10,000 years to date. Other major Internet VoD publishers include MSN Video, Google Video, Yahoo Video, CNN, and a plethora of copycat YouTube sites [19]. VoD wipes out the necessity to go to your video store to buy films and offers access to a large collection of materials. With VoD, users will have the flexibility of choosing the content, as well as scheduling the program they desire to watch [20].

There are two major ways to implement the VoD architecture: *centralized architecture* and *distributed architecture*. In the centralized architecture, clients are directly connected to the video server through the network, as shown in Figure 2.2. A video server has access to the video content storage and is responsible for the delivery of the video content in uninterrupted streams. Even though centralized VoD systems are simple to manage, the major problem of this architecture lies in the poor scalability, as the service capacity is well defined by server limitations. A system expansion may lead to huge costs in resource increment. When the number of clients increases, the number of streams needed may be enormous resulting in additional channel bandwidth. The performance of centralized VoD systems can be improved by adding local servers. The local sites

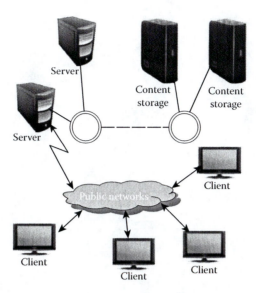

Figure 2.2   VoD centralized architecture.

do not maintain media archives; however, they can store popular movies in their video buffers. The contents of the buffers can be delivered to clients more quickly without accessing the central server. Videos that are not buffered at local sites can be delivered to clients from the central archive when they are requested.

In the *distributed architecture*, multiple video servers are distributed throughout the network infrastructure. Each video server controls and manages a subset of the content storage and is responsible for a subset of the video streams. Figure 2.3 shows a typical layout for the distributed architecture. Ideally, all the popular content is replicated at the video servers connected to each exchange. This significantly decreases traffic between the servers and, as a result, settles down the bandwidth requirements between the main hubs. If a local server does not have a requested video title, it searches through a list of all the video servers that have that title and picks the one with the least network load. The distributed architecture is a viable option as it relaxes the bandwidth requirements on the network.

In the VoD system, popular content can attract a large number of viewers, who performs requests asynchronously. As there are dedicated channels to service each user request, the bandwidth requirements are increased significantly as more and more streams are requested. The fundamental challenge of VoD service is how to meet the on-demand expectation of users without consuming a large amount of bandwidth at the content server. A number of schemes have been proposed that focus on the efficient bandwidth usage of the content server. A common thread in all schemes is the use of *multicasting*. Since a user cannot watch the video at once, broadcasting protocols can only provide near-VoD service. Multicast transmission sends exactly one copy of the stream not over the whole network but only down the branches of the network where one or more viewers are tuned in. In this way, the available network bandwidth can be used more efficiently. For multicast in a VoD system, there is a possibility that more than one user requests the same video at the same time. The probability of more than one user requesting the same video will be high if the number of videos available is small when compared to the number of users. Even if there is a large video archive, there will be a set of popular videos that are requested by many users, thus

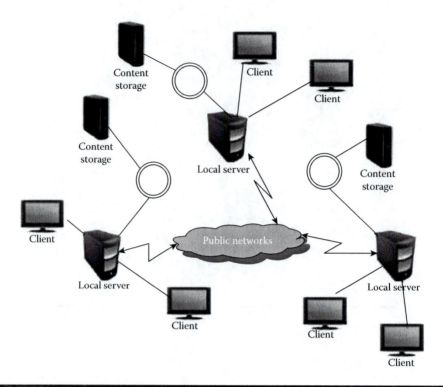

Figure 2.3   VoD distributed architecture.

increasing the chance of multicasting. If a requested video is multicast, all the users in the multicast group will be served by one channel, thus saving the network bandwidth [20]. Some multicasting schemes propose how to provide efficient and practical multicasting, whereas others assume the availability of multicasting to all participating users. Multicasting is mainly implemented in three ways: *IP multicast, overlay network-based multicast,* and *application-layer multicast* (ALM).

IP multicast implements the service at the IP layer and offers efficient group communication. IP multicast requires fairly sophisticated router software that allows the server to replicate streams as required by the clients. The user of a multicast has no control over the media presented. Like in *broadcast,* the choice is simply to watch or not to watch. The user's host communicates with the nearest router to get a copy of the stream. Four classes of IP multicasting approaches have been proposed to overpass the gap between synchronous IP multicast and asynchronous VoD streaming: *batching, patching, periodic broadcasting,* and *merging* [21].

The basic idea of "batching" is to delay the requests for the different videos for a certain amount of time (batching interval), so that more requests for the same video arriving during the current batching interval may be serviced using the same stream. Thus, requests that are made by many different viewers for the same video can share a common video stream if these requests are spaced closely enough [22]. Batching can only be used with popular videos since unpopular videos are unlikely to receive multiple requests during the delay interval. While clients' requests are not instantly granted, the batching technique in fact offers a near-VoD service but not a true VoD service.

In the "patching scheme," an existing multicast can expand dynamically to serve new clients. Most of the communication bandwidth of the server is organized into a set of logic channels, and each is capable of transmitting a video at the playback rate. The remaining bandwidth of the server is used for control messages such as service requests and service notifications [23]. A channel is either a regular channel in which the server multicasts the entire video or a patching channel in which the server multicasts only the leading part of the video. When a client requests a video from the server, the server instructs the client to download from a regular channel and a patching channel. The client exits the patching channel after it downloads the leading part of the video but remains in the regular channel until the end of the video [21]. Patching is very simple and does not require any specialized hardware. Since all requests can be served immediately, the clients experience no service delay, and true video on-demand can be achieved. Patching is very effective in reducing the bandwidth and storage requirements if the number of requests from the users is within a certain limit. Beyond that, patching loses its competence as it results in starting multiple patches of the same video and augments the bandwidth needs.

The idea behind "periodic broadcasting scheme" is to divide the video into a series of segments and broadcast each segment periodically on dedicated server channels. Clients wait for the beginning of the first segment and download the data of the next segment while watching the current segment. User waiting time is usually the length of the first segment [24]. In [24], the periodic broadcasting protocols are divided into three groups: *pyramid broadcasting, harmonic broadcasting,* and *hybrid broadcasting. Pyramid-like schemes* such as those discussed in [25] and [26] have increasing size segments and equal bandwidth channels. The segment size of the videos in this protocol follows a geometrical series, and different videos are mingled together in each logical channel. In *pyramid broadcasting,* the system requires that the video data be transferred at a rate much higher than it is consumed to provide on-time delivery of the videos. In this scheme, video segments are of geometrically increasing sizes, and the server network bandwidth is evenly divided to periodically broadcast one segment in a separate channel. This solution requires expensive client machines with enough bandwidth to cope with the high data rate on each broadcast channel. *Harmonic-like schemes* such as that discussed in [27] have equal-size segments and decreasing bandwidth channels. They divide the video into equal-size segments and transmit them in logical channels of decreasing bandwidth. This requires much less server bandwidth than pyramid broadcasting protocols. A new family of the "hybrid broadcasting protocol" includes Pagoda broadcasting [28] and New Pagoda [29] broadcasting schemes. These protocols are hybrid of pyramid-based protocols and harmonic-based protocols. They partition each video into fixed-size segments and map them into a small number of data streams of equal bandwidth and use time-division multiplexing to ensure that successive segments of a given video are broadcast at the proper decreasing frequencies. The result is that they do not require significantly more bandwidth and, at the same time, do not use more logical streams.

The common issue among batching and patching is that they require twice or more bandwidth at the user system than the nominal playback rate since users must establish multiple streaming sessions concurrently. They also require a substantial amount of disk space in order to store segments from one of the streams while the other is being played out. In addition, they all assume the availability of multicasting capability at all participating nodes. In the *stream-merging scheme* [30], the key idea is to encode the media at a bit rate just slightly less than the client's receive bandwidth. The receive bandwidth that is left unused during viewing is used to perform near-optimal hierarchical stream merging. This technique has been shown to be highly effective in reducing server bandwidth. However, it may take a long time to complete the merging process or may never be realized when the time gap is large between two sessions. It also requires an encoder/decoder system that dynamically changes the rate of stream.

Centralized VoD systems such as CNN Pipeline, YouTube, and Uitzending gemist have drawbacks due to the limited scalability of these systems. Batching and patching would increase the scalability of these systems a lot; however, there is no support for broadcasting or multicasting in the Internet backbone. IP multicast requires routers to maintain per group state. However, very few routers on the Internet can support IP multicast. Overhauling the Internet with IP multicast capable routers is a task considered not feasible in the near future. The routing and forwarding table at the routers now needs to maintain an entry corresponding to each unique multicast group address. This increases the overheads and complexities at the routers. Another issue is that there is lack of experience with additional mechanisms such as reliability and congestion control on top of IP multicast, which makes the ISPs wary of enabling multicasting at the network layer [31]. For these and other reasons, researchers have looked at other ways to achieve an efficient and effective group communication such as *overlay network-based multicasting* and ALM.

In *overlay network-based multicasting*, a network dedicated for the purpose of multicasting is created on top of an existing IP network. Only those routers, that is, overlay nodes that are equipped with multicasting functionality, participate in multicast-specific service; other routers simply forward packets in multicast sessions as regular unicast flows. An example of overlay multicast implementation is the overlay multicast network infrastructure (OMNI) [32], which offers an overlay architecture for media-streaming applications.

In OMNI, service providers deploy multicast service nodes that run the routing and forwarding of information to a group of clients. OMNI follows a two-tier approach to overlay multicast (Figure 2.4). The lower tier contains a set of service nodes that are distributed throughout a CN infrastructure such as the Internet and provides data distribution services to any host connected to an OMNI node. An end host subscribes with a single OMNI node to receive multicast data service. The OMNI nodes organize themselves into an overlay that forms the multicast data delivery backbone. For the second layer, the data delivery path from the OMNI nodes to its clients is independent of the data delivery path used in the overlay backbone. This path can be built using network layer multicast, ALM, or a set of unicast paths [33]. The strengths of overlay network-based multicasting include the ability to deploy a large-scale multicast network without needing

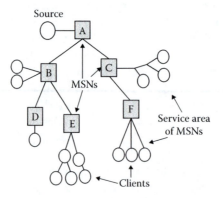

Figure 2.4 OMNI architecture. (From Banerjee, S., and Bhattacharjee, B., A comparative study of application layer multicast protocols. Retrieved 8 April 2010 from http://citeseerx.ist.psu .edu/, 2001. With permission.)

to upgrade all IP routers, support virtually unlimited number of multicast groups, and provide a practical solution for the deployment of group communication infrastructure on the Internet. However, it typically requires semipermanently installed overlay nodes that will remain in service for an extended period of time or at least for the duration of the multicast session. For this reason, it is difficult to construct and maintain such a network within an environment where network nodes are highly dynamic, such as ad hoc and peer-to-peer (P2P) networks.

Owing to the drawbacks presented by "overlay multicast" and the slow deployment of "IP multicast" technology on the global Internet, an application-layer solution has been adopted; this approach is referred to as ALM. In ALM, the multicasting functionality is implemented at the application layer. ALM protocols do not change the network infrastructure; instead they employ multicast forwarding functionality exclusively at end hosts. Unlike network layer multicast where data packets are replicated at routers inside the network, in ALM, data packets are replicated at end hosts. In this multicast strategy, group membership, multicast tree construction, and data forwarding are solely controlled by participating end hosts; thus, it does not require the support of intermediate nodes such as routers or dedicated servers. The P2P approach has ALM premises [21].

2.3.3 Live Video Streaming

Internet streaming technology also brings in more interesting applications such as transmission of traditional TV content in a much more flexible manner. Due to cost considerations, conventional TV networks normally offer channels only if there are enough user bases. For example, a TV network may be willing to offer Hindi programs in New York City, where a large Indian population live, but not in many other parts of the country [34]. The introduction of live streaming services enables users to watch several TV channels through the Internet simultaneously. In live streaming, video streams are being generated at the same time as it is being downloaded and viewed by the clients. Thus, we are dealing with the distribution of a file of unknown and unpredictable length in which the data are only available for a small period of time. In this case, the most important challenge is the play-out delay, namely, the time elapsing between the content production and its play-out. The lone action that the client should be able to carry out is to switch channels. The end-user experience is similar to a live TV broadcast as all of the users will intend to watch the most recently generated content. The user requires a download speed not less than or equal to the playback speed if data loss is to be evaded. The popular live video streaming service is *IP television* (IPTV). With the extensive acceptance of broadband residential access and the progress of video compression technologies, IPTV may be the next popular Internet application [35].

IPTV is a system where a digital television service is delivered using IP over a network infrastructure, which may comprise delivery by a broadband Internet connection. Thus, IPTV offers digital television services over IP for residential and business users at a lesser cost. The official definition approved by the International Telecommunication Union focus group on IPTV (ITU-T FG IPTV) is as follows: "…multimedia services such as television/video/audio/text/graphics/data delivered over IP based networks managed to provide the required level of QoS and experience, security, interactivity and reliability." IPTV also makes it easier for users to access ostracized VoD content, such as well-known movies from decades ago, which are no longer offered in any important TV channel [34].

IPTV is a union of computing, communication, and content, as well as an amalgamation of broadcasting and telecommunication technologies. It enables triple play of *voice, data*, and *video*. The triple-play idea is that clients can subscribe to one service that offers voice, data, and video—all three brought into the home or office over one line and by one service provider. The use of IP as a video delivery mechanism is omnipotent. An IPTV service system does not change the structure of content and channel production of the original television network. However, it just amends the controlled mode of transmission, that is, it makes use of pure IP signaling to change channels and control other functions. In this fashion, selection space of content has been significantly expanded for users [36]. IPTV has a different infrastructure from TV services, which makes use of a push metaphor in which the entire content is pushed to the clients. IPTV has two-way interactive communications between operators and users, for example, streaming control functions such as pause, forward, and rewind, which traditional cable television services lack [37].

A typical IPTV system consists of four main components, as shown in Figure 2.5 [38]. The video head end (VH) captures all programming content, including linear programs and VoD content. The VH receives the content through satellite or terrestrial fiber networks. It is also responsible for encoding the video streams into MPEG-2 or MPEG-4 formats. VH encapsulates the video streams into a transport format, which are then sent to the CN, using IP multicast or IP unicast. The CN groups the encoded video streams into their respective channels. The CN is unique to the service provider and often includes equipment from multiple vendors. At this stage, IPTV traffic can be protected from other Internet data traffic to guarantee a high level of QoS. The broadband remote access server is responsible for maintaining user policy management, such as subscriber authentication and accounting, IP address assignment, and service advertisement. In the reverse direction, traffic from multiple end users is aggregated and routed to the CN by digital subscriber line access multiplexers. The home network connects both the home computer(s) and the IPTV set-top boxes (STBs) to a broadband service to offer the data, voice, and video services in subscribing homes [38]. The STB converts a scrambled digital compressed signal into a signal that is sent to the TV.

Globally, many of the world's major telecom providers are exploring IPTV as a new revenue opportunity from their existing markets. Two major U.S. telecommunication companies, AT&T and Verizon, have invested significantly to replace the copper lines in their networks with fiber-optic cables for delivering many IPTV channels to residential customers [39]. The world's leading markets for IPTV for now are France, South Korea, Hong Kong, Japan, Italy, Spain, Belgium, China, Switzerland, and Portugal. TV2 Sputnik is an IPTV service provider that uses the public Internet for content distribution. It is offered by one of the public service broadcasters in Denmark. Optimal Stream is an IPTV service provider that delivers IPTV to Danish households over the public Internet. The United Kingdom launched IPTV early but it has been slow to grow. IPTV is just beginning to grow in Central and Eastern Europe; now it is growing in South Asian countries such as Sri Lanka and, especially, India. Major vendors for IPTV in India include UTStarcom, Alcatel Lucent, SeaChange, Harmonic, Cisco, Irdeto, Harris, Viaccess, NDS, Conax, Verimatrix,

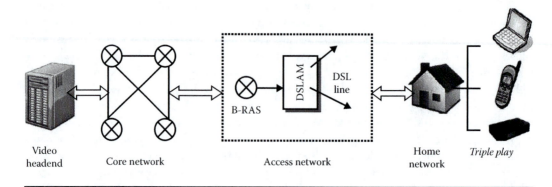

Video headend   Core network   Access network   Home network   *Triple play*

B-RAS   DSLAM   DSL line

Figure 2.5 IPTV system.

Oracle, and Sun Microsystems. In [40], it is projected that consumer IP traffic will grow annually at 57%, driven and dominated by video traffic.

According to Multimedia Research Group, the number of global IPTV subscribers will grow from 44 million at the end of 2010 to 111.5 million in 2014, which is a compound annual growth rate of 26%. The forecast shows that Europe will be the regional leader, with 42% of the world-wide IPTV subscribers total in 2014, maintaining its lead mostly because of the sheer number of large Tier-1 service providers and because of continued strong IPTV growth in some countries. According to a recent study by CISCO, the global IP traffic will continue to be dominated by video, exceeding 91% of global consumer IP traffic by 2014. The study expects that, together, all types of file-sharing traffic will nearly triple by 2014, still accounting for 27% of all Internet traffic. Internet video is predicted to account for 46% of all traffic in the same year.

2.4 Failure of Traditional Streaming Techniques

The traditional client/server-based streaming provides good performance and high availability rates if the number of clients is limited. However, the deployment and maintenance costs of these schemes are usually very high. The current estimation of YouTube's costs is $1 million per day, and these costs could increase extremely if more videos continue to be switched to greater qualities [41]. The high bandwidth required by live streaming video greatly limits the number of clients that can be served by a source. In fact, many streaming services today offer relatively low resolution in order to save bandwidth. The quality of those streaming services is typically not comparable with that in traditional TV networks. Resource management is thus a key issue in Internet streaming deployment. In client/server-based media-streaming systems scenario, on the one hand, the processing power, storage capacity, and input/output (I/O) throughput of the server may become the bottleneck; on the other hand, a large number of long-distance network connections may also lead to traffic congestion. Hence, the system cannot meet the performance requirements of large-scale real-time media-streaming applications [42].

Consider a situation in an on-demand video service offered by Akamai to Doordarshan Online (http://www.dd.now.com). In April 2001, Doordarshan Online used Akamai's content distribution network to webcast an India–Australia cricket match. The company provisioned certain bandwidth from Akamai with its average number of clients in mind. As the match approached an exciting finish, the number of clients demanding the feed increased to surpass the provisioned bandwidth. The servers went down, leading to disrepute of the site and annoyance among end users. The above example shows that unicast schemes scale badly for flash crowds. In spite of growing the server resources, the surge in the number of requests leads to saturation of server resources. Current trends indicate that such problems will exacerbate in the near future as the potential for demand intensifies; more events will be webcast live by various companies to feed an increasing client base. It is rational to anticipate that, as the edge bandwidth increases, the size of flash crowds will also increase, corresponding to taller spikes in traffic [43]. However, the IP multicast technique being offered to address these problems needs support from special hardware, and the costs of infrastructure setup and administration are expensive. In essence, the traditional techniques cannot resolve the problems of video streaming effectively [42]. Hence, an alternate mechanism is required to prevail over the resource saturation.

The emerging distributed information sharing architecture, P2P networks, has been widely accepted as a means to address the resource problem with Internet streaming applications such as VoD and live streaming and to provide an alternative for client/server computing. Still in its infancy, both live and on-demand P2P streaming have the potential of altering the way we watch TV, providing ubiquitous access to an enormous number of channels.

2.5 P2P Networks

The WWW can be viewed as a massive distributed system consisting of millions of clients and servers for accessing associated documents. Servers preserve collections of objects, whereas clients provide users a user-friendly interface for presenting and accessing these objects. The inadequacy of the client–server model is evident in WWW. Because resources are concentrated on one or a small number of nodes and in order to provide 24/7 access with satisfactory response times, complicated load-balancing and fault-tolerance algorithms have to be employed. The same holds right for network bandwidth, which adds to this tailback situation. These two key problems inspired researchers to come up with schemes for allocating processing load and network bandwidth among all nodes participating in a distributed information system [44].

P2P networks are a recent addition to the already large number of distributed system models. P2P networking has spawned immense attention worldwide among both Internet users and computer professionals. P2P computing takes advantage of existing computing power, computer storage, and networking connectivity, allowing users to leverage their collective power to the "benefit" of all. The P2P system is defined as "a self organizing system of equal, autonomous entities (peers) which aims for the shared usage of distributed resources in networked environment avoiding central services." Nodes in a P2P network usually play equal roles, so these nodes are also called peers. In this chapter, the terms "peer" and "node" are used interchangeably in the context of P2P networks. The peers cooperate in a distributed manner to achieve the desired objective. The most important characteristics of P2P technology are direct interaction and data exchange between peer systems rather than through a central server. This is the basis for decentralized distributed computing. P2P networks are self-organized and adaptive. Peers may come and go freely. P2P systems handle these events automatically [45].

One of the main features in a P2P system is that each node contributes resources including bandwidth, storage space, and CPU power, and consequently, the entire system capacity can, in fact, increase as more nodes enter the system. This is piercingly contrary to the client and server architecture, in which the addition of clients always degrades the overall performance. Another benefit that P2P brings is the robustness in case of failure as each node does not rely on any centralized server for content retrieval. In a P2P system, participating nodes mark at least part of their resources as "shared," allowing other contributing peers to access these resources. Thus, if node A publishes something and node B downloads it, then when node C asks for the same information, it can access it from either node A or node B. As a result, as new users access a particular file, the system's capability to provide that file increases [46]. P2P networks have the prospective of diminishing the user perceived latency by pushing the data and computation to a location closer to the users.

Among the various P2P systems in different application domains, file-sharing systems, where files are exchanged among peers, dominate the applications of P2P systems. P2P systems normally form, at the application level, a decentralized overlay network with its own routing mechanism. Until now, a few most important categories of P2P systems have been introduced with their own merits and demerits. Generally, P2P systems are categorized as *centralized, decentralized structured,* and *decentralized unstructured* P2P systems. In the beginning, a P2P system started out with a centralized index system, where file locations are indexed in a number of selected servers for speedy searches. In the centralized model, such as Napster [47], central index servers are used to maintain a directory of shared files stored on peers with the intention that a peer can search for the location of a desired content from an "index server." Conversely, this design makes a single-point failure, and its centralized nature of the service generates systems vulnerable to denial of service (DoS) attacks. The subsistence of a central authority introduced numerous legal problems such that the centralized index method was replaced by the decentralized index system. Subsequently, P2P systems have become completely decentralized in all their functions. Decentralized P2P systems have the advantages of eliminating reliance on central servers and providing freedom for participating users to exchange information and services directly between each other. Decentralized P2P systems can be categorized into two major systems: *unstructured* and *structured*.

In decentralized unstructured P2P systems, such as Gnutella, there is neither a centralized index nor any strict control over the network topology or file placement. By and large, the peers self-configure into an overlay network with no particular intended topology. Distribution of files is probably managed in an ad hoc way not designed to result in any particular arrangement. Since these systems have no coupling between the network topology and data placement, locating a desired file is not simple. Nodes joining the network, following some loose rules, form the network. In these systems, data are stored anywhere in the system and searched for by broadcasting queries to all peers within a specified distance. These methods are simple and highly robust to alteration in the overlay network topology. Conversely, the ineffectiveness of broadcasting raises doubts about their scalability.

In decentralized structured models, such as Chord [48], Pastry [49], and CAN [50], the shared data placement and topology characteristics of the network are robustly controlled on the basis of distributed hash functions. The index is distributed in a precise way across the overlay network topology. The result is that queries are directed resourcefully toward the exact index location, solving the scalability problem of unstructured methods. On the other hand, structured schemes have troubles of their own: complexity, high maintenance overhead, a rigid structure that is somewhat fault intolerant, and inability to support range and keyword queries, which are quite popular in P2P applications. Due to these drawbacks, structured methods have not so far been deployed on any broad scale [51].

Until recently, Internet P2P systems assumed all peers are identical and uniform in resources. Functionality is thus distributed without considering real-world heterogeneity of peer capabilities. For example, some peers may have smaller disk and slower processor speed than others. However, they perform the same role and responsibility as other peers with greater capabilities. This results in instances of inefficiency and bottlenecks in performance due to very limited capabilities of these peers. To account for and even exploit the existence of such heterogeneity of peer capabilities, the notion of superpeers, which are well provisioned in terms of resource capacity, has recently been introduced. A superpeer often plays the role of a server that manages the queries and responses for a subset of ordinary peers [52]. It acts as a server to a set of clients in the system, and the whole set of superpeers is regarded as a centralized server to the clients just like Napster. Thus, this approach basically forms a hierarchical overlay network, where the top layer contains the superpeers, and the bottom layer consists of the peers. KaZaA is a P2P file-sharing application, which employs the idea of "superpeers." The notion of superpeers has been proposed in a recent version of the Gnutella protocol to perk up the scalability of its original system.

2.5.1 Challenges in P2P Streaming

Over the past few years, P2P networks have appeared as an auspicious method for the delivery of multimedia content over a large network. The intrinsic characteristics make the P2P model a potential candidate to solve various problems in multimedia streaming over the Internet [53]. The P2P streaming is more elegant for two reasons. First, P2P does not need support from Internet routers and is thereby cost effective and simple to deploy. Second, a peer simultaneously acts as a client as well as a server, thus downloading a video stream and at the same time uploading it to other peers watching the program. Consequently, the P2P streaming significantly decreases the bandwidth needs of the source [54]. The objective of P2P streaming mechanisms is to maximize delivered quality to individual peers in a scalable fashion in spite of the heterogeneity and irregularity of their access link bandwidth. The aggregate available resources in this approach physically grow with the user population and can potentially scale to any number of participating peers [55]. Each peer should continuously be able to offer suitable content to its connected peers in the overlay by making use of an outgoing bandwidth of participating peers [56]. However, providing P2P video streaming services for a large number of viewers creates very difficult technology challenges on both system and networking resources.

While traditional P2P file distribution applications target flexible data transfers, P2P streaming focuses on efficient delivery of audio and video content under stiff timing requirements. Stream data are instantaneously received, played, and passed to other associated peers. For example, the P2P file-sharing application BitTorrent permits peers to interchange any segment of the content being distributed since the order in which they arrive is not important. In contrast, such techniques are not viable in streaming applications [57]. Video files are directly played out while they are being downloaded. Therefore, pieces, which are received after their play-out time, degrade user experience. This degradation is visible either as missing frames or as a playback stop, which is also denoted by stalling. While redundancy schemes might be suitable for streaming because they do not require further communication between a sender and a receiver, retransmission might not be possible because of the strict timing requirements. In addition, peers have limited upload capacities, which stems from the fact that the Internet was designed for the client/server paradigm and applications. Furthermore, the streaming systems suffer from packet drop or delay due to network congestion [41].

In a P2P streaming, the end-to-end delay from the source to a receiver may be excessive because the content may have to go through a number of intermediate receivers. The behavior of receivers is unpredictable; they are free to join and leave the service at any time, thus discarding their successor peers. Receivers may have to store some local data structures and exchange state information with each other to preserve the connectivity and to perk up the effectiveness of the P2P network. The control overhead at each receiver for satisfying such purposes should be small to keep away from excessive use of network resources and to overcome the resource limitation at each receiver. This is important to the scalability of a system with a large number of receivers [58].

Organizing the peers into high-quality overlay for disseminating the video stream is a challenging problem for broadcasting video in P2P networks. The constructed overlay must be effective both from the network and the application outlooks as broadcasting video concurrently requires high bandwidth and low latencies. On the other hand, a start-up delay of a couple of seconds is abided for applications that are in real time. The system should be able to accommodate tens of thousands of receivers at a time. At the same time, the overheads associated must be reasonable even at large scales. The construction of overlay must take place in a distributed fashion and must be robust to dynamic changes in the network. The system must be self-improving, in that the overlay should incrementally progress into a better structure as more information becomes available [54].

2.5.2 Approaches for Overlay Construction

Existing streaming techniques in the P2P approach can be categorized into schemes supporting *P2P live video streaming* and those supporting *P2P on-demand video streaming*. Some techniques

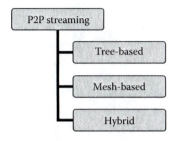

Figure 2.6   P2P streaming types.

can offer both services. Several P2P streaming systems of the two aforementioned categories have been deployed to provide on-demand or live video streaming services over the Internet.

Based on the overlay network structure, P2P streaming systems are broadly classified into three categories: *tree-based, mesh-based*, and *hybrid* schemes (Figure 2.6). The tree-based approaches use push-based content delivery; however, the mesh-based approaches use swarming content delivery. Several P2P live streaming and VoD applications are built on these schemes. This section briefly discusses all the three categories of overlays along with example applications.

2.5.2.1 Tree-Based Overlay

Similar to an IP multicast tree formed by routers at the network level, users participating in a video streaming session can form a tree at the application layer that is rooted at the video source server. Tree-based overlays implement a tree distribution graph, rooted at the source of content (Figure 2.7). In principle, each node receives data from a parent node, which may be the source or a peer. The tree-based systems typically distribute video by actively pushing data from a peer to its children peers [59].

A common approach to P2P streaming is to organize participating peers into a *single tree–structured overlay*, over which the content is pushed from the source toward all peers (e.g., [60]). In

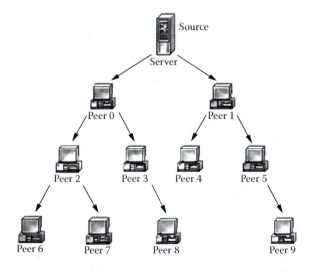

Figure 2.7   Single-tree model. (From Liu, J. et al., *Proc. IEEE*, 96(1), pp. 11–24, 2008. With permission.)

this way, organizing peers is called *single-tree streaming.* In these systems, peers are hierarchically organized in a tree structure where the root is the stream source. The content is spread as a continuous flow of information from the source down to the tree. Each user joins the tree at a certain level. All the load is supported by the interior nodes of the tree, whereas leaves are just receiving data. Systems belonging to this category mainly differ in the algorithms used to create and maintain the tree structure. Given a set of peers, there are many possible ways to construct a streaming tree to connect them up. The goal of tree construction algorithm is to maximize the bandwidth to the root of all nodes. Since these systems are very close to IP multicast, trying to emulate its tree structure, they are able to achieve data paths that do not differ too much from IP multicast paths.

Tree construction and maintenance can be done in either a *centralized* or a *distributed fashion* in single-tree streaming systems. In a centralized solution (Figures 2.8 and 2.9), a central server

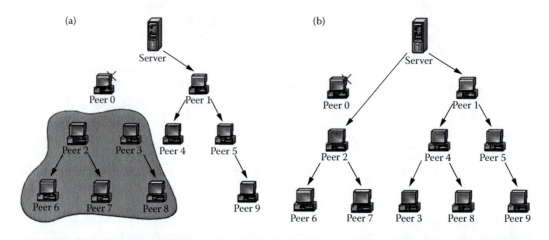

Figure 2.8 Streaming tree reconstruction. (a) Peer 0 departs. (b) Tree recovery after churn. (From Liu, J. et al., *Proc. IEEE*, 96(1), pp. 11–24, 2008. With permission.)

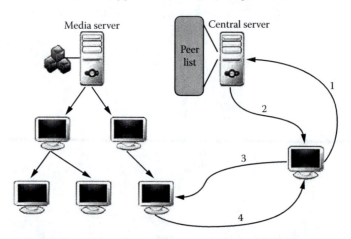

Figure 2.9 Centralized solution for tree construction and maintenance. (1) The peer sends its join request to the central server. (2) The central server sends proper providers for the peer and informs the peer about them. (3) The peer contacts them. (4) Then, the provider(s) sends data to peer.

controls the tree construction and recovery. When a peer joins the system, it contacts the central server. Based on the existing topology and the characteristics of the newly joined peer, such as its location and network access, the server decides the position of the new peer in the tree and notifies it to which parent peer to connect. The central server can detect peer departure through either a graceful sign-off signal or some type of time-out-based inference. In both cases, the server recalculates the tree topology for the remaining peers and instructs them to form the new topology [61]. For a large streaming system, the central server might become the performance bottleneck and the single point of failure. To address this, various distributed algorithms, for example, ZigZag [62], have been developed to construct and maintain streaming tree in a distributed way. If peers do not change too often, single tree-based systems require little overhead as packets are forwarded from peer to peer without the necessity of additional messages. However, in high-churn environments, the tree would be frequently damaged and reconstructed. This process requires considerable control message overhead. Consequently, peers must buffer data for at least the time required to repair the tree in order to evade packet loss [59].

2.5.2.2 Tree-Based Live Streaming Systems

The most popular system using a single-tree approach is NICE [63]. NICE is an acronym that stands for *NICE the Internet Cooperative Environment*. NICE was initially designed for low-bandwidth data-streaming applications with a large number of receivers. The protocol arranges the set of end hosts into a hierarchy based on round-trip-time information between hosts (Figure 2.10). The basic operation of the protocol is to create and maintain the hierarchy. The hierarchy implies the routes. Logically, each member keeps a detailed state about other members that are near in the hierarchy and only has limited knowledge about other members in the group. The hierarchical structure is also important for localizing the effect of member failures. While constructing the NICE hierarchy, members that are "close" with respect to the distance metric are mapped to the same part of the hierarchy: this produces trees with low stretch.

SpreadIt [43] builds an application-level multicast tree over the set of clients. Nodes are organized into different levels (Figures 2.11 and 2.12). For each node n, at level $l + 1$ ($l = 0, 1, 2,...$), there is a node called its parent p at level l; n is called a child of p. All nodes in the subtree rooted at p are called its descendants. Each peer within the tree is responsible for forwarding the data to its descendants. Each client node needs to be enabled with a basic peering layer between the application and transport layers. Peering layers at different nodes coordinate among themselves to

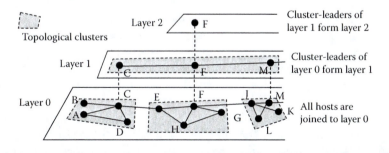

Figure 2.10 Hierarchical arrangement of hosts in NICE. (From Banerjee, S. et al., Scalable application layer multicast. *Proc. ACM SIGCOMM Conf.***, ACM Press, New York, 2002. With permission.)**

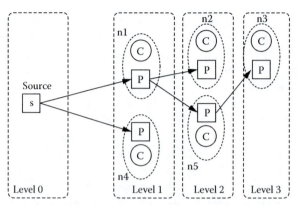

N = {n1, n2, n3, n4, n5}, P: Peering layer, C: Application client

Figure 2.11 SpreadIt: an application-level multicast tree built on the peers. (From Rowston, A. and Druschel, P. Pastry: Scalable, distributed object location and routing for large-scale peer-to-peer systems. *Proc. IFIP/ACM*, Middleware, Heidelberg, Germany, 2001. With permission.)

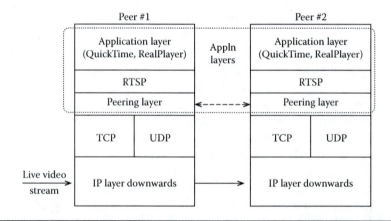

Figure 2.12 SpreadIt: a layered architecture of a peer. (From Deshpande H. et al. Streaming live media over peer-to-peer network, Technical report, Stanford University, 2001. With permission.)

establish and maintain a multicast tree. The application (RealPlayer, Windows Media Player, etc.) gets the stream from the peering layer on their local machines. SpreadIt uses only a single distribution tree and hence is vulnerable to disruptions due to node departures.

End System Multicast (ESM) [64] is an infrastructure for media broadcasting implemented by Carnegie Mellon University. ESM allows broadcasting audio/video data to a large pool of users. The ESM system employs a structure-based overlay protocol that constructs a tree rooted at the source. The information is delivered following a traditional single-tree approach, which implies that any given peer receives streams from only one source. Each ESM node maintains information about a small random subset of members, as well as information about the path from the source to itself. A new node joins the broadcast by contacting the source and retrieving a random list of members that are currently in the group. It then selects one of these members as its parent using the parent selection algorithm. To learn about members, a gossip-like protocol is used. Each node also maintains the application-level throughput it is receiving in a recent time window. If its performance is significantly below the source rate, then it selects a new parent as described in the parent selection algorithm. When a node joins the broadcast or needs to make a parent change, it probes a random subset of nodes it knows. The probing is biased toward members that have not been probed or have low delay.

Figure 2.13 shows an example of the ESM task. The end receivers could play the role of parent or children nodes. The parent nodes perform the membership and replication process. The children nodes are receivers who are getting data directly from the parent nodes. There is one central control server and one central data server residing in the same root source. Any receiver can play the role of parent to forward data to its children. Each client has two connections: *a control connection* and *a data connection*.

One advantage of ESM is that it resolves the deployment problems of IP multicast. However, doing multicasting at end hosts incurs some performance penalties. Generally, end hosts do not handle routing information as routers do. In addition, the limitation in bandwidth and the need of forwarding messages from host-to-host using unicast connection and, consequently, incrementing the end-to-end delay of the transmission process contribute to the price to pay for this approach. These reasons make end-system multicast less efficient than IP multicast.

ZigZag [62], which was proposed by the University of Central Florida, improves the NICE protocol. The tree organization is very close to that proposed by NICE. The algorithms for structure building and maintenance are quite similar to NICE, and all the NICE structure's properties are still valid. ZigZag organizes receivers into a hierarchy of clusters and builds the multicast tree atop this hierarchy according to a set of rules called C-rules (Figure 2.14). A cluster has a head that is responsible for monitoring the memberships of the cluster and an associate head that is responsible for transmitting the content to cluster members. Therefore, the failure of the head does not affect the service continuity of other members, or in case the associate head departs, the head is still working and can designate a new associate head quickly. Whereas in NICE, everything is forwarded by the cluster leader (CL); here, the one responsible for data forwarding is the associate head. The ZigZag protocol control overhead is low. A receiver needs to exchange control information to $O (\log N)$ other receivers in the worst case. On average, it communicates with, at most, a constant number of other receivers. ZigZag is best applicable to streaming applications such as

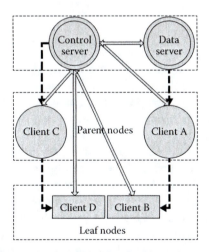

Figure 2.13　Example of ESM.

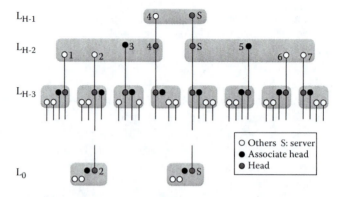

Figure 2.14 Administrative organization of peers in ZigZag. (From Tran, D. A. et al. ZIGZAG: An efficient peer-to-peer scheme for media streaming. *Proc. IEEE INFOCOM,* **2, pp. 1283–1292, 2003. With permission.)**

a single media server broadcasting a live long-term sport event to many clients, each staying in the system for a long-enough period. As ZigZag is focused on single-source media streaming, it is not suitable for media-streaming applications, where multiple sources are presented. The main drawback of ZigZag is that it does not consider the upload bandwidth capacity of peers in join procedure. Also, because ZigZag creates single-tree connection between peers, it has the general problems of single-tree model, such as not using upload bandwidth of leaves and vulnerability to failure of interior nodes.

2.5.2.3 Multitree-Based Overlay

Single-tree-based solutions are perhaps the most natural approach and do not require sophisticated video-coding algorithms. However, one concern with single-tree-based approaches is that the failure of nodes, particularly those higher in the tree, may disrupt delivery of data to a large number of users and potentially result in poor transient performance. If an interior node does not have the required computational or bandwidth resources to serve all its children, peers in its subtree will suffer high delays in data reception or will never receive the stream. The amount of data lost varies from one system to another and depends on the repairing mechanism being adopted. These systems do not seem to exploit very well all the available peers' resources and particularly the available bandwidth. For instance, the leaf nodes account for a large portion of peers in the system, and they do not contribute their uploading bandwidth, which greatly degrades the peer bandwidth utilization efficiency. In response to these concerns, researchers have been investigating more resilient structures for data delivery. In particular, one approach that has gained popularity is the *multitree-based* approach [54].

In the multitree-based approach (Figure 2.15), an overlay construction mechanism organizes participating peers into multiple trees. Each peer determines a proper number of trees to join based on its access link bandwidth. Each peer is placed as an internal node in only one tree and as a leaf node in other participating trees. When a peer joins the system, it contacts the bootstrapping node to identify a parent in the desired number of trees. In multiple-tree-based P2P live streaming systems, the video is encoded into multiple substreams, and each substream is delivered over one tree. The quality received by a peer depends on the number of substreams that it receives [61]. To keep the population of internal nodes balanced among different trees, a new node is added

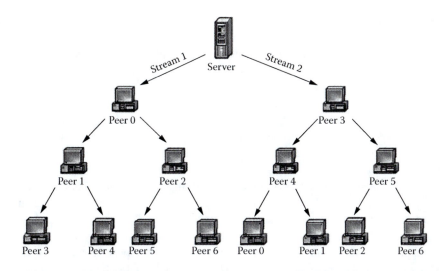

Figure 2.15   Multiple tree-based streaming.

as an internal node to the tree that has the minimum number of internal nodes. To maintain short trees, a new internal node is placed as a child for the node with the lowest depth that can accommodate a new child or has a child that is a leaf. In the latter case, the new node replaces the leaf node, and the partitioned leaf should rejoin the tree similar to a new leaf. When an internal node of a tree departs, each one of its child nodes and the subtree rooted at them are partitioned from the original tree and thus should rejoin the tree. Peers in such a partitioned subtree initially wait for the root of the subtree to rejoin the tree as an internal node. If the root is unable to join the subtree after a certain period of time, individual peers in a partitioned subtree independently rejoin the tree with the same position as a leaf or internal node. The content delivery is a simple push mechanism where internal nodes in each tree simply forward any received packets for the corresponding description to all of their child nodes. Therefore, the main component of the tree-based P2P streaming approach is the tree construction algorithm [65].

There are two key advantages for the multiple-tree solution. First, if a peer fails or leaves, all its children lose the substream delivered from that peer, but they still receive the substreams delivered over the other trees. For this reason, all of its children would receive video streams in case of loss of a substream. Second, a peer plays different roles as an internal node and as a leaf node in various trees. The upload bandwidth of an internal node can be utilized to upload the substream delivered over that tree. At the same time, in order to provide high bandwidth utilization, a peer with a high upload bandwidth can supply substreams in several trees [61].

If peers do not change too often, multitree streaming systems require little overhead since packets are forwarded from node to node without the need for extra messages. However, in high-churn environments, the tree must be continuously destroyed and rebuilt. This process requires considerable control message overhead. Hence, nodes must buffer data for at least the time required to repair the tree to avoid packet loss [63,66].

2.5.3 Multitree-Based Live Streaming Systems

Few applications built on multitree concept are available today. Examples are SplitStream and CoopNet.

SplitStream [67] is a multitree streaming system proposed in 2003 by the Microsoft Research Center. The technique is designed to overcome the inherently unstable forwarding load in conventional tree-based multicast systems. The main idea of SplitStream is to split the stream into dissimilar independent stripes and multicast each stripe using a separate tree. To ensure that the forwarding load can be spread across all participating peers, a forest of stripe trees is constructed in such a way that a node is an interior node in at most one stripe tree and is a leaf node in all the other ones. Such a set of trees is called *interior-node-disjoint*. Figure 2.16 illustrates how SplitStream balances the forwarding load among the participating peers. In this example, the original content is split into two stripes and multicast in separate trees. Each peer, other than the source, receives both stripes and is an internal node in only one tree and forwards the stripe to two children. When an overloaded node receives a request from a prospective child, it either rejects this child or accepts it and rejects one of its existing children, which is less desirable than the new child. A node is more desirable if its node id is closer to its parent node id. In both cases, the rejected child contacts one of the children of the overloaded node.

SplitStream builds the multicast trees for the stripes while respecting the inbound and outbound bandwidth constraints of the peers. It offers resilience to node failures and unannounced departures, even while the affected multicast tree is repaired. One of the main problems with SplitStream is the impact of nodes with heterogeneous bandwidth on its efficiency. Another problem is that in an interior-node-disjoint, nodes receive distinct stripes with different latencies as nodes are decisively placed in different distances from the root of multiple trees. This is undesirable for a live media-streaming application, which involves strict timing constraints. The problem is augmented when the system scales to trees with larger depth, and nodes are placed in diverse distances from the source. The former will either increase the source-to-end delay or disrupt the continuity of the media, whereas the latter wastes the bandwidth of both the sender and the receiver and unnecessarily burdens the network.

CoopNet (cooperative networking) [68] combines aspects of infrastructure-based and peer-to-peer content distributions. It adopts multiple description coding to carry on media data-layering treatment and then transmits media data in different layers along different tree paths. A resourceful server plays a central role in constructing and managing the distribution trees, whereas the bandwidth for forwarding the media data stream is still contributed by the distributed set of peers (Figure 2.17). The system builds multiple distribution trees spanning the source and all the receivers. When a node wants to join, it contacts the central server, which responds with a designated

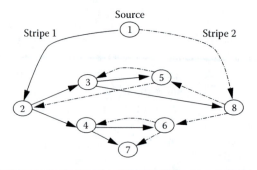

Figure 2.16 Simple example illustrating the basic approach of SplitStream [67]. The original content is split into two stripes. An independent multicast tree is constructed for each stripe such that a peer is an interior node in one tree and a leaf in the other.

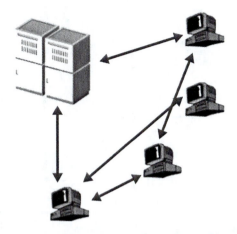

Figure 2.17   Streaming media content using CoopNet.

parent node in each tree. When a node leaves gracefully, it informs the central server, which will find a new parent for the children of the departed node and notifies the children of the identities of their new parent. CoopNet supports both the live streaming as well as on-demand streaming services. The CoopNet approach is good, where a lower quality content presentation is preferred over loss of or delayed quality content presentation. However, since the central server needs to maintain full knowledge of all the distribution trees, it will put a heavy control overhead on the server. Thus, the scalability is not very good. Another problem is that the central server does constitute a single point of failure [4].

2.5.3.1 Mesh-Based Overlays

To combat the peer dynamics, many recent P2P streaming systems use the mesh-based streaming approach. In the mesh-based approach, participating peers form a randomly connected overlay or a mesh. In these overlays, the original media content from a source is distributed among different peers. Each node knows every other node in the system. As a result, each node maintains connections with quite a few other nodes (neighbors) in the network. Each peer exchanges the data with a set of neighbors. If one neighbor leaves, the peer can still download the video from the remaining neighbors. Meanwhile, the peer will add other peers into its neighbor set. Unlike single-tree systems in mesh-based systems, each peer can receive data from multiple supplying peers. Thus, mesh-based streaming systems are robust against peer dynamics. The major challenges in mesh-based P2P live streaming systems are neighborhood formation and data scheduling [61].

Upon arrival, a peer in the network contacts a bootstrapping node (tracker) to receive a set of peers that can potentially serve as parents. This approach is very similar to BitTorrent (Figure 2.18). The main advantage of this swarming content delivery is the ability to effectively utilize the outgoing bandwidth of participating peers as the group size grows [65]. The tracker provides a list of peers containing the information of a random subset of active peers available. Using this list, the peer attempts to initiate peering connections, and if successful, it starts exchanging video content with its neighbors. To handle unexpected peer departures, peers regularly exchange keep-alive messages. At the same time, depending on the system's peering strategies, a peer connects to new neighbors not only in response to peer departures but also when better streaming performance can be achieved.

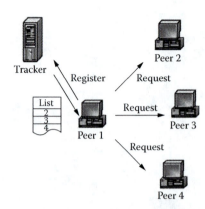

Figure 2.18   Peer list retrieval from Tracker Server. (From Liu, Y. et al., *Journal of Peer-to-Peer Networking and Applications*, 1(1), pp. 18–28, 2008. With permission.)

In mesh-based systems, the concept of video stream becomes invalid due to the mesh topology. The basic data unit in mesh-based systems is a video chunk. The multimedia server divides the media content into small media chunks of a small time interval, each of them with a unique sequence number that serves as a sequence identifier. Later, each chunk is disseminated to all peers through the mesh (Figure 2.19). Since chunks may take different paths in order to reach a peer, they may arrive to destination in a nonsequential order. To deal with this matter, received chunks are normally buffered into memory and sequentially rearranged before delivering them to its media player, ensuring continuous playback [61].

Mainly there are three major flavors of data exchange designs in mesh-based systems: *push, pull*, and *hybrid push–pull* (Figure 2.20). In a *mesh–push* system, a peer actively pushes a received chunk to its neighbors who have not obtained the chunk yet. There is no clearly defined parent–child relationship in a mesh-based system. A peer might blindly push a chunk to a peer already having the chunk. It might also happen that two peers push the same chunk to the same peer. Peer-uploading bandwidth will be wasted in redundant pushes. To address that problem, chunk push schedules need to be carefully planned between neighbors, and the schedules need to be reconstructed upon neighbor arrivals and departures.

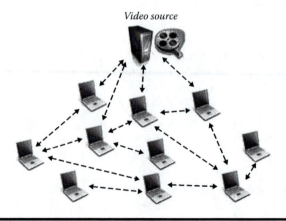

Figure 2.19   P2P live video streaming.

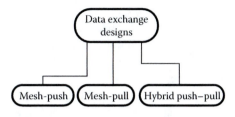

Figure 2.20   Data exchange designs in mesh-based systems.

Another method for data delivery is the *pull* method. The main idea of the pull method is that each peer explicitly requests the required chunk from other peers. Each peer has a neighbor set, and it periodically exchanges data availability information (buffer maps) with its neighbors. A buffer map contains the sequence numbers of chunks currently available in a peer's buffer. Whenever a peer receives such information from other peers, it learns about the chunks it has not yet received. It then requests the missing chunks from the peers in the neighbor set that possess it. Redundancy is avoided, as the node pulls data only if it does not already possess them. Furthermore, since any chunk may be available at multiple partners, the overlay is robust to failures; departure of a node simply means that its partners will use other partners to receive data segments. Finally, the randomized partnerships imply that the potential bandwidth available between the peers can be fully utilized [69]. A disadvantage of the pull technique is that both frequent buffer map exchanges and pull requests produce more signaling overhead and introduce additional delays while retrieving a chunk.

The pull mode in the unstructured overlay, which is inherently robust, can work well with the high churn rate in the P2P environment, whereas the push mode can efficiently reduce the accumulated latency observed at user nodes. The pure pull method cannot meet the demands of delay-sensitive applications because of the striking latency accumulated hop by hop. Additionally, strong buffer capacities at each node are needed to store the exchanging data. The *hybrid push–pull* [70] streaming can greatly reduce the latency and inherit most good features such as simplicity and robustness of the pure pull method. Each node uses the pull method as a start-up, and after that, each node will relay a chunk to its neighbors as soon as the packet arrives without explicit requests from the neighbors. The streaming packets are classified as *pulling packets* and *pushing packets*. A pulling packet of a node is delivered by a neighbor only when the packet is requested, whereas a pushing packet is relayed by a neighbor as soon as it is received. Each node works under the pure pull mode in the first time interval when just joining. After that, based on the traffic from each neighbor, the node will subscribe the pushing packets from its neighbors accordingly at the end of each time interval. A simple roulette wheel selection scheme is employed to allocate pushing packets in the next time interval to each neighbor. The selection probability of a neighbor is equal to the percentage of traffic from that neighbor in the previous time interval. Meanwhile, the lost packets induced by the unreliability of the network link or the neighbor's failure will be pulled as well from the neighbors, where the roulette wheel selection scheme is also used to select the suppliers of each packet from neighbors. Thus, most of the packets received will be pushing packets from the second time interval.

2.5.4 Popular Mesh-Based Live Streaming Systems

Several applications are developed by researchers for various categories of mesh-based P2P streaming. AnySee is a push-based streaming application, whereas CoolStreaming, Chainsaw, PPLive,

PPStream, and SopCast are examples of pull-based streaming applications. GridMedia and PRIME are applications developed based on the hybrid scheme called the push–pull approach.

AnySee [71] is a mesh-push-based streaming system in which resources are assigned based on their locality and delay. The basic workflow of AnySee is as follows: initially, a mesh-based overlay is constructed. Every peer, with a unique identifier, first connects the bootstrapping peers and selects one or several peers to construct logical links. Each peer thus maintains a group of logical neighbors. A location detector-based algorithm is employed to match the overlay with the underlying physical topology. Initially, all streaming paths are managed by the *single overlay manager*, which deals with the join/leave operations of peers. The *interoverlay optimization manager* explores appropriate paths, builds backup links, and cuts off paths with low QoS for each end peer. The manager maintains two sets of active streaming paths, including the current streaming path and the precomputed backup paths, of all the peers in the network. Thus, when a peer fails or leaves the network selfishly, a new path is selected from the backup sets to replace the broken link, thus restoring the connectivity of the network. The system diagram of an AnySee node is shown in Figure 2.21. This mechanism is advantageous because the neighboring peers are not swarmed with requests due to a peer's departure; instead, the overlay manager just replaces a lost link by referring to the backup set and thus replacing a peer efficiently. Hence, AnySee restores the connectivity of the network very quickly [72] since the restoration plan for the descendant peers of the missing peer is carried out beforehand. The *key node manager* allocates the limited resources, and the *buffer manager* manages and schedules the transmission of media data. The goal of the key node manager is to determine the number of requests that a peer should have. Videos are partitioned into chunks, each with a fixed playing time of 1 s. Peers fetch chunks from sources or peers and cache them in local memory. The weakness of AnySee is that the media quality cannot be guaranteed since a group of randomly selected peers may not have enough resources to provide the desired media quality.

Chainsaw [73] is a mesh-pull-based system that does not rely on a rigid network structure. In this scheme, peers are notified of new packets by their neighbors and must explicitly request a packet from a neighbor in order to receive it. This way, duplicate data can be eliminated, and a peer can ensure it receives all packets. Every peer maintains a *window of interest*, which is the range of sequence numbers that the peer is interested in acquiring at the current time. It also maintains and informs its neighbors about a *window of availability*, which is the range of packets that it is willing to upload to its neighbors. The window of availability will typically be larger than the window of

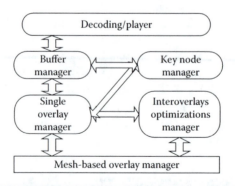

Figure 2.21 The system diagram of an AnySee node. (From Liao, X. et al., AnySee: Peer-to-peer live streaming. *Proc. IEEE INFOCOM*, Barcelona, Spain, 2006. With permission.)

interest. For every neighbor, a peer creates a list of *desired packets*, that is, a list of packets that the peer wants, and is in the neighbor's window of availability. It will then apply some strategy to pick one or more packets from the list and request them via a request message. A peer keeps track of what packets it has requested from every neighbor and ensures that it does not request the same packet from multiple neighbors. It also limits the number of outstanding requests with a given neighbor to ensure that requests are spread out over all neighbors. Nodes keep track of requests from their neighbors and send the corresponding packets as bandwidth allows. The system does not provide a mechanism to enforce a fair resource contribution as Chainsaw allows peers to define their own maximum uploading bandwidth and fails to deter free riding [74]. Chainsaw can potentially invite high network and CPU overheads due to per packet announcements.

PPLive is a mesh-pull P2P streaming platform that distributes both live and prerecorded contents. The major difference with BitTorrent is that, in PPLive, packets must meet the playback deadline. In January 2008, the PPLive application provided almost 500 channels with 1 million daily users on average. The number of channels in December 2008 was reported to be equal to 1775. The PPLive platform consists of multiple overlays. A single overlay corresponds to a PPLive channel. Each peer in an overlay is identified by the pair (IP address and port number). Figure 2.22 shows the basic actions of a PPLive peer. At first, the PPLive peer downloads channel list from the *channel list server* via http. After that, for the selected channel, the peer collects a small set of peers involved in the same overlay by querying the *membership servers* via UDP. The peer communicates with the peers in the list to obtain additional lists, which it aggregates with its existing peer list via UDP. In this manner, the peer maintains a list of other peers watching the same channel [75]. In order to relax the time requirements, to have enough time to react to node failures, and to smooth out the jitter, packets flow through two buffers: one is managed by PPLive and the second by the media player. A downside of such architecture is the long start-up delay [76]. The working of PPStream is very similar to PPLive.

DONet (or CoolStreaming) [77] is another successful mesh-pull P2P streaming system implemented by the Universities of Hong Kong and Vancouver. In DONet, every node periodically exchanges data availability information with a set of partners and retrieves unavailable data from

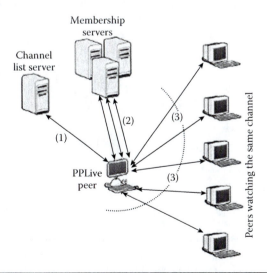

Figure 2.22 PPLive basic architecture.

one or more partners or supplies available data to partners. A node consists of three key modules (Figure 2.23): a *membership manager*, which helps the node maintain a partial view of other overlay nodes; a *partnership manager*, which establishes and maintains partnership with other nodes; and *a scheduler*, which schedules the transmission of video data. The *scheduler* determines which segment should be obtained from which partner and downloads segments from partners and uploads their wanted segments. CoolStreaming requires newly joining nodes to contact the origin server to obtain an initial set of partner candidates. Each node also maintains a partial subset of other participants in the group. CoolStreaming employs Scalable Gossip Membership protocol (SGAM) to distribute membership messages. A CoolStreaming node can depart either gracefully or accidentally due to crash. In either case, the departure can be easily detected after an idle time, and an affected node can quickly react through rescheduling using the buffer map information of the remaining partners. CoolStreaming also lets each node periodically establish new partnerships with nodes randomly selected from its local membership list. This operation helps each node maintain a stable number of partners in the presence of node departures and explore partners of better quality, for example, those constantly having a higher upload bandwidth and more available segments [54]. CoolStreaming supported several different types of media players, such as Windows Media Player and Real Player. Using the scheduling algorithm and a strong buffering system, CoolStreaming achieves a smooth video playback and a very good scalability, as well as performance. The overall streaming rate and playback continuity of the CoolStreaming system is proportional to the amount of peers online at any given time [78]. One of the disadvantages of DONet is that notifying peers and, afterward, requesting segments possibly results in long delays before any data are exchanged. Similarly, due to the random selection algorithm, the QoS cannot be assured. Moreover, DONet assumes that all the peers can cooperate in the replication of the stream; it is likely to have selfish peers in systems that do not want to share their upload bandwidth.

SopCast is a free BitTorrent-like P2PTV application, born as a student project at Fundan University in China. SoP is the abbreviation for Streaming over P2P. In SopCast, the channels can be encoded in Windows Media Video, Video file for Realplayer (RMVB), Real Media (RM), Advanced Streaming Format, and MP3. A client has multiple choices of TV channels, each of which forms its own overlay. Each channel streams either live audio–video feeds or loop-displayed movies according to a preset schedule. The viewer tunes into a channel of his or her choice, and

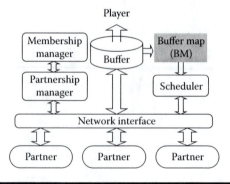

Figure 2.23   Generic system diagram for a DONet node. (From Zhang, X. et al., *Coolstreaming/ DONet: A Data-driven Overlay Network for Efficient Live Media Streaming*. IEEE INFOCOM, 2005. With permission.)

SopCast starts its own operations to retrieve the stream. After some seconds, a player pops up, and the stream can be seen. It also allows users to broadcast their own channel. SopCast provides a low overall frame loss ratio. However, it suffers from peer lags, that is, peers watching the same channel might not be synchronized. Moreover, the zapping time is extremely high [79].

GridMedia [80] adopts a push–pull streaming mechanism to fetch data from the partner nodes. The pull mode in the unstructured overlay can work well with the high churn rate in a P2P environment, whereas the push mode can reduce the accumulated latency at the user side. A well-known rendezvous point (RP) tracker server is deployed to assist the construction of the overlay. As a start-up, a participating node first contacts the RP to get a list of part of the nodes already in the overlay called a login process. Then, the participating node will randomly select some nodes in this list as its neighbors. GridMedia mainly consists of *multisender-based overlay multicast protocol (MSOMP)* and *multisender-based redundancy retransmitting algorithm (MSRRA)*. The MSOMP originates from the streaming server, which is a node at the root. It deploys a mesh-based two-layer structure and groups all the peers into clusters with multiple distinct paths from the source root to each peer. Then, with one or several leaders in each cluster, all the leaders construct the backbone of the overlay to build the upper layer. The MSOMP provides each leader with multiple parents to receive distinct streams simultaneously. It utilizes the existing IP multicast service, which is available in many local area networks (LANs). The IP multicast domain (IMD) is a local network of any size that supports IP multicast. An IMD could be a single host, a LAN, etc. In each IMD, there is a header peer that is responsible for disseminating streaming content to other peers in the same IMD. As soon as the header leaves, a new header will be elected to replace the original role. The MSOMP connects the IMDs by a unicast tunnel altogether. MSOMP-based GridMedia architecture is shown in Figure 2.24.

To address the problem of long-burst packet loss, the MSRRA is proposed at the sender peers to patch the lost packets by using receiver peer loss pattern prediction. In the MSRRA, each receiver peer obtains streaming packets simultaneously from multiple senders. Every sender peer transmits part of the streaming content. As soon as there is congestion occurring on one link, the

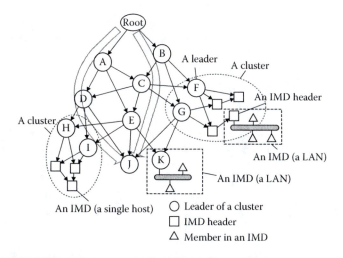

Figure 2.24 GridMedia architecture based on MSOMP. (From Zhang, M. et al., *Gridmedia: A practical peer-to-peer based live video streaming system*. IEEE 7th Workshop on Multimedia Signal Processing, pp. 287–290, 2005. With permission.)

receiver will take notice of this congestion, and subsequently, it notifies other senders who will continuously patch the episode of lost packets. The MSRRA efficiently relieves the impact of node failure, network congestion, and link switch operations.

PRIME [81,82] is a scalable push–pull mesh-based P2P streaming mechanism for live content. The foremost design goal of PRIME is to diminish bandwidth bottleneck and content bottleneck. PRIME incorporates swarming content delivery, which combines *push* content reporting by parents with *pull* content requesting by children. Each peer simultaneously receives content from all of its parents and provides content to all of its children. Given the available packets at individual parents, a packet scheduling scheme at each peer periodically determines an ordered list of packets that should be requested from each parent. Parents simply deliver requested packets by each child in the provided order and at the rate that is determined by the congestion control mechanism. Each segment of the content is delivered to individual participating peers in two phases: *diffusion phase* and *swarming phase*. During the diffusion phase, each peer receives any piece of a new segment from its parent in the higher level. Therefore, pieces of a newly generated segment are progressively pulled by peers at different levels. During the swarming phase, each peer receives all the missing pieces of a segment from its parent in the same or lower levels. These parents are called *swarming parents*. Each piece of any new segment is diffused through a particular diffusion subtree during the diffusion phase of that segment. Then, the available pieces are exchanged between peers in different diffusion subtrees through the swarming mesh during the swarming phase of the segment. The application of the two different phases for content delivery leads to effective utilization of available resources to accommodate scalability and also minimizes content bottleneck. The disadvantage of PRIME is that, if content bottleneck happens, nodes have to wait long in order to find their required data units after a few swarming phases, when those data become available in their neighborhood. Hence, there is no assurance for a reasonable level of streaming quality. Moreover, the algorithm does not consider the behavior of the P2P system in presence of a churn.

HyPO [83] is a hybrid P2P overlay for live media streaming. The scheme optimizes the overlay by organizing peers with similar bandwidth ranges in geographical area into a mesh overlay and forms a tree overlay by selecting peers that are determined as stable. Figure 2.25 illustrates a two-layer mesh/tree overlay in HyPO. Depending on the tree optimization mechanism, the peers that have a large bandwidth will be near the media source node in the tree overlay and evenly distributed in the tree with branches of a similar depth. Consequently, tree optimization reduces the

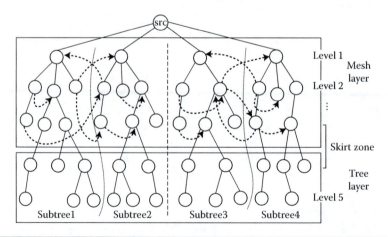

Figure 2.25 Two-layer mesh/tree overlay in HyPO.

average depth of the tree, thus enhancing the scalability. The mesh in HyPO is not an auxiliary connection since the peers in the mesh member always deliver the data in mesh style until the mesh member becomes a tree member. However, since all procedures of HyPO rely on a bootstrap server, a discontinuous period may occur if the server fails. Furthermore, the HyPO does not mention how data are delivered in its mesh overlay [84].

mTreebone [85] is a collaborative tree mesh design that leverages both mesh and tree structures. The key idea is to identify a set of stable nodes to construct a tree-based backbone, called *treebone*, with most of the data being pushed over this backbone. These stable nodes, together with others, are further organized through an auxiliary mesh overlay, which facilitates the treebone to accommodate node dynamics and fully exploit the available bandwidth between overlay nodes. Other nonstable nodes are attached to the backbone as outskirts. Figure 2.26 shows an mTreebone framework. In this scheme, the mesh connection is invoked only if there is an isolated node affected by parent departure or failure. The treebone maintenance and optimization only happen at the treebone nodes, and there is no extra overhead for the outskirt peers. Normally, the streaming quality is much better for the treebone nodes due to the better stability of their data delivery paths from the source. The key challenge is that we need to identify the set of stable overlay nodes and position them at appropriate locations in the tree. Such a requirement can conflict with the bandwidth and delay optimization in tree construction. An additional complication when discussing stability is that this depends on human behavior, that is, on how long the user decides to stay [69]. Locality-based clustering was not considered in mTreebone. On the other hand, CliqueStream [86], which is a hybrid overlay similar to mTreebone, exploits the properties of a clustered P2P overlay to achieve the locality properties (Figure 2.27). CliqueStream elects one or more stable nodes of maximum available bandwidth in each cluster and allocates special relaying role to them. To maintain transmission efficiency, a content delivery tree is constructed out of the stable nodes using the structure in the underlying routing substrate, and content is pushed through them. Less stable nodes within a given cluster then participate in the content dissemination and pull the content creating a mesh around the stable nodes.

The emerging hybrid push–pull P2P streaming overlays present a viable alternative for the traditional way of overlay construction such as tree and mesh since the hybrid design greatly simplifies the

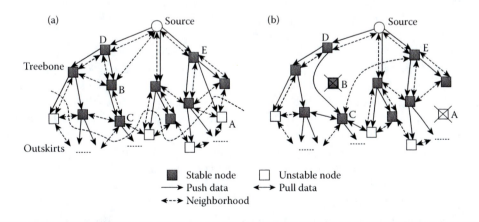

Figure 2.26 mTreebone framework. (a) Hybrid overlay. (b) Handling node dynamics. (From Wang, F. et al., *IEEE Transactions on Parallel and Distributed Systems*, 21(3), pp. 379–392, March 2010. With permission.)

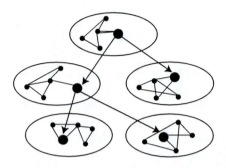

Figure 2.27 Streaming topology in CliqueStream. (From Asaduzzaman, S. et al. CliqueStream: An efficient and fault-resilient live streaming network on a clustered peer-to-peer overlay. *Proc. Eighth International Conference on Peer-to-Peer Computing,* **pp. 269–278, 2008. With permission.)**

overlay construction and maintenance processes and, at the same time, largely retains its efficiency and achieves fine-grained control over load.

2.6 P2P on Demand Video Streaming

The existing VoD schemes pose several issues such as infeasibility of multicast, server crashes, and high maintenance and deployment costs of dedicated overlay routers. However, P2P-based video streaming provides an alternative architecture for VoD services. In a P2P VoD system, all peers are Internet-connected hosts, which store and stream the video to the requesting clients. The cost of these peers and Internet access would be borne by the clients rather than by providers of VoD service. Because there is an abundant supply of potential supplying peers with underutilized resources such as bandwidth and storage, P2P-based architectures should have costs that are significantly less than the traditional client–server and CDN solutions [87].

Applying P2P live streaming techniques directly into VoD streaming is not a trivial task due to several reasons. Like P2P live streaming systems, the P2P VoD systems also deliver the content by streaming. However, peers can watch different parts of a video at the same time, hence thinning their ability to help each other and relieve of the server [88]. A VoD capability would enable users to start watching a video after waiting for a small start-up time while downloading the video in parallel. Even though the shorter end-to-end delay makes live streaming more lively for the users, since, in VoD streaming, the video stream is previously recorded, the liveness is irrelevant. Hence, a short tree rooted at the video server and spanned over peers is not desirable in VoD streaming. The users should be able to watch the video at an arbitrary time, unlike in live streaming where they need to synchronize their viewing times. The users should also be able to perform control operations such as rewind and forward on the video [89]. In addition, the relationship between various variables is different for the two types of streaming. For example, a peer will likely stop watching a VoD stream when its QoS degrades, but the peer may not do the same thing for a live stream because he or she does not have an option of watching it again in the future. Therefore, it is expected that, if the QoS of the video stream reduces, there will be many more peers leaving the system in the VoD streaming case than the case of live streaming. This stretches the significance of a strong failure recovery protocol in a VoD streaming system. The protocol reconnects the abandoned peers efficiently, so that there are no loss of frame and no long delay at a client's playback.

Another important requirement of a VoD service is *scalability*. A typical video stream imposes a heavy burden on both the network and the system resources such as the disk I/O of the server. A VoD system should permit a new peer to join the system fast. The shorter the joining time, the shorter the start-up delay for a peer. The joining requests of peers arrive to the system at different times. It is expected that the system must deliver the video in full length to every peer without making the server become a bottleneck [90]. P2P VoD systems usually require users to contribute a larger amount of storage as these systems need huge buffer sizes in order to satisfy the diversified request from peers on different kinds of video programs. This storage space is usually 1 GB in PPLive. In effect, after a user installs PPLive and runs the system for the first time, he or she could see an unknown kind of file of 1 GB existing in secondary storage [88].

Like video streaming systems, P2P VoD systems are generally classified as *tree-* and *mesh-based* systems.

2.6.1 Tree-Based VoD Systems

The users using tree-based overlay is synchronized and receive the content in the order the server sends it out. This is fundamentally different from the requirement imposed by VoD service. The major issue in tree-based systems is the design of appropriate procedure for accommodating asynchronous users into the system. P2Cast and P2VoD are examples of tree-based P2P VoD systems.

P2Cast [91] is an early patching scheme for VoD service. It is founded on the patching scheme proposed to support VoD service using native IP multicast. P2Cast addresses two key technical issues such as constructing an application overlay appropriate for streaming and providing a continuous stream playback in the face of disruption from an early departing client. The clients arriving within a threshold form a session. For each session, the server, together with the P2Cast clients, forms an application-level multicast tree over the unicast-only network. The clients in P2Cast can forward the video stream to other clients and also cache and serve the initial portions of a video to other clients. Every client actively contributes its bandwidth and storage space to the system while taking advantage of the resources located at other clients. The entire video is streamed over the application-level multicast tree, so that it can be shared among clients. For clients who arrive later than the first client in the session and thus miss an initial segment of the video, the segment can be retrieved from the server of other clients that have already cached that initial segment. P2Cast can serve many more clients than the traditional client–server unicast service. The recovery scheme in P2Cast lets peers receive data from server directly when parent departure occurs. However, this increases the workload of the server.

P2VoD [92] is a tree-based P2P VoD scheme that tries to solve the problems of quick join, provides fast and localized failure recovery without jitter, effectively handles clients' asynchronous requests, and provides small control overhead, compared to P2Cast. Each client in P2VoD has a *variable-size* first-in–first-out buffer to cache the most recent content of the video stream it receives. Existing clients in P2VoD can forward the video stream to a new client as long as they have enough outbound bandwidth and still hold the first block of the video file in the buffer. The failures are managed with the concept of *generation* and a *caching scheme*. The caching scheme allows a group of clients, arriving to the system at different times, to store the same video content in the prefix of their buffers. Such groups form a generation. When a member of a generation leaves the system, any remaining member of that generation can provide the video stream without jitter to the abandoned children of the leaving member, provided that outbound bandwidth is sufficient. In P2VoD, a streaming connection is assumed to be of constant bit rate, which equals to the playback rate of the video player. The recovery process in VoD is more complicated [93]. In addition, P2VoD does not consider the heterogeneous bandwidth of peers.

Cache and relay is a tree-based approach in which a VoD client commonly relies on the content that resides in its parents' buffers. In this scheme, routers do not carry multicast functionalities. Hence, end hosts are in charge for the caching and allocation of streaming media. The end hosts may be client machines or proxies thereof, and these systems maintain retrieved media objects in their local caches provisionally. If another client requests the media objects later on, the original server can forward the request to those end hosts who are physically closer to the client.

oStream [94] takes advantage of buffering capabilities of end hosts by employing cache and relay approach. The scheme employs a spanning tree algorithm for peers to construct an overlay for media streaming. oStream reduces the topological inefficiencies such as link stress and stretch introduced by using ALM.

A framework called *DirectStream* [95] allows clients to take advantage of the benefits of interval caching and VoD service with VCR operation support. DirectStream comprises a directory server, content servers, and clients. The directory server works as a central administrative point. It maintains a database that keeps track of all servers and clients participating in DirectStream and helps new clients to locate the required service. The content servers provide the same functionality as in the traditional client–server service model storing contents in their repository and serving clients' requests so long as sufficient bandwidth is available. Thus, the clients in DirectStream function as P2P nodes. A peer caches a moving window of the latest received content and serves latecomers by continuously forwarding the cached content. A set of active clients, among which a P2P streaming overlay is established, is called a cluster. Clusters in DirectStream evolve over time, and each client in a cluster shares the same stream. The service search process for a new request consists of four steps, as indicated in Figure 2.28. First, the new client sends a request to the directory server to ask for the video starting at position. The directory server then looks into its database and returns a list of candidate nodes, including both the content server and clients that have the

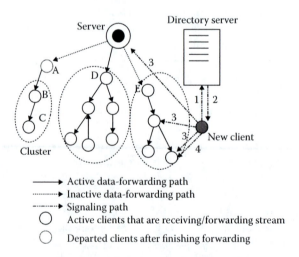

Figure 2.28 DirectStream architecture. (From Suh, K. et al., A peer-to-peer on-demand streaming service and its performance evaluation. *Proc. 2003 IEEE International Conference on Multimedia and Expo (ICME 2003)*, July 2003. With permission.)

content to serve this request. The new client determines from which node to retrieve the stream using the QoS parent selection algorithm. Using this algorithm, a client selects a parent node that has sufficient bandwidth. The new client contacts the selected candidate node and asks to forward the stream. After the connection is successfully set up, the new client signals back to the directory server and registers itself into the database. DirectStream significantly reduces the workload posed on the server. Another advantage is that it scales well as the popularity of the video increases even if participating clients behave noncooperatively. DirectStream has two drawbacks. The centralized management presents a single point of failure. When numerous different ancestors fail, a peer can quickly starve its buffer.

In tree-based VoD systems, peers on the upper layer always play an important role in the whole overlay network. Their departure will lead to the lower layer network fluctuation. Moreover, each peer has only one data supplier, which will cause inefficient utilization of available bandwidth in a heterogeneous and highly dynamic network environment. At the same time, in cache- and relay-based systems, if a parent jumps to another play point in the video, it starts to receive media data, which is of no interest for its children and those need to search for a new parent.

2.6.2 Mesh-Based VoD Systems

In mesh-based VoD systems, no specific topology is created. Peers in the network, based on the design rules, connect to several parents to receive video packets. Mesh-based VoD systems have lower protocol overhead, are much easier to design, are more resilient to high rates of churn, and hence are more popular. Current P2P mesh-based systems have been shown to be very proficient for large-scale content distribution with few server resources. However, such systems have been designed for generic file distribution and provide a limited user experience for viewing media content. However, in VoD systems, the difficulty lies in the fact that users want to receive blocks "sequentially" in order to watch the movie while downloading. In addition, in VoD services, the users may be interested in different parts of the movie and may compete for system resources. Over all, the main challenge resides in designing systems that ensure that users can start watching a movie at any point in time, with small start-up times and sustainable playback rates [96].

BitTorrent (BT) is one of the most successful mesh P2P mechanisms for distributing huge volumes of content over the Internet. It is a scalable file-sharing protocol that also incorporates swarming data transfer mechanism. There are several limitations of original BT strategy in providing video streaming. In BT, files are segmented on space. Although the default piece selection mechanism of BitTorrent is very efficient in minimizing the probability for rare pieces to become extinct and in providing peers with rare pieces, it fails despondently in case of time-sensitive traffic. The reason is that, with time-sensitive data, each piece must be received within a certain time limit. This factor is not taken into consideration in the original piece selection mechanism of BT, and thus, it cannot provide time-sensitive distribution services since pieces are requested based on their rareness and not by their deadline. Consequently, the current piece selection mechanism needs modifications in order to support a time-sensitive service such as VoD [97]. BASS and BiToS are examples of BitTorrent-based mesh P2P VoD systems.

BitTorrent-Assisted Streaming System (BASS) [98] extends the current BitTorrent system to provide a near-VoD service. Since BASS uses the assistance of BT for streaming, it utilizes the service of an external server, which stores all of the publisher's videos and guarantees that the users can play back the video at the playback rate without any quality degradation, with the only modification to BitTorrent being that it should not download any data prior to current playback point. It is allowed to use the rarest-piece-first and tit-for-tat policies. In the rarest-piece-first policy, the client requests a piece based on the number of copies it sees available and chooses the least common one. In tit-for-tat, a leecher (one who downloads) reciprocates to other leechers that send it pieces by giving higher priority to their requests. From the media server, BASS downloads pieces in order, skipping over pieces that have already been downloaded by BitTorrent, or are currently in the process of being downloaded and are expected to finish before their playback deadline arrives. The system overview of BASS is given in Figure 2.29. Even though BASS reduces the load at the server by a significant amount, the design of the system is still server oriented, and hence, the bandwidth requirements at the server increase linearly with the number of users [96].

Kangaroo [99] is a system focused on providing both P2P VoD services as well as live streaming content. Kangaroo resembles a typical mesh-based P2P system in that it consists of *peers* coordinated by a *tracker*. Kangaroo handles DVD operations with minimum delays, network overhead, and server resources. Kangaroo implements a hybrid scheduling policy that combines selfish (sequential segment downloads for continuous playback) with altruistic (local rarest to improve segment diversity) behavior. A peer consists of several subcomponents, that is, the segment scheduler, the peer selection scheduler, and the neighborhood manager. The segment scheduler decides what segment should be scheduled for download next, whereas the peer selection scheduler decides which neighbor peer(s) to schedule the download from and the neighborhood manager that constantly revisits the peer's neighborhood and decides which are the best peers to get/push data from/ to. Each peer in Kangaroo downloads data in parallel from a small number of neighbors through data connections. Peers also maintain a number of control connections that are used to exchange information about available segments in a neighborhood, thus enabling the peer to infer the popularity and location of the segment for scheduling. Kangaroo resembles a gossip-based overlay in

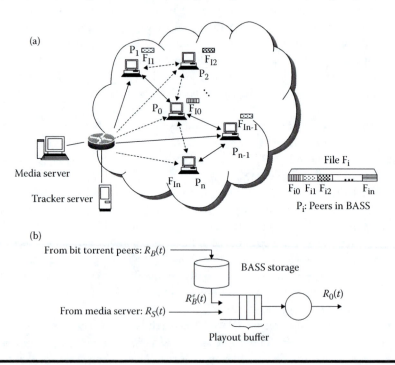

Figure 2.29 BASS. (a) System overview. (b) Client model. (From Dana, C. et al., *Bass: Bittorrent assisted streaming system for video-on-demand*. International Workshop on Multimedia Signal Processing (MMsP) © 2005 IEEE.)

which a smart tracker is used to implement peer coordination. Kangaroo also provides low buffering times and high swarming throughput under user VCR-like operations. However, it does not consider user viewing behavior. Thus, VCR-like operations may cause long response time [100].

The *BiToS system* [101] is also based on BitTorrent. The main idea is to divide the missing blocks into two sets: "high-priority set" and "remaining piece set," and request with higher probability blocks from the high-priority set (Figure 2.30). The high-priority set contains all the pieces that are quite close to be reproduced. Thus, a peer desires to download these pieces earlier, in contrast with the remaining pieces set, which contains pieces that will not be needed in the near future. After the initiation of the player, the player buffer requests the needed pieces from the received pieces buffer. In BiToS, the major emphasis is given for the careful scheduling of the video blocks. The pieces that miss their playback deadline are simply dropped. Hence, this may lead to degradation in video playback quality. In addition, due to the asymmetric nature of the Internet connections and heterogeneity of the peers, the system cannot guarantee that pieces requested are always available for playback on time [97].

COCONET [102] is a novel and efficient way of organizing peers to form an overlay network for supporting streaming and neighbor lookup for continuous playback or VCR operations. The scheme utilizes a cooperative cache-based technique where each peer contributes a certain amount of storage to the system in return for receiving video blocks. The system uses this cooperative cache to organize the overlay network and serve peer requests, thereby reducing the server bottleneck supporting VCR-related operations. In several P2P VoD systems, peers share video segments only with neighboring peers based on its playing position. However, in highly skewed viewing patterns, most of the peers are clustered around a particular playing position, and very few peers are distributed at different positions throughout the video length. Hence, the peers may not find any or very few neighbors to satisfy their demand. COCONET avoids this situation. In order to find new supplier peers at different parts of the movie length, P2P VoD systems maintain an updated index of the live peers with their available video segments. Unlike other P2P VoD systems, COCONET does not use indexing at the tracker. Instead, the tracker only maintains a small subset of live peers, which is queried only once as an RP when a new peer joins the system. Each COCONET peer

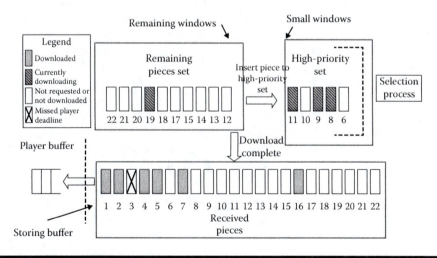

Figure 2.30 BiToS approach. (From Vlavianos, A. et al., BiTos: Enhancing BitTorrent for Supporting Streaming Applications. *Proc. IEEE Global Internet Symposium*, **2006. With permission.)**

builds an index based on the cooperative cache contents, which helps in finding any supplier peer for any video segment throughout the entire video length. The control overhead of COCONET is also low in order to maintain the overlay structure, even during a heavy churn. COCONET also has better load balancing and fault tolerance properties. The distributed contributory storage caching scheme helps in spreading the query load uniformly through the overlay and organizes the overlay in a uniform and randomized fashion, which makes the content distribution independent from playing position.

2.7 Mobile VoD

Streaming video to mobile users is rapidly emerging as a crucial multimedia service. With the emergence of wireless technologies such as IEEE 802.11 and Bluetooth, mobile users are enabled to connect to each other directly without any networking infrastructure such as the Internet and infra-structure-based wireless LANs. In other words, the users form a *mobile ad hoc network* (MANET). Due to the increasing popularity of wireless networks, mobile VoD systems have found many practical applications. For example, airlines can now provide VoD services in airport lounges to entertain the waiting passengers on their laptops or personal digital assistants (PDAs). Universities can install mobile VoD systems that allow students to watch important video lectures anywhere anytime on campus. In mobile VoD systems, the equipment used to watch video broadcasting falls in a wide spectrum of heterogeneous capabilities, ranging from powerful laptops to primitive PDAs. One of the main issues in mobile streaming is the heterogeneity found in mobile devices: diverse display size, computing power, memory, and media capabilities. The wireless bandwidth is limited, whereas a video is typically large. A video server enabled with 802.11g could not deliver more than 36 1.5-Mb/s MPEG1 video streams at once to its wireless clients. However, 802.11b can only sup-port at most seven concurrent such video streams. The load of a VoD system is usually distributed unevenly; it is heavy only over a short period of time. For instance, in the airport lounge example, the system would have a heavy load only during 1 h or 2 h before a flight departure. Therefore, the system should be able to adapt to different loads and make necessary adjustment to the broadcast-ing schedule so as to minimize the total bandwidth usage. Load adaptivity is also important to the mobile VoD system because of its energy consumption. Since the coverage of wireless transmission is limited, we often need multiple hosts to cover a large-enough service area. For this reason, a sig-nificant amount of energy for the intermediate mobile hosts is consumed. In order to save energy, the system should use smaller total bandwidth when the load is light [103].

PatchPeer [104] is a VoD technique for the wireless environment. The basic idea of PatchPeer is to take advantage of the distinct features of the hybrid wireless network (Figure 2.31) to overcome the scalability issue associated with the original patching technique in a traditional wireless net-work. Figure 2.32 shows the typical interactions between a requesting peer with its neighboring peers and the server in PatchPeer. The requesting peer first sends the identification of the requested video to the server. It receives the starting time of the latest regular channel that is streaming the video from the server. The requesting peer then requests a patching stream from a neighboring peer to compensate for the initial missing part of the video. If a neighboring peer can provide the patching stream, the requesting peer receives the patching stream from the neighboring peer and the regular stream from the base station. Otherwise, the requesting peer receives both the patch-ing stream and regular stream from the base station. The PatchPeer scales better than the original patching, as most of the patching streams in PatchPeer are provided by mobile clients themselves, leaving the base station with more downlink bandwidth to serve more clients.

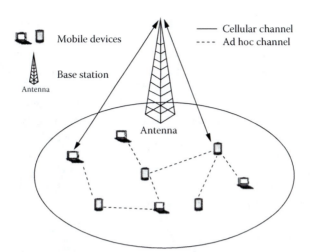

Figure 2.31   Unified cellular and ad-hoc network architecture.

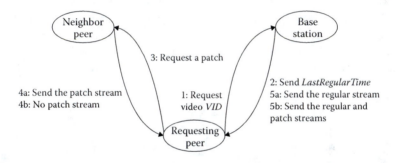

Figure 2.32   Collaboration diagram of PatchPeer. (From Do, T. T. et al., *Peer-to-Peer Network Applications*, 2, pp. 182–201, 2009. With permission.)

MobiVoD [105] is a mobile VoD system that employs a periodic broadcast protocol to achieve maximum scalability. The clients leverage an ad hoc network caching technique to minimize the service delay. The system consists of three components: *video server*, *clients*, and *local forwarders*. Due to the limitation of wireless transmissions, a video server cannot transmit a video to clients located in a wide geographic range. Thus, a scatter of stationary and dedicated computers called *local forwarders* is provided to relay the service to client's transmission coverage area. This area is called a local service area. If a client is within the service area of a local forwarder, the former can receive the video packet broadcast from the latter. The server and set of local forwarders form a service backbone. The service backbone is interconnected either via a wired area network/LAN or via an infrastructure-based wireless network. A video is divided into segments, each broadcasts on a separate communication channel. When a new client joins the system, it waits until the next broadcast of the first segment starts to download the first segment. After playing the first segment, the client immediately switches to the broadcast of the second segment to download it and so on until all segments have been downloaded. Periodic broadcasting makes the system scalable with an increase in the number of clients. However, as the period that a new client must wait before it starts the VoD service is significant, MobiVoD employs two caching policies: random cache and dominating-set cache (DSC). Random cache permits a client to cache the first segment with some probability. Even if a new client finds some clients in its neighborhood, the chance of keeping cache by these clients may be low in random cache. Hence, as an alternative, DSC, which maintains a dominating set of the clients D_{set}, is used. A client belonging to D_{set} caches the first video segment. Using any one of the caching schemes, when a new client requests the VoD service, it joins the current broadcast immediately and downloads the video packets broadcast into a playback buffer. As for the beginning portion that was already transmitted by the current broadcast, the new client downloads and plays it immediately from a nearby cache. The new client switches to the playback buffer to play the rest of the video. MobiVoD focuses on popular videos, whereas PatchPeer handles videos with diverse popularity. Moreover, MobiVoD is using only the wireless local area networks, whereas PatchPeer operates in a hybrid wireless environment.

MOVi (mobile opportunistic VoD) [106] is a mobile P2P VoD application based on ubiquitous WiFi-enabled devices such as smartphones and ultramobile PCs. MOVi addresses challenges such as limited wireless communication range, user mobility, and variable-user population density by exploiting the opportunistic mix use of downlink and direct P2P communication for improving the overall system throughput. It exploits sparingly distributed access points, user mobility, unstable channel conditions, and population density to provide a high bit rate on-demand video streaming service. MOVi is composed of two logical components: a mobile client node called MOVi Client (MC) and a network of servers known collectively as MOVi Server (MS) (Figure 2.33). MS maintains three key functions. It derives a connectivity map of link quality between MCs, schedules the direct transfer of content segments between pairs of MCs based on the map, and tracks the content delivery and caching status of all MCs within its domain. A MOVi Client maintains

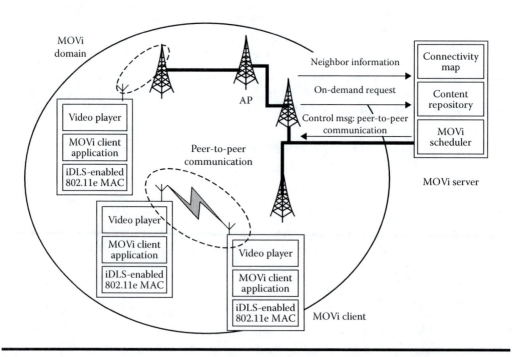

Figure 2.33 MOVi architectural components. (From Yoon, H. et al. On-demand video streaming in mobile opportunistic networks. *Proc. Sixth Annual IEEE International Conference on Pervasive Computing and Communications*, pp. 80–89, 2008. With permission.)

two key functions: serves as a temporal cache to help content diffusion inside the MOVi network and acts as a channel state monitor by periodically observing link quality to its neighboring MCs and updates any changes to MS. The content is stored at a central repository and is fragmented into multiple equal segments prior to distribution. Each segment is normally mapped to several packets. Upon receiving requests from MCs, MS delivers content segments over the downlink path and schedules direct segment exchanges between MCs. If direct P2P communication is not possible with other MCs, the requested segment is delivered from MS to the MC via the access point path. An MC has no knowledge of which content segments reside in its neighboring MC, and it simply waits for direct communication triggers from MS. Once the MC receives all segments that make up a video frame segment, the frame is handed to the media player. The player then decodes the frame for play-out. If there are missing segments within yet to be played out video frame segment, MC sends immediate on-demand request to the MS to recover the missing segments. Neighboring peer discovery in MOVi is carried out by evaluating the signal-interference-to-noise ratio value between MCs and the active duration of neighboring MCs. MOVi is able to increase the number of supported concurrent users twofold, compared with unicast-based on-demand video streaming, as well as reduce video start-up delay by half.

   P2P mobile VoD (P2MVOD) [107] allows a moving client to receive streaming data on demand from other moving clients in P2P architecture by utilizing multicast VoD technology. P2MVOD divides video content into the same-sized segments. The segments are then broadcasted to eliminate the mobile routing overhead of unicast and multicast routing protocols. Segmenting the content enables multiple clients to share the accountability for providing all video content. A control server is used to control the segmentation of video content and the delivery of each segment. The control server does not store or deliver any video content. It only possesses information on the segments that can be provided by the clients that are storing them. Each client knows the control server address and submits a request for video delivery to it. On receiving a request, the server searches for a client with segments of the required content and forwards the request to that client. The control server holds and maintains a schedule that describes the time at which individual segments must be sent. By referring to the schedule, the control server determines the segments that must be sent to the new client and searches for other clients that can provide these segments. Furthermore, it queries them regarding the possibility of sending at a time determined by it. The receiving clients do not need to know the identities of the clients. P2MVOD reduces the traffic on both the links compared to the patching technique, although it adds to traffic when the request rate is low.

2.8 Mobile Live Streaming

Delivering media to large numbers of mobile users presents challenges due to the stringent requirements of streaming media, mobility, wireless, and scaling to support large numbers of users. Live streaming to mobile devices is thus a challenging task [108]. P2P-based near-live video streaming is becoming more and more popular with users of fixed-line broadband network access, but it is mostly unavailable to mobile users.

   In [109], a real-time P2P streaming system for the mobile environment is presented. Peers are grouped into clusters according to their proximity in order to proficiently exchange data between peers. Clusters also help with scalability issues of peer maintenance. The architecture of the overlay network with three clusters sharing a certain streaming service is shown in Figure 2.34. Peers exchange actual media data between each other using RTP. RTP sessions are split into a number

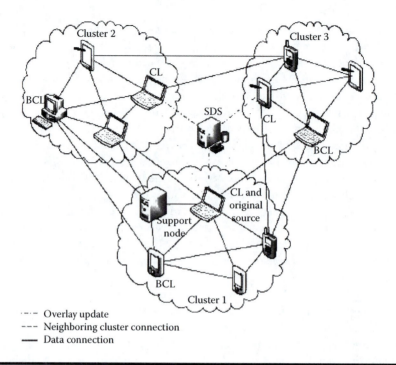

Figure 2.34 Overlay architecture in a mobile environment. (From Peltotalo, J. et al., *International Journal of Digital Multimedia Broadcasting*, Article ID 470813, 15 pages, 2010. With permission.)

of partial streams in such a way that it allows reassembling the original media session in real time at the receiving end. There is one cluster leader (CL) assigned to each cluster with the possibility for one or more backup CLs (BCLs). CLs are used to manage peers inside the cluster and to connect new arriving peers. Each ordinary peer performs periodic keep-alive messaging to inform its existence to the CL and all other peers from which it has received RTP packets. This helps to avoid unnecessary data transmission because RTP uses the UDP, and the sending peer does not otherwise know that the receiving peer is still in the network. A new arriving peer selects a suitable cluster according to its best knowledge of locality using round-trip time values between CLs and itself. At the time of joining a cluster, a peer receives an initial list of peers from which the actual media data can be acquired. The corresponding CL inserts joined peers into its peer list. A peer finally selects its sources for the stream. When the cluster grows too large to be handled by a single CL, the cluster should be split into two separate clusters. The existing CL assigns one of its BCLs to become a new CL for the new cluster and redirects a number of existing peers to the new cluster. Merging of two clusters must be done when a cluster becomes too small. All peers in the streaming network are forming a nonhierarchical mesh structure. The system offers very low initial buffering times.

LocalTree [110] is a scalable algorithm that minimizes the energy used in packet transmission. Peers are organized into two tiers: the base tier and the tree tier. Peers are first connected in a simple unstructured mesh in the base tier. The base tier provides a network for further optimization of the energy consumption. In the base tier, peers utilize only local neighbor information to make independent distributed decisions on whether to rebroadcast a packet or not. The base-tier mesh is further optimized by the tree–tier algorithm. In the tree tier, groups of relatively stable nodes are then identified based on node and link conditions. They are then connected, following a greedy tree construction algorithm. With the two-tier operation, LocalTree is able to adapt different network dynamics.

2.9 P2P Streaming and Cloud

Cloud computing refers to both the applications delivered as services over the Internet and the hardware and systems software in the data centers that provide those services. Cloud computing offers different service models as a base for successful end-user applications. Due to the elastic infrastructure provided by the cloud, it is suitable for delivering VoD and live video streaming services. A few schemes for media streaming that integrate the benefits of P2P and cloud technologies have been proposed recently.

A streaming mechanism that merges P2P and cloud computing technologies for achieving efficient media streaming is proposed in [111]. Figure 2.35 represents an overall view of the proposed cloud–P2P architecture. The cloud contains multimedia streaming servers. The service has first level or directly connected clients (C_1, C_2, C_3, and C_4) and higher level clients (HP_{11}, HP_{12},...). The first-level clients after login consult and choose one among the three types of price packets. The price packets for customers are defined by considering three types of QoS parameters such as jitter, latency, and bandwidth. Similarly, higher level clients acquire an information list with QoS status for all connected clients. The higher level clients contact the first or higher level customers instead of the provider for streams. There exists option for peers (P_{111}, P_{112}, P_{121},...) to connect to higher level clients who want to offer their service for free. The streaming network is thus organized into a P2P tree overlay. The service provider has direct centralized management for managing the contract policies among all types of customers.

AngelCast [112] is a cloud-based live stream-acceleration service with optimized multitree construction that combines both P2P and cloud technologies. In P2P-based live streaming systems, the play-out rates are constrained by the upload bandwidth of clients. Usually, the upload bandwidth is lower than the download bandwidth for the participating peers. This limits the

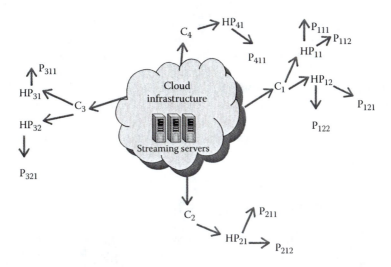

Figure 2.35 P2P streaming cloud architecture.

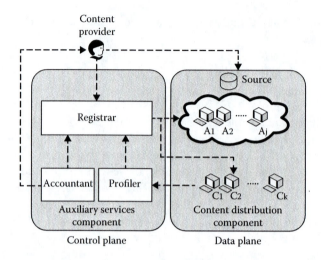

Figure 2.36 Architectural elements of AngelCast. (From Sweha, R. et al., AngelCast: Cloud-based Peer-Assisted Live Streaming Using Optimized Multi-Tree Construction. *Proc. ACM Multimedia Systems (MMSys)*, **2012. With permission.)**

quality of the delivered stream. Therefore, to leverage P2P architectures without sacrificing the quality of the delivered stream, content providers use additional resources to complement those available through clients. In AngelCast, a content provider is guaranteed that its clients would be able to download the stream at the desired rate without interruptions while extremely utilizing the benefits from P2P delivery. AngelCast achieves this by employing special servers from the cloud called *angels*. The angels can supplement the gap between the average client upload capacity and the desirable stream bit rate. Angels download only the minimum fraction of the stream that enables them to fully utilize their upload bandwidth.

Figure 2.36 shows the architectural elements of AngelCast. The Registrar collects information about clients, making fast membership management decisions that ensure smooth streaming. When a new node joins the stream, it contacts the Registrar and informs it of its available upload bandwidth. The Registrar uses a data structure representing the streaming trees and assigns the new client to a parent node in each tree. The Registrar also decides how many future children the new node can adopt in each tree. Content providers contact the Registrar to enroll their streams. The Registrar uses the profiler to estimate the uplink capacity of clients. The Accountant uses the estimated gap between the clients' uplink capacity and the stream play-out bit rate to give the content provider an estimate of how many angels it will need. The service of angels is utilized to achieve QoS. AngelCast makes sure that any parent has at least two children, except for the nodes in the second to last level, ensuring a logarithmic depth of all trees.

2.10 Security in P2P Video Streaming

The open and anonymous nature of the P2P network makes it an ideal medium for attackers to spread malicious content. As a result, widespread and unrestricted deployment of P2P systems exposed a number of security vulnerabilities. In a P2P environment, the collaboration of all peers is very important for the correct functioning of the system. Every peer is soaking up network bandwidth. If too many users access the same network resource, the network bandwidth may be used up, resulting in a DoS. A malicious node would continuously issue queries with high time-to-live values on the network, thus generating a huge amount of network traffic, rendering the network unusable by other honest peers. The peer who offers a resource may go offline, whereas other fellow peers are downloading from it. A malicious peer may just simply route a query to a nonexistent peer or an unreliable peer with long latency.

Securitywise, P2P streaming systems are more challenging than other P2P applications because they are more vulnerable to QoS fluctuations. Live streaming protocols are most sensitive to delay and delay jitter. If a user is not receiving packets in time, he or she may grow dissatisfied with the quality of the delivery and leave the system altogether. Due to this, the peers connected to that machine may also be affected. Even minor quality variations cause the viewing experience to lose appeal and the user to drop the service. P2P streaming is vulnerable to manipulation and threats at the transport and network layers. Clever attacks can compromise selectively the guarantees that a streaming session should provide, rendering some channels unusable, or making the broadcast unavailable in particular locations [113].

Traditional security mechanisms typically protect resources from malicious users by restricting access to only authorized users. However, the problems in P2P systems relate more with trustworthiness rather than security. Therefore, there is demand for mechanisms to maintain the trust of P2P systems. Trust management is a successful approach that helps maintain the overall credibility level of the system, as well as encourage honest and cooperative behavior. The inspiration of trust management is that since, in a P2P system, there is no central authority that can authenticate and guard against the actions of malicious peers, it is up to the peer to protect itself and to be responsible for its own actions. Consequently, each peer in the system needs to somehow assess information received from another peer in order to determine the trustworthiness of both the information, as well as the sender. This can be attained in many ways such as relying on direct experiences or obtaining reputation information from other peers [114].

2.10.1 Common Attacks and Solutions

2.10.1.1 DoS Attacks

DoS attacks decrease or cease total capable network activity. The goal of such an attack is to exhaust key resources at the target, diminishing the target's capacity to either provide or receive service. Resources that can be exhausted include the target's downstream bandwidth, upstream bandwidth, CPU processing, or TCP connection resources. Compared with the widely used file-sharing networks, P2P streaming networks are more vulnerable to DoS attacks. There are several reasons for DoS attacks in P2P streaming. Video streaming requires high bandwidth. Hence, a certain amount of data loss could make the whole stream useless. Since streaming applications require their data to be delivered in a timely fashion, data with a missed deadline are useless. Usually a streaming network consists of a limited number of data sources. Hence, the failure of the data source could bring down the whole streaming system [115]. For example, malicious nodes send excessive amounts of requests or duplicate packets intended for their peers. Thus, a fair node would be flooded with useless messages or too many requests for it to handle. Consequently, the ability to bring a contribution to the streaming session is compromised. In this way, the resources of the system are exhausted with a relatively small effort on the attacker side.

Ripple stream [115] is a DoS resilience framework that employs a credit system to allow peers to evaluate other peers' behaviors. The overlay is organized according to a credit-constrained peer

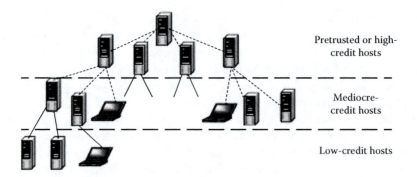

Figure 2.37   Example of a ripple-stream-based overlay.

selection mechanism. The peers share the credit information with each other. Peers with high trustworthiness are kept in the central part of the overlay structure. Malicious nodes, with low reliability, are pushed to the peripheral of the network. The higher the credit is, the closer a peer can be to the data source. Credit management component in the system translates a user's behavior to its credit value. In ripple stream, when a new peer A joins the overlay, it will first obtain a list of peers with mediocre credit from a bootstrap mechanism. After joining the overlay, A accumulates credit by fulfilling its duties. These credit-related operations are handled by the credit component included in ripple stream. In the meantime, A also tries to find upstream peers that can provide better service based on some overlay optimization principles. If A discovers malicious behaviors of other peers, it disconnects from these peers and reports its discovery to the credit system. An example of a ripple-stream-based overlay is shown in Figure 2.37. Ripple-stream achieves DoS resilience with the credit system, and during attacks, the ripple stream stabilizes the overlay and substantially improves the streaming quality.

2.10.1.2 Free Riding

Nodes that consume services offered by other nodes but that do not themselves contribute services to the P2P network are known as free riders. In P2P networks, free riding is a familiar problem. A free rider guzzles more resources than it contributes. In the case of P2P streaming, a free rider is a peer who downloads data but uploads little or no data in return. The encumbrance of uploading is on the unselfish peers, who may be too few in number to provide all peers with a satisfactory QoS. In both live streaming and VoD, peers require a minimal download speed to sustain playback. Hence, free riding is very harmful as the unselfish peers alone may not be able to provide all the peers with sufficient download speeds.

A classification of free-riding techniques for file-sharing applications is presented in [116]. The schemes are categorized as monetary-, reciprocity-, and reputation-based approaches. Monetary-based approaches charge peers for the services they receive. Because these services are still very low cost, such approaches are also called micropayment-based solutions. The technique proposed in [117] is an example of a monetary-based approach. The main disadvantage is that the proposed solutions require some centralized authority to monitor each peer's balance and transactions. This can cause scalability and single-point-of-failure problems. In reciprocity-based approaches, a peer monitors other peers' behaviors and evaluates their contribution levels. The well-known P2P application BitTorrent implements a reciprocity-based approach by adjusting a peer's download speed according to its upload speed. Reciprocity-based approaches face several implementation issues such as fake services published by peers. Since a peer itself provides contribution-level information, the credibility is in question. In reputation-based approaches, peers with good reputations are offered better services. These approaches construct reputation information about a peer on the basis of feedback from other peers. Reputation-based approaches store and manage long-term peer histories. XRep [118] is an example of an autonomous reputation system. Reputation sharing is achieved in XRep through a distributed algorithm by which resource requestors can evaluate the consistency of a resource offered by a participant before beginning the download.

In [119], two policies to limit the number of free riders in a P2P streaming system are proposed: *block-and-drop* (BD) and *block-and-wait (BW) policies*. With the BD policy, free riders that would like to join the streaming session are blocked if the free upload capacity in the overlay is less than the streaming rate. Under the BW policy, free riders are blocked if the overlay does not have enough available upload capacity. The same users can be temporarily disconnected and have to wait to reconnect if there is not enough capacity to serve all peers. Under the BD policy, both the blocking and the dropping probabilities can be high. Therefore, the free riders already admitted to the system are frequently dropped. Under the BW policy, the number of free riders waiting to be reconnected is very low for all parameter settings. As a result, free riders receive the stream without interruption with high probability. This feature makes the BW policy a good option to control free riders.

In [120], a mechanism for give-to-get free-riding resilient for P2P VoD systems is proposed. In give-to-get, peers have to forward the chunks received from a peer to others in order to get more chunks from that peer. By preferring to serve good forwarders, free riders are excluded in favor of well-behaving peers. When the bandwidth in the P2P system becomes scarce, the free riders will experience a significant drop in the experienced QoS. Free riders will thus be able to obtain video data only if there is spare capacity in the system.

In [121], a rank-based peer-selection mechanism for peer-to-peer media streaming systems is proposed. The mechanism provides incentives for cooperation through service differentiation. Contributors to the system are rewarded with flexibility and choice in peer selection to provide high-quality streaming sessions. Free riders are given limited options in peer selection and hence receive low-quality streaming. The contribution of a user is converted into a score, then the score is mapped into a rank, and the rank provides flexibility in peer selection. Cooperative users earn higher rank by contributing their resources to others and eventually receive high-quality streaming. Free riders have limited choice in peer selection and hence receive low-quality streaming. The incentive mechanism reduces the data redundancy required during a streaming session to tolerate packet loss.

A payment-based incentive mechanism for P2P live media streaming is proposed in [122]. In a payment-based system, the P2P network is treated as a market. Every overlay node plays the double role of service consumer and provider. Consumers try to buy the best possible service from service providers at a minimum price, whereas the providers strategically decide their respective prices in a pricing game, in order to maximize their economic revenues in the long run. A peer earns points by forwarding data to others. The data streaming is divided into fixed-length periods, during which peers compete with each other for good data suppliers for the next period in a first-price auction-like procedure using their points. Once a peer finds an ideal parent, it takes part in a competition for that parent. If it wins, it becomes a child of that parent; otherwise, it gets a list of the winner peers, from which it attempts to find a new best parent. It again takes part in the competition for that new parent and continues this process until it wins a parent or has no parents to choose. In the latter case, it tries to find a parent in a best effort manner.

2.10.1.3 Pollution Attacks

In a P2P live video streaming system, a polluter can introduce corrupted chunks. Explicitly, an attacker can join a current video channel and create partnerships with other peers watching the channel. The attacker can then announce to its partners that it has a large number of chunks for the current video stream. When the neighbors request advertised chunks, the attacker sends bogus polluted chunks in place of legitimate chunks. Each receiver mixes into its playback stream the polluted chunks it receives from the attacker, along with other chunks it receives from its other neighbors. The polluted chunks damage the quality of the rendered video at the receiver. Polluted chunks received by an unsuspicious peer not only affect that single peer but also, since the peer also forwards chunks to other peers and those peers, in turn, forward chunks to more peers, and so on, the polluted content can potentially spread through much of the P2P network. If the amount of polluted data is very important, users might ultimately get unsatisfied and completely stop using the system [123]. Figure 2.38 shows [124] a P2P network where a polluter sends bogus chunks to other peers, falsely marking these chunks as legitimate. These corrupted chunks get propagated through the P2P network. Content pollution attacks can severely impact the QoS in P2P live video streaming systems. There are solutions proposed in the literature for managing the pollution attacks in P2P file-sharing applications. However, few schemes are available for fighting pollution attacks in P2P streaming applications.

In [125], a lightweight nonrepudiation protocol called Malicious node Identification Scheme (MIS) for network-coding-based P2P streaming networks is introduced. With a network-coding technique, instead of merely relaying the packets of information they receive, the nodes of a network will take several packets and combine them together for transmission. However, the "combination" nature of network coding makes it vulnerable to pollution attacks. MIS employs an

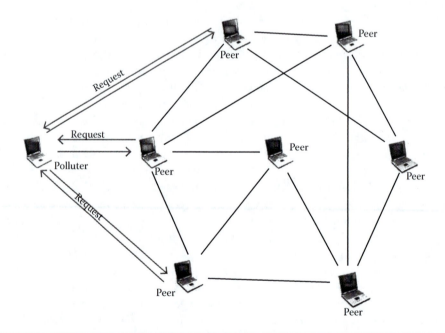

Figure 2.38 Pollution attack in a P2P live video streaming system.

approach for detecting the existence of malicious nodes. Each decoding node detects corrupted blocks by checking if the decoding result matches the specific formats of video streams. Any node having an unreliable decoding result will send an alert to the servers to trigger the process of recognizing malicious nodes. The servers then compute a checksum based on the original blocks and distribute it to the nodes using the streaming overlay. The checksum helps the nodes to detect which neighbor has sent it a corrupted block. The accuracy of MIS is based on the condition that no node can lie when reporting a suspicious node that has sent a corrupted block, and evidence associated with the corrupted block is necessary to demonstrate to the servers that the reported node has really sent the block. A nonrepudiation transmission protocol is used to achieve this. MIS has good computational efficiency and the ability of managing a large number of corrupted blocks and malicious nodes. However, this scheme requires a distribution of multiple checksums to all the peers when an attack is detected, which incurs significant communication overhead. In [126], a trust management system is proposed that identifies attackers and excludes them from further sharing of multimedia data to resist pollution attacks in P2P live streaming.

2.11 Summary

Video-over-IP applications have recently attracted a large number of users on the Internet. With streaming, a user does not have to wait to download a file to play it and can watch the video in real time. The basic solution for streaming video over the Internet is the traditional client–server service model. A client sets up a connection with a video source, and the video content is streamed to the client directly from the server. However, the client–server design harshly restricts the number of concurrent users in video streaming due to the bandwidth bottleneck at the server side. Another model, CDNs, overwhelmed the same bottleneck issue by adding dedicated servers at physically dissimilar locations. This results in expensive deployment and maintenance.

P2P networking is a very promising model to construct various distributed applications. Recently, quite a few P2P streaming systems have been deployed to provide live and VoD streaming services on the Internet. Compared with conventional approaches, the main benefit of P2P streaming is that each peer contributes its own resources to the streaming session. Administration, maintenance, and responsibility for operations are hence dispersed among several users, instead of focusing on a few servers. For this reason, there is a rise in the quantity of resources in the network. Accordingly, the usual bottleneck problem of the client–server systems is further reduced. The P2P architecture thus extends extremely well with large user population and also provides a scalable and economical alternate to traditional streaming services. Mainly, there are two well-known schemes for P2P video streaming: tree-based and mesh-based. A hybrid of the two schemes is also emerging. This chapter reviewed the architecture of these models and also briefly explained few systems built on these approaches. The application of P2P technology in video streaming on MANETs is known as mobile P2P streaming. The issue of live video streaming on MANETs is still a real challenge due to frequent changes in network topology and the sensitiveness of radio links. A few schemes for media streaming, which integrate the benefits of P2P and cloud technologies, have been proposed recently. Security has significant impact on P2P-based streaming applications. Media streaming is inherently more prone to attacks as it is very difficult to monitor the participating peers in the overlay. The network consists of thousands of nodes but not all can be trusted. Security forms one of the most critical issues in a streaming system. This chapter also reviewed various security issues and mechanisms for preventing such attacks.

References

1. INTERNAP: The next big wave in convergence (Building the right foundation for Internet TV). *Streaming Media Magazine*. Retrieved May 5, 2010 from www.internap.com.
2. Kozamernik, F. *Streaming Media Over the Internet: An Overview of Delivery Technologies*. Retrieved May 23, 2010 from www.ebu.ch/trev_index-xz.html.
3. Wiggins, R. W. Overview of streaming audio and video. *Building a Community Information Network: A Guidebook*, 1–10. Retrieved February 2, 2010 from www-personal.umich.edu/~csev/citoolkit/content/book/chapter8.pdf, 1999.
4. Okuda, M. Enabling large-scale peer-to-peer stored video streaming service with QoS support. Unpublished doctoral dissertation, University of Pittsburgh, 2006.
5. Peng, G. *CDN: Content distribution network*. Retrieved May 28, 2010 from http://arxiv.org/abs/cs.NI/0411069, 2004.
6. Ho, K. M., K. T. Lo, and J. Feng. Multimedia Streaming on the Internet: Compression, QoS Control/Monitor, Streaming Protocols, Media Synchronization. In Furht. B (Eds), *Encyclopedia of Multimedia Second Edition*. USA: Springer, 2008.
7. Beggs, J., and D. Thede. *Designing Web Audio*. USA: O'Reilly Media, 2001.
8. Telecom, ATIS Telecom Glossary 2007, American National Standard for Telecommunications. Retrieved June 1, 2010 from http://www.atis.org, 2007.
9. Dutson, B. *The European Streaming Industry: Clearing the Barriers to Growth*. Retrieved 12, May 2010 from http://www.streamingmedia.com/.
10. Apostolopoulos, J., W. Tan, and S. Wee. *Video Streaming: Concepts, Algorithms, and Systems*. (HPL-2002-260). California: Hewlett-Packard Laboratories, September 2002.
11. Wu, D., Y. T. Hou, W. Zhu, Y. Zhang, and J. M. Peha. Streaming video over the internet: Approaches and directions. *IEEE Transactions on Circuits and Systems for Video Technology*, vol. 11(3), pp. 282–300, 2001.
12. Wang, Y., Y. Wenger, J. Wen, and A. Katsa. Error resilient video coding techniques: Real-time communications over unreliable networks. *IEEE Signal Processing Magazine*, vol. 17(4), pp. 61–82, July 2000.
13. Rabinovich, M., and O. Spatscheck. *Web Caching and Replication*. USA: Addison-Wesley Longman Publishing Co., 2002.
14. Bekker, H., and E. Verharen. *State-of-the-art research into streaming media caching and replication techniques*. Retrieved 17, April 2010 from https://doc.novay.nl/dsweb/.
15. Baentsch, M., L. Baum, G. Molter, S. Rothkugel, and P. Sturm. *Caching and replication in the World Wide Web or a European perspective on WWW-connectivity*. Retrieved 12, May 2010 from http://research.cs.ncl.ac.uk.
16. Pathan, M., and R. Buyya. *A Taxonomy and Survey of Content Delivery Networks*. (GRIDS-TR-2007-4). Grid Computing and Distributed Systems Laboratory, The University of Melbourne, Australia, February 2007.
17. Lazar, I. and W. Terrill. Exploring content delivery networking. *IT Professional*, vol. 3(4), pp. 47–49, 2001.
18. Dilley, J., B. Maggs, J. Parikh, H. Prokop, R. Sitaraman, and B. Weihl. Globally distributed content delivery. *IEEE Internet Computing*, vol. 6(5), pp. 50–58, September 2002.
19. Huang, C., J. Li, and K. Ross. Can internet video-on-demand be profitable? *ACM SIGCOMM Computer Communication Review*, vol. 37(4), pp. 133–144, October 2007.
20. Kalva, H., and B. Furht. Techniques for Improving the Capacity of Video-on-Demand Systems. In *29th Annual Hawaii International Conference on System Sciences (HICSS-29)*, pp. 308–315, USA: IEEE Computer Society, 1996.
21. Zhang, X., and H. Hassanein. Video-on-demand streaming on the Internet: A survey. In *25th Biennial Symposium on Communications*, pp. 88–89, USA: IEEE Press, doi:10.1109/BSC.2010.5472998, 2010.
22. Aggarwal, C. C., J. L. Wolf, and P. S. Yu. On Optimal Batching Policies for Video-on-Demand Storage Servers. In *International Conference on Multimedia Computing and Systems (ICMCS '96)*, pp. 253–258, USA: IEEE Computer Society, 1996.

23. Hua, K. A., Y. Cai, and S. Sheu. Patching: A multicast technique for true video-on-demand services. In *ACM Multimedia International Conference*, pp. 191–200, USA: ACM, 1998.

24. Hu, A. Video-on-demand broadcasting protocols: A comprehensive study. In *Infocom 2001*, pp. 508–517, USA: IEEE Press, 2001.

25. Viswanathan, S., and T. Imiehnski. Metropolitan area video-on-demand service using pyramid broadcasting. *Multimedia Systems*, 4(4), pp. 197–208, 1996.

26. Hua, K., A., and S. Sheu. Skyscraper broadcasting: A new broadcasting scheme for metropolitan video-on-demand systems. *ACM SIGCOMM Computer Communication Review*, 27(4), pp. 89–100, 1997.

27. Juhn, L. S., and L. M. Tseng. Harmonic broadcasting for video-on-demand service. *IEEE Transactions on Broadcasting*, 43(3), pp. 268–271, 1997.

28. Paris, J. F., S. W. Carter, and A. Long. Hybrid broadcasting protocol for video on demand. In *Multimedia Computing and Networking Conference (MMCN'99)*. *San Jose*, pp. 317–326, 1999.

29. Paris, J. F. A simple low-bandwidth broadcasting protocol for video-on-demand. In *8th Int'l Conference on Computer Communications and Networks (IC3N'99), Boston-Natick, MA*, pp. 118–123, USA: IEEE Press, 1999.

30. Eager, D., M. Vernon, and J. Zahorjan. Bandwidth skimming: A technique for cost-effective video-on-demand. In *Multimedia Computing and Networking (MMCN)*, 2000.

31. Banerjee, S., and B. Bhattacharjee. *A comparative study of application layer multicast protocols*. Retrieved 8 April 2010 from http://citeseerx.ist.psu.edu/, 2001.

32. Banerjee, S., C. Kommareddy, B. Bhattacharjee, and S. Khuller. Construction of an Efficient Overlay Multicast Infrastructure for Real-time Applications. In *IEEE Infocom 2003, San Francisco*. pp. 1521–1531, USA: IEEE Press, 2003.

33. Moen, D. M. *Overview of Overlay Multicast Protocols*, Retrieved 11 April 2010 from http://bacon.gmu.edu/XOM/pdfs/Multicast%20Overview.pdf, 2004.

34. Xiao, Z., and F. Ye. *New Insights on Internet Streaming and IPTV*. In *2008 International Conference on Content-Based Image and Video Retrieval (CIVR'08)*, Canada, pp. 645–654, USA: ACM, 2008.

35. Hei, X., Y. Liu, and K. W. Ross. IPTV over P2P streaming networks: The Mesh-Pull Approach. *IEEE Communications Magazine*, vol. 46(2), pp. 86–92, 2008.

36. Yiding, H., H., Nanyang, W. Juan, and L. Jian. *Applications and Development Prospects of IPTV*. International Conference on Measuring Technology and Mechatronics Automation, China, USA: IEEE Computer Society, pp. 687–689, 2010.

37. Xiao, Y., X., Du, J., Zhang, F., Hu, and S. Guizani. Internet Protocol Television (IPTV): The killer application for the next-generation internet. *IEEE Communications Magazine*, vol. 45(11), pp. 126–137, November 2007.

38. Shihab, E., L. Cai, F. Wan, and A. Gulliver. Wireless mesh networks for in-home IPTV distribution. *IEEE Network*, vol. 22(1), pp. 52–57, January 2008.

39. Chen, Y., Y. Huang, R. Jana, H. Jiang, and Z. Xiao. *When is P2P Technology Beneficial for IPTV Services?* 17th International workshop on Network and Operating Champaign, IL, USA, 2007.

40. Cisco Systems, Cisco Report on the Exabyte Era. Retrieved 17, April 2010 from www.hbtf.org/files/cisco_ExabyteEra.pdf, 2008.

41. Abboud, O., K. Pussep, A. Kovacevic, K. Mohr, S. Kaune, and R. Steinmetz. Enabling Resilient P2P video streaming—Survey and analysis, *Multimedia Systems*, vol. 17(3), pp. 177–197, June 2011.

42. Gao, W., H. Longshe, and F. Qiang. Recent advances in peer-to-peer media streaming system. *Journal of China Communications*, pp. 52–57, October 2006.

43. Deshpande H., M. Bawa, and H. Garcia-Molina. Streaming live media over peer-to-peer network, Technical report, Stanford University, 2001.

44. Aberer, K., and M. Hauswirth. An overview on peer-to-peer information systems. *Proc. Workshop on Distributed Data and Structures*, pp. 171–188, March, 2002.

45. Li, X. and J. Wu. Searching techniques in peer-to-peer networks. Edited by J. Wu, *Handbook of Theoretical and Algorithmic Aspects of AdHoc, Sensor, and Peer-to-Peer Networks*, Boston: Auerbach Publications, 2005.

46. Tewari, S., Performance Study of Peer-to-Peer File Sharing, Doctoral Dissertation, University of California at Los Angeles, 2007.

47. Kim, A., and L. Hoffman. Pricing Napster and other Internet peer-to-peer applications, Technical report, George Washington University, 2002.

48. Stoica, I., R., Morris, D. Karger, Kaashoek, and H. Balakrishnan. Chord: A scalable peer-to-peer lookup service for internet applications. *Proc. SIGCOMM*, 2001.

49. Rowston, A., and P. Druschel. Pastry: Scalable, distributed object location and routing for large-scale peer-to-peer systems. *Proc. IFIP/ACM*, Middleware, Heidelberg, Germany, 2001.

50. Ratnasamy, S., P. Francis, M. Handley, and R. Karp. A scalable content-addressable network, *Proc. SIGCOMM*, 2001.

51. Pyun, Y. J., and D. S. Reeves. Constructing a balanced, (log(N)/1oglog(N))-diameter super-peer topology for scalable P2P systems. *Proc. Fourth International Conference on Peer-to-Peer Computing (P2P2004)*, IEEE Computer Society, August 2004, pp. 210–218, 2001.

52. Seet, C., C. Lau, W. Hsu, and B. Lee. A Mobile System of Super-Peers Using City Buses. *Proc. 3rd Int'l Conf. on Pervasive Computing and Communications Workshops*, IEEE Computer Society, USA, pp. 80–85, 2005.

53. Meddour, D., M. Mushtaq, and T. Ahmed. Open Issues in P2P Multimedia Streaming, *MULTICOMM 2006 Proceedings*, pp. 43–48, 2006.

54. Liu, J., S. G. Rao, B. Li, and H. Zhang. Opportunities and challenges of peer-to-peer internet video broadcast. *Proc. IEEE*, vol. 96(1), pp. 11–24, 2008.

55. Hoong, P., K. and H. Matsuo. A two-layer super-peer based p2p live media streaming system. *Journal of Convergence Information Technology*, vol. 2(3), pp. 47–57, 2007.

56. Magharei, N. and R. Rejaie. Understanding Mesh based Peer-to-Peer Streaming. *Proc. International Workshop on Network and Operating Systems Support for Digital Audio and Video (NOSSDAV '06)*, Newport, Rhode Island, USA, 2006.

57. Silverston, T., O. Fourmaux, A. Botta, A. Dainotti, and A. P. G. Ventre. Traffic analysis of peer-to-peer iptv communities. *Computer Networks*, vol. 53(4), pp. 470–484, 2009.

58. Tran, D. A., K. A. Hua, and T. T. Do. A peer-to-peer architecture for media streaming. *IEEE Journal on Selected Areas in Communications*, vol. 22(1), pp. 121–133, 2004.

59. Marfia G., G. Pau, P. Di Rico, and M. Gerla. P2P Streaming Systems: A Survey and Experiments. *Proc. 3rd STMicroelectronics STreaming Day (STreaming Day'07)*, Genoa, Italy, 2007.

60. Chu, Y., S. G. Rao, S. Seshan, and H. Zhang. Enabling conferencing applications on the internet using an overlay multicast architecture. *Proc. ACM SIGCOMM*, 2001.

61. Chen, C. W., Z. Li, and S. Lian. *Intelligent Multimedia Communication: Techniques and Applications*, Berlin, Heidelberg: Springer, pp. 195–215, 2010.

62. Tran, D. A., K. Hua, and T. Do. ZIGZAG: An efficient peer-to-peer scheme for media streaming. *Proc. IEEE INFOCOM*, vol. 2, pp. 1283–1292, 2003.

63. Banerjee, S., B. Bhattacharjee, and C. Kommareddy. Scalable application layer multicast. *Proc. ACM SIGCOMM Conf.*, New York: ACM Press, 2002.

64. Chu, Y., S. Rao, and H. Zhang. A Case for End System Multicast. *Proc. ACM Sigmetrics*, International Conference on Measurement and Modeling of Computer Systems, 2000.

65. Magharei, N., R. Rejaie, and G. Yang. *Mesh or Multiple-Tree: A Comparative Study of Live P2P Streaming Approaches*. 26th IEEE International Conference on Computer Communications-INFOCOM2007, pp. 1424–1432, 2007.

66. Yang, F., X. Dai, and X. Ru-zhi. *The Common Problems on the Peer-to-Peer Multicast Overlay Networks*. 2008 International Conference on Computer Science and Software Engineering, pp. 98–101, 2008.

67. Castro, M., P. Druschel, A. M. Kermarrec, A. Nandi, A. Rowstron, and A. Singh. SplitStream: High-bandwidth multicast in cooperative environments. *Proc. 19th ACM Symposium on Operating Systems Principles*, New York: ACM Press, pp. 298–313, 2003.

68. Padmanabhan, V., H. Wang, P. Chou, and K. Sripanidkulchai. Distributing streaming media content using cooperative networking. *Proc. ACM/IEEE NOSSDAV*.

69. Liu, Y., Y. Guo, and C. Liang. Video streaming systems. *Journal of Peer-to-Peer Networking and Applications*, vol. 1(1), pp. 18–28, 2008.

70. Zhang, M., J. Luo, L. Zhao, and S. Yang. A peer-to-peer network for live media streaming: Using a push-pull approach. *Proc 13th Annual ACM International Conference on Multimedia*, 2005.

71. Liao, X., H. Jin, Y. Liu, L. M. Ni, and D. Deng. AnySee: Peer-to-peer live streaming. *Proc. IEEE INFOCOM*, Barcelona, Spain, 2006.

72. Ghoshal, J., L. Xu, B. Ramamurthy, and M. Wang. *Network Architectures for Live Peer-to-Peer Media Streaming*. University of Nebraska-Lincoln CSE Technical reports, TR-UNL-CSE-2007-020, 2007.

73. Pai, V., K. Tamilmani, V. Sambamurthy, K. Kumar, and A. B. Mohr. Chainsaw: Eliminating trees from overlay multicast. *Proc. 4th Int. Workshop Peer-to-Peer Systems (IPTPS)*, February 2005.

74. Shah, P., J. Rasheed, and J. Paris. *Performance study of unstructured P2P overlay streaming systems*. 17th International Conference on Computer Communications and Networks, ICCCN 2008, pp. 608–613, August 3–7, 2008.

75. Spoto, S., R. Gaeta, and M. Grangetto. *Analysis of PPLive through active and passive measurements*. IEEE International Symposium on Parallel and Distributed Processing, Rome, Italy, pp. 1–7, 2009.

76. Chen, Y., C. Chen, and C. Li. *A Measurement Study of Cache Rejection in P2P Live Streaming System*. 28th International Conference on Distributed Computing Systems Workshops, IEEE Computer Society, 2008.

77. Zhang, X., J. Liu, B. Li, Y. Yum. *Coolstreaming/DONet: A Data-Driven Overlay Network for Efficient Live Media Streaming*. IEEE INFOCOM, 2005.

78. Venot, S., and L. Yan. *Peer-to-peer media streaming application survey*. International Conference on Mobile Ubiquitous Computing, Systems, Services and Technologies, IEEE Computer Society, pp. 139–148, 2007.

79. Fallica, B., L. Yue, A. Fernando, R. Kuipers, E. Kooij, and P. Van Mieghem. Assessing the quality of experience of SopCast. *Int. J. Internet Protocol Technology*, vol. 4(1), pp. 11–23, 2009.

80. Zhang, M., Y. Tang, L. Zhao, J. G. Luo, and S. Q. Yang. *Gridmedia: A practical peer-to-peer based live video streaming system*. IEEE 7th Workshop on Multimedia Signal Processing, pp. 287–290, 2005.

81. Magharei, N., Y. Guo, and R. Rejaie. *Issues in offering live P2P streaming service to residential users*. IEEE CCNC, 2007.

82. Magharei, N., and R. Rejaie. PRIME: Peer-to-peer receiver-driven mesh-based streaming. *IEEE/ACM Transactions on Networking*, vol. 17(4), pp. 1052–1065, 2009.

83. Byun, H., and M. Lee. *HyPO: A Peer-to-Peer Based Hybrid Overlay Structure*. IEEE ICACT 2009, Feb. 2009.

84. Awiphan, S., S. Zhou, and J. Katto. Two-layer mesh/tree overlay structure for live video streaming in P2P networks. *Proc. 7th IEEE Consumer Communications and Networking Conference (CCNC)*, pp. 1–5, 2010.

85. Wang, F., Y. Xiong, and J. Liu. mTreebone: A collaborative tree-mesh overlay network for multicast video streaming. *IEEE Transactions on Parallel and Distributed Systems*, vol. 21(3), pp. 379–392, March 2010.

86. Asaduzzaman, S., Y. Qiao, and G., Bochmann. CliqueStream: An efficient and fault-resilient live streaming network on a clustered peer-to-peer overlay. *Proc. Eighth International Conference on Peer-to-Peer Computing*, pp. 269–278, 2008.

87. Liuy, Z., Y. Sheny, S. Panwary, K. W. Rossz, and Y. Wang. *Efficient Substream Encoding and Transmission for P2P Video on Demand*. Packet Video 2007, pp. 143–152, 2007.

88. Huang, Y., T. Z. J. Fu, D. Chiu, J. C. S. Lui, and C. Huang. *Challenges, design and analysis of a large-scale P2P-VoD system*, SIGCOMM'08, USA, August 17–22, 2008.

89. Annapureddy, S., C. Gkantsidis, P. R. Rodriguez, and L. Massoulie. *Providing video-on-demand using peer-to-peer networks*. Microsoft Research Technical Report, MSR-TR-2005-147, October 2005.

90. Do, T. T., K. A. Hua, and M. A. Tantaoui. Robust video-on-demand streaming in peer-to-peer environments. *Computer Communication*, vol. 31(3), pp. 506–519, 2008.

91. Guo, Y., K. Suh, J. Kurose, and D. Towsley. P2cast: Peer-to-peer patching scheme for VoD service. *Proc. 12th World Wide Web Conference (WWW-03)*, 2003.

92. Do, T. T., K. A. Hua, and M. A. Tantaoui. *P2VoD: Providing fault tolerant video-on-demand streaming in peer-to-peer environment*. IEEE International Conference on Communications, Paris, France, pp. 1467–1472, 2004.

93. Roh, J.-H., and S.-H. Jin. *Video-on-Demand Streaming in P2P Environment*. IEEE International Symposium on Consumer Electronics (ISCE 2007), pp. 1–5, 2007.

94. Cui, Y., B. Li, and K. Nahrstedt. *IEEE Journal on Selected Areas in Communications*, vol. 22(1), pp. 91–106, 2004.

95. Suh, K., Y. Guo, J. Kurose, and D. Towsley. A peer-to-peer on-demand streaming service and its performance evaluation. *Proc. 2003 IEEE International Conference on Multimedia and Expo (ICME 2003)*, July 2003.

96. Guha, S., S. Annapureddy, C. Gkantsidis, D. Gunawardena, and P. Rodriguez. Is high-quality VoD feasible using P2P swarming? *Proc. WWW*, pp. 903–911, 2007.

97. Pandey, R. R., and K. K, Patil. Study of BitTorrent based Video on Demand Systems. *International Journal of Computer Applications*, vol. 1(11), pp. 29–33, 2010.

98. Dana, C., D. Li, D. Harrison, and C. Chuah. *Bass: Bittorrent assisted streaming system for video-on-demand*. International Workshop on Multimedia Signal Processing (MMsP) IEEE Press, 2005.

99. Yang, X., M. Gjoka, P. Chhabra, A. Markopoulou, and P. Rodriguez. Kangaroo: Video Seeking in P2P Systems. *Proc. IPTPS*, 2009.

100. Xu, T., W. Wang, B. Ye, W. Li, S. Lu, and Y. Gao. Prediction-based Prefetching to Support VCR-like Operations in Gossip-based P2P VoD Systems. *Proc. IEEE ICPADS'09*, Shenzhen, China, 2009.

101. Vlavianos, A., M. Iliofotou, and M. Faloutsos. BiTos: Enhancing BitTorrent for Supporting Streaming Applications. *Proc. IEEE Global Internet Symposium*, 2006.

102. Bhattacharya, Z. Yang, and D. Panuse. *COCONET: Co-operative Cache Driven Overlay NETwork for P2P VoD Streaming*. QSHINE, pp. 52–68, 2009.

103. Regant, Y., S. Hung, and H. F. Ting. An optimal broadcasting protocol for mobile video-on-demand. *Proc. Thirteenth Australasian Symposium on Theory of Computing*, vol. 65, 2007.

104. Do, T. T., K. A. Hua, N. Jiang, and F. Liu. PatchPeer: A scalable video-on-demand streaming system in hybrid wireless mobile peer-to-peer networks. *Peer-to-Peer Network Applications*, vol. 2, pp. 182–201, 2009.

105. Tran, D. A., L. Minh, and K. A. Hua. *Proc. 2004 IEEE International Conference on Mobile Data Management*, pp. 212–223, 2004.

106. Yoon, H., J. Kim, F. Tan, and R. Hsieh. On-Demand Video Streaming in Mobile Opportunistic Networks. *Proc. Sixth Annual IEEE International Conference on Pervasive Computing and Communications*, pp. 80–89, 2008.

107. Sato, K., M. Katsumoto, and T. Miki. *P2MVOD: Peer-to-peer mobile video-on-demand*. Sixth Annual IEEE International Conference on Pervasive Computing and Communications, pp. 80–89, 2008.

108. Noh, J., M. Makar, and B. Girod. Streaming to Mobile Users in a Peer-to-Peer Network. *Proc. 5th International ICST Mobile Multimedia Communications Conference*, 2009.

109. Peltotalo, J., J. Harju, L. Vaatamoinen, I. Bouazizi, and I. D. D. Curcio. RTSP-based Mobile Peer-to-Peer Streaming System. *International Journal of Digital Multimedia Broadcasting*, vol. 2010, Article ID 470813, 15 pages, 2010.

110. Zhang, B., S. G. Chan, G. Cheung, and E. Y. Chang. LocalTree: An Efficient Algorithm for Mobile Peer-to-Peer Live Streaming. *Proc. IEEE International Conference on Communications (ICC2011)*, pp. 1–5, 2011.

111. Trajkovska, I., J. Salvachua, and A. M. Velasco. A novel P2P and cloud computing hybrid architecture for multimedia streaming with QoS cost functions. *Proc. MM'10 International Conference on Multimedia*, 2010.

112. Sweha, R., V. Ishakian, and A. Bestavros. AngelCast: Cloud-based Peer-Assisted Live Streaming Using Optimized Multi-Tree Construction. *Proc. ACM Multimedia Systems (MMSys)*, 2012.

113. Gheorghe, G., R. Lo Cigno, and A. Montresor. Security and Privacy Issues in P2P Streaming Systems: A survey. *Journal of Peer-to-Peer Networking and Applications*, pp. 1–17, 2010.

114. Ding, C., C. Yueguo, and C. Weiwei. *A Survey Study on Trust Management in P2P Systems*. Available: http://citeseerx.ist.psu.edu/viewdoc/summary?doi = 10.1.1.137.997, 2004.

115. Wang, W., Y. Xiong, and Q. Zhang. *Ripple-Stream: Safeguarding P2P Streaming against DoS Attacks*. IEEE International Conference on Multimedia and Expo, pp. 1417–1420, 2006.

116. Karakaya, M., I. Korpeoglu, and O. Ulusoy. Free riding in peer-to-peer networks. *IEEE Internet Computing*, vol. 13(2), pp. 92–98, 2009.

117. Vishnumurthy, V., S. Chandrakumar, and E. G. Sirer. KARMA: a secure economic framework for P2P resource sharing. *Proc. Workshop on the Economics of Peer-to-Peer Systems*, 2003.

118. Damiani, E., S. De Capitani Di Vimercati, S. Paraboschi, P. Samarati, and F. Violante. A reputation-based approach for choosing reliable resources in peer-to-peer networks. *Proc. 9th ACM Conference on Computer and Communications Security*, pp. 207–216, New York: ACM Press, 2002.
119. Chatzidrossos, I., and V. Fodor. On the effect of free-riders in P2P streaming systems. *Proc. International Workshop on QoS in Multiservice IP Networks*, 2008.
120. Mol, J., J. Pouwelse, M. Meulpolder, D. Epema, and H. Sips. Give-to-Get: Free-riding-resilient video-on-demand in P2P systems. *Proc. SPIE, Multimedia Computing and Networking Conference*, vol. 6818, 2008.
121. Habib and J. Chuang. Incentive mechanism for peer-to-peer media streaming. *Proc. International Workshop on Quality of Service*, 2004.
122. Tan, G., and S. A. Jarvis. A payment-based incentive and service differentiation mechanism for peer-to-peer streaming broadcast. *Proc. the 14th International Workshop on Quality of Service*, 2006.
123. Dhungel, P., X. Hei, K. W. Ross, and N. Saxena. The pollution attack in P2P live video streaming: Measurement results and defenses. *Proc. Peer-to-Peer Streaming and IP-TV workshop (P2P-TV'07)*, 2007.
124. Seedorf, J. *Security issues for P2P-based voice and video streaming applications*. iNetSec 2009 Open Research Problems in Network Security, vol. 309, IFIP Advances in Information and Communication Technology, Boston: Springer, pp. 95–110, 2009.
125. Wang, Q., L. Vu, K. Nahrstedt, and H. Khurana. Identifying malicious nodes in network-coding-based peer-to-peer streaming networks. *Proc. IEEE Mini INFOCOM'10*, 2010.
126. Hu, B., and H. V. Zhao. Pollution-resistant peer-to-peer live streaming using trust management. *Proc. 16th IEEE International Conference on Image Processing*, pp. 3057–3060, 2009.

Chapter 3

P2P Streaming over Cellular Network: Issues, Challenges, and Opportunities

Mohammad Zulhasnine, Changcheng Huang, and Anand Srinivasan

Contents

3.1 Introduction

The peer-to-peer (P2P) network is exploding with great interest that aims to overcome the most of the main limitations of the traditional client/server architecture. In a P2P system, logically connected users called peers form an application-level overlay network on top of the physical network. Few notable characteristics of the P2P system are scalability, dynamic, self-organizing, user driven, and decentralized control. Media content sharing over P2P networks is promising because of the distributed nature of the P2P system.

As more appealing applications and services are offered, we are being less wired on the contrary. Fourth-generation (4G) mobile technology aims at meeting the growing demands of supporting data applications at a faster speed. Forecast conducted by Cisco [1] suggests that mobile video traffic will exceed 1 Exabyte per month by 2012–2013, 20 years after the first short message service (SMS) was sent. On the other hand, online web reached the same milestone in 2004, more than 30 years after the first email was sent. This figure indicates the tremendous impact of wireless video contents in the near future. Applications such as live video streaming, Internet radio, and video conferencing have proliferated by riding over the P2P network. Enabling these P2P features to cellular users is still in limbo due to limitations caused by heterogeneity, mobility, and the time-varying capacities of the wireless channel. In addition, cellular links are expensive and inadequate than that of the Internet link, and the paradoxical reality is that peers scatter randomly, which leads to needless traffic traversal through multiple links within a provider's network.

The streaming source is usually located on the Internet. One major challenge is to design a scheme that would disseminate streaming contents among the peers efficiently. For cellular peers, all streaming contents come through the base station, raising the issue of congestions, and the system is limited by the capacity of the base station. It is very common that smartphones and tablets are equipped with multiple wireless interfaces: a cellular carrier that provides Internet access and a wireless local area network (WLAN) in ad hoc mode. The latter provides an opportunistic use of resources within a group free of charge. Cellular peers may collaborate to save the expensive wireless bandwidth and to avoid congestion at the base station. Prolonged viewing time demands the cellular streaming system to be energy efficient.

Peer selection strategy allows requesting peers to choose sending peers from the potential senders. If sending peers are chosen randomly, the overlay network exhibits robustness. However, that causes a substantial amount of traffic flow through the underlying physical network. Choosing neighbors of close proximity is more important in case of video streaming due to high volume of data transmission. Peer selection strategy is one of the major challenges toward an efficient cellular P2P system.

A file-sharing application does not need to meet any time constraint or sequentiality requirement; only the total time to download an entire file is significant. On the other hand, live streaming is delay sensitive; users need to download a certain range of segments timely. A segment arriving after its scheduled playback time is useless. Meeting the time constraint is therefore critical for the streaming service. Two categories of streaming applications are live streaming and video-on-demand (VoD). In a live streaming application, a source progressively generates new segments of contents. The content is therefore disseminated in realtime, the buffer size is relatively small, and playbacks are loosely synchronized among all peers. Delay and jitter may freeze the live streaming contents; however, playback can skip contents to overcome the problem. The streaming goal over the wireless channel is to enhance the user-perceived quality and save the scarce wireless bandwidth at the same time. The scope of this work is limited to the user-driven P2P live streaming applications running over cellular devices. The base station is not aware of such applications and does not broadcast contents.

With an exploding number of handheld devices, from smartphones, ePads, gaming consoles, to portable media players, designing a P2P streaming system over a cellular network is an important consideration, especially in the context of next-generation converged networks. In this work, we provide a comprehensive overview of the recent advancement in the area of wireless P2P streaming. We present architectures, protocols, and issues toward enabling P2P streaming on wireless devices. Section 3.2 describes the mechanism of constructing efficient overlay in details. Section 3.3 illustrates the cellular bandwidth saving technique through collaboration among peers. Section 3.4 presents the essence and strategy of peer selection in the wireless P2P streaming system. Section 3.5 discusses various issues on realizing a wireless streaming system. Section 3.6 concludes our work with future direction to enable P2P streaming over the cellular network.

3.2 Designing Efficient Overlay Network

In a P2P system, the users (peers) form an application-level overlay network on top of the physical network. Based on the overlay formation technique, P2P systems are classified into unstructured P2P systems such as Gnutella [2] and KaZaA [3] and structured P2P systems such as Chord [4], Pastry [5], and content-based networks [6]. Unstructured lookup protocols in [2] and [3] flood the queries with restriction and therefore accrue considerable message overhead and lessen the search accuracy. Structured lookup protocols in [4–6] construct an overlay network using a distributed hash table (DHT) and thereby improve search efficiency and accuracy.

Streaming over a P2P network relies on the overlay network and yields some penalties in terms of bandwidth usage due to the mismatch between the physical and overlay networks. The P2P streaming system needs to feed the peers with current contents continuously. Because of the magnitude of contents per unit time and the nature of playing the same contents throughout the entire system, efficient overlay construction is the key to realize a cellular streaming system. Few important aspects of an efficient overlay network are scalability, proximity awareness, and low maintenance overhead. Here, we will discuss few efficient overlay systems.

3.2.1 Proximity-Aware Topology Construction

Several researchers presented topology-aware clustering schemes where peers partition into groups based on proximity in terms of a network metric (e.g., latency or round-trip time) [7–8]. This way, peers belonging to the same group are close to each other.

3.2.1.1 Network Metric-Based Clustering

The method proposed in [7] makes use of information derived from Border Gateway Protocol (BGP) routing table snapshots, whereas the one in [8] exploits landmark nodes to measure the proximity. The authors in [9] presented a topology-aware overlay named as mOverlay. In mOverlay, peers with close proximity form a group using a dynamic landmark procedure. Unlike [8], mOverlay does not require any extra deployment of landmark nodes; rather, groups within the mOverlay serve as the landmark nodes for the joining peer. A joining peer first contacts a peer of an arbitrary group. The peer then measures the distance from that group. That group also provides proximity information from other groups. The peer then joins the group with the closest distance. There is overhead to guide a joining node in finding its closest existing cluster.

3.2.1.2 Network Prefix-Based Clustering

Huang et al. [10] proposed a P2P scheme where peers form clusters according to their network prefix divisions. This divides the network into multiple network-aware clusters and helps peers to search files in nearby peers first. One super peer from each cluster maintains indexes of shared file and peers' location. The bootstrap peer maintains an up-to-date cluster routing table to direct new peers joining the network to the appropriate clusters. The bootstrap peer requires routing information from a BGP router to update the cluster routing table. The authors conducted the simulations with WiFi-connected peers. However, peers belonging to the same cluster and with the same network prefix division can be far away from each other, let alone in meeting the WiFi communication range. The presence of anonymous routers, whose existence can be detected but not their address, also leads to distorted and inflated inferred topology [13].

3.2.1.3 Base Station-Aware Clustering

The work in [11] divides the wireless users according to their WiMAX base stations. Peers from the same base station form a local mesh network exploiting WiFi connectivity, and the base stations construct a DHT-based overlay network. Peers first search contents within the local mesh and then on the DHT-based system if required. Results show that this method can greatly reduce the number of lookup messages in physical networks. However, base stations' participation to the P2P system is not realistic. The method also requires knowledge of the physical layer and therefore is unsuitable for application-driven P2P systems.

3.2.1.4 Physical Layer-Dependent Proximity

Wang and Ji [12] proposed a topology-aware P2P protocol based on Chord for the mobile ad hoc network where physically closed nodes get adjacent *ids* exploiting nodes' position information.

Entering node broadcasts request to join the existing wireless P2P network. Any node, upon receiving the entering application, calculates the relative distance from the energy level of the received signal. From the received energy strength and *id* space margin information, the entering node gets a node *id* adjacent to the physically close nodes. The method is only suitable for the infrastructure less wireless network and does not consider the Internet gateway in the design. The cost of device discovery should not outweigh the benefits of proximity-aware communication.

3.2.1.5 Geographical Location-Based Proximity

Canali et al. [13] proposed a location-aware overlay network named MeshChord for wireless mesh networks. To acquire location awareness, the algorithm assigns adjacent *id*'s to physically closed peers. However, this method requires location information possibly using the Global Positioning System receivers. The users also may not feel comfortable to release geographical location due to privacy and security reasons. The DHT-basedP2P system "GeoRoy" [14] is a location-aware variant of Viceroy [15]. GeoRoy is based on a two-tier architecture where mesh routers (super peers) form the DHT-based upper tier, and the mobile mesh clients (leaf peers) provide contents to the system. The use of DHT in the upper tier helps in the indexing of the contents by the lower-tier users. Thus, the super peers only provide a distributed catalog of the resources held by the leaf peers, and the leaf peers publish and request resources with the aid of the super peers. The peers' *id*'s are computed from geographical information in both GeoRoy [14] and MeshChord [12]. However, MeshChord is built on top of Chord, whereas GeoRoy is built on top of Viceroy. The use of a geographically aware hash function reduces the mismatch between the overlay network and the underlying network and thereby makes the resource lookup more efficient in terms of hop distance.

3.2.2 Hierarchical Overlay Construction

The hierarchical DHT system improves the lookup performance by utilizing super peers. However, load balancing among super peers is a crucial problem of the hierarchical DHT system. In addition, such a system must avoid isolation of regular peers if a super peer fails.

3.2.2.1 Super Peer-Based Hierarchy

The P2P system by nature exhibits heterogeneity, and "super peers," which are more powerful and stable, do exist. Joung and Wang [16] proposed a two-level hierarchical Chord, named Chord^2, where super and regular peers join in two separate tiers. The motivation therein is to reduce the maintaining cost of the DHT-based P2P system. Regular peers, which form the outer tier, perform periodical maintenance of their successor links only. They inform the super peers if a change is detected due to the peer joining/leaving process. The super peers send finger table update notification to the responsible regular peers only. This way, Chord^2 avoids periodical refreshing of the finger table by the regular peers and reduces maintenance cost. In [17], the networked divided into several groups. Each group consists of one stable and several unstable peers (who join/leave very frequently). A stable peer gets higher bits on the left side, covering the entire *id* space, and plays a more important role in routing and replication. The unstable peer gets a node *id* with lower bits on the left side, which makes its *id* region as small as possible. The unstable peers only assist the reliable peers. The authors claimed that such a system minimizes the maintenance overhead. Super peers should be uniformly distributed throughout the overlay network and meet some other properties such as load balancing, adaptability to peer churn, and leveraging peer heterogeneity.

3.2.2.2 DHT-Based Hierarchy

Zoels et al. [18] proposed a load-balancing algorithm for the two-tier DHT system with disjoint groups. To achieve an equal load for every super peer, a distributed load-balancing algorithm determines to which super peer the joining peer would connect. Such a load-balancing approach may create mismatch between the physical layer and the logical layer of the overlay network.

3.2.3 DHT-Based Proximity-Aware Overlay

In "eQuus" [19], a DHT-based P2P system, peers that are close in terms of a proximity metric are grouped into the same cliques. The bootstrap peers of such a system maintain the address of one node of each clique in its routing table. The joining node contacts an arbitrary bootstrap node to determine the closest clique. The bootstrap node replies with the address of one node of each clique. The joining node needs to contact all these nodes to determine the proximity. However, such a system fails to balance load when the underlying nodes cluster in an unbalanced manner. Cramer and Fuhrmann [20] proposed a topology-aware DHT system based on Chord. In this scheme, *id* assignment is relaxed, and logical successors are chosen according to the physical proximity in the ad hoc network. However, discovering physically nearby neighbors is expensive for larger networks and often appears impractical.

3.2.4 Example of an Application-Driven Efficient Overlay System

In [21], we proposed an efficient overlay network named cellular Chord (C-Chord) that integrates the cellular users into the P2P network in a topology-aware fashion. The hierarchical overlay system "C-Chord" conceptually separates stable wired peers and wireless peers. The stable Internet peers form the large main Chord (*m*-Chord); on the contrary, cellular users form several (equal to the number of base stations) auxiliary Chord (*a*-Chord). The cellular peers under the same base station belong to a particular a-Chord. The base station does not participate in any of the Chord; cellular users only utilize the base station's unique numeric identification number (Cell-ID) to form the *a*-Chord. The *a*-Chord organizes the physically close cellular peers together. The Cell-ID serves as a rendezvous point for peers under the same access point. Figure 3.1 shows the relationship between physical and logical networks in the proposed C-Chord system. Here, the *m*-Chord consists of four-wired peers, and each *a*-Chord consists of two cellular peers. The cellular peers under the same base station belong to a particular *a*-Chord. The base station does not participate in any of the Chord. Cellular users have the option to choose between Internet peers and cellular peers. The users of the cellular network can download either diverse content from the stable Internet peers at a faster rate or social contents from the peers within the same base station, avoiding Internet data penalty. Our proposed system reduces the high management cost of routing information with fewer entries in the routing table for the cellular peers.

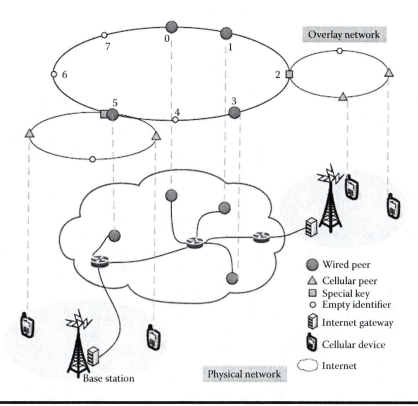

Figure 3.1 **Relationship between physical and logical networks in the proposed C-Chord system. For the clarity of the picture, only the special key and part of the network are shown.**

3.3 Collaborative Streaming over Cellular Network

Because of the proliferation of wireless handheld devices, there are many important scenarios when mobile users in close physical proximity may prompt to watch the same media contents such as live streaming of a popular game show, election night results, and concerts. Realizing P2P streaming applications are entirely user driven, Internet service provider (ISP)/mobile operator's cooperation is not feasible. That is, the base station is not aware of the applications running on mobile devices and does not broadcast contents. Cellular channels are limited in number and expensive; multimedia streaming demands huge bandwidth and does not scale to the number of users. To avoid bottleneck at the base station and to reduce streaming cost, few wireless peers may download contents from the Internet peers and then share the contents with other wireless peers by broadcasting. To realize such collaborative P2P streaming on cellular devices, we need to address the following issues:

- How to select agents in a distributive manner
- How to find assisted peers in close vicinity
- How to disseminate streaming contents from the agents avoiding collisions
- How to deal with the dynamics of network size
- How to ensure fairness of streaming cost sharing
- How to ensure video quality

3.3.1 Collaborative Streaming System

Recently, there have been quite a few collaborative P2P streaming systems proposed such as COSMOS [22], LocalTree [23], CHUMCAST [24], COMBINE [25], WiMA [26], and MOVi [27]. All of the above-mentioned streaming systems have the same motivation of reducing streaming costs through collaborative sharing. Few cellular peers pull streams from the base stations and then share the contents with the remaining users using a free broadcast channel such as WiFi or Bluetooth. These systems, however, differ in implementations; we will discuss the important details of such systems. Peers that pull streaming contents and then broadcast are called agents, whereas peers that receive contents through broadcasting are termed as assisted peers. Figure 3.2 shows the underlying physical architecture of a collaborative P2P streaming system.

In COSMOS [22], each agent that randomly selects one video description still enjoys the full quality of video through sharing. This feature saves valuable cellular bandwidth. However, an agent requires other agents with remaining video descriptions in its communication range. The agents and the assisted peers interchange their role alternatively to ensure the cost of fairness. Peers' density always plays a crucial role in determining the rebroadcast scope. To avoid duplication of packets and the contention of the wireless medium, peers also exchange neighboring information and streaming buffer map periodically. COSMOS employs dynamic broadcast, instead of fixed broadcast, where peers determine the broadcasting scope, depending on its local density, and thereby reduce flooding and channel redundancy. In LocalTree [23], few cellular peers download streaming contents from the Internet and then redistribute the contents at multihop distance over the WiFi-connected unstructured mesh network. Each peer acquired receives states of packets of its neighbors. To avoid collision, peers wait in proportion to the value of received packet per neighbor before rebroadcasting. The peer with the least value of received packet per neighbor is the most preferred to rebroadcast. Each time a peer rebroadcasts, it also decreases time-to-leave (TTL) by "1." Peers in such a system may calculate the distance from the puller in terms of the hop count from the TTL field. Any peer, experiencing stable hop count, may join to construct a tree for further reducing the battery consumption. The TTL determines the depth of the tree. The base-mesh structure proves resilient against a peer's connection failure. The proposed method minimizes battery consumption by reducing the rebroadcasting scope and constructing stable tree on top of base-mesh overlay. Peers with better connection with the base station may fetch contents

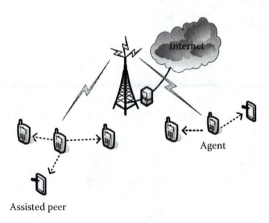

Figure 3.2 Collaborative P2P streaming system where cellular peers share streaming cost.

from the Internet peers in a more energy optimal way. CHUMCAST [24] also follows a collaborative streaming approach where nearby wireless peers form a group employing a WLAN. Each peer plays the role of an agent by turn and distributes the contents by a broadcasting technique.

The agent in the CHUM network transmits service beacon periodically. The beacon contains its own *id* and the group *id* along with some other information. The arriving wireless peer transmits the joining request with the information of the *id* of the beaconing peer. Any group member, upon receiving the joining request, compares the *id* and drops the request if it is a mismatch. Only the beaconing peer finds the match and accepts the joining request. Each wireless peer gets support from one wired peer (CHUMCAST server) in the Internet and uploads one-hop neighboring wireless peers' information to its supporting server. The agent's server (active server) collects membership information from other supporting wired peers and selects a set of rebroadcasting peers. The active server also schedules the next proxy and delivers all membership and topology information to that agent's server. Wireless peers in the CHUM network rely on CHUMCAST servers. The authors did not mention any kind of incentives for such a server's role. If these servers are common wired peers (which sounds more reasonable in a P2P network), then the failure of the CHUMCAST servers could severely affect CHUM network performance.

COMBINE [25] differs from other collaborative streaming systems on how it forms the collaborative groups. In COMBINE, a requester (termed "initiator") identifies a set of collaborators from its vicinity and forms a collaborative group in an energy-efficient way. To join the group, each potential member wakes up periodically to broadcast a short message and keeps their WLAN cards in a low-energy mode in the remaining time. The initiator always keeps its WLAN card switched on to discover such a message. The message contains the cost of download (in terms of the monetary and the battery energy) and expected WWAN speeds of the corresponding potential member. The initiator then chooses a set of collaborators based on received messages. WiMA [26] suggests that wireless peers with greater abilities serve as agents. However, WiMA does not describe how the comparison is made to select agents from the wireless peers. To find arriving peers, each agent broadcasts a heart beating (HB) packet periodically. The joining peer, upon receiving the HB packet, requests streaming contents from the agent. Under stable condition, the agent simply broadcasts contents to the assisted peers. The assisted peer downloads missing contents from some other peers. WiMA also does not describe how an agent delivers contents without collision in the presence of another agent. The streaming quality of the wireless peers heavily depends on the availability of streaming contents in the agent. Collaborative streaming system MOVi [27] deploys a centralized MOViserver located at the back-haul network. The server manages the connectivity maps of shared devices, tracks the media segments, and also schedules the agents to disseminate streaming contents to the assisted peers. The shared users serve as the cache for the media contents and are responsible for data transfer under the control of the MOVi server. In [28], wireless peers compete among them to become an agent based on their *id*s. Each peer broadcasts a special "hello" message along with its *id* over the shared link to claim the role of an agent. If he or she hears a "hello" massage from another peer with lower *id*, the receiver gives up its claim. This way, the peer with the lowest *id* becomes the agent. The agent may also assign a chunk download task to the assisted peers [29]. The assisted peer periodically monitors its Internet access link bandwidth, estimates the assigned chunk download probability within the deadline, and reports it to the agent. The agent identifies the chunks with lower probability of successful download and reassigns the chunk download task to other peers enjoying good Internet connections. Both collaborative streaming systems in [28–29] differ from others that the assisted peers also download contents through the Internet access link. The authors suggested deploying cache at the mesh access point (MAP) that has inherent gateway to the Internet [30]. Client forms a

P2P network within itself and with the cached MAPs and downloads/shares contents aided by a P2P tracker. However, this scenario requires ISP cooperation, which is most unlikely in the context of the P2P network. The cache placement problem is an active area of research.

3.3.1.1 Feedback Mechanism to Recover Missing Chunks

Multicasting in a collaborative streaming system exploits the inherent broadcast nature of a shared wireless channel. Ironically, the same physical property poses a challenge on incorporating a feedback mechanism while ensuring quality. The feedback mechanism may overwhelm the sender, especially in a dense network. In addition, without proper feedback method, it is also difficult to estimate the optimum transmission rate.

We just opened a Pandora's box by stating that the feedback mechanism is essential to ensure quality while multicasting over the wireless channel. In a traditional automatic repeat request (ARQ) error control method, an acknowledgment (ACK) is sent by the receiver to the sender to inform that a correct frame has been received.

The sender retransmits the corresponding frame if no ACK is received. When the traditional ARQ method is implemented in multicasting over the wireless channel, high bandwidth and extreme coordination are required to process ACKs by each receiver. This leads to a feedback implosion problem. An alternative solution to this problem is to implement negative ACK (NACK), where a receiver only sends NACK feedback to the sender upon receiving an erroneous frame. Even with a NACK-based protocol, significant bandwidth is consumed and overhead accrues if the sender needs to identify receivers that did not receive a correct frame. One way to circumvent the disadvantages and provide scalability is to combine the NACK messages. Still, the sender has to compromise the enumeration of the number of receivers with the NACK message and is unable to determine the suitable rate of transmission.

In the wireless multicast method, the sender has to transmit at the lowest rate sustainable to the receiver at the worst-condition environment. This way, valuable capacity is wasted because of the prolonged channel occupancy. Piamrat et al. [31] proposed a dynamic rate-adaptation mechanism, where the sender adapts a multicast rate based on the quality of experience (QoE) feedback by the receivers. If the receivers experience lower QoE, the sender reduces the multicast transmission rate. On the other hand, if the receivers experience higher QoE, the sender increases the multicast transmission rate. This way, the sender transmits at a rate higher than the lowest instant sustainable rate of the receiver with the worst channel condition.

For real-time traffic like user datagram protocol (UDP)-based streaming, in-time delivery is more crucial than providing a hard-core guarantee of a correct packet arrival. It is more acceptable to allow few erroneous packets than to wait for retransmission as long as the target level of user satisfaction is achieved. Selecting multiple senders is a viable alternative to the feedback mechanism and quality assurance solution.

3.3.2 Leveraging P2P Concept to Operator's Multimedia System

Multimedia broadcast/multicast service (MBMS) in today's cellular networks is capable of broadcasting rich multimedia contents to a large number of mobile peers. However, packet loss is inevitable as the optimal rate selection is challenging due to the time-varying nature of wireless links. The individual feedback mechanism for the retransmission request is not scalable and may overwhelm the multimedia server.

Mobile users may form the P2P network based on secondary wireless interfaces such as 802.11 and repair packet losses. In PatchPeer [32], the base station, with multicast capability, delivers VoD to a set of mobile users. Any user, with missing parts, downloads the remaining parts from the neighboring user employing the unicast method. Selecting an appropriate neighboring peer is challenging; downloading from a peer at multihop distance may cause delay. In [33], the base station broadcasts multimedia contents in blocks to the mobile users. To repair the packet losses, mobile users are divided into several clusters. The connected dominating set algorithm determines the dominator among the cellular users. The dominator is selected as the clusterhead, and the remaining users join one of their neighboring dominators. The normal users wait for a random time between "0" and "0.005" s and then send packet to their dominators. The scheme assumed that one dominator identifies the lost packets on the whole network and coordinates with the other dominators to trigger the repair process. Each dominator then encodes all of its packets with a random linear network coding technique and then broadcasts that coded packet around its neighborhood. The proposed method thus reduces the burden of the base station's downlink channels.

Mobile video broadcast service may employ multiple transmissions from the base station to ensure video quality to the scattered mobile users of variable channel quality. Hua et al. [34] proposed to broadcast different layers of video using different channels with different coverage ranges. The base layer is broadcasted throughout the entire cell to maintain the minimum quality for all viewers. Other layers of video are broadcasted with decreasing covering range as the enhanced layers increase. In order to improve the perceived video quality further, mobile users that received enhancement layers forward those layers to users with weaker channel condition through ad hoc links. Local broadcasters are selected from the average cellular data rate.

3.4 Peer Selection Strategy

The P2P streaming system is dynamic by nature and exhibits heterogeneity in terms of bandwidth. A receiving peer in such a system relies on multiple sending peers to attain the best possible streaming quality. The receiver maintains a contact list of potential senders from which it chooses an active set for streaming. This process is known as the peer selection mechanism. The receiver first ranks the potential senders based on some application-level properties such as the contribution level, partnership willingness, upload bandwidth, and energy levels. Appropriate peer selection from the potential senders plays a vital role especially for wireless networks.

3.4.1 Related Peer Selection Modules

3.4.1.1 Topology-Aware Peer Selection

In [35], the receiver finds an optimal set of active peers by following one of the three approaches: random, end-to-end, or topology-aware. In the random approach, the receiver hardly achieves minimal rate requirement from the aggregate rate of all connected peers because of the peers' low availability and the congestion at the shared path. In the end-to-end approach, a receiver selects peers based on the individual end-to-end expected bandwidth and availability information and does not consider the shared segments. In the topology-aware approach, the receiver selects peers to maximize the expected aggregated rate at the receiver. The topology-aware peer selection technique performs best as it considers shared paths. This scheme applies to large-scale P2P networks and does not address the wireless-related issues. In [36], the authors addressed the issue of increased contention on the shared wireless channel when multiple nodes try to access the same file simultaneously. They also proposed a cooperative P2P file transfer protocol that increases the aggregate throughput by selecting potential download peers with a minimum interference on the download paths. However, this protocol selects potential download peers based on the current load and interference on the download paths for a single receiver at a time. Li et al. [37] also emphasized the importance of appropriate peer selection from the discovered potential sender peers. In their proposed algorithm, the candidates for the senders send an evaluation score based on several factors such as energy, link quality, movement, lingering time, and security. The receiver peer waits a period for the evaluation scores from several candidates before ranking them. However, the evaluation score changes repeatedly in the time-varying wireless environment, and the price paid to acquire evaluation scores is not worthwhile most likely.

3.4.1.2 Load-Aware Peer Selection

Zhang et al. [38] claimed that locality-aware peer selection might degrade the performance of the mobile P2P network over 3G cellular networks due to load mismatch. When a cellular peer selects multiple peers from the same base station as partners, the underlying traffic still traverses through the gateway GPRS support node, costing two-hop expensive cellular channels. Choosing such a local peer also might cause load variation among the cell base stations. The authors suggested that a peer should select a partner from the cell with least traffic load and better upload bandwidth. The authors did not mention how the peer selection module collects a base station's traffic load.

Peer selection methods, as proposed in [35–37], select potential senders optimal for a single receiving peer. Peering decision, only at the receiver end, is always suboptimal, as that particular receiver does not consider other receivers' optimality. An efficient peer selection module therefore should consider the scenario where multiple requesting peers download contents from multiple potential senders. Both stable marriage (SM) algorithm and game theory are strong candidates to model such a problem where multiple peers are involved with their own selfish interests.

3.4.1.3 Matching Algorithmic Peer Selection

The SM algorithm finds a stable matching between men and women through a sequence of proposals from men to women. Both men and women have preference lists. A stable assignment is called optimal when there is no man/woman pair, of which both have stimulus to elope. Gale and Shapley [39] first introduced this algorithm, and Gusfield and Irving [40] later extended the algorithm to many variations. Our work [41] presented a peer selection module for cellular peers based on the SM problem that chooses the appropriate candidate from the discovered potential senders. In that peer selection problem, a man is analogous to the requesting peer, and a woman is analogous to the potential sender. As a man proposes to a woman, the requesting peer sends request to the potential senders. We provided an alternate heuristic algorithm to solve the problem. The solution offers cellular users a choice of downloading contents either from the Internet peers at a faster rate or from other cellular users from the same base station avoiding the Internet data penalty. The results showed that streaming quality improved due to an intelligent selection of peers among the potential senders.

3.4.1.4 Cooperative Peer Selection

A multiarmed bandit can be thought of as a slot machine with multiple levers. When pulled, each lever earns a reward. A gambler's objective is to maximize the sum of the rewards earned from sequential lever pulls. The gambler needs to explore all levers to acquire a reward value associated with each lever but should not lose too much rewards through exploration at the same time. The restless bandit algorithm is an extension of the multiarmed bandit problem. The work in [42] proposed a multiple sender selection scheme to improve the performance in terms of data rate and energy consumption in the context of the multihop wireless network.

The sender selection process is based on the restless-bandit algorithm, where the potential senders' states (active or passive) are determined in a distributed and cooperative manner. The lever is equivalent to a sender, and pulling the lever is equivalent to setting the sender's status as active. The formulation is *restless* as the reward, associated with each sender, changes independent of the sender selection process due to the time-varying nature of the wireless channel. Peer selection has crucial influence on the network's overall throughput and traffic distribution as well. Gurses and Kim [43] proposed a peer selection scheme that attempts to maximize the overall throughput of the wireless adhoc network. The scheme also allocates the optimal rate to the receiving peers by controlling the MAC layer parameters. Such control of the underlying parameters is not possible for real-world P2P applications.

3.4.1.5 Game-Theoretic Peer Selection

The game theory analyzes the attitude of the decision makers ("players") whose rational decisions affect each other. The players interact among themselves and form coalitions to gain (often-conflicting) rewards through their decisions. In [44], the authors proposed a peer selection process where peers play a strategic game to determine the parent–child relationship. Each parent forms a coalition with its children. Each combination of coalition offers a specific value. When a peer joins a coalition, it brings an additional value to the existing coalition and gets its share of utility. Every player tries to maximize its share of utility by making an attempt to join a coalition that offers higher utility. A stable coalition is formed when no member of the coalition has incentive to leave. A child usually accepts more than one parent until it meets its incoming traffic demand. Any peer with higher outgoing bandwidth gets few utility from a coalition that drives that peer to form coalition with more parents. This way, peers of high outgoing bandwidths are resilient to peer failure, and the system is less susceptible to the peers' dynamics.

3.4.2 ISP Cooperation on Peering Process and the Associated Risks

Recently, few researchers suggested that ISPs and P2P application provider could collaborate for peering decision [45–49]. In [45], Aggarwal et al. proposed to install an ISP-owned facility called *Oracle* that would assist a peer client on locality-aware peering decision. P2P users provide a list of possible peers to the *Oracle*. The *Oracle* ranks the candidate peers for the client based on BGP information. Such guidance on peering decision reduces cross-domain traffic. *Oracles* from different ISPs may exchange information to make the peer ranking more efficient [46]. Figure 3.3

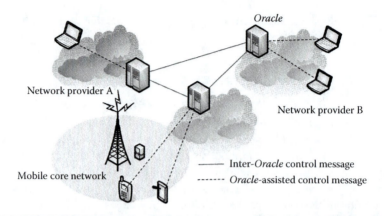

Figure 3.3 ISP-controlled *Oracle* provides assistance on peering decision. For clarity, we only show logical connection between peers and *Oracles* and among *Oracles* themselves.

shows that the ISP-controlled *Oracle* assists peers to select appropriate partners. Xie and Yang [47] proposed to introduce *iTracker* to the ISP that performs similar function as the *Oracle* [45], [46]. iTracker collects information from both the P2P provider (application tracker) and ISP and provides peering decision based on distance, load, or peer's bandwidth heterogeneity. In [48], the application tracker provides peering instructions to the clients from end-to-end path utilization information reported by the ISPs.

The objective of the application tracker is to balance P2P traffic by minimizing the maximum link utilization through optimum partner selection. A content distribution network (CDN) connects clients with low-latency replica servers (located in ISPs). Peers can make use of information collected by the CDN to select partners. For example, peers close to the same server select each other as partners to reduce unwanted traffic flows [49]. CDN-assisted peer selection eliminates the need for additional entity such as *Oracle*. In [50], Kaya et al. proclaimed the potential risks coming out of cooperation between P2P providers and ISPs. The authors emphasized that ISP-owned *Oracle* should be avoided for inter-ISP peering decisions as ISPs have conflicting preferences for inter-ISP links. A centralized *Oracle* monitored by a third party is needed to ensure fairness among multiple ISPs.

3.4.3 Design Criteria of Peer Selection Modules

We propose the following recommendations to design an efficient peer selection module:

- The peer selection should leverage the underlying network conditions. However, the price paid to infer topology should not exceed the reward achieved from it.
- The peer selection technique should sustain the peer churn rate, with the peer joining/leaving the system all the time.
- The receiver should experience improved streaming quality by choosing potential senders with better load distribution.
- When collaborative streaming is not an option, a cellular receiver should prefer a wired sender to a cellular sender to save wireless bandwidth when avoiding two-hop wireless communication.

- When a cellular receiver finds more than one cellular sender with a close data rate, the one with better energy state is more preferred.
- P2P streaming is always user driven; ISP cooperation is only possible for the CDN-based streaming system.

3.5 Miscellaneous Issues on Realizing Cellular P2P Streaming System

3.5.1 P2PSIP for Cellular Peers

The Session Initiation Protocol (SIP) is a signaling protocol that may establish, modify, and terminate a multimedia session. A dedicated proxy server maintains the registration and provides SIP message routing functionality. As a text-based application-layer protocol, SIP attracted researchers to explore its uses in P2P applications. Li and Wang [51] proposed a P2P file-sharing application using SIP as a control protocol for cellular networks. The work also suggested few modifications to the SIP in order to meet the requirements of cellular P2P application. However, the SIP requires server-centric architecture and exhibits difficulties in NAT/firewall traversal. P2P SIP (P2PSIP) overcomes the limitations of the centralized SIP proxy server and allows discovering peers and resources in a distributive manner. Davies and Gardner [52] proposed a P2P-based framework that locates mobile devices and initiates streaming services between them without the aid of a centralized SIP server. Several Internet peers serve as the virtual switches and coordinate the connections between mobile devices. The responsible Internet peer of a particular mobile device is allocated based on the mobile device's *id*. The work in [53] measured the performance of mobile phone devices in a Chord-based P2PSIP overlay network. The authors concluded that the mobile peer's memory and CPU are sustained without impairments. However, the message overhead in traditional Chord increases wireless network overload and battery consumption and thereby degrades the performance of mobile peers. Wu and Womack [54] emphasized that the maintenance of the overlay network is challenging for cellular peers due to limited battery life and mobility. The work proposed a mobile-initiated connectivity check that exhibits efficiency due to reduced cellular paging load and lessened wake-up duration.

3.5.2 Barriers Posed by NATs and Firewalls

A next-generation cellular network is expected to offer all-IP-based mobile broadband access to the clients. Mobile network operators (MNOs) generally deploy various types of middle boxes such as a network address translator (NAT) and firewall. A NAT hides a huge number of users behind it and provides them Internet access, even with limited public IP address space. Firewalls protect mobile users by isolating them from malicious activities on the Internet. However, cellular peers located behind NATs should possess the ability to initiate connections to other peers and accept connections initiated by other peers. Figure 3.4 illustrates how a NAT appears as a black box to the laptop host and prohibits the smartphone from receiving P2P contents.

Few NAT traversal schemes that are being standardized in the Internet are the Engineering Task Force such as Session Traversal Utilities for NAT (STUN) [55], Traversal Using Relays around NAT (TURN) [56], and Interactive Connectivity Establishment (ICE) [57]. The work in [58] studied the NAT boxes to predict their port allocation methods. The authors claimed that the cellular peers could utilize existing NAT traversal techniques without modifications. For

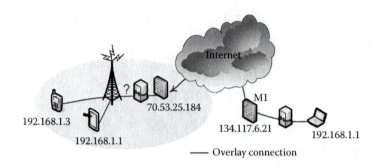

Figure 3.4 NAT IP address of the smartphone is not addressable to the laptop.

example, with STUN, a network client may acquire its external address and port information. Skype, aP2P-based Internet telephone system, employs STUN or a very similar method to solve the NAT traversal problem. P2PSIP may also utilize existing NAT traversal schemes. The work in [59] compared the delay on-call setup and connection establishment for P2PSIP and traditional centralized SIP-oriented overlay network.

Wang et al. [60] developed a tool that is composed of client software running on each mobile device and a dedicated server located on the Internet. They found that operators' firewalls are still vulnerable to attacks such as battery draining and service denials because of the lack of cooperation between the mobile application developers and cellular network operators.

Timeout for idle TCP connections set by the middle box also appears to be a problem for the mobile devices when running P2P applications [60]. Most of the user-driven applications of smartphones (e.g., SMS and push-based email) require maintenance of occasionally idle connections to save battery energy. Live streaming application requires long-lived connection with Internet peers, and peers running such application may appear idle to the network operator when they share contents among themselves. NAT boxes and firewalls aggressively set timeouts for idle TCP connections to recycle the resources held by inactive connections. The extra radio activities for reconnection exacerbate energy consumption on mobile devices.

3.5.3 Copyright and Legal Issues

P2P services and applications are exposed to legal challenge due to frequent sharing of copyrighted materials by the end users. The early P2P system Napster was built over a single-server tracker system that provides content lookup services to the peers. Eventually, Napster faced accusation of violating copyright and ceased its operation [61]. Today's real-world P2P streaming systems, such as PPLive [62], SopCast [63], and Coolstreaming/DONet [64], employ "trackers" only to connect peers sharing common media contents. The peers themselves do the streaming chunk discovery utilizing gossip or DHT-based queries. It is controversial whether to hold such a "tracker" to account for copyright infringement.

3.5.4 Opportunistic Use of WiFi Hotspots

Smartphones usually have multiple access interfaces and may access the Internet at WiFi hotspots and thereby off load the base station's traffic load. In [65], the authors presented experimental results on 3G mobile traffic offloading through WiFi hotspots. From statistical analysis, the authors claimed that WiFi hotspot connections could offload about 65% of the total mobile data and save 55% of battery power. Collins et al. [66] reiterated the WLAN-based universal coverage for mobile devices ranging from home to public places such as university and business campuses, sport complex, and entertainment parks.

Despite the rapid growth of WiFi hotspots, their coverage is still quite limited when it is compared to a large footprint of the cellular network. Moreover, many WiFi hotspot providers only allow a limited number of applications to their subscribers and prohibit them from getting rich multimedia contents. Connection timeout and frequent reconnection authentication may also hinder playing live streaming media.

3.5.5 Scalable Video Coding

Two major scalable video codings that cope with heterogeneity and jitter are media description coding (MDC) and layered coding (LC). Each description in MDC contributes a different amount of enhancement toward the final reconstructed video quality. With MDC, each peer adjusts the number of description for playback according to its bandwidth and device capability. In LC, video is encoded into one base layer and multiple enhancement layers. The base layer ensures the minimum quality of video as perceived by the viewers. The viewer experiences improved quality of video as the number of enhanced layers increases. The implementation of MDC or LC gives reality to live streaming on cellular devices.

3.5.6 Decoder Compatibility and Screen Resolution Heterogeneity

P2P users from different networks (e.g., WiFi, ADSL, WiMAX, and cellular networks) may have decoder and image resolution limitations. The work in [67] proposed a content management system that offers adaptation functionalities to the peers from diverse networks. The proposed architecture introduces two media-aware network elements (MANEs): a seamless home media gateway (sHMG) at the edge of the extended home environment and a seamless network media gateway (sNMG) at the edge of the cellular access networks. Terminal-aware content adaptation and content protection are now distributed to the deployed media gateways. Figure 3.5 shows the proposed architecture with MANE.

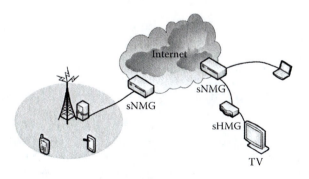

Figure 3.5   MANEs for decoding and image resolution compatibility.

3.6 Conclusion

P2P live streaming is becoming extremely popular to users who enjoy Internet via wired broadband access or WiFi access that is connected to a wired backbone network. This feature is still unavailable to the mobile users, although they have access to the Internet. Cellular channels are limited in number and expensive. Because of the magnitude of contents per unit time and the nature of playing same contents throughout the entire system, topology-aware overlay construction is the key to an efficient P2P streaming system. Few important aspects of an efficient overlay network are scalability, proximity awareness, and low maintenance overhead. Appropriate peer selection from the potential senders also plays a vital role for P2P streaming. The peer selection module should be application-driven. A good peer selection module interacts with other peers for peering decision, increases system throughout, and improves individual user experience. For cellular peers, all streaming contents come through the base station, raising the issue of congestion, and the system is limited by the capacity of the base station. A collaborative streaming system saves the expensive wireless bandwidth and avoids congestion at the base station.

The influence of a mobile streaming system from a social network point of view is enormous. One can be a broadcaster and even build his or her own TV station without the help of a commercial server using minimal resources. Imagine you are enjoying a movie online on your tablet at home and you like to share it with your family. Rather than huddling around the small screen, you can share the contents with a WiFi-enabled TV by employing P2P streaming protocol instantly. In the world we live in today, people want entertainment anytime, anywhere. Whether you are in an airport, stadium, shopping center, subway station, traffic jam, or in any other crowded area, you would not miss any on-air TV shows. Social and community network exhibits high locality characteristics. In a long-distance train, a good number of people might be interested in watching the same live content (e.g., the coverage of an Olympic event). Multiplayer gaming and proximity video advertising could also benefit from the mobile P2P streaming concept. The rise of tablet usages has boosted the consumer demand for the cellular P2P streaming service. Realizing the impact, industries are working on proximity-aware P2P applications. The next step is toward live P2P streaming service on handheld devices. Qualcomm's recent innovation, "Flashlinq" technology allows devices to communicate directly in the operator spectrum [68]. 3GPP is also working on standardizing D2D communication that would provide a breakthrough to the collaborative streaming approach. SopCast, a P2P streaming system over the Internet, already launched its mobile counterpart for Android devices [63]. Other similar online streaming services will follow suit.

References

1. Cisco, "Visual networking index: Forecast and methodology: Global mobile date traffic update, 2011–2016," *White Paper*, 2012.
2. "Gnutella protocol development." [Online]. Available: http://rfc-gnutella.sourceforge.net/index.html/.
3. "KaZaA." [Online]. Available: http://www.kazaa.com.
4. I. Stoica, R. Morris, D. Karger, M. F. Kaashoek, and H. Balakrishnan, "Chord: A scalable peer-to-peer lookup service for internet applications," in *Proc. (SIGCOMM) Conf. on Applications, technologies, architectures, and protocols for computer communications*, 2001, pp. 149–160.
5. A. I. T. Rowstron and P. Druschel, "Pastry: Scalable, decentralized object location, and routing for large-scale Peer-to-Peer systems," in *Proc. of the IFIP/ACM International Conference on Distributed Systems Platforms Heidelberg*, 2001, pp. 329–350.

6. S. Ratnasamy, P. Francis, M. Handley, R. Karp, and S. Shenker, "A scalable content addressable network," in *Proc. (SIGCOMM) Conf. on Applications, Technologies, Architectures, and Protocols for Computer Communications. ACM*, 2001, pp. 161–172.

7. B. Krishnamurthy and J. Wang, "On network-aware clustering of web clients," in *Proc. of the conference on Applications, Technologies, Architectures, and Protocols for Computer Communication (SIGCOMM)*, 2000, pp. 97–110.

8. Ratnasamy, M. Handley, R. Karp, and S. Shenker, "Topologically-aware overlay construction and server selection," in *IEEE Joint Conference of the Computer and Communications Societies (INFOCOM)*, vol. 3, 2002, pp. 1190–1199.

9. X. Y. Zhang, Q. Zhang, Z. Zhang, G. Song, and W. Zhu, "A construction of locality aware overlay network: mOverlay and its performance," *IEEE Journal of Selected Areas in Communications*, vol. 22, no. 1, pp. 18–28, Jan. 2004.

10. C.-M. Huang, T.-H. Hsu, and M.-F. Hsu, "Network-aware P2P file sharing over the wireless mobile networks," *IEEE Journal of Selected Areas in Communications*, vol. 25, no. 1, pp. 204–210, Jan. 2007.

11. J. Li, L. Huang, W. Jia, M. Xiao, and P. Du, "An efficient implementation of file sharing systems on the basis of WiMAX and Wi-Fi," in *Proc. IEEE International Conference on Mobile Ad hoc and Sensor Systems*, Oct. 2006, pp. 819–824.

12. S.Wang and H. Ji, "Realization of topology awareness in peer-to-peer wireless network," in *International Conference on Wireless Communications, Networking and Mobile Computing*, Sept. 2009, pp. 1–4.

13. C. Canali, M. E. Renda, P. Santi, and S. Burresi. "Enabling efficient peer-to-peer resource sharing in wireless mesh networks," *IEEE Transactions on Mobile Computing*, vol. 9, no. 3, pp. 333–347, Mar. 2010.

14. L. Galluccio, G. Morabito, S. Palazzo, M. Pellegrini, M. E. Renda, and P. Santi, "Georoy: A location-aware enhancement to Viceroy peer-to-peer algorithm," *Elsevier*, vol. 51, no. 8, pp. 1998–2014, June 2007.

15. D. Malkhi, M. Naor, and D. Ratajczak, "Viceroy: A scalable and dynamic emulation of the butterfly," in *Proceedings of the Symposium on Principles of Distributed Computing. ACM*, 2002, pp. 183–192.

16. Y.-J. Joung and J.-C. Wang, "Chord2: A two-layer chord for reducing maintenance overhead via heterogeneity," *Computer Networks*, vol. 51, no. 3, pp. 712–731, 2007.

17. K. Kim and D. Park, "Mobile NodeID based P2P algorithm for the heterogeneous network," in *International Conference on Embedded Software and Systems*, 2005, pp. 1–8.

18. S. Zoels, Z. Despotovic, and W. Kellerer, "Load balancing in a hierarchical DHT-basedP2P system," in *International Conference on Collaborative Computing: Networking, Applications and Worksharing (CollaborateCom)*, 2007, pp. 353–361.

19. T. Locher, S. Schmid, and R. Wattenhofer, "eQuus: A provably robust and locality aware peer-to-peer system," in *IEEE International Conference on Peer-to-Peer Computing*, Sept. 2006, pp. 3–11.

20. C. Cramer and T. Fuhrmann, "Proximity neighbor selection for a DHT in wireless multihop networks," in *IEEE International Conference on Peer-to-Peer Computing*, 2005, pp. 3–10.

21. M. Zulhasnine, C. Huang, and A. Srinivasan, "Topology-aware integration of cellular users into the P2P system," in *IEEE Vehicular Technology Conference Fall (VTC) Fall*, 2011, pp. 1–5.

22. M.-F. Leung and S. H. G. Chan, "Broadcast-based peer-to-peer collaborative video streaming among mobiles," *IEEE Transactions on Broadcasting*, vol. 53, no. 1, pp. 350–361, March 2007.

23. B. Zhang, S. G. Chan, G. Cheung, and E. Y. Chang, "LocalTree: An efficient algorithm for mobile peer-to-peer live streaming," in *IEEE International Conference on Communications (ICC)*, June 2011, pp. 1–5.

24. S.-S. Kang and M. W. Mutka, "A mobile peer-to-peer approach for multimedia content sharing using 3G/WLAN dual mode channels," *Wireless Communications and Mobile Computing*, vol. 5, no. 6, pp. 633–645, 2005.

25. G. Ananthanarayanan, V. N. Padmanabhan, L. Ravindranath, and C. A. Thekkath, "COMBINE: Leveraging the power of wireless peers through collaborative downloading," in *Proc. of the International Conference on Mobile Systems, Applications and Services (MobiSys)*, 2007, pp. 286–298.

26. W. Jiang, X. Liao, H. Jin, and Z. Yuan, "WiMA: A novel wireless multicast agent mechanism for live streaming system," in *International Conference on Convergence Information Technology*, 2007, pp. 2467–2472.

27. H. Yoon, J. Kim, F. Tan, and R. Hsieh, "On-demand video streaming in mobile opportunistic networks," in *IEEE International Conference on Pervasive Computing and Communications (PerCom)*, March 2008, pp. 80–89.

28. M. Stiemerling and S. Kiesel, "A system for peer-to-peer video streaming in resource constrained mobile environments," in *Proc. of ACM Workshop on User-Provided Networking: Challenges and Opportunities (U-NET)*, 2009, pp. 25–30.

29. M. Stiemerling and S. Kiesel, "Cooperative P2P video streaming for mobile peers," in *Proc. of International Conference on Computer Communications and Networks (ICCCN)*, Aug. 2010, pp. 1–7.

30. Y. Zhu, W. Zeng, H. Liu, Y. Guo, and S. Mathur, "Supporting video streaming services in infrastructure wireless mesh networks: Architecture and protocols," in *IEEE International Conference on Communications (ICC)*, 2008, pp. 1850–1855.

31. K. Piamrat, A. Ksentini, J. M. Bonnin, and C. Viho, "Q-DRAM: QoE-based dynamic rate adaptation mechanism for multicast in wireless networks," in *IEEE Global Telecommunications Conference (GLOBECOM)*, 2009, pp. 1–6.

32. T. Do, K. Hua, N. Jiang, and F. Liu, "PatchPeer: A scalable video-on-demand streaming system in hybrid wireless mobile peer-to-peer networks," *Peer-to-Peer Networking and Applications*, vol. 2, no. 3, pp. 182–201, 2009.

33. Y. Liu, B. Guo, C. Zhou, and Y. Cheng, "A CDS based cooperative information repair protocol with network coding in wireless networks," in *IEEE Global Telecommunications Conference (GLOBECOM)*, Dec. 2010, pp. 1–5.

34. S. Hua, Y. Guo, Y. Liu, H. Liu, and S. S. Panwar, "Scalable video multicast in hybrid3g/ad-hoc networks," *IEEE Trans. Multimedia*, vol. 13, no. 2, pp. 402–413, April 2011.

35. M. Hefeeda, A. Habib, B. Botev, D. Xu, and B. Bhargava, "PROMISE: peer-to-peer media streaming using collect cast," in *Proc. ACM International Conference on Multimedia*. ACM, 2003, pp. 45–54.

36. S. M. ElRakabawy and C. Lindemann, "Peer-to-peer file transfer in wireless mesh networks," in *Proc. Wireless on Demand Network System and Services (WONS)*, 2007, pp. 114–121.

37. X. Li, H. Ji, R. Zheng, Y. Li, and F. R. Yu, "A novel team-centric peer selection scheme for distributed wireless P2P networks," in *IEEE (WCNC)*, 2009, pp. 1–5.

38. Y. Zhang, X. Zhou, F. Bai, and J. Song, "Peer selection for load balancing in 3Gcellular networks," in *International Conference on Information Science and Engineering (ICISE)*, Dec. 2010, pp. 2227–2230.

39. D. Gale and L. S. Shapley, "College admissions and the stability of marriage," *The American Mathematical Monthly*, vol. 69, no. 1, pp. 9–15, 1962.

40. D. Gusfield and R. W. Irving, *The stable marriage problem: structure and algorithms*. Cambridge, Mass.: MIT Press, 1989.

41. M. Zulhasnine, C. Huang, and A. Srinivasan, "Towards an effective integration of cellular users to the structured peer-to-peer network," *Peer-to-Peer Networking and Applications*, vol. 5, no. 2, pp. 178–192, 2012.

42. P. Si, F. R. Yu, H. Ji, and V. C. M. Leung, "Distributed multisource transmission in wireless mobile peer-to-peer networks: A restless-bandit approach," *IEEE Transactions on Vehicular Technology*, vol. 59, no. 1, pp. 420–430, Jan. 2010.

43. E. Gurses and A. N. Kim, "Maximum utility peer selection for P2P streaming in wireless ad hoc networks," in *IEEE GLOBECOM*, 2008, pp. 1–5.

44. M. K. H. Yeung and Y.-K. Kwok, "On game theoretic peer selection for resilient peer-to-peer media streaming," *IEEE Transactions on Parallel and Distributed Systems*, vol. 20, no. 10, pp. 1512–1525, Oct. 2009.

45. V. Aggarwal, A. Feldmann, and C. Scheideler, "Can ISPs and P2P users cooperate for improved performance?" *ACM SIGCOMM Computer Communication Review*, vol. 37, no. 3, July 2007.

46. Z. Dulinski, M. Kantor, W. Krzysztofek, R. Stankiewicz, and P. Cholda, Optimal choice of peers based on BGP information," in *IEEE International Conference on Communications (ICC)*, May 2010, pp. 1–6.

47. H. Xie, Y. R. Yang, A. Krishnamurthy, Y. G. Liu, and A. Silberschatz, "P4P: Provider portal for applications," in *ACM SIGCOMM Computer Communication Review*, 2008, pp. 351–362.

48. C. Wang, N. Wang, M. Howarth, and G. Pavlou, "An adaptive peer selection scheme with dynamic network condition awareness," in *IEEE International Conference on Communication (ICC)*, 2009, pp. 1–5.

49. D. R. Choffnes and F. E. Bustamante, "Taming the torrent: A practical approach to reducing cross-ISP traffic in peer-to-peer systems," *ACM SIGCOMM Computer Communication Review*, pp. 363–374, 2008.

50. A. O. Kaya, M. Chiang, and W. Trappe, "P2P-ISP cooperation: Risks and mitigation in multiple-ISP networks," in *IEEE Global Telecommunications Conference (GLOBECOM)*, 2009, pp. 1–8.

51. L. Li and X. Wang, "P2P file-sharing application on mobile phones based on SIP," in *International Conference on Innovations in Information Technology*, 2007, pp. 601–605.

52. S. E. Davies and S. Gardner, "A novel and non SIP-based framework for initiating a multimedia session between mobile devices," in *International Conference on Internet Multimedia Services Architecture and Applications (IMSAA)*, 2008, pp. 1–6.

53. J. Mänpää and J. J. Bolonio, "Performance of REsource LOcation; Discovery (RELOAD) on mobile phones," in *IEEE Wireless Communications and Networking Conference (WCNC)*, 2010, pp. 1–6.

54. W. Wu and J. Womack, "Efficient connectivity maintenance for mobile cellular peers in a P2PSIP-based overlay network," in *IEEE Consumer Communications and Networking Conference (CCNC)*, Jan. 2011, pp. 252–256.

55. J. Rosenberg, "Interactive connectivity establishment (ICE): A protocol for network address translator (NAT) traversal for offer/answer protocols," IETF, Internet Draft work in progress, Tech. Rep., Feb. 2010.

56. R. M. J. R. P. Matthews, "Traversal using relays around NAT (TURN): Relay extensions to session traversal utilities for NAT (STUN)," IETF, Internet Draft work in progress, Tech. Rep., Feb. 2010.

57. P. M. J. Rosenberg, R. Mahy, and D. Wing, "Session traversal utilities for NAT (STUN)," RFC 5389, IETF, Tech. Rep., 2009.

58. L. Makinen and J. K. Nurminen, "Measurements on the feasibility of TCP NAT traversal in cellular networks," in *Next Generation Internet Networks (NGI)*, 28–30, 2008, pp. 261–267.

59. J. Maenpaa, V. Andersson, G. Camarillo, and A. Keranen, "Impact of network address translator traversal on delays in peer-to-peer session initiation protocol," in *IEEE Global Telecommunications Conference (GLOBECOM)*, Dec. 2010, pp. 1–6.

60. Z. Wang, Z. Qian, Q. Xu, Z. Mao, and M. Zhang, "An untold story of middle boxes in cellular networks," *ACM SIGCOMM Computer Communication Review*, vol. 41, no. 4, pp. 374–385, Aug. 2011.

61. "A & M Records, Inc. v. Napster," vol. 239 F.3d 1004 (9th Cir.), 2001.

62. "PPLive." [Online]. Available at http://www.pplive.com/.

63. "SopCast." [Online]. Available at http://www.sopcast.com/.

64. X. Zhang, J. Liu, B. Li, and Y. S. P. Yum, "CoolStreaming/DONet: A data-driven overlay network for peer-to-peer live media streaming," in *INFOCOM*, vol. 3, March 2005, pp. 2102–2111.

65. K. Lee, I. Rhee, J. Lee, S. Chong, and Y. Yi, "Mobile data offloading: How much can WiFi deliver?" in *International Conference on emerging Networking EXperiments and Technologies (CoNEXT)*. ACM, 2010, pp. 1–12.

66. K. Collins, S. Mangold, and G. M. Muntean, "Supporting mobile devices with wireless LAN/MAN in large controlled environments," *IEEE Commun. Mag.*, vol. 48, no. 12, pp. 36–43, 2010.

67. L. Garcia, L. Arnaiz, F. Alvarez, J. M. Menendez, and K. Gruneberg, "Protected seamless content delivery in P2P wireless and wired networks," *IEEE Wireless Communications*, vol. 16, no. 5, pp. 50–57, Oct. 2009.

68. Qualcomm, "Flashlinq," *GSMA Mobile World Congress*, 2012.

Chapter 4

Peer-to-Peer Video-on-Demand in Future Internet

Mostafa M. Fouda, Zubair Md. Fadlullah,
Al-Sakib Khan Pathan, and Nei Kato

Contents

4.1 Background

Have you noticed how fast Internet traffic is growing lately? It is an amazing statistic for those who have not. If you just go about your way of hooking up with the Internet and send emails, read news, chat, watch YouTube videos, and interact with your e-buddies on social networks such as Facebook and Twitter, get ready to be surprised. The volume of the Internet traffic has increased eight times than that it was only 5 years ago. And there is no sign that this explosive growth will end any time soon. For the more techno-savy bunch, the visual network index (VNI) [1] forecast from CISCO should be enough. The VNI indicates that Internet traffic will continue to rise fourfold in the course of the next 5 years. The VNI further projects another fascinating figure: the Internet video may comprise 40% of the entire consumer Internet traffic, which may reach up to a whooping 62% by the end of 2015.

Let us ask ourselves for the reason behind this groundswell in video traffic. As the prime reason of this surge in video traffic, we may certainly point fingers at the vast improvement in both access and core network technologies and the deployment of fiber optical communication networks. Advances in wide-band access networks and their large-scale deployment are promising data delivery at a remarkable pace. Even the simplest of Internet users would attest to this. Ask anyone who used the Internet a decade ago. The painfully slow dialing-up days to connect to the Internet are certainly over, the dial-up modems are almost extinct, too—perhaps our kids and grandkids would enjoy knowing about them in a museum or an archive. Unlike the early days of a staggeringly slow Internet, today's Internet users do not restrict themselves to only web browsing and text/image data transfer. Instead, they usually enjoy and share a wide variety of multimedia services, including audio and video streaming. As the resources and technologies to deliver multimedia contents continue to evolve toward 40/100 Gb networks [2], multimedia content distribution techniques also need to evolve significantly. This necessity arises from the tremendous pressure of meeting the quality-of-experience (QoE) requirements of the users who subscribe to these services. Many of you, perhaps, have heard of the term "quality of service," which is simply dubbed as "QoS." QoE is a similar term for multimedia content delivery. Put yourself in a user's shoe. Then, answer the following questions: When you want to watch a video content on your device of choice (which may range from a personal computer to a laptop, to a smartphone or a tablet device), how fast could the software or application load the video? When you watch the video, does it play smoothly or does it break while playing? In the latter case, how often does the video freeze? At the end, how satisfied or happy were you with the experience of availing of this service? QoE gives a measure to this level of satisfaction. As the multimedia industry continues to grow with millions of wireless and mobile devices and tons of apps, it is becoming ever so difficult to satisfy the QoE of most of the—let alone all—customers.

In order to catch up with the growth of multimedia applications, many service architectures have appeared over the last few years. While some belong to the basic client/server model, others are more sophisticated. Consider Apple's iCloud or Google's cloud technologies as the more sophisticated distributed approaches. To date, one of the most promising multimedia service architectures is, however, the peer-assisted content delivery technology. Who or what is a peer? Let us try to explain from our real-world experience. A peer is typically a friend or a colleague who helps us in accomplishing some task, right? We can extend this concept of "peers" in the networking context. "Peers" refer to user nodes (such as computers, laptops, and smartphones) that help one another in achieving some common task. Now, what is the common task in the networking concept? It depends on the nature of the application. A very popular and common peer-assisted application is sharing files without the need to have any centralized server. Many of us are already familiar with the BitTorrent [3] file-sharing protocol. For example, "μTorrent" is one of the most popular BitTorrent applications available today for transferring files among peers. Refer to an example scenario, which is illustrated in Figure 4.1, which shows how a distribution of Ubuntu is being downloaded usingthe "μTorrent" program from a number of seeds/peers. Note that the "μTorrent" program classifies the connected hosts into two types, namely, seeds (hosts having full content) and peers (hosts having only part of the content). In the figure, the user is connected to 36 seeds out of 880. It is worth mentioning that, throughout the chapter, we refer to seeds as peers also. BitTorrent has certainly been popular with home users and often been under the heat from law enforcers due to legal issues, including copyright breach. BitTorrent is, however, a basic peer-assisted technology. If we investigate further, we may find more sophisticated peer-assisted multimedia delivery techniques including Video-on-Demand (VoD) systems [4], where a number of peers help the content source by providing other users with the video chunks, which they have already downloaded.

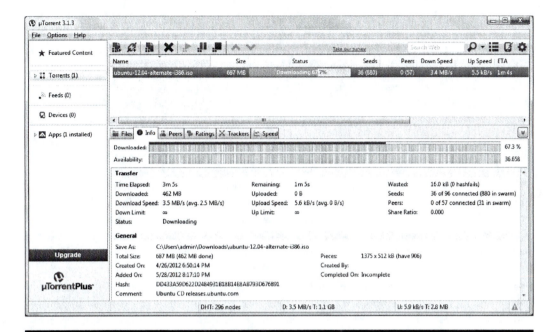

Figure 4.1   File download example using "µTorrent," one of the most popular clients for P2P client using the BitTorrent protocol.

While designing peer-assisted VoD is still a challenging issue in the current Internet setting, imagine the challenges it may lead to from the context of future Internet. Global Environment for Network Innovations (GENI) [5] and Future Internet Design (FIND) [6] initiated by the National Science Foundation (NSF) of the United States are two projects that have emphasized on addressing the need for cross-layer Internet design, network virtualization, dynamic switching of optical circuits, service discovery and composition, service management, traffic, and routing engineering, along with a number of other directions toward the future Internet. The Europeans, meanwhile, have come up with their own project. The European edition of the initiative, which was conducted within the 7th Framework Program (FP7), was given an interesting name: FIRE," which stands for the Future Internet Research and Experimentation [7]. In the far east, Japan was also not far behind. They launched a New Generation Network architecture through the National Institute of Information and Communications Technology [8], in which much focus was given on generating future Internet concepts and technologies by the year 2015. So, thus far, we have different future Internet concepts from the Americans, Europeans, and Japanese. These projects have featured different design considerations and objectives to formulate how the actual future Internet would be like and what prospect this new paradigm would hold. In spite of their differences, all the projects had a common aspect: they all imply that the existing Internet architecture can no longer deal with the huge volume of multimedia traffic. Currently, the network operators are having a hard time to cope with the growth of multimedia traffic as they are required to offload traffic at optimal points of the current Internet framework. For example, in case of mobile networks, the optimal traffic-offloading points are the radio access networks and base stations. The operators, therefore, need to ensure that such entities in the network are not overwhelmed by the sheer volume of data traffic. Thus, future Internet initiatives are simply suggesting that it is the end of the line for the current Internet as far as next-generation multimedia traffic handling is concerned.

As a multimedia distribution technology, live video streaming aims at providing the endusers with real-time feeds such as television news and sport channels, live events, etc. Different paradigms, ranging from client/server-based solution to P2P technologies, can be used for distributing multimedia content to the users. The client/server solution can use either unicast or multicast modes. In the unicast mode, the server distributes the content to each user indivdiually. This consumes lots of bandwidth and results in a bottleneck in the server. As a remedy to this issue, the multicasting solution offers distributing the same content to multiple users at the same time. While the multicasting solution holds immense prospect, it requires especial routers (capable of multicasting) to deliver the content to the target users. However, this requires replacing all conventional routers with multicast-capable ones, and this seems to be far away from what we have now. On the other hand, P2P solutions can be used easily for the live video streaming. A commercial example of a P2P live video streaming program is illustrated in Figure 4.2, featuring "SopCast." In "SopCast," many users are able to watch a certain live video feed by cooperating with each other. This cooperation takes place by sharing the pieces or chunks of videos that the users already have received. By doing so, it is possible to reduce the streaming load on the main server to some extent. It is worth noting, however, that the main challenge of this technique is to come up with a system that can ensure that all the peers may successfully receive the streamed content and play it in a smooth fashion.

On the other hand, VoD streaming, which happens to be one of the largest contributors to the multimedia traffic, is more popular for Internet users since it provides user-interactive facilities. The user-interactive options include pausing the video, jumping to a different section of the content, performing rewind/forward options, watching the content in a slow motion, etc. In fact, YouTube [9] is a great example of VoD streaming, where Internet users can watch any part of a video of their liking at any time, as shown in Figure 4.3. With the more sophisticated interactive

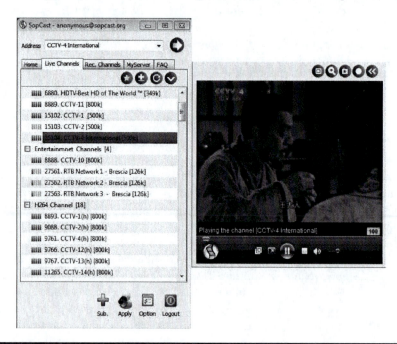

Figure 4.2   Example of a popular video streaming application called SopCast, which is available on the Internet for watching TV channels.

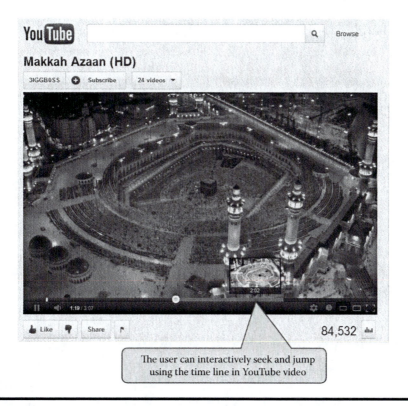

Figure 4.3   Interactive video streaming on YouTube.

features of VoD over live video streaming, multicasting cannot be used since different users may be interested to watch different parts of a video. In conventional VoD techniques, the server sends separate chunks of the video to individual customers, but this is an expensive process in terms of bandwidth cost. From the future Internet perspective, it is important to design a cheaper and reliable VoD technology, which needs to be carefully designed to achieve low cost, decent QoE, and high scalability. In fact, there exist a number of strategies for delivering VoD content. The first technique consists of the traditional client/server architecture, in which a single server sends the video to its viewers. The second technique is the content distribution network (CDN) paradigm, which aims at avoiding the bottleneck near that server by hosting duplicated contents in different regional servers. The other approach is to use P2P overlay networks for VoD streaming. In the P2P VoD strategy, the client-side resources are best utilized for uploading and sharing video content with other users or peers, and this eliminates the need for a centralized content server. Also, it is possible to combine the P2P VoD approach with the CDN paradigm to offer more flexibility and reliability. However, note that the cheapest VoD solution is P2P-based because it uses the already available client-side resources at the peers. To quickly compare these different VoD paradigms, please refer to Table 4.1.

The remainder of this chapter is structured as follows. At first, we present recent trends in Internet traffic burst and the evolution of multimedia technologies. Then, a survey of several future Internet initiatives is provided, indicating that they all aim at acheiving a common target, despite having different visions. Next, the chapter investigates whether conventional P2P VoD schemes fit future Internet initiatives. This requires us to understand peer-assisted VoD systems,

Table 4.1 Comparison of Different VoD Paradigms

Paradigm	Central Server Needed?	User Scalability	Performance Improvement Needs Infrastructure Change?	Cost
Client/Server	√	Low	Yes	Relatively high
CDN	√	Low per server	May need replicated content servers, broadband access, routers, etc.	Relatively high
P2P	x	High	No	Relatively low

and therefore, an overview of the state-of-the-art P2P VoD technology is presented in the following section. Directions toward how P2P-based VoD schemes can be improved to accommodate the need of future Internet is then delineated. The chapter ends with final caveats, future guidelines, and concluding remarks.

4.2 Recent Trends in Internet Traffic Burst

Let us try to pinpoint the main reasons behind the recent trend of the traffic increase in the Internet. The increasing number of broadband subscribers is certainly a main contributor. Everyone wants to enter the digital realm these days—everybody wants to be "connected," so to speak. Thus, the increase in cyber population means that there will be an increase in digital traffic. Improvement in bandwidth per subscriber has also contributed to the overall traffic increase. For example, let us consider Japan. In this island country, the number of asymmetric digital subscriber line subscribers started to decrease significantly because they have opted to get hooked to a faster and more bandwidth-intensive fiber to the home, which is commonly known as FTTH connection. The FTTH subscribers in Japan is now over 16 million [10]. This trend of increasing bandwidth per subscriber naturally resulted in the increasing use of video-related services, such as YouTube. The average data size per video content is also increasing because the increased bandwidth per user allows easy transmission of high definition (HD) videos, swiftly making a transition away from the traditional 480p Standard Definition technology. This talk on HD videos takes us to HD televisions (HDTVs). Is watching television part of your favorite pastime or maybe you are an occassional viewer enjoying sports channels or news? In either case, you will find that flat-screen liquid-crystal display television sets are quickly replacing their older counterpart—the rock-heavy cathode-ray-tube monitor TVs. The resolution of the television has increased so much recently, and the digital transmission of the television programs on the HDTVs through pay-per-channel and on-demand video streaming is providing so much better viewing quality. The technology is becoming more sophisticated with the buzzing appearance of the cutting-edge Ultra HD Television (UHDTV). Some also call this Super Hi-Vision. This technology is developed by the NHK Science and Technology Research Laboratory, Japan, and can generate 16 times higher resolution than that possible in the current 1080p HDTV format. The picture in UHDTV technology will comprise four times as many pixels in both horizontal and vertical directions (a whooping 7680 × 4320 pixels in contrast with 1920 × 1080 in 1080p HDTV, as demonstrated in Figure 4.4). If you are a music lover and appreciate surround sound systems, this technology has

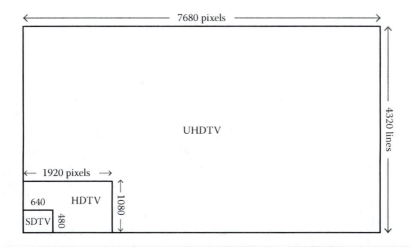

Figure 4.4 Recent trend of video resolution improvement. Note the huge leaps in resolution from SDTV to HDTV and eventually to UHDTV technology.

something incredible to offer you: it supports 22 speaker-based surround sounds. Given recent trends, it is expected to have UHDTV available for residential users between the years 2016 and 2020. Meanwhile, technical factors such as high storage requirement for these audio and video contents and their effective distribution mechanisms are being investigated.

Imagine how much bandwidth will be required for accommodating UHDTV videos. For real-time UHDTV quality video transmission, a bandwidth of more than 70 Gb/s may be consumed. The evolution of UHDTV technology, therefore, dictates the trend of persistent Internet video traffic growth for decades to come. Even for an overoptimist, it is difficult to assume that existing Internet and Internet Protocol (IP)–based technology can deal with this enormous pressure.

4.3 Future Internet Initiatives: One Goal, Different Visions

In this section, let us review a few important concepts that have been recently introduced to facilitate future Internet. This will help the readers to know the specific requirements of future Internet.

The Internet has, indeed, survived for nearly four decades since its concept came into reality in the 1970s. The initial concept, which resulted in the phenomenal rise in Internet communications, is quite limited now; we are no longer restricted to just connecting our computers to the Internet and share some files. We do not see ourselves confined to only sit at a wired workstation and just be content by surfing the Internet for news and information or by chatting with our friends and colleagues. The old-school concept of Internet services based on text/image browsing is no longer sufficient to satisfy current user requirements, which demonstrate high demand for multimedia services. This actually makes us wonder how the Internet has managed to absorb the heightened pressure from a wide range of networked and mobile users and applications (also not to mention their ever-increasing number), state-of-the art business models, latest networking devices and equipment, etc. Recently, researchers have realized the importance to note the limitation of the prevailing Internet architecture and replace it with a futuristic Internet framework.

Toward this end, a number of research projects have been proposed to design and develop the future Internet. The first notable future Internet initiative, which comes into mind, is FIND

[6]. FIND was envisioned by the NSF in the United States, and it focuses on developing network architecture, security, advanced wireless and optical properties, economical principles, and, in general, ways to effectively construct a global network in two decades from now. The second project, which is worth citing, is the GENI program [5]. GENI was launched to provide the missing elements in the current Internet such as security, reliability, evolvability, and manageability.

Meanwhile, Europeans are running their own projects to develop a futuristic Internet. The notable future Internet initiative from the European side are the FIREworks (Future Internet Research and Experimentation—Strategy Works) project [7] and the Evolved Internet Future for European Leadership (EIFFEL) [11] Support Action (SA) for the 7th Framework Program (FP7). The NanoDataCenter project (NaDa) [12] offers next-generation data hosting and content distribution paradigm. By enabling a distributed hosting edge infrastructure, the NaDa architecture aims at enabling next-generation interactive services and applications. Similarly, the P2P-Next integrated project [13] expects to construct a next-generation P2P content delivery platform by combining high-profile academic and industrial players.

The Network-Aware P2P-Television Application over Wise Networks (NAPA-WINE TV) services [14] over the Internet brought a new dimension to peer-assisted streaming. NAPA-WINE TV services can be provided by either exploiting IP multicast functionalities or relying on a pure P2P approach. The concept of NAPA-WINE television demonstrates that the P2P approach has been successfully exploited to overcome the limitations of IP multicast functionalities concerning a single point of failure and can potentially offer a scalable infrastructure on the global level. Recently, several P2P-TV systems have started to appear, with the last generation offering high-quality TV (P2P-HQTV) systems. These systems provide a ubiquitous access to the service. However, at the same time, they are raising concern for network carriers because of the potential out-of-control growth of their traffic.

VITAL++ [15] project aims at combining and experimenting with the best Internet Protocol Multimedia Subsystem (IMS)-like control plane functionality and P2P technology to come up with a new communication paradigm. In this vein, a pan-European test bed with distributed test sites integrated by IMS technology has been set up for VITAL++. By doing so, it is possible to test the reference content applications and services using the P2P approach to determine which optimization algorithms can distribute network resources in the best way to satisfy the quality-of-service demands of P2P users. Note that these different projects agree on the common goal, namely, develop the Internet for the future. However, while they are at their common target, their implementation strategies are not necessarily the same. In other words, the differences in design objectives among these projects are quite remarkable. In fact, similar contrast in design objectives can also be seen even among the Internet Engineering Task Force (IETF) work groups on this issue. For example, one IETF group proposed the Low Extra Delay Background Transport protocol in order to minimize additional latency due to sophisticated networking applications. On the other hand, another IETF group introduced the differentiated services (Diffserv) code point for bulk traffic for identifying, marking, and managing congestion events because of P2P traffic under the Diffserv framework. Furthermore, to manage the P2P traffic volume within an acceptable level at the network operators' expensive links, two IETF groups were formed. They were called Decoupled Application Data Enroute (DECADE) [16] and Application Layer Traffic Optimization (ALTO) [17] working groups. DECADE identifies the problem with current Internet caches (e.g., P2P and web caches), which attempt at providing adequate storage capabilities within the network for reliable access of resources. However, DECADE is not capable of explicitly supporting individual P2P application protocols or user access to the content providers' caches. ALTO, on the other hand, offers a simple yet effective way to provide the network information to the P2P applications. This helps P2P applications to reduce the overhead associated with measuring topology information such as path performance metrics computation. In addition, instead of random initial peer selection, adoption of traffic localization for P2P applications was also done by introducing the ALTO service.

In the next section, we describe a number of existing works on P2P VoD, which may fit the aforementioned future Internet architectures.

4.4 Existing P2P VoD Schemes: Do They Fit Future Internet Initiatives?

In order to satisfy the requirements of future Internet, several research works have appeared in recent literature, which aimed at combining P2P strategies with the server–client streaming architecture. BitTorrent Assisted Streaming System (BASS) for VoD is such an example [18]. BASS consists of an external media server and a modified BitTorrent [3] protocol. The peers in BASS cannot download contents prior to the current playback time. Instead, they receive chunks from the media server sequentially. In addition, they do not require obtaining the already downloaded or currently being downloaded chunks by the BitTorrent. The BASS framework is, however, not suited to support a large population of peers. In other words, it is not scalable and cannot fit the future Internet scope.

For solving the scalability problem with BASS, the BitTorrent Streaming (BiToS) [19] framework was proposed. BiToS does not consider any external media server for streaming the on-demand video. Instead of using the "rarest first" strategy, BiToS exploits a selection mechanism based on discarding the missing chunks (i.e., the chunks unable to meet their playback time). However, BiToS exhibits poor performance when the peers fail to download more and more chunks. In other words, the higher the missing chunk rate gets, the more disrupted the video playback becomes at the BiToS peers.

Annapureddy et al. [20] offered a different approach altogether for P2P VoD streaming in one of their works. Their work focused on providing small start-up delays in peer-assisted VoD systems by network coding integrated segment scheduling. This way, they demonstrated that it is possible to highly utilize the available resources. From practical considerations, however, this approach is not applicable for future Internet VoD environment because the peers cannot share real-time information quickly enough. To overcome this problem, a scheduling mechanism fitting the future Internet requirements that is able to instruct the concerned peers to help each other in real-time streaming is essential. In this vein, VoD scheduling schemes over P2P mesh-based overlay networks with user-scalability features need to be proposed with consideration given to future Internet requirements.

4.5 Overview of Peer-Assisted VoD Systems

With the advances in broadband access technologies, multimedia content distribution in the form of interactive on-demand video streaming services has gained immense popularity in recent time (e.g., YouTube [9]). In the conventional client/server paradigm, upon users request to watch videos, the centralized entity/server delivers the videos to the users. Two of the notable shortcomings of the client/server-style VoD approach are given as follows.

1. *Scalability:* The system cannot support the increase in content subscribing population.
2. *Limited streaming rate:* The bandwidth of the server is constrained. Thus, the subscribers cannot enjoy good streaming rate if many people are watching the video.

To overcome these shortcomings, peer-assisted video streaming has become an attractive alternative to its server-based counterpart (e.g., YouTube) for VoD provisioning. It already appeared as a technology for the future Internet by targeting global users connected to heterogeneous networks. From the broadcaster/content provider view point, the P2P approach provides an additional incentive because it can serve a large number of peers (i.e., many more subscribers in contrast with the conventional client/server-based approach) without investing into additional resources. If we consider the users' perspective, they also experience improved delivery rate of the multimedia content. At the same time, the users can also upload their already received content to other peers and thus may participate in the VoD streaming process, even without having any extra feature or resources.

What is the implication of using peer-assisted VoD in the future Internet? The implication is significant and not to be underestimated since this will allow only minimal change to the existing Internet infrastructure without having to place additional content servers. Furthermore, effective peer-assisted VoD approaches in the future Internet may also help overcome bandwidth/ processing load bottlenecks, mitigate start-up delay, reduce end-to-end latency, and, at the same time, improve the video playback rate.

We can broadly classify existing peer-assisted VoD systems into two categories, that is, tree-based and mesh-based systems. In tree-based VoD systems, every node receives data from a source or parent node following a tree-like structure. Tree-based peer-assisted VoD systems are not suitable for the future Internet setting. The reason behind this is that the nodes of the trees (i.e., peers) are expected to frequently change, resulting in frequent tree reconstruction events. As a consequence, the child nodes may fail to obtain the video-streaming feed, and they may need to wait for the tree to be reconstructed. On the other hand, in the mesh-based overlay network, a new peer (that has just joined the overlay) contacts the tracker to receive a list of currently active peers, which are able to contribute to the video streaming. After receiving the active peer list, the new peer begins sending requests to the other peers in the list to obtain the necessary video chunks. Also, it is worth noting that a peer is connected with a small subset of active peers at any time, and it is permitted to exchange video chunks and control messages only with those peers. In Figure 4.5, the overall architecture of the mesh-based peer-assisted VoD strategy is demonstrated. The mesh-based peer-assisted VoD networks are more resilient to node failures in contrast with their tree-based counterpart. The reason behind this is that the changes in the aforementioned mesh-based infrastructure can be dealt with far greater ease compared to the tree-based reconstruction.

Most of the prevailing peer-assisted VoD streaming strategies follow a rather naive assumption: a user selects a video, starts watching it from the beginning, and continues to watch it until the end. This assumption is not practical as it does not consider user-interaction events. What if the user wants to pause the video or replay it from the beginning? What if the user is bored with the current scene of a movie and wants to skip this by fast forwarding? The aforementioned assumption is simplistic in nature to keep the system design simple. This sacrifices the practical aspect: instead of continuously playing the video content, VoD subscribers are often habituated to jump to a more interesting scene either due to lack of time or interest. This phenomenon (frequent user interaction with peer-assisted VoD applications) presents a significant challenge to the continuity of video playback in terms of "seeking delay." Seeking delay indicates the time needed since the request for a video segment until the segment becomes available. To ensure an almost zero seeking delay to watch the video without any disruption, every video segment needs to be prefetched by a peer before the playback of the segment. This improves the continuity of the playback. Unfortunately, as mentioned earlier, most conventional peer-assisted VoD systems, based on their rather simplistic assumption that users watch the video sequentially without any interactive "seek" operation, perform "prefetching" of the video segments in a sequential fashion.

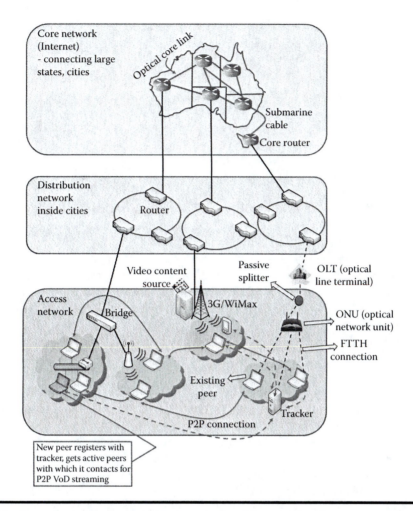

Figure 4.5 Overview of the peer-assisted meshed architecture for the VoD streaming process.

This explains why many of you, at one point of time or another, have tried to randomly jump to another section of a YouTube video and got stuck for a while. To overcome this problem, it is necessary to formulate a strategy that is able to prefetch random video segments. Toward this end, a prefetching scheme for obtaining appropriate video segments *a priori* at the peers in a mesh-P2P VoD environment is covered as a case study in the next section.

One of the design objectives of a peer-assisted VoD system is to maximize the aggregate throughput of all the peers. While throughput maximization in P2P applications has not been studied well in literature, by attempting to maximize the throughput, a peer may end up receiving duplicate packets (i.e., the same video segments or chunks) because it obtains streams from multiple sources. To overcome this problem, what the peer can do is to negotiate with the source node(s) and request for specific chunks. The conventional strategies may also be prone to inefficient utilization of network resources in large and heterogeneous environments, the presence of which is highly expected in the future Internet. As a consequence, they may not be able to maximize the aggregate throughput. Therefore, improved peer-assisted VoD strategies are required to fit the future Internet scope.

Next, we present a case study in which we demonstrate an improved P2P-based VoD scheme, to accommodate the requirements of the future Internet.

4.6 How to Improve P2P-Based VoD Schemes: A Case Study to Fit the Need of Future Internet

So far, we validated the reason behind the need to design an effective peer-assisted VoD system for overcoming the shortcomings associated with conventional VoD strategies under the context of the future Internet architecture. By designing the future Internet-oriented VoD system, it is possible to meet the high expectation of QoE from the user point of view. In this section, we present a case study on a P2P VoD mechanism, which exploits the overlay network resources by taking into account the upload bandwidth available at the source node, as well as all active peers [4,21]. This mechanism also combines an efficient prefetching scheme with a scheduling strategy to obtain the appropriate video segments *a priori* at the receiving peer's terminal. This strategy demonstrates that, by enhancing the conventional peer-assisted VoD design, it is possible to deal with the new trends and challenges of peer-assisted VoD streaming in the future Internet.

The considered system for our discussion consists of a tracker, which contains a list of existing peers in the P2P overlay. In the tracker, the information regarding the peers is maintained. Peer-related information includes their IP addresses, port numbers, available upload bitrate, and chunk bitmap. Let S, R, and P denote the initial source peer or the initial content streaming peer, receiving peer (who wants to view the video content), and the list of selected peers to serve the receiving peer, respectively. When a new peer appears at the P2P mesh overlay, it expects to be served by other nodes, which are already in the active peer list of the overlay tracker. In other words, the new peer becomes the receiving peer R. The tracker provides the list of active neighboring peers P to R. In doing so, the tracker follows the following policies, instead of randomly selecting P to serve R.

1. The peers, which possess the video segments following the last playable chunk of R, are selected as potential peers in the list P.
2. The tracker further filters the potential P by choosing only the peers with sufficient upload capacity and the relatively low appearance in the neighbor list of the other currently served peers. By doings so, the system distributes the load over the participating peers according to their available resources.
3. If P is sufficient to serve R with the multimedia content, S is omitted in R's neighbor list. S is included in R's neighbor list only under critical conditions. For example, when there is no peer to construct list P to serve R or when all the serving peers even altogether do not contain the required video content, S is included in R's list of neighboring peers.
4. If P is not complete to adequately serve R, then random peers are considered to construct P.

Now that we have explained the peer selection strategy, let us move to the operational phases of the peers. The peers operate in two phases: prefetching and scheduling phases. The prefetching phase is implemented in R's terminal. On the other hand, the scheduling phase is implemented in each of the peers in P. The algorithms used in these two phases are described in the following.

The prefetching phase is further divided into two steps. In the first prefetching step, R requests each peer in P for specific bandwidth slots. It is worth noting that each bandwidth slot is adequate for transferring a single chunk of the target video. To efficiently exploit the available bandwidth, the number of bandwidth slots requested by R is set to be directly proportional to the upload capacity of each peer in *P*. This is not allowed to exceed the maximum number of needed chunks available at each peer in *P*. Along with sending the request of specific bandwidth slots, *R* transmits its own time-to-freeze (*TTF*) information to *P*. The parameter *TTF* represents the remaining time for the video playback to enter the freezing state. Figure 4.6 demonstrates this through an example whereby a peer in *P* exploits the received *TTF* value for prioritizing requests during the course of the scheduling process.

Now, we move to the scheduling phase, which is implemented at each of the peers in *P*. The available upload bandwidth of every peer in *P* is assigned to the currently requesting *R*'s. The bandwidth assignment is performed according to the following algorithm.

1. Each peer in *P* allocates up to *K*% of its available upload bandwidth to *R* with ($TTF < TTF_{th}$), where TTF_{th} is a predefined threshold. The highest priority is assigned to *R* with the minimum possible value of *TTF*. The objective of this priority-based assignment is to ensure a smooth playback at *R* to avoid freezing of the video streaming.
2. *R* with relatively higher upload capacity is allocated more bandwidth slots. This is done to ensure that a given *R* with a higher upload capacity is assigned more bandwidth by peers in *P*, so that *R* may be able to fill its video chunks as soon as possible to exploit its own upload capacity to serve the content to other requesting peers.
3. *R*, which has just joined the overlay and has not yet received any video chunk, is assumed to have a *TTF* of zero. By doing so, the system sets the highest priority for new peers. Also, this minimizes the start-up delay for playing the video at *R*.

Following the scheduling process, each peer in *P* transmits a specific number of bandwidth slots to each requesting *R*. As a result, the second phase of prefetching is triggered, which allows *R* to determine the required video chunks, to be requested from the peers in *P*.

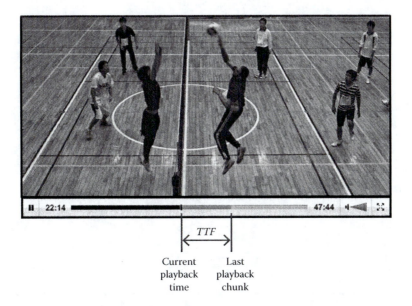

Figure 4.6 Example showing the concept of the *TTF* parameter.

If you have been following carefully, by now you should realize that the main objective of the work highlighted in this case study is to ascertain smooth video playback. To fulfill this objective, the second step of prefetching is executed as follows: When R requests the peers in P for video chunks, priority is given on the chunks, which appear after the playback time with a maximum interval equal to a tunable time window T. After serviced by peers in P with all the chunks requested within T, R makes an assumption on the quality of the VoD streaming as not to degrade at least over the next T. From this point, R does not need to request the chunks sequentially. Then, R can randomly request its uplink peers in P to increase the diversity of chunks in the overlay.

Next, we highlight the ability of the tracker to select the peers in more detail. The peer-assisted VoD scheme in this case study assumes that the tracker has network topology information. In other words, the tracker is considered to have knowledge about the different network domains. Another important assumption is made: the tracker may acquire link congestion information over different network domains. This is actually possible from the point of view of network providers and operators, if they deploy monitoring agents at different points of the network and allow the tracker to receive information regarding network configuration and traffic dynamics. The domain-based localization denotes the selection of nearby peers, which belong to the same domain. Note that choosing a nearby peer is not always an optimal decision. As an example to this case, consider the simple scenario in which the link to a nearby peer is experiencing congestion. This implies that link congestion should be also taken into consideration in the selection of peers.

In Figure 4.7, there are two illustrated cases for your careful consideration. In the scenario in this figure, there are three domains, namely, D_1, D_2, and D_3. Peers P_1, P_2, and P_3 belong to D_1. Let $d_{i,j}$ denote the end-to-end propagation delay between two neighboring peers P_i and P_j. In the example, let us consider ($d_{2,3} < d_{2,1}$). When P_2 searches for its neighboring peers, it may be assigned to either P_1 or P_3 based on the domain-based localization scheme. However, when there is congestion in the link between P_2 and P_3, there should be congestion awareness in the peer selection scheme, which needs to select P_1 based on the network domain, as well as the link congestion.

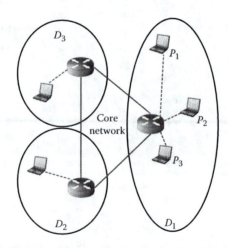

Figure 4.7 Example for the comparison between domain-based localization and congestion awareness strategies for peer selection.

4.7 Conclusion

In this chapter, we focused on the importance of having effective P2P streaming of multimedia contents over the future Internet. In order to ensure high QoE for peer-assisted VoD users, it is essential to ascertain short start-up delay, smooth video playback, and scalability. In this chapter, we have pointed out that the upload bandwidth utilization of the involved peers may affect these in a significant way, and this will remain a key concern in the future Internet architecture. We presented a case study showing how this issue can be dealt with by designing an effective prefetching mechanism for obtaining the necessary video chunks, an efficient scheduling algorithm for assigning the upload bandwidth, and an efficient peer selection strategy for selecting the most appropriate peers that serve the requesting peer in the best possible manner. Also, we highlight the importance of integrating the prefetching and scheduling mechanism with domain-based and congestion-aware peer selection strategies. Combined strategies such as these exhibit encouraging performance, even when the system is operating under the worst case scenarios, that is, servicing a potential flash crowd of peers. Also, this way, if an adequate scheduling scheme can be designed for P2P VoD streams, the content replication technology used by many of the conventional P2P solutions may no longer be required.

A note of caution for the interested readers (or potential researchers in this field): the peer-assisted VoD systems may face further challenges in the future Internet. A crucial point is mobility; a high number of users in future Internet may be expected to be mobile. It will also be interesting to study the impact of mobility of peers on the selection of peers and the QoE of the VoD subscribers. Another note of caution is regarding the selfish behavior of the peers. In the current Internet setting, peers with low-bandwidth resources may not contribute by refusing to become uplink nodes to participate in the VoD streaming. Admittedly, the future Internet may improve bandwidth constraints in the core networks, and FTTH solutions may solve the last mile access bandwidth limitations. However, mobile users are expected to form the significant majority of next-generation Internet users, and they may not necessarily have the processing power or time to act as content-streaming entities. Whether they may be exploited to act as nonselfish peers in facilitating VoD services also remains to be seen.

References

1. CISCO, "CISCO Visual Networking Index: Forecast and Methodology, 2010–2015," published in June 1, 2011. Available online at http://www.cisco.com/en/US/solutions/collateral/ns341/ns525/ns537/ns705/ns827/white_paper_c11-481360.pdf.
2. A. Weissberger, "ComSocSCV Meeting Report: 40/100 Gigabit Ethernet—Market Needs, Applications, and Standards," October, 2011. Available online at http://community.comsoc.org/blogs/ajwdct/comsocscv-meeting-report-40100-gigabit-ethernet-%E2%80%93-market-needs-applications-and-standar.
3. B. Cohen, "The BitTorrent Protocol Specification," Jan. 2008. Available online at http://www.bittorrent.org/beps/bep_0003.html.
4. M. Fouda, T. Taleb, M. Guizani, Y. Nemoto, and N. Kato, "On Supporting P2P-based VoD Services over Mesh Overlay Networks," in *Proc. IEEE GLOBECOM'09*, Hawaii, USA, December 2009.
5. J. B. Evans and D. E. Ackers, "Overview of GENI and Future Internet in the U.S." May 22, 2007. Available online at http://www.geni.net.
6. NSF NeTS FIND Initiative. Available online at http://www.nets-find.net.
7. CORDIS, "Future Internet Research and Experimentation: An Overview of the European FIRE Initiative and Its Projects," September 1, 2008. Available online at http://cordis.europa.eu/fp7/ict/fire.
8. T. Aoyama, "A New Generation Network: Beyond the Internet and NGN," *IEEE Communications Magazine,* vol. 47, no. 5, May 2009, pp. 82–87.

9. YouTube, Available online at http://www.youtube.com.
10. S. Namiki, T. Kurosu, K. Tanizawa, J. Kurumida, T. Hasama, H. Ishikawa, T. Nakatogawa, M. Nakamura, and K. Oyamada, "Ultrahigh-Definition Video Transmission and Extremely Green Optical Networks for Future," *IEEE Journal of Selected Topics in Quantum Electronics*, vol. 17, no. 2, March–April 2011, pp. 446–457.
11. EIFFEL Support Action, Dec. 2006. Available online at http://www.fp7-eiffel.eu.
12. NanoDataCenter (NaDa) Project. Available online at http://www.nanodatacenters.eu.
13. The P2P-Next project. Available online at http://www.p2p-next.org.
14. The NAPA-WINE TV Services, 2007. Available online at http://napa-wine.eu/cgi-bin/twiki/view/Public.
15. ICT-VITAL++ Project—Embedding P2P Technology in Next Generation Networks: A New Communication Paradigm and Experimentation Infrastructure. Available online at http://www.ict-vitalpp.upatras.gr.
16. DECADE Working Group Charter. Available online at http://trac.tools.ietf.org/bof/trac/wiki/decade.
17. Application-Layer Traffic Optimization (ALTO) Working Group. Available online at http://www.ietf.org/html.charters/alto-charter.html.
18. C. Dana, D. Li, D. Harrison, and C. Chuah, "BASS: BitTorrent assisted streaming system for video-on-demand," in *Proc. International Workshop on Multimedia Signal Processing (MMSP)*, Shanghai, China, November 2005.
19. A. Vlavianos, M. Iliofotou, and M. Faloutsos, "BiToS: Enhancing BitTorrent for supporting streaming applications," in *Proc. IEEE Global Internet*, Barcelona, Spain, April 2006.
20. S. Annapureddy, S. Guha, C. Gkantsidis, D. Gunawardena, and P. Rodriguez, "Exploring VoD in P2P Swarming Systems," in *Proc. IEEE INFOCOM'07*, Alaska, USA, May 2007.
21. M. M. Fouda, Z. M. Fadlullah, M. Guizani, and N. Kato, "Provisioning P2P-based VoD Streaming for Future Internet," *Journal of Mobile Networks and Applications (MONET)*, Aug. 18, 2011.

Chapter 5

IPTV Networking: An Overview

Georgios Baltoglou, Eirini Karapistoli, and Periklis Chatzimisios

Contents

5.1 Introduction

Over 30 years ago, networking experts from military and educational environments introduced Transmission Control Protocol (TCP)/Internet Protocol (IP), giving birth to packet-switched networks to circumvent any single point of failure, which is common in circuit-switched networks such as public switched telephone networks (PSTNs). The design of this new network type had been governed with one and only concept: best-effort data delivery. By this, it was implied that the network would make its best attempt to transfer data, yet no guarantee about that could be given and early applications of this scheme such as HTTP and email needed nothing more. In the recent past though, a variety of other applications have been developed, the use of which has become (or is gradually becoming) quite popular, such as IP Television (IPTV), which is the newest hot topic in computer networks.

IPTV is a service with great expectations. It thrives to replace the former methods of TV broadcasting such as over-the-air (OTA), cable, or satellite TV distribution, making use of the packet-switched Internet infrastructure. In order to succeed, it also provides a variety of new services such as video-on-demand (VoD) and videocassette recording-like functionality for the distributed TV media, together with the ability to coexist with high-rate data surfing and voice-over-IP (VoIP) traffic in what is called a *triple-play* service. Yet, the IPTV streams, which are highly maladaptive to network deficiencies, pose great demands in all separate layers of the network architecture. As such, the network must facilitate a high-bandwidth physical infrastructure to be able to allow the traversal of multiple standard-definition (SD) or even high-definition (HD) channels. Even so, a service better than best effort or quality of service (QoS) must be implemented to allow prioritization and ensure the timely delivery of IPTV packets. At the same time, the different existing standards need to converge, allowing the IPTV service to be utilized in different types of devices, from TV set-topboxes (STBs), home computers, or even 3G and upcoming 4G mobile smartphones.

While different interpretations of IPTV exist, they all contain a common denominator. As stated in [1], IPTV is defined as "an alternate method of distributing television content over IP by a telecom carrier or an Internet service provider (ISP)." What diversifies IPTV from the common TV broadcast mechanism over air, cable, or satellite, is the support for a limitless number of services.

To begin with, IPTV is an interactive technology. It offers personalization of viewing experience by allowing the viewer to search for content by title or actor's name, change and control camera angles of the video playback, retrieve statistics such as player performance in sports games, use picture-in-picture functionality that allows "channel surfing" without needing to leave the program, and real-time participation in shows. In addition, it provides digital video recording (DVR) functions, with capabilities of pausing, fast forwarding, and rewinding not only the recoded but also the currently playing (live) multimedia content, in conjunction with the commodity of recording multiple channels simultaneously.

Another important feature of IPTV is VoD. VoD permits a customer to browse through an online program or film catalog and, upon request, to watch trailers and then select specific content for playback. Furthermore, this feature provides the customer with a large set of DVD-like functionalities such as the ability to pause and jump to next movie sections. However, IPTV's greatest advantage is the fact that it can seamlessly support integration and convergence with other services. As such, it offers the ability of using VoIP communication with on-screen caller ID, *anytime* and *anywhere* access to the multimedia content, over all types of devices [personal computers (PCs), television sets, and cell phones], and the manipulation of DVR and VoD functions with these devices. However, as bandwidth is a limited resource and the IP infrastructure is based on a best-effort delivery method, certain heavy demands are imposed that need to be fulfilled in order for this highly inelastic traffic to gain assured quality and IPTV to function as prospected. This is where operators need to future-proof their infrastructure and delivery platforms to be able to accommodate all new distribution mechanisms without quality loss while managing complex billing and content scheduling operations and securing content across multiple platforms.

In this chapter, effort has been made in providing a detailed background on IPTV, its market proliferation, and in identifying its demands. Since IP can and will be used to deliver video over all sorts of networks, both fixed and wireless, IPTV solutions are at the microscope of this chapter. By assessing the performance of these networks, we discuss issues that can deteriorate the quality of IPTV or refrain its deployment. Finally, we try to address whether deploying a problem-free IPTV network is feasible and explain what the future holds for this technology.

5.2 IPTV on the Rise

Over the last years, IPTV has emerged from being a new service trend to one of the hottest topics of computer and communication networks for various reasons. To start with, as a technology, it has provided the means for telecommunication companies (or telcos) to refine their economy by entering a new entertainment platform. At the same time, as an infrastructure, it necessitates that new manufacturers/suppliers are formed for both delivering the service and manufacturing and supporting the content, respectively. Specifically, aside from telcos, suppliers for multicast and VoD content are of need along with Electronic Program Guide (EPG) data suppliers. At the same time, new companies to manufacture head-end and access systems, IPTV software platforms, and STBs, as well as manufacturers of measurement equipment and IPTV network monitoring devices, are all major participants. Aside from the key players previously mentioned, there is a vast audience that awaits for services that are not currently implemented in simple TV distribution. These points have led to the exponential growth of the IPTV market, appointing it as a very promising one.

While this flourishing market started with a slow pace, it achieved 4.3 million subscribers in 2005 and more than 40 million by the end of 2010. Currently, it holds a global market of approximately 70 million subscribers, which is a number that is forecasted to reach or surpass the 110 million mark by 2014, resembling a compound annual growth rate of 26% [2–4].

As shown in Figure 5.1, the highest growth rate is presented in the European region, with a noteworthy number of strong IPTV deployments in France, Italy, and Spain and very strong competitive offerings from the Scandinavian countries. An additional boost is created as remaining eastern European countries adopt the IPTV technology. At the same time, the Asian region is closely following and is expected to surpass Europe in the number of IPTV subscribers in the near future, as China and India will ultimately be the largest markets worldwide.

Equivalently, global IPTV service revenue is constantly growing, having exceeded 18 billion euros by the end of 2011 and forecasted to surpass 35 billion in 2014, yielding a compound annual

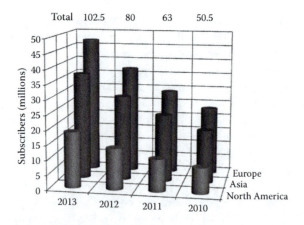

Figure 5.1   IPTV subscribers per region.

growth rate of 28%. Europe and North America are generating the greatest amount of this revenue due to the low average prices in the Asian region.

Based on these current and forecast reports, the IPTV arena is getting increasingly competitive and is achieving constant year growth, despite downturn of economy that scourges several countries worldwide. As IPTV is proliferating, its market presents solid opportunities for IPTV operators in Europe, Asia, and North America to establish a new service platform with high revenue.

5.3 IPTV Architecture

While IPTV provides a variety of features, it does so posing great demands on physical and network infrastructures in order for its reliable service delivery to be ensured. The following sections provide an overview of the characteristics of the IPTV service such as the available media coding formats, the protocols, and the access networks that can guarantee its prospected performance.

5.3.1 Media Encoding and Compression

Since IPTV is offered as a competitive technology to cable, satellite, or OTA TV broadcasting, it should also maintain the characteristics of these technologies in terms of content quality. Transmission of uncompressed video, however, imposes a major problem: that of excessive bandwidth. AnnSD video in an uncompressed format would require 270 Mb/s, whereas a full HD content of 1920 × 1080 resolution, 16-bit color depth for its color space, and a frame rate of 25 frame/s would require approximately 3 Gb/s including its multichannel audio and the protocol overhead needed to traverse the Internet. These numbers of data rates appoint uncompressed video transmission inexpedient and compression of eminent use.

In an effort to standardize video compression for all uses, ISO created the Moving Pictures Expert Group (MPEG). MPEG handled video as moving pictures and tried attaining a high compression factor by exploiting redundancy between pictures. Specifically, this type of encoding, which is also referred to as temporal-redundancy or interframe compression, uses preceding and posterior frames in a sequence to compress the current frame. Thus, the decoder adds the differences to the previous picture to produce the new picture. In this scheme, three types of frames are defined, that is, intra-frames (*I*-frames), bidirectional frames (*B*-frames), and predicted frames (*P*-frames). *I*-frames are encoded completely and are handled as reference frames for the compression process. In *B*-frames, both past and future pictures are used as a reference in compression. In these frames, only the motion of the section of the frame that has moved is encoded. Finally, *P* frames are coded with respect to the nearest previous *I*- or *P*-frames. Thus, like *I*-frames, *P*-frames can also serve as a prediction reference for *B*-frames and future *P*-frames. Figure 5.2 shows the prediction of an object in its new position by shifting pixels from the previous picture using motion vectors. Any prediction errors are eliminated by comparing the predicted picture with the actual picture. The coder sends the motion vectors and the errors. The decoder shifts the previous picture by the vectors and adds the errors to produce the next picture. Typically, a reference *I*-frame is coded every 15–18 frames, and *B*- and *P*-frames follow. This sequence of a reference frame and consecutive *B*- and *P*-frames forms a group of pictures (GOP).

While MPEG-1, which was the first standard of video compression to be released, provided resolution of just 352 × 240 pixels at 1.5 Mb/s and only stereo audio, its successor MPEG-2 was designed to support a wide range of applications and services of varying bit rate, resolution, and quality.

Additionally, it specified a transport stream (TS), which is a communications protocol for audio, video, and data. MPEG-TS provides a special format for transmitting MPEG video multiplexed with other streams specially designed for delivering data in real time over unreliable transport media to a device that is assumed to start reading data from some point after the beginning of transmission, as in Internet-streaming video and digital video broadcasting (DVB). Because of its features, MPEG-2 became an internationally recognized standard and the core of most digital television and DVD formats.

However, in order to support HD or full-HD content over packet-switched networks, even further compression that would maintain the quality of the video would be desirable. The H.264 project group created a standard that is capable of providing good video quality at substantially lower bit rates than previous standards. H.264 at a given rate offers a much better quality, whereas, at a specified quality, it requires roughly 50% of the bit rate that MPEG-2 needs. Furthermore, H.264 offers better motion prediction than MPEG-2 by incorporating variable macroblock sizing and supports multiple deblocking filters presenting fewer artifacts than MPEG-2. Finally,

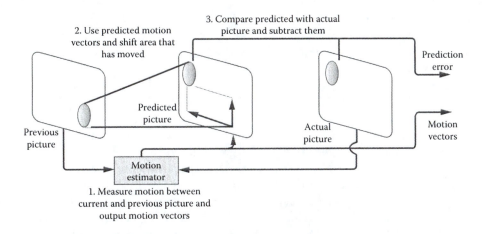

Figure 5.2　Image motion prediction.

it provides easier packetization than MPEG-2 and a maximum resolution of 4096 × 2304. As expected, however, it also has some great disadvantages. H.264 has greatly increased computational requirements. HD decoding on an average PC is difficult in real time, whereas real-time encoding is not even possible. Additionally, its complexity appoints difficult and expensive the design and optimization of encoders and decoders. In general terms, however, the international community is migrating to this standard, especially for full-HD formats such as Blu-ray.

Another encoding algorithm that aspires to be standardized for HD IPTV video delivery is Video Codec 1 (VC-1), which is an enhancement to Windows Media Video 9 (WMV-9). VC-1, as stated in [5], is particularly optimized for high-bit-rate content of high resolution. VC-1 offers comparable quality to H.264 for high bit rates and resolutions, but it does lack the ability to maintain a low bit rate at lower resolutions.

The evolution of IPTV content undeniably attracts viewers. The SD content offered both by OTA and IPTV broadcasting is steadily giving its place to HD channels. More than that, 3D material is in its commercial infancy, but it is progressively growing as 3DTV sets gain of a larger market share. These new media contents are quickly embraced by customers who, in turn, expect IPTV content to feature the same multimedia characteristics as their cable and satellite TV sets. Additionally, consumers are focusing even more on VOD and online video rentals (iVOD). Services such as NetFlix have conditioned the viewing audience to the ease of renting on-demand streaming content. Subsequently, the multicast networking scheme that helped alleviate the network from extravagant traffic is appointed inadequate, and unicast traffic traversal is again utilized. Though this continuously changing media content environment augments the viewers' quality of experience (QoE), it also threatens the end-to-end quality of IPTV and poses further demands on all infrastructure layers.

5.3.2 Protocols Utilized in IPTV

TV video content as broadcasted now multiplexes all channels and broadcasts them all in continuous availability. If all the number of channels were to be broadcasted in a unicast fashion to every user, even with H.264 or VC-1 compression schemes, the bandwidth requirement would still be extravagant. Thus, high-bandwidth applications, such as MPEG video, can be sent to more than one receiver simultaneously only by using IPmulticast. IP multicast is a bandwidth-conserving technology for simultaneously delivering a single stream of data to multiple users [6]. All alternatives require the source to send more than one copy of the data. Some even require the source to send an individual copy to each receiver. The Internet Assigned Numbers Authority controls the assignment of IP multicast addresses, which have been allocated in the range of 224.0.0.0–239.255.255.255.

5.3.2.1 IGMP

As multicast is based on the concept of a group that expresses interest in receiving a particular data stream, a membership for this multicast group has to be established. For IP networks, this membership is provided by the Internet Group Management Protocol (IGMP). IGMP is an integral part of the IP multicast specification, operating above the network layer. It dynamically registers individual hosts in a multicast group on a particular local area network (LAN). Hosts identify group memberships and indicate their interest in joining a group by sending IGMP messages called *membership reports* to their local multicast router. Routers listen to IGMP messages and periodically send out *membership queries* to discover which groups are active or inactive on a particular subnet. These queries verify that at least one host of a specific subnet is still interested in receiving the multicast stream of that group. In IGMP version 1, after no reply for three consecutive IGMP queries, the router times out the group by stopping traffic forwarding toward that group. In its second development, the hosts can actively communicate with the local multicast router, showing their intention of leaving the group. After that, the router sends a group query to determine whether other hosts with interest to the multicast data remain. In case of no response, the group is timed out. With this approach, unnecessary traffic is stopped sooner, reducing the overall latency.

As IGMP messages are transmitted as multicast packets, they are indistinguishable from the multicast traffic at Layer 2. For this reason, switches examine some of Layer 3 information in the IGMP packets sent between the hosts and the router. This process is called IGMP snooping. When the switch listens to an IGMP host report for a multicast group, it adds the host's port number to the associated multicast table entry, and when hearing an IGMP leave group message, it removes the port from the table. IGMP snooping requires that every multicast packet be filtered for control information, which can often bottleneck the switch performance. Because of that, it is usual that IGMP snooping is used on high-end switches, with special integrated circuitry that can perform IGMP checks in hardware.

5.3.2.2 Transfer Protocols

A transfer protocol must be implemented to deliver the video content of IPTV services since IGMP does not act as a transport protocol itself. Real-time applications such as commercial live video distribution do not (and should not) make use of a retransmission mechanism as that utilized in TCP. Thus, in most cases, User Datagram Protocol (UDP) is used, defying the risk that exists for audio and video broadcasts to undergo content degradation due to packet loss. To alleviate these risks, in some cases, IPTV makes use of Real-Time Transport Protocol (RTP) and Real-Time Transport Control Protocol (RTCP), which is a transfer protocol relying on UDP and its associated control protocol. RTP can provide data transfer of inelastic real-time traffic, whereas RTCP can provide a feedback mechanism on the transferred data, allowing packet-loss detection in conjunction with other delay measurements. It should also be noted that, for the on-demand functions of DVR and VoD in IPTV previously mentioned, the Real-Time Streaming Protocol (RTSP) can be used. RTSP [7] is a protocol that finds broad usage in streaming media systems by allowing the client to control the streaming media server in a remote fashion, issuing commands such as "play" and "pause" and giving the option of time-based access to the server's files.

5.3.2.3 Multicast Routing and Forwarding

Finally, a multicast network requires a routing protocol to build forwarding paths between the subnet of the content source and each subnet containing members of the multicast group. The usual way that this can be accomplished is by Protocol-Independent Multicast (PIM). PIM, as implied by its name, is IP routing protocol-independent. Thus, it can leverage whichever unicast routing protocols are used to populate the unicast routing table, including Enhanced Interior Gateway Routing Protocol, Open Shortest Path First, Border Gateway Protocol, or static routes, and it uses this routing information to perform the multicast forwarding function. PIM supports two particular modes: (1) dense mode (PIM-DM) and (2) sparse mode (PIM-SM), depending on how many routers in the network will need to distribute multicast traffic for each multicast group [8]. Specifically, in PIM-DM, it is assumed that, when a source starts sending, all downstream systems

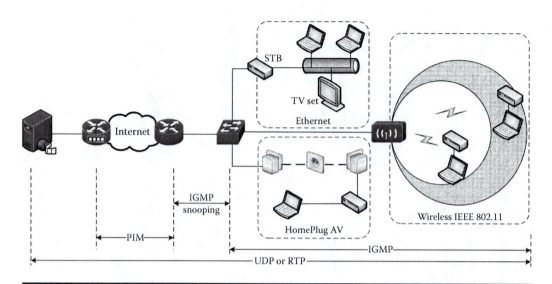

Figure 5.3   IPTV sample network model.

want to receive multicast datagrams; thus, initially, multicast datagrams are flooded to all areas of the network. If some areas of the network do not have group members, PIM-DM will prune off the forwarding branch by instantiating prune state. As a result, PIM-DM is mainly designed for multicast LAN applications, whereas the PIM-SM is for wide-area interdomain networks.

In contrast to unicast routing, where traffic is routed through the network along a single path from source to destination, caring only of the destination address, in multicast routing, the source sends traffic to an arbitrary group of hosts represented by a group membership. As such, the router must determine which direction is upstream, directing toward the source, and which one is downstream. As multiple downstream paths may exist, the router must replicate the packet and forward all traffic down the appropriate downstream paths only, which is not necessarily all paths. This process is referred to as reverse path forwarding (RPF). In RPF, when a multicast packet arrives at the router, the latter performs a check using the existing unicast routing table to determine the upstream and downstream neighbors and forwards the packet only if it is received on its upstream interface. RPF guarantees that the multicast distribution is loop-free.

A network topology designed to deliver a multicast service such as IPTV using the aforementioned protocols is depicted in Figure 5.3. Evidently, the number of protocols utilized and the networking infrastructure mechanism insert a series of further delays that have to be kept under strict limits to ensure comparability to a traditional distribution.

5.4 Standardization Landscape

While the proper technology has to be in place, a key success factor specifically for IPTV operators is not to rely solely on technology as an innovator but to enable cross-platform portability and uniform user interface standards to improve user friendliness. Although there are no established standards at this stage, much effort has been made toward IPTV standardization during the last couple of years. The technical areas of IPTV that are being addressed include service and functional requirements, architecture (NGN and non-NGN), QoS/QoE, traffic management mechanisms, performance monitoring, security aspects, end systems and home networking, middleware, metadata, applications, and content platforms, as well as protocols and control plane aspects.

The most important efforts toward IPTV standardization are the following [9,10]:

■ *DVB IPI:* DVB over IP networks [11] provides a set of technical specifications, which is also referred to as the DVB-IPTV handbook, that covers the delivery of DVB MPEG-2-based services over IP-based networks. The key functionalities specified by the DVB-IPTV handbook are the delivery of DVB MPEG-2 TS-based services over IP networks for Live Media Broadcast services and content-on-demand (CoD) services, the Service Discovery and Selection mechanism for MPEG-2-based audio/video transport over IP, the use of command and control application-level protocol Real-Time Streaming Protocol (RTSP) to control CoD services, as well as two optional protocols for Application Layer Forward Error Correction (AL-FEC) protection of streaming media and for providing protection against packet loss through a retransmission mechanism. Due to the fact that operators need flexibility on deploying IPTV services (depending on the current market requirements and business models), it is considered that a small set of service-oriented profiles is enough for low-cost and differentiated services that do not require full implementation of the DVB-IPTV handbook. For this reason, the *technical specification* ETSI TS102 826 [12] defines four service-oriented profiles:
  1. A basic profile to accommodate existing IPTV deployments of live TV services
  2. Live media broadcast profile to build live IPTV services carried over multicast transport
  3. CoD profile to build on-demand IPTV services carried over unicast transport
  4. Content download profile to build services of content available for download

■ *ITU-T FG IPTV:* The International Telecommunications Union—Telecommunication Standardization Sector (ITU-T) has formed the Focus Group (FG) IPTV [13] in order to coordinate the development of global IPTV standards, taking into account the existing work of the ITU-T study groups (SGs). The main goals of FG IPTV were definition of IPTV, coordination, interoperability and gap analysis of existing standardization activities and ongoing works, and harmonization of the development of new standards. The FG IPTV ended in December 2007, and the Global Standards Initiative (IPTV-GSI) working group was established to, via related SGs, harmonize standards on a global level and coordinate with other standardization-developing organizations. In particular, SG13 (future networks including mobile and next-generation networks) is working on details about IPTV architectural design issues, SG16 (multimedia coding, systems, and applications) primarily focuses on the IPTV terminal device basic models and multicast function support, whereas SG12 (performance, QoS, and QoE) is acting mainly toward the completion of new recommendations on QoE for IPTV.

■ *ETSI TISPAN IPTV:* Telecoms and Internet converged Services and Protocols for Advanced Networks (TISPAN) IPTV is the standards group within the European Telecommunications Standards Institute (ETSI), which addresses several specifications regarding the IPTV architecture. TISPAN defines the functional architecture and reference points of a customer network device for both IP multimedia subsystem (IMS)-based and non-IMS-based IPTV services. ETSI TISPAN employed the IMS concept in its NGN Release 1 (NGN R1) within the NGN architecture framework but did not address IPTV at all. In the TISPAN NGN R2, IMS was used to support IPTV services, and several specifications addressed IPTV regarding service requirements [14] and architectures with a non-IMS IPTV subsystem [15] and IMS-based IPTV [16]. Two philosophies were developed on how those requirements should be fulfilled. The first one, namely, *Dedicated IPTV,* was to adapt existing Internet Engineering Task Force protocols to an NGN environment, and the second one, namely, *IMS-Based IPTV,* was designed to take advantage of the benefits of the 3GPP IMS for providing IPTV services. In Release 3 (R3), the two families of specifications support new features described like access to third-party content, P2P distribution, and more features that will be added progressively. The R3-dedicated IPTV specification is discussed in [17], and the IMS-based IPTV specification is discussed in [16].

■ *ATIS IIF:* The Alliance for Telecommunications Industry Solutions (ATIS) initiated in July 2005 the IPTV Interoperability Forum (IIF), which develops standards and specifications to enable the interoperability, interconnection, and implementation of IPTV systems and services, including VoD and interactive TV services. The scope of the work within IIF includes the following areas [18]:

1. Coordinate standards activities that relate to IPTV technologies
2. Develop self-consistent ATIS standards, such as interoperability requirements, specifications, guidelines, and technical reports
3. Provide a venue for interoperability activities
4. Provide a venue for the development of standards and other documents for IPTV systems/services deployed over a managed IP and/or NGN infrastructure

   The *Committee of IIF* is composed of five committees: (1) architecture, (2) IPTV security solutions, (3) QoS metrics, (4) IIF testing and interoperability, and (5) metadata and transaction delivery.

■ *3GPP MBMS:* The Third Generation Partnership Project (3GPP) specifies Multimedia Broadcast/Multicast Service (MBMS) specifications [19] mainly to define an efficient way to deliver and control multicast and broadcast innovative services over 3G networks.

■ *OMA BCAST:* The Open Mobile Alliance (OMA) is the mobile IPTV-related standardization body, and it introduces the concept of the mobile broadcast services enabler to address the ITU-T FG IPTV requirements related to mobile IPTV.

As there is no unique established standard at this stage, bringing together all these different initiatives into one flexible architecture is a real challenge that has to be fulfilled for IPTV to successfully reach its market potential. Another prerequisite for the IPTV standardization bodies is to develop standards that ensure secure and reliable delivery of entertainment video and related services to subscribers. For a comprehensive survey on published specifications and ongoing activities on IPTV security in main standards organizations, including ITU-T IPTV FG, ETSI TISPAN, DVB, ATIS, Open IPTV Forum, and China Communications Standards Association, the reader may refer to [20]. As stated in the work of Lu et al. [20], toward creating the trustworthy environment needed to broadcast IPTV data, IPTV standards must primarily comply with two different categories of security, namely, *confidentiality* and *integrity.*

■ *Confidentiality* refers to the control and authorization of access to the content. Unauthorized access to IPTV may result in content theft or excessive congestion that will lead to denial of service (DoS) of legitimate subscribers. Both wired and wireless networks can face security threats and attacks including eavesdropping, session hijacking, message replay, and DoS, with wireless connections, being more susceptible to it as they lack the inherent security provided by the physical infrastructure of wired networks [21].

■ *Integrity* is nowadays referred to as Digital Rights Management, and it is a form of rules restricting the illegal free content redistribution, or its alteration, which is in violation of the copyright laws. Failure in compliance to this security measure will eventually cause the reluctance of content distributors to cooperate with the apparently unreliable telecom operators.

5.5 Wired and Wireless IPTV Networks

5.5.1 Wired

As IPTV requires high data rates, the wired distribution networks that can be utilized are variants of asymmetrical digital subscriber line 2+ (ADSL2+), very high speed digital subscriber line (VDSL), and optical fiber networks.

An ideal choice for an IPTV access network would be fiber-to-the-premises (FTTP). FTTP makes use of fiber-optic communication in which an optical fiber is run directly onto the customers' premises. FTTP can be further categorized according to where the optical fiber ends:

- Fiber-to-the-home (FTTH) is a form of fiber optic communication delivery in which the optical signal reaches the end user's living or office space.
- Fiber-to-the-building (FTTB) is a form of fiber optic communication delivery in which the optical signal reaches the private property enclosing the home or business of the subscriber or set of subscribers. In contrast to FTTH, the optical fiber terminates before reaching the home, with the path extended from that point up to the user's space to run over a physical medium other than the optical fiber, that is, copper cable.

While FTTH and FTTB could provide great benefit to IPTV, the migration to it is not mandatory for IPTV to function properly. To further reduce costs, another technology can be of use. Fiber-to-the-node (FTTN), which is also called fiber-to-the-cabinet, is a fiber-optic architecture where its fibers are run to a cabinet serving an area that is usually less than 1500 m in radius and can obviously contain several hundred customers. This technology can allow the maximum exploitation of current broadband technologies, and it is growing in availability.

Another—and certainly the most common—method of broadband Internet connection is ADSL. In its current implementation ADSL2+, it is offering data rates of up to 24 Mb/s. The maximum theoretical throughput of ADSL2+ is enough to carry one HD channel and two SDTV channels and still conserve room for voice and high-speed data connection.

Finally, an access network with desirable characteristics for IPTV is the very high bit rate digital subscriber line (VDSL). VDSL, like ADSL, also makes use of twisted pair cables that are already installed in PSTNs. The current implementation VDSL2 (ITU recommendation G.993.2) specifies eight profiles that address a range of applications including up to 100 Mb/s symmetric transmission on links of 100 m long (using a bandwidth of 30 MHz) and asymmetric operation with downstream rates in the range of 10–40 Mb/s on links of lengths ranging from 1 to 3 km (using a bandwidth of 8.5 MHz). Evidently, the deployment of VDSL2, along with FTTN, will result in situating VDSL2 nodes close to subscribers. Thus, operators will be able to boost capacity enough to support multiple HDTV streams to a household without having to replace the entire copper infrastructure with fiber as in FTTH. However, this is not the only advantage of VDSL2 as an IPTV access network. A very important aspect of the VDSL2 standard is that it uses Ethernet as multiplexing technology in the first mile. As a result, the access architecture can be simplified into an end-to-end Ethernet network that uses virtual LANs (VLANs) as the service-delivery mechanism across the entire access network [22].

Currently, both fiber optics and VDSL are penetrating markets worldwide, creating access networks with strong foundation for deployment and further proliferation of IPTV.

5.5.2 Wireless

In order to provide *anywhere–anytime* internet connection, wireless technologies have proven to be a key component in the success of home and industrial networking. These wireless communication technologies like IEEE 802.11 (with its standards 802.11a/b/g/e/n), IEEE 802.16 (Worldwide Interoperability for Microwave Access, WiMAX), and 3GPP Long-Term Evolution (LTE) can enable consumers to enjoy IPTV services without the need of rewiring their homes.

The IEEE 802.11 protocols have become the dominant standard for wireless LANs (WLANs) mainly because of the advantages that they offer such as interoperability, mobility, flexibility, and cost-effective deployment. The IEEE 802.11 standards include IEEE 802.11b, specifying the operation of up to 11-Mb/s data rate [23]; the IEEE 802.11g [24] and 802.11a [24] standards, enhancing the operation to 54-Mb/s data rate (on the 2.4- and 5-GHz band, respectively); and the recently standardized IEEE 802.11n [25], enhancing the operation that can reach more than 540 Mb/s. Furthermore, the IEEE 802.11e standard [26] has been developed and standardized, aiming at providing QoS support for various kinds of wireless applications. While it decreases delay and jitter, it still does not guarantee that IPTV data will always arrive and in a timely manner. Also, to be effective, networks using 802.11e must propagate their QoS requirements from source to destination, and that propagation depends on all network components being 802.11e-compliant. As such, IEEE is working on amendments to the 802.11 established standards that specify further augmentations. IEEE 802.11ac [27] and 802.11ad [28] Task Groups are aiming to standards that can provide data rates exceeding 1 Gb/s (for the 5- and 60-GHz bands, respectively), whereas IEEE 802.11aa Task Group [30] is working on enhancements that could enable transporting audio video streams with robustness and reliability while at the same time allow for the graceful and fair coexistence of other types of traffic. A more detailed analysis of the IEEE 802.11 standards (currently active and amendments under development) can be found in [31]. It is therefore evident that wireless networking architectures that clearly aim at delivering high-quality IPTV content are of major concern.

The last mile in the access network providing high-capacity broadband wireless connectivity can also utilize WiMAX technology based on IEEE 802.16 standards that include physical (PHY) and medium access control (MAC) layer specifications covering several different frequency ranges for a variety of existing protocol stacks [32]. The theoretical coverage radius of the first IEEE 802.16 a/d/e standards could reach 50 km, and data rates up to 75 Mb/s could be achieved by supporting either point-to-multipoint or mesh mode topologies. IEEE 802.16 standards define five data delivery services, namely, unsolicited grant service, real-time variable rate (RT-VR), nonreal-time variable rate (NRT-VR), best effort (BE), and extended real-time variable rate (ERT-VR). (The latter was introduced in IEEE 802.16e [33].) The rtPS and Extended rtPS services are designed to support real-time service flows that generate variable-size data packets for real-time data services that require guaranteed data rate and delay such as MPEG video and VoIP. The recent IEEE 802.16-2009 [34], as well as the 802.16 j/h/m [35–37] standards, provides enhanced data rates (aiming to make available up to 1 Gb/s peak throughput) that will allow for an improved user experience and support for multimedia services. In particular, the IEEE 802.16m Advanced Air Interface provides a more flexible and efficient QoS framework to support emerging and evolving mobile Internet applications by introducing a new scheduling service, adaptive granting and polling (aGP) service, quick access, delayed bandwidth request, and priority-controlled access [38]. Based on the above, WiMAX appears to have the required QoS features needed to support the inelastic video of IPTV. Even more importantly, it can provide IPTV services to both fixed and mobile wireless stations, which greatly contributes to the realization of anytime–anywhere access to TV channel broadcasts [39,40].

A technology that aspires to provide mobile internetworking is LTE. Deployment of mobile broadband systems based on the LTE radio access technology (based on the 3GPP LTE Release 8 [41]) is currently ongoing on a broad scale and provides high data rates (up to 100 Mb/s in downlink and 50 Mb/s in uplink). At the same time, LTE-Advanced (3GPP Release 10) is being developed mainly under the close partnership between network operators and equipment vendors [42]. LTE specifications support multiple QoS requirements by means of different bearers, which are packet flows established between the packet data network gateway and the user terminal, each one associated with a QoS. Broadly speaking, bearers can be classified into two categories based on the nature of the QoS they provide: guaranteed bit rate (GBR) and non-GBR. The traffic running between a particular client application and a service can be differentiated into separate service data flows (SDFs). SDFs mapped to the same bearer receive a common QoS treatment. A bearer is assigned a scalar value referred to as a QoS class identifier (QCI), which specifies the class to which the bearer belongs. Aside from QCI, there are several QoS attributes associated with the LTE bearer such as allocation and retention priority, maximum bit rate (MBR), and aggregate MBR (AMBR) [43]. Overall, LTE networks have the necessary characteristics and QoS attributes to support IPTV distribution for a large set of mobile devices.

5.5.3 In-Home Networking

Aside from the access networks that are utilized in IPTV distribution, it is now becoming even more common for the end users to distribute the IPTV traffic that reaches the customer premiss equipment to other in-house locations via wired, wireless, and even powerline communications (PLC). A typical WLAN consists of an access point (AP) connected via wireless link to a number of stations deployed within a range of up to 100 m. The most widely deployed technology used in home networking is IEEE 802.11. This technology, which has reached full maturity as a frequent home-networking method, is able to deliver traffic in bit rates approximating 54 Mb/s, which are adequate enough for the majority of applications that are deployed in Fast-Ethernet internetworking. Furthermore, the new IEEE 802.11n protocol specification can reach 540 Mb/s, ensuring bandwidth abundance for even more demanding applications. For these reasons, it seems that the high-speed IEEE 802.11n technology can easily cover a typical house with sufficient bandwidth to support video, gaming, data, and voice applications and thus provide IPTV service as the access network. However, less than optimum conditions can easily decrease the data rate to prohibitive values. Furthermore, the often poor connection between a wireless user and an AP results in severe packet losses and very poor jitter levels, which are usually not even suited for VoIP applications as these standards are mainly designed to provide Internet access and network file transfer. The IEEE 802.11e specification enhances the QoS of the legacy 802.11 by introducing priorities for traffic types to overcome certain QoS issues for real-time traffic. Enhanced distributed channel access (EDCA) is utilized in order to support prioritized QoS by defining four access categories (ACs), namely, voice, video, best effort, and background. Each AC is characterized by specific values for a set of access parameters that statistically prioritize channel access for one AC over another. While it decreases delay and jitter, IEEE 802.11e still does not guarantee that IPTV data will always arrive and in a timely manner. Also, to be effective, networks using IEEE 802.11e must propagate their QoS requirements from source to destination, and that propagation depends on all network components being 802.11e-compliant. Furthermore, the only QoS mechanism broadly implemented so far in wireless adapters available on the market is prioritization of traffic known as Wi-Fi Multimedia (WMM) [44]. WMM is a part of Wi-Fi certification program that implements a subset of the IEEE 802.11e standard and provides the means for traffic prioritization in order to satisfy the most urgent needs for a QoS solution for Wi-Fi networks. The defined IEEE 802.11e admission control procedures are also included in WMM specifications [45], which are based on the EDCA method. However, admission control is not fully implemented in the currently available APs [46]. Another disadvantage of WMM is that, to take advantage of its functionality in a Wi-Fi network, it requires that both the AP and the clients running applications that have a need for QoS are Wi-Fi-certified for WMM and have WMM enabled devices [47]. Moreover, WMM-enabled devices can take advantage of their QoS functionality only when using applications that support WMM and can assign the appropriate priority level to the traffic streams they generate.

The main characteristics of in-home networks include a small network size, short distance between stations, and the fact that all network devices belong to and are managed by the same owner. An alternative wired type of in-home networking technology that utilizes the existing electrical power supply network to provide various broadband services for in-building networking and the last-mile access is PLC. This type of network is considered as the most widely available wired medium since power lines exist in almost every residential or industrial building, making the installation easier and less expensive [48]. However, power lines often suffer from high noise levels, multipath fading, and interference from various appliances. In particular, HomePlug Powerline Alliance initially developed the HomePlug 1.0 standard [49], which supported bit rates of up to 14 Mb/s and had high market penetration [50]. The most recent PLC standard is HomePlug AV (HPAV) [51] and can provide higher data rates up to 200 Mb/s. While early adoptions were usually capable of lower than fast-Ethernet networking speeds, latest developments are capable of delivering from several megabits per second to several hundred megabits per second over the main electrical wires installed in home and small-office environments. Recently, IEEE and ITU-T have announced the 1901 [52] and G.hn [53] PLC standards, respectively. Both standards specify MAC- and PHY-layer specifications and provide solutions to major issues and limitations such as interoperability among PLC devices. Although PLC offers a promising solution (due to the ease of deployment and the available infrastructure), it seems that PLC implementations are not always a good choice for in-house distribution of IPTV services [54]. Thus, a comprehensive performance analysis for high-quality video distribution capabilities of IEEE 1901 and ITU-T G.hn standards is needed.

5.6　Measuring Performance of IPTV

5.6.1　Network Performance Metrics

5.6.1.1　Packet Loss

Due to the fundamental design of IP networks as a best-effort delivery platform independent of application requirements, packets and data are susceptible to loss. If the network is unable to properly transmit data, it consequently discards them, leaving it to the application to properly handle the side effects of the lost packets [55].

The sources of packet losses can be summarized as follows:

1. Analog and electromagnetic interference, such as impulse noise, usually caused by external factors including electrical devices close to the proximity of the network equipment as well as weather conditions. These interferences exceed the error correction schemes of the physical layer, resulting in packet losses.
2. Short-term transient changes in bandwidth, which usually arise from equipment tolerances in conjunction with the configuration selection. QoS constraints involve defining bandwidth limits that, if not handled properly, will result in packet loss. Input buffers and packet processing can be exceeded by the bursty nature of traffic, resulting in packet loss or packets being received too late to be of any use.
3. Equipment issues, failures, and incompatibilities, including bad fiber connections or network media, as well as standardized hardware that are not fully interoperable with other vendors' models. Because different pieces of equipment perform differently, a "constant bit rate" (CBR) stream from one device may not be sufficiently a CBR for another, resulting in the possibility of discarded packets due to exceeding packet buffers over very short times. This phenomenon is quite common during mode conversion from Ethernet to Asynchronous Transfer Mode (ATM), which is a typical scenario of ADSL environments.

In contrast to the other metrics described hereafter, measuring loss over long-term periods and analyzing the average is inappropriate. To exemplify, an impulse noise that usually occurs in bursts of 10–50ms would be hidden in a sample of 1 h as a 0.001% packet loss. From the perspective of network monitoring, this concealed value would require no special attention and would be identified as a good measurement. However, the end user might, in fact, have an undesirable viewing experience. As such, packet loss measurement, and especially the analysis of the obtained measurements by the operators, should be conducted with regard to both the occurring duration and the frequency of occurrences over a time period.

For the inelastic nature of the IPTV service, packet losses are a major issue. Since it is fairly common for IPTV deployments to transport packets via UDP, no protection to packet loss is provided. Packet loss of the audio stream can be exhibited as dropouts, squeaking noises, variations in sound volume, named chirping, or skipping. For the video stream, the impact is varying, depending on the video frame that was affected. As explained in Section 5.3.1, *I* frames act as a reference for all frames in a GOP. Consequently, loss of part or all of an *I*-frame propagates and can persist for the entire GOP, with a typical duration of 0.5–1 s. Similarly, as *P*- and *B*-frames can be referenced by others, loss issues can also persist but usually to a lesser extent and with less duration. The more flexible interpicture prediction of MPEG-4 compression that is now even more frequently used for delivering HD content can worsen this effect. Packet loss of as little as 10^{-4} (one packet per minute) is generally considered unviewable, and one lost packet per hour (or 2×10^{-6}) is the baseline per the DVB standard. Packet loss effects result into mild pixelization on the viewed content, blocked pixels for a duration of a few frames, unnecessary elongation or repetition of frames (stuttering), and frame freezes; in the worst case, it can cause the STB to crash or reboot.

Additionally, packet loss can occur during channel change. MPEG decoders have to wait until the next reference frame before presenting the image to the viewer. Packet loss at this frame can cause the decoder to wait until the next good frame, significantly increasing the zapping delay. Telcos can make GOP times shorter, as they can be as far as 2 s apart. However, shorter GOPs for improved zapping would directly result in a higher bandwidth requirement. Therefore, a tradeoff has to be made between a short zapping delay and the added cost of higher bit rate.

5.6.1.2 One-Way Propagation Delay

One-way propagation delay is the time it takes for a packet to traverse from the source to the destination, as it propagates through the medium [55].

In packet-based networks, one-way propagation delay is measured by sending a precisely time-stamped packet across the network under test. To have an accurate one-way propagation delay measurement, the packet is tagged with the same frame attributes as the packets carrying the service, such as VLAN, QoS, and destination address. Timestamping is performed by the originating probe.

Two key factors affect the resolution and accuracy of these one-way measurements: (1) synchronization error between the test units' reference clocks and (2) the intrinsic error of the measurement device itself. Both of these must be minimized to provide a meaningful one-way delay measurement—if the error approaches even a tenth of a millisecond, the accuracy will be insufficient to reliably detect SLA performance issues.

5.6.1.3 Packet Delay Variation

Due to the fact that one-way propagation delay is not constant over time, packet latency across a network is varying. The metric that measures this variability is named packet delay variation (PDV) or "jitter" [56]. Jitter is calculated as the difference in end-to-end delay between successive packets in a flow, ignoring packets that might have been lost.

As the Ethernet frames arrive at the STB at varying rates determined by the network conditions, buffering is required to help smooth out the variations. Based on the size of the buffer, there are delivery conditions that can make the buffer overflow or underflow, which results in degradation of the perceived video. Similarly, knowing the characteristics of a specific STB, the service provider might be able to characterize the maximum jitter supported by the IPTV network before noticing a considerable video degradation. This value will be a decisive factor when monitoring or analyzing the video QoS at the customer premises.

The effects of PDV on video content when a buffer overrun is caused are similar to the ones demonstrated by mild packet loss. To be specific, pixelization of areas of the viewed content and other visible distortions such as horizontal or vertical lines are typical. In the case of buffer underrun, frame freeze is the most usual experienced issue. In both cases, the channel change time is also affected.

5.6.1.4 Out-of-Order and Reordered Packets

As IP packets traverse through a number of heterogeneous network types and over dynamic topologies, they are not bound to reach their destination at a constant rate, thus arriving out of order. Since not all devices support packet reordering before presenting them to the decoder, out-of-order packets can cause distortions similar to the ones presented by packet loss.

5.6.2 QoE Metrics

5.6.2.1 MOS and R-Factor

The E-Model is another rating system that provides an objective measurement of quality based on packet loss, jitter, and delay. The E-Model reports results as *R*-values. The following is an approximate relationship between the mean opinion score (MOS) and E-Model quality ratings.

The E-model [57] is designed as an objective method of quantifying the quality based on a series of impairments, such as packet loss, jitter, and delay. It is a computational model that can be useful to transmission planners to help ensure that users will be satisfied with end-to-end transmission performance. The primary output of the model is a scalar rating of transmission quality. A major feature of this model is the use of transmission impairment factors that reflect the effects of modern signal processing devices. The result of any calculation with the E-model in a first step is the transmission rating factor R, which combines all transmission parameters relevant for the considered connection.

This rating factor R is defined as

$$R = R_o - I_s - I_d - I_e + A \tag{5.1}$$

where R_o represents, in principle, the basic signal-to-noise ratio, including noise sources such as circuit noise and room noise.

Factor I_s is a combination of all impairments that occur more or less simultaneously with the voice signal. Factor I_d represents the impairments caused by delay, and the equipment impairment factor I_e represents impairments caused by low-bit-rate codecs. The advantage factor A allows for compensation of impairment factors when there are other advantages of access to the user. The term R_o and the I_s and I_d values are further subdivided into more specific impairment values.

The MOS is a subjective measurement indication, which ranks the perceived audio/video quality based on user feedback. In MOS, users determine the video quality by rating displayed video sequence on a scale of 1 (very bad) to 5 (excellent). The mean averages of multiple users are taken, and so, the MOS is computed. As such, the MOS provides a numerical value for the quality that the end user experiences. The MOS has been synonymous to QoE for the assessment of networks delivering voice, and this fashion has continued for the video streams as well. This QoE indicator is connected to the R-value metric using the graph illustrated in Figure 5.4.

Based on previous studies [58,59], multimedia QoE tends to have "Good". "Acceptable," or "Poor" grades, subjective of user perception for certain levels of delay, jitter, and packet loss values. As such, MOS can be given as the function MOS = f {delay, jitter, loss}, mapping QoE to QoS levels for different resolutions. Subsequently, a set of 27 network conditions can be defined for

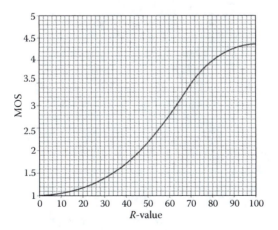

Figure 5.4   Relation of MOS and R-value.

evaluating a network's performance, with each condition denoted by a triplet as follows: ⟨[GGG], [GGA], [GGP], ..., [PPP]⟩. When comparing MOSs, it is of vital importance to consider that certain video types inherently produce a higher level of quality than others. For example, HDTV delivers a higher resolution and picture size than regular SDTV; therefore, maintaining all other factors identical, the MOS for an HD video stream will be higher than the MOS for the same sequence delivered in SD. Equivalently, multimedia content delivered on large TV screens will, by default, achieve a higher MOS than the same content being displayed on mobile or hand devices such as tablets. As a result, relying solely on the absolute MOSs can be misleading when comparing these dissimilar types of video service, as viewers tend to form expectations of quality based in part on the perceived capabilities of the medium. As presented in [60], a video viewed on a cellular handset might receive an absolute MOS of 3.1 when little or no quality degradation is evident, whereas, for an HDTV video sequence, a MOS of 3.1 would suggest that there were significant impairments present.

5.6.2.2 MDI

Another QoE measurement metric that finds frequent use in SLA supervision and network monitoring is the media delivery index (MDI). MDI passively and actively monitors audio/video IP flows and provides an indication of jitter and loss. It is expressed in the form of MDI = {DF; MLR}, where DF denotes the delay factor and MLR denotes the media loss rate. The delay factor is a value of time in milliseconds, showing the amount of data the buffers must be able to contain for this millisecond time in order to normalize jitter, computed at specific regular intervals. Accordingly, MDI denotes the packets lost during that time frame.

In contrast to MOS and *R*-value, MDI is a QoE measure from the perspective of the telecom operator or service provider, and thus, it involves characteristics that they have control over.

5.6.3 IGMP and IPTV Service-Specific Metrics

The performance indicators dealing with channel changing are those that greatly affect user-perceived QoE. Inherently, IPTV cannot achieve changing of channels instantaneously as supported by OTA and cable TV broadcasting. In order to explain this, the process that occurs when changing to a new channel has to be analyzed.

In IP multicasting, each TV channel is assigned with a multicast address. When an end user wants to watch a channel, a request is sent to join the corresponding multicast group. Usually, the nearest router belonging to the requested multicast group replicates and forwards the video stream to the end user. As a result, zapping delay is created since the channel change information has to travel upstream, in a worst-case scenario through the whole network to the head end [61].

Channel zapping can be decomposed into several parts:

1. Processing delay of the remote control request (d_1), which is the time interval between the remote control request and the transmission of the join message
2. Network delay (d_2), which is the time interval between the transmission of the join/leave message and the reception of the first video frame of the requested channel
3. STB layer delay (d_3), which is the time needed by the STB IP stack to process incoming packets and deliver the content to the MPEG decoder engine
4. STB jitter buffer delay (d_4), which is the time until the STB jitter buffer reaches the fullness set point prior to the forwarding of the video signal to the decoder function

As a result, total delay $D = d_1 + d_2 + d_3 + d_4$. Evidently, d_1, d_3, and d_4 are dependent on the design of the STB, and only d_2 is the IP network-dependent delay.

When the user changes the current channel from Ch1 to Ch2, the STB generates an IGMP leave message for channel Ch1 and sends it to the home gateway (HG, which might be a server, switch, or router). After receiving the IGMP leave message, the HG generates an IGMP group-specific query message to the home network in order to see if any host has membership to the group in the Leave message. The sent group-specific query message has its max response time set to last member query interval. If no reports are received after the response time of the last query expires, the HG assumes that the group has no local members, stops consequently the forwarding of the multicast stream corresponding to the group, and sends its upper router the IGMP leave message for the group. After the STB sends the IGMP leave message to the HG, it then sends an IGMP join message for the new channel. HG receives it and sends the IGMP join message to the upper router if no other local hosts have membership for that specific group. Last-hop router (LHR), which is the first router connected to HG, receives the join message and sends PIM join message to the other multicast routers in the access network. Then, the multicast stream for the group corresponding to the new channel selected can be transmitted through several routers and HG and reach the STB. As a result, in the network-dependent delay, the channel zapping time is a sum of the IGMP leave processing time, IGMP join processing time, PIM processing time, multicast stream forwarding time from FHR to LHR, multicast stream forwarding time from LHR to HG, and multicast stream forwarding time from HG to IP STB. The whole process of channel changing and the IGMP queries is depicted in Figure 5.5. The "leave-process" not only creates delays in channel changing time but also causes a substantial increase in bandwidth. As previously

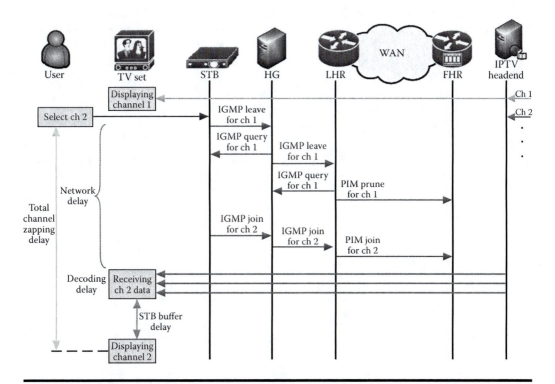

Figure 5.5 "Zapping" IGMP queries.

described, although IGMP leave packets are sent at a specific time point, an additional time has to pass in order for the multicast stream to stop. As channels are sequentially changed, the bandwidth increases and may even reach the network limit, which will result in severe packet loss.

In order to measure the network performance from the perspective of channel changing, the following metrics are defined:

1. *Join and leave delay:* The join delay is defined to be the elapsed time from issuing the IGMP join request until arrival of the first packet in the multicast packet stream. The join operation is said to fail if no packet has arrived when the time has come to issue the IGMP leave request.

   The leave delay is defined to be the elapsed time from issuing the IGMP leave request until reception of the last packet in the multicast packet stream. The leave operation is said to fail if no packet could be stated to be the last packet, that is, the packet stream was not stopped and was still arriving when it was time to join the channel again.

2. *Channel and link bandwidth:* The channel bandwidth is determined by monitoring the number of packets/bytes received during the "zap" interval. Similarly, the total link bandwidth is sampled at the time of issuing the IGMP join request.

   It should be noted that the link bandwidth may be useful to examine and detect if multiple multicast channels are received at the same time (overlapping channels) as depicted in Figure 5.6. This could happen if an IGMP leave request fails or the switch in use is misconfigured and forwards unwanted multicast traffic to the probe.

5.6.4 Quality Margins for IPTV Services

From the above descriptions of the QoS/QoE metrics, it is evident that IPTV quality is not equally affected by them. As a result, defining its tolerance margins and exact requirements is more than a difficult task. While QoS classes have been standardized [62], IPTV is composed of different services, varying from a simple streaming video to interactive applications, as already described. As such, no standardization has been made so far, providing the exact limits beyond which IPTV service quality is not acceptable. Aside from the explanations given on the previous section and Table 5.1, IPTV can be described as very bandwidth demanding, very sensitive to packet loss, and simply sensitive to latency and jitter.

Based on [63], some margins for the QoS/QoE metrics can be derived. As such, latency should be kept less than 200 ms, whereas PDV or jitter that can deteriorate the video quality at greater extent should be kept under 50 ms. Packet loss for IPTV, which has the greatest impact in video

Table 5.1 QoS Classes

Service Characteristics	QoS Class	Delay	Jitter	Loss
Real-time, jitter sensitive, very interactive	0	100 ms	50 ms	10^{-4}
Real-time, jitter sensitive, interactive	1	400 ms	50 ms	10^{-3}
Transaction data, highly interactive	2	100 ms	–	10^{-3}
Transaction data, interactive	3	400 ms	–	10^{-3}
Low loss (bulk data, video streaming)	4	1 s	–	10^{-4}
Traditional applications of IP networks	5	–	–	–

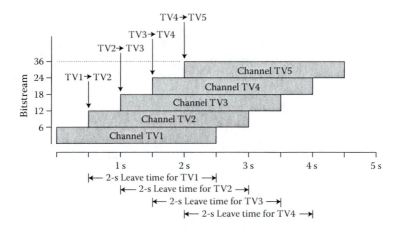

Figure 5.6 Channel "zapping."

quality, should be kept below 10^{-4}. To further specify, as losses are bound to happen in bursts, the bandwidth stream of the IPTV video should be related to the loss rate tolerance. Data from [63] provide packet loss rate values for streams of bit rates varying from 3 to 5 Mb/s. As specified on the same report, the maximum duration of every single error should not exceed 16 ms.

Establishing margins of satisfactory performance for the "zapping" function is more complicated, as explained in the previous section. Methods of reducing "zapping" delay have been proposed in [64–66]. In [63], the acceptable delay time is between 500 ms and less than 2 s, whereas in [61], the measured time for channel zapping was 2.1 s.

Generally, a time less than 1 s can be safely called satisfactory [67], whereas a value more than 2 s seems to exceed being characterized as simply annoying by end users [68]. Even in specific cases, the 1-s channel change delay is regarded a nuisance for the user, as his or her expectations have been formed by the traditional TV broadcasting. As a result, currently, there is no knowledge about the explicit relation between the channel zapping time and the user experience (QoE), as expressed with MOS, but only approximate criteria are given for its evaluation [61]. Based on measurement equipment vendors [69], join and leave times for different QoE classes are presented in Table 5.2.

Finally, for the MDI QoE measure, any MLR greater than 0 implies that packets are either lost or delivered out of order, whereas an acceptable margin for DF is within 9 and 50 ms. The gap between the two values is attributed to the broad differences in the processing power and overall quality of available STBs, since the DF must be fine-tuned according to the highest amount of jitter that the STB can handle before the appearance of distortions [70].

Table 5.2 Zap Join and Leave Experience Margins

Experience	Join Delay	Leave Delay
Excellent	50	250
Good	150	750
Fair	300	1500
Poor	500	2500
Bad	1000	5000

5.6.5 Direct, Active, and Passive Measurements

Direct, active, and passive measurements have become the common practice for most of the ISPs' networks, and the measurement data collection provides means for studying end-to-end performance bottlenecks in the network path and also for understanding Internet traffic characteristics [71]. Especially for the inelastic nature of IPTV, traffic measurements and monitoring can provide the needed information in order to assess the performance of networks in delivering this service, the impact that IPTV can have on other services delivered through the same medium, and the expected level of experience that the end users can enjoy. More than that, each type of measurement is particularly suited for different types of networks so that traffic and network characteristics remain unaffected when conducting them.

5.6.5.1 Direct Measurement

Direct measurements are based on a device's capability to maintain information about its own performance. The stored performance statistics can be collected using a mechanism such as Simple Network Management Protocol (SNMP) that extracts them from a management information base. The very large number of MIBs that exist can support more than proportional amounts of performance information, which would make direct measurements one of the best sources of performance monitoring. However, the limits of the capacity and the rate required to store such information, in conjunction with the inefficient SNMP polling mechanism, appoint them as the worst.

5.6.5.2 Passive Measurements

The approach of passive measurements uses devices to inspect the traffic as it passes by. This is achieved through the use of special purpose devices or equivalent software implementations, with Remote Monitoring (RMON)-, MRTG-, and Netflow-capable devices being some of them. Passive measurements are using multiple measurements on the same packet to infer traffic performance.

As passive measurements can monitor the performance on a single network element, such as a router, or a network interface, they can collect essential measurements for network management, such as link utilization, router load, errors, and queue drops. However, they also have limits:

- As packets are collected so that performance calculations can be made, the amount of data is substantial and can reach up to enormous sizes for high payload traffic.
- To be able to use passive measurements for end-to-end performance, prior knowledge of the path needs to be known.
- As passive measurements might involve inspecting of all packets that traverse a network, security implications might arise.

It should be noted that the issue of a large amount of collected data can be alleviated by collecting data at intervals, namely, packet sampling. However, this type of monitoring requires a very high level of synchronization of the clocks so that there is no time deviation. Although technically feasible, time synchronization demands additional hardware or software installation such as usage of either dedicated Network Time Protocol servers, Global Positioning System–enabled systems, or proprietary-engineered algorithms. Still, packet sampling as a method to decrease the large volume of passive measurements is not very efficient as traffic is known to fluctuate dynamically and unpredictably especially in an Internet environment.

5.6.5.3 Active Measurements

The approach of active measurements requires inserting test packets into the network for measurement purposes. As the generated traffic will traverse the same network infrastructure, it will eventually be affected by it, with the same manner that the real traffic is. Introducing artificial (emulated) packets to the network provides the advantage of having absolute control of the nature of the traffic parameters. As such, parameters can be adjusted to have traffic of an exact amount of volume, generation intervals, sampling frequency, scheduling, packet sizes, and types. However, active measurements do create extra traffic as the packets are inserted into the network. If the extra generated volume is too high, problems arise, appointing the obtained measurements inaccurate. To avoid phenomena like that, traffic frequency generation can be regulated, so as not to overcome specific network-dependent boundaries.

The most important issue of active measurements is synchronization, as described earlier. However, most traffic monitoring and supervision systems have proprietary synchronization methods implemented on their probes to keep measurement accuracies in the vicinity of microseconds.

Active measurements are frequently preferred especially for monitoring of the inelastic IPTV multicast traffic. Unlike the passive measurements that need to capture all types of packets, including packets that do not need to be monitored, active measurements can generate traffic to reflect only the data of the studied application service, which, in this case, is IPTV. The excessive capturing of a large number of packets, which is demonstrated by passive measurements, can create additional central processing unit (CPU) usage to the measuring devices and network equipment, leading to network performance deterioration or even network unresponsiveness. The unnecessary delays that might be caused highly affect the IPTV traffic that requires minimal delay. Also, new probing devices released by various vendors allow the real data stream to be included in the generated traffic stream so that hybrid measurement data are obtained, verifying that both emulated and real data are equivalently affected by network conditions.

However, it is worth mentioning that active measurements have limitations of their own:

■ With active measurements, we can only measure how the network affects the generated packets as they traverse it. It is impossible, however, to identify how specific network devices or interfaces handle the traffic themselves. For instance, with passive or direct measurements, it is easy to collect link utilization, server response, router load, errors, and queue drops, which are infeasible to accomplish with active measurements. As such, it is right to claim that active measurements are best suited for path segments, whereas other types of measurements are more applicable to specific points in the network.

■ Finally, the traffic generation of active measurements can cause subsequent problems in the case that the generated data are affecting the real data transmission, which can easily occur as we reach bandwidth limitations. This appoints active traffic measurements ideal only for networks that have capacity abundance, so that artificial data do not impact network performance.

5.7 Performance Evaluation of Networks

But what is expected from networks offering IPTV? Performance wise, current wired access networks can easily adapt to the QoS level needed by IPTV service to meet viewing audience expectations in terms of QoE. With fiber-optic prices dropping, VDSL, FTTN, and FTTH technologies are reaching the end consumer, bringing along an ideal level of network performance. Extensive measurement sessions have been held even in real-life networks utilizing active performance measurements to verify the QoS/QoE levels that can be achieved while IPTV service is multicasted on a large number of end users and for various available video codec standards. Such a sample network is presented in [72], with MPEG-2 TS IPTV traffic being generated using the same attributes as the real IPTV media that traverses the network. Equivalent access networks to the aforementioned network that are engineered to distribute IPTV content, can easily maintain close to zero packet losses and very low levels of delay and jitter for all services that comprise the triple-play bundle, as shown in Table 5.3. Even for HD content delivery, the measurement sessions conducted exhibited the same results: zero packet losses, a mean delay of 555–675 μs, and a mean jitter of 112–131 μs. Subsequently, the QoE that such access networks can deliver to a demanding viewing audience is of a very high level. QoE metrics such as the *R*-value are nominal, yielding an overall "excellent" QoS chart for all types of channels (a chart that shows the amount of time spent in the different QoS levels, in percent of the complete test period), as shown in Figure 5.7. The pie-slices that appear with 0%, such as "Poor" and "Bad," denote a percentage less than 0.05. Instead, QoE classes that would have an absolute zero value would not be displayed on the pie chart.

As stated, IPTV is highly maladaptive to network frailties, and a common point of such weakness is usually the end user's home network. With the ability of using a large variety of networking technologies, it is usual for IPTV content reaching the customer premiss equipment to be distributed via an in-home network to other in-house locations utilizing Ethernet, wireless, and even PLC. Although these may be well suited for best effort Internet access, it is needed to further verify if they support the delivery of IPTV content, at what extent and the overall level of QoS and QoE that can be achieved for "triple-play" services. A primary concern in dimensioning and comparing network technologies is to identify the maximum capacity that can be attained. In the measurement sessions that were held for in-home network monitoring, the Ethernet network exhibited a constant capacity of 97 Mb/s, whereas the HomePlug AV network with a nominal speed of 200 Mb/s failed to reach Ethernet capacity values, fluctuating within the range of 43–49 Mb/s. It should be noted that, in the real-life situation where the network capacity was tested, other devices including routers, switches, computers, and home appliances were connected on the main grid as typically happens in a home network, deteriorating though the maximum capacity that could be attained. Accordingly, the capacity of the wireless networks had major fluctuations in both IEEE 802.11g and IEEE 802.11n standards as the distance from the AP inside the testing facilities varied. As such, the capacity of the IEEE 802.11g wireless network ranged between 7 and 14 Mb/s and that of the IEEE 802.11n between 11 and 85 Mb/s, as illustrated in Figure 5.8a, along with the capacities obtained while stressing the rest network types. Note that the leftmost values were those that are more distant from the AP (with beacon strength of 70 dBm), whereas the rightmost and higher values were those obtained when the distance between the client and the AP was between 2 and 3 m within the same room (beacon strength of 40 dBm).

Table 5.3 IPTV and VoIP Delay and Jitter Results

Metric	Delay (μs)	Jitter (μs)	Delay (μs)	Jitter (μs)
Min	471–586	0–58	188–312	0–0
Mean	554–667	111–125	214–333	1–17
Median	548–662	73–126	208–330	1–2
Max	719–8185	232–3125	285–470	17–169

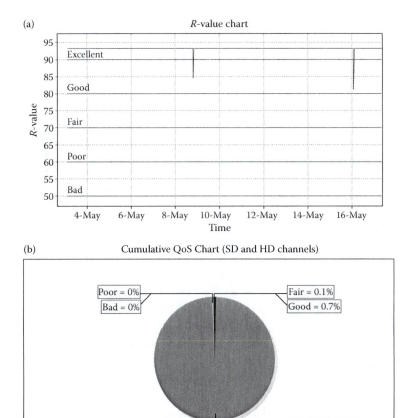

Figure 5.7   (a) *R*-value chart. (b) Cumulative QoS chart.

In the passive measurement sessions held, MPEG-4 H.264 codec (part 10: AVC) traffic was monitored and provided the performance values depicted in Table 5.4. Based on these QoS performance values, the overall MOS is depicted in Figure 5.8b, together with the upper and lower boundaries as they were defined in Section 5.6.2.1.

Progressing from network performance parameters to QoE-specific metrics, like IGMP join and leave delays, the inherited shortcomings of IPTV start to emerge. Over the series of active measurement sessions held on the fiber-optic access network, a large number regarded the "zapping" join and leave delays of six IPTV channels with average data rates between 6.2 and 6.8 Mb/s. Indicative charts of the join and leave delays are depicted in Figure 5.9.

It should be noted that the network of study was engineered in order to circumvent possible QoE degrading issues. As such, *fast-leave* has been implemented on the access network switches. This important addition to the original IGMP, which was included in the second version of RFC2236, allows the switch to remove an interface from the forwarding-table entry without the need of previously sending out group-specific queries to the interface. According to the original IGMP version, when a switch with IGMP snooping enabled receives an IP group-specific IGMP leave message, it sends a group-specific query out the interface where the leave message was received to determine if there are any other hosts attached to that interface that are interested in

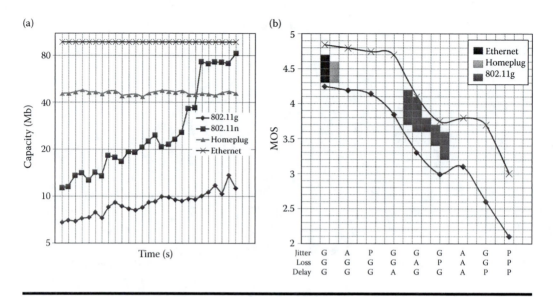

Figure 5.8 (a) Network capacity. (b) Overall MOS chart.

the MAC multicast group. If the switch does not receive an IGMP join message within the query response interval and none of the other IP groups corresponding to the MAC group are interested in the multicast content for that MAC group, then the interface is removed from the port list of the (MAC-group, VLAN) entry in the Layer-2 forwarding table. However, in the second version of RFC2236, the important addition of fast-leave was made. With fastleave enabled on the VLAN, an interface can be removed immediately from the port list of the Layer-2 entry upon reception of the IGMP leave message, unless a multicast router was learned on that specific port. Fast-leave processing ensures optimal bandwidth management for all hosts on a switched network, even when multiple multicast groups are in use simultaneously. This IGMP feature, however, can only be used in VLANs where exactly one host is connected to each interface, as it occurred in the studied network. As observed in Figure 5.9a, expected zap Leave times were indeed kept at very low levels. In most cases, the zap leave delay was in the vicinity of 100 ms, with very low values in the proximity of 50 ms being frequent. Values as low as 10 ms, although rare, still existed in the measurement samples, illustrating how the fast-leave IGMP snooping implementation played an

Table 5.4 Home Networking QoS Results

Metric	Delay (ms)	Jitter (ms)	Packet Loss (%)
Ethernet	Mean: 1.5–2.79	Mean: 1.35–1.45	Mean: 10^{-6}
	Max: 3.21–9.77	Max: 3.2–4.13	Max: 10^{-5}
HomePlug	Mean: 2.86–3.12	Mean: 1.68–2.32	Mean: 10^{-5}
	Max: 12.52–14.3	Max: 3.25–4.69	Max: 10^{-4}
802.11g	Mean: 2.1–3.3	Mean: 1.72–2.89	Mean: 0.13
	Max: 220–348	Max: 21.5–23.96	Max: 1.7

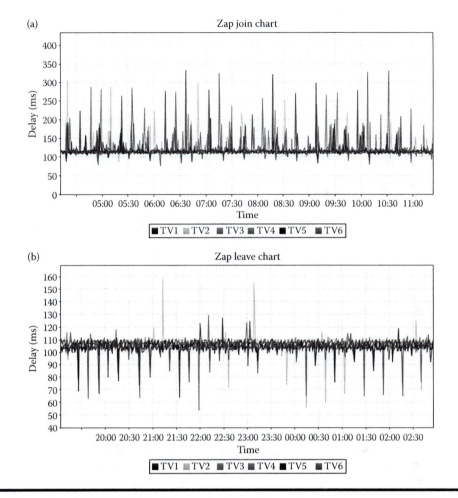

(a)

Figure 5.9 (a) Zapping join chart. (b) Zapping leave chart.

important role in decreasing zapping leave delay. At the same time, isolated cases of leave failures also existed, yet very few in proportion to the total zapping measurement sessions.

While zapping leave times were sufficiently low for the channel leave experience to be mostly characterized as "excellent", the same cannot be said about the channel join delays. Evidently from Figure 5.9b, join times frequently surpassed 150 ms, yielding an overall "good" level of experience. It should be noted that the media encoding scheme of IPTV further deteriorates the zapping time since further delay is introduced after the zap join until the first *I*-frame reaches the decoder.

In the event that leave failure was observed, the link bandwidth peaked, illustrating what was described in Section 5.6.3. Such a case is depicted in Figure 5.10. In networks where there is capacity abundance, this growth in link bandwidth as exhibited in Figure 5.10b after a leave fail might impact the QoS performance metrics, but it will not highly affect IPTV QoE. In contrast, in case that this increase occupied all the network capacity, something that can easily happen in in-home networks such as IEEE 802.11g, packet losses would start to occur, massively affecting the QoE for the viewing audience. Furthermore, both STBs and CPE (especially low-end models) cannot handle the high bandwidth increase. With a large number of prioritized packets traversing through them, high CPU utilization occurs, causing packets to either be delayed or dropped, and

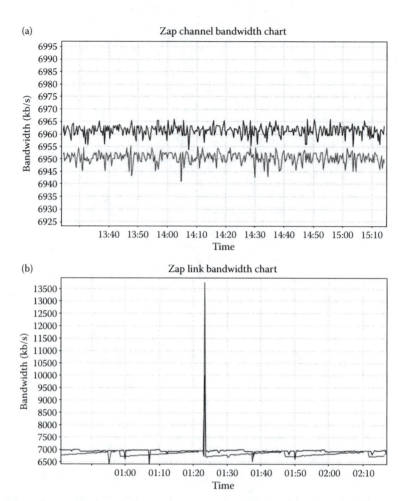

Figure 5.10 **(a) Channel bandwidth. (b) Zap link bandwidth.**

even freezes and crashes on the devices handling them, greatly deteriorating the projected video content.

The cumulative MOS concerning the zap join and leave delays is illustrated in Figure 5.11. As expected, zap leave delays are "excellent" at the most extent, with very few portions of "bad" experience caused by the IGMP leave fails that have infrequently occurred. Accordingly, zap join delays are mostly limited to a "good" overall experience for the end user. As a result, the overall zapping experience (join and leave) should be considered to be ranging between "good" and "excellent" levels but definitely not "excellent" in total as it would be expected from a network with the QoS performance as this. It is therefore mandatory for other mechanisms to exist in order to further decrease zapping times and make them more comparable to traditional TV broadcasting low zap delays.

Works including [64] provide a method for drastically decreasing the zapping join time by creating adjacent multicast channel groups to the channel being played at the present time and multicast their content to the HG in advance. To do so, when the STB sends an IGMP join request, the HG also sends a join request for the adjacent channel group; thus, when the channel is changed to

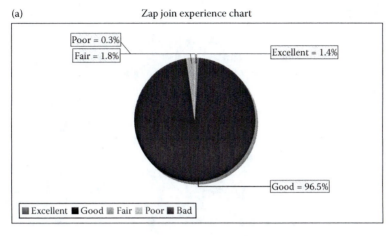

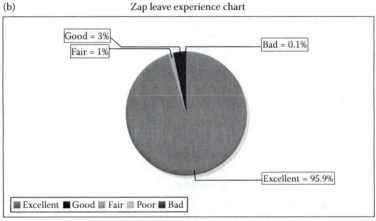

Figure 5.11 **(a) Cumulative zap join MOS. (b) Cumulative zap leave MOS.**

one that belongs to the adjacent channel group, it is joined with very improved delay. Mechanisms such as that can adequately be implemented in networks as that previously mentioned due to the absence of strict capacity constraints. Multicasting a series of channels on a link can, as described, consume the available bandwidth and cause QoE degradation issues. To achieve better zapping experience from networks lacking large capacity, other algorithms have been utilized, such as that discussed in [73]. This work models the users' zapping behavior to prejoin channels that are highly likely to be selected next. More than that, mechanisms like this are capable of determining the most efficient number of channels that should be prejoined by estimating expected channel zapping time and expected bandwidth usage with a semi-Markov process. As such, a channel zapping time is improved while maintaining a minimized bandwidth usage, something that might be of great importance for a particular network architecture.

Evidently, current network infrastructures are capable of delivering a high level of IPTV QoE to the viewing audience. Points of frailty such as redistributing the IPTV content to other in-home locations and inherited IPTV problems such as suboptimal zap delay can be circumvented by carefully deploying in-home networks and utilizing current network technologies and special mechanisms to successfully deliver a service comparable to and better than traditional OTA, satellite, and cable TV.

5.8 Conclusions

This chapter presents an overview of the evolutionary IPTV approach, its services, and its key features in terms of wired and wireless access networks. While IPTV has demands in network performance and capabilities, open issues such as security, QoS, and QoE that need to be addressed, as well as a standardization process that needs eventually to converge, are being quickly embraced by both users and telcos. With the interactive and personalized services it targets to offer, IPTV can become the new wave of home entertainment for its clients. This is because IPTV provides telcos a new architecture on which they can build new service platforms and offer innovative applications refining their economy. Taking advantage of existing network architectures that guarantee exceptional IPTV traffic delivery and utilizing recent network infrastructures such as the IMS architectural framework, telcos can offer the end user the ability to experience the IPTV service in its full potential. Having achieved that, no matter whether IPTV will be delivered on TV sets via PCs and STBs or on 4G wireless mobile devices, either as a separate service or included in a triple-play bundle, there is little doubt that it will have the ability to forever change the way video content is consumed in daily life.

References

1. Palmer, S. *Television Disrupted: The Transition from Network to Networked TV*. Focal Press, 2006.
2. Multimedia Research Group. "Semiannual IPTV Global Forecast Report 2010–2014," Multimedia Research Group (MRG), Tech. Rep., Jun. 2010.
3. Companies and Markets. "Global IPTV Market Forecast to 2014," Global Market Report, Feb. 2011.
4. B. Forum. "IPTV Q4 2010," Broadband Forum, Tech. Rep., Jan. 2011.
5. Loomis, J., and M. Wasson. "VC-1 Technical Overview," Online Article, Oct. 2006.
6. Williamson, B. "Developing IP Multicast Networks," Cisco Press, 2000.
7. Schulzrinne, H., A. Rao, and R. Lanphier. "Real time streaming protocol (RTSP)," *IETF RFC 2326*, Apr. 1998.
8. Estrin, D. et al. "Protocol independent multicast-sparse mode (pim-sm)," *IETF RFC 2362*, Jun. 1998.
9. Maisonneuve, J., M. Deschanel, J. Heiles, W. Li, H. Liu, R. Sharpe, and Y. Wu. "An overview of IPTV standards development," *IEEE Transactions on Broadcasting*, vol. 55, no. 2, pp. 315 –328, Jun. 2009.
10. Mikoczy, E., D. Sivchenko, B. Xu, and J. Moreno. "IPTV systems, standards and architectures—Part II: IPTV services over IMS: Architecture and standardization," *IEEE Communications Magazine*, vol. 46, no. 5, pp. 128–135, May 2008.
11. ETSI TS 102 034 V1.2.1 tech. spec., "DVB: Transport of MPEG 2 Based DVB Services over IP-Based Networks," Sep. 2006.
12. ETSI TS102 826, tech. spec., "Digital Video Broadcasting (DVB); DVB-IPTV Profiles for TS 102 034," Nov. 2009.
13. ITU-T FOCUS GROUP ON IPTV, wkg. doc. IPTV-DOC-0084, "IPTV Architecture," 4th FG IPTV mtg., Bled, Slovenia, 2007.
14. ETSI TS 181 016 V3.3.1, tech. spec., "TISPAN; Service Layer Requirements to Integrate NGN Services and IPTV," Jul. 2009.
15. ETSI TS 02049 V0.0.8, tech. spec., "TISPAN; IPTV Architecture: Dedicated Subsystem for IPTV Functions in NGN," Sep. 2007.
16. ETSI TS 182 027 V3.5.1, tech. spec., "TISPAN; IPTV Architecture; IPTV Functions Supported by the IMS Subsystem," Mar. 2011.
17. ETSI TS 182 028 V3.3.1, tech. spec., "TISPAN; NGN integrated IPTV Subsystem Architecture," Oct. 2010.
18. ATIS IPTV High Level Architecture Standard (ATIS-0800007), "ATIS IPTV Interoperability Forum (IIF)," 2007.

19. 3GPP TS 23.246 V8.0.0, tech. spec., "Multimedia Broadcast/Multicast Service (MBMS); Architecture and Functional Description," rel. 8, Sep. 2007.
20. Lu, T., F. Xie, Y. Peng, and J. Xie. "Analysis of security standardization for IPTV," in Advanced Computer Control (ICACC), 2011 3rd International Conference on, Jan. 2011, pp. 219–223.
21. Johnston, D., and J. Walker. "Overview of IEEE 802.16 security," *IEEE Security Privacy*, vol. 2, no. 3, pp. 40–48, May–Jun. 2004.
22. Eriksson, P.-E., and B. Odenhammar. "VDSL2: Next important broadband technology," *Eriksson Review*, Apr. 2006.
23. "IEEE Std 802.11b Part 11: Wireless LAN Medium Access Control (MAC) and Physical Layer (PHY) Specifications: Higher-Speed Physical Layer Extension in the 2.4 GHz Band," IEEE Std 802.11b-1999, 2000.
24. "IEEE 802.11g-2003 Part 11: Wireless LAN Medium Access Control (MAC) and Physical Layer (PHY) Specifications: Further Higher Data Rate Extension in the 2.4 GHz Band," IEEE 802.11g-2003, 2003.
25. "IEEE 802.11g-2003 Part 11: Wireless LAN Medium Access Control (MAC) and Physical Layer (PHY) Specifications: Further Higher Data Rate Extension in the 5 GHz Band," IEEE Std 802.11a-1999, 1999.
26. "IEEE 802.11n-2009, ŞPart 11: Wireless LAN Medium Access Control (MAC) and Physical Layer (PHY) Specifications Amendment 5: Enhancements for Higher Throughput," IEEE 802.11n-2009, Oct. 2009.
27. "IEEE 802.11e-2005 Part 11: Wireless LAN Medium Access Control (MAC) and Physical Layer (PHY) Specifications Amendment 8: Medium Access Control (MAC) Quality of Service Enhancements," IEEE 802.11e-2005, 2005.
28. "IEEE 802.11ac/D2.0 Draft Standard Part 11: Wireless LAN Medium Access Control (MAC) and Physical Layer (PHY) Specifications Amendment: Enhancements for Very High Throughput for Operation in Bands Below 6GHz," IEEE 802.11ac/D2.0, Jan. 2012.
29. "IEEE 802.11ad/D7.0 Draft Standard Part 11: Wireless LAN Medium Access Control (MAC) and Physical Layer (PHY) Specifications Amendment 3: Enhancements for Very High Throughput in the 60 GHz Band," IEEE 802.11ad/D7.0, Apr. 2012.
30. "IEEE 802.11aa/D9.0 Draft Standard Part 11: Wireless LAN Medium Access Control (MAC) and Physical Layer (PHY) Specifications Amendment 3: MAC Enhancements for Robust Audio Video Streaming," IEEE 802.11aa/D9.0, Jan. 2012.
31. Hiertz, G., D. Denteneer, L. Stibor, Y. Zang, X. Costa, and B. Walke. "The IEEE 802.11 universe," *IEEE Communications Magazine*, vol. 48, no. 1, pp. 62–70, Jan. 2010.
32. Ahson, S., and M. Ilyas. "WiMAX Technologies, Performance Analysis and QoS." Taylor and Francis, 2008.
33. "IEEE Standard for Local and Metropolitan Area Networks Part 16: Air Interface for Fixed and Mobile Broadband Wireless Access Systems Amendment 2: Physical and Medium Access Control Layers for Combined Fixed and Mobile Operation in Licensed Bands and Corrigendum 1," IEEE Std 802.16e-2005 and IEEE Std 802.16-2004/Cor 1-2005 (Amendment and Corrigendum to IEEE Std802.16-2004), pp. 1–822, 2006.
34. "IEEE Standard for Local and metropolitan area networks Part 16: Air Interface for Broadband Wireless Access Systems," IEEE Std802.16-2009 (Revision of IEEE Std 802.16 - 2004), pp. C1–2004, 2009.
35. "IEEE Standard for Local and metropolitan area networks Part 16: Air Interface for Broadband Wireless Access Systems Amendment1: Multiple Relay Specification," IEEE Std 802.16j-2009 (Amendment to IEEE Std 802.16-2009), pp. C1–290, 12 2009.
36. "IEEE Standard for Local and metropolitan area networks Part 16: Air Interface for Broadband Wireless Access Systems Amendment 2: Improved Coexistence Mechanisms for License-Exempt Operation," IEEE Std 802.16h-2010 (Amendment to IEEE Std 802.16-2009), pp. 1–223, 30 2010.
37. "IEEE Standard for Local and metropolitan area networks Part 16: Air Interface for Broadband Wireless Access Systems Amendment 3: Advanced Air Interface," IEEE Std 802.16m-2011(Amendment to IEEE Std 802.16-2009), pp. 1–1112, 5 2011.
38. Alasti, M.,B. Neekzad, J. Hui, and R. Vannithamby. "Quality of service in WiMAX and LTE networks [topics in wireless communications]," *IEEE Communications Magazine*, vol. 48, no. 5, pp. 104–111, May 2010.

39. She, J., F. Hou, P.-H. Ho, and L.-L. Xie. "IPTV over WiMAX: Key success factors, challenges, and solutions [advances in mobile multimedia]," *IEEE Communications Magazine*, vol. 45, no. 8, pp. 87–93, Aug. 2007.

40. Oyman, O., J. Foerster, Y.-J. Tcha, and S.-C. Lee. "Toward enhanced mobile video services over WiMAX and LTE [WiMAX/LTE update]," *IEEE Communications Magazine*, vol. 48, no. 8, pp. 68–76, Aug. 2010.

41. Sesia, S., I. Toufik, and M. Baker. "LTE, The UMTS Long Term Evolution: From Theory to Practice," Wiley, 2009.

42. "3GPP LTE - 23.303 v8.31 (release 8)," http://www.3gpp.org.

43. H. Ekstrom. "QoS control in the 3GPP evolved packet system," *IEEE Communications Magazine*, vol. 47, no. 2, pp. 76–83, Feb. 2009.

44. "Wi-Fi certified for WMM," http://www.wi-fi.org, 2008.

45. "Wi-Fi Alliance WMM (including WMMTM Power Save) Specification," version 1.1, 2003.

46. Spenst, A., K. Andler, and T. Herfet. "A post-admission control approach in wireless home networks," *IEEE Transactionson Broadcasting*, vol. 55, no. 2, pp. 451–459, Jun. 2009.

47. "Wi-Fi Alliance/Wi-Fi Certified for WMM: Support for Multimedia Applications with Quality of Service in Wi-Fi Networks," White Paper, Sep. 2004.

48. Galli, S., and O. Logvinov. "Recent developments in the standardization of power line communications within the IEEE," *IEEE Communications Magazine*, vol. 46, no. 7, pp. 64–71, July 2008.

49. HomePlug Powerline Alliance, "HomePlug 1.0 specification," Jun. 2001.

50. Lee, M. K., R. E. Newman, H. A. Latchman, S. Katar, and L. Yonge. "HomePlug 1.0 powerline communication LANs: Protocol description and performance results," *International Journal of Communication Systems*, vol. 16, issue 5, pp. 447–473, 2003.

51. HomePlug Powerline Alliance, "HomePlug AV Specification," 2005.

52. "IEEE Standard for Broadband over Power Line Networks: Medium Access Control and Physical Layer Specifications," IEEE Std1901-2010, pp. 1–1586, 30 2010.

53. Oksman, V., and S. Galli. "G.hn: The new ITU-T home networking standard," *IEEE Communications Magazine*, vol. 47, no. 10, pp.138–145, Oct. 2009.

54. Luby, M., M. Watson, T. Gasiba, and T. Stockhammer. "High-Quality Video Distribution using Power Line Communication and Application Layer Forward Error Correction," in IEEE International Symposium on Power Line Communications and Its Applications (ISPLC'07), Mar. 2007, pp. 431–436.

55. Almes, G., S. Kalidindi, and M. Zekauskas. "A One-Way Packet Loss Metric for IPPM," IETF RFC 2680, Sep. 1999.

56. Demichelis C., and P. Chimento. "IP Packet Delay Variation Metric for IP Performance Metrics (IPPM)," RFC 3393, Nov. 2002.

57. ITU-T. "The e-model, a computational model for use in transmission planning," ITU-T G.107 Recommendation, May 2000.

58. Calyam, P., M. Sridharan, W. Mandrawa, and P. Schopis. "Performance measurement and analysis of h.323 traffic," in *Passive and Active Measurement Workshop*, vol. 3015, 2004, pp. 137–146.

59. Takahashi, A., D. Hands, and V. Barriac. "Standardization activities in the ITU for a QoE assessment of IPTV," *IEEE Communications Magazine*, vol. 46, no. 2, 2008, pp. 78–84.

60. Telchemy Incorporated. "Understanding IP Video Quality Metrics," http://www.telchemy.com/appnotes/Understanding, Feb. 2008.

61. Luo, X., Y. Jin, Q. Zeng, W. Sun, and W. Hu. "Channel Zapping in IP over Optical Two-Layer Multicasting for Large Scale Video Delivery," in *6th International Conference on Information, Communications and Signal Processing*, Dec. 2007, pp. 1–4.

62. ITU-T Recommendation. "Network Performance Objectives for IP-Based Services," ITU-T Y.1541 Recommendation, Feb. 2003.

63. D. Forum. "Triple-Play Services Quality of Experience (QoE) Requirements," Technical Report TR-126, Dec. 2006.

64. Cho, C., I. Han, Y. Jun, and H. Lee. "Improvement of channel zapping time in iptv services using the adjacent groups join-leave method," in *The 6th International Conference on Advanced Communication Technology*, vol. 2, 2004, pp. 971–975.

65. Joo, H., H. Song, D.-B. Lee, and I. Lee. "An effective iptv channel control algorithm considering channel zapping time and network utilization," *IEEE Transactions on Broadcasting*, vol. 54, no. 2, pp. 208–216, Jun. 2008.
66. Jennehag, U., T. Zhang, and S. Pettersson. "Improving transmission efficiency in h.264 based IPTV systems," *IEEE Transactions on Broadcasting*, vol. 53, no. 1, pp. 69–78, Mar. 2007.
67. Kozamernik, F., and L. Vermaele. "Will Broadband TV shape the future of broadcasting?" EBU Technical Review, Apr. 2005.
68. Fuchs, H., and N. Farber. "Optimizing channel change time in IPTV applications," in *IEEE International Symposium on Broadband Multimedia Systems and Broadcasting*, 31 2008–April 2 2008, pp. 1–8.
69. Prosilient Technologies AB, "PT-Analyzer Technical Manual," 2006.
70. Agilent Technologies Inc. "IPTV QoE: Understanding and interpreting MDI values," Tech. Rep., Sep. 2008.
71. Calyam, P., D. Krymskiy, M. Sridharan, and P. Schopis. "Active and passive measurements on campus, regional and national network backbone paths," in *Proceedings on 14th International Conference on Computer Communications and Networks* (ICCCN 2005), Oct. 2005, pp. 537–542.
72. Baltoglou, G., E. Karapistoli, and P. Chatzimisios. "Real-world IPTV network measurements," in *2011 IEEE Symposium on Computers and Communications (ISCC)*, 28 2011–July 1 2011, pp. 830 –835.
73. Lee, C. Y., C. K. Hong, and K. Y. Lee. "Reducing Channel Zapping Time in IPTV Based on User's Channel Selection Behaviors," *IEEE Transactions on Broadcasting*, vol. 56, no. 3, pp. 321–330, Sep. 2010.

SAFETY AND SECURITY IN NETWORKS

II

Chapter 6

A Walk through SQL Injection: Vulnerabilities, Attacks, and Countermeasures in Current and Future Networks

Diallo Abdoulaye Kindy and Al-Sakib Khan Pathan

Contents

6.1 Introduction

In recent years, the World Wide Web (WWW) has witnessed a staggering growth of many online web applications that have been developed for meeting various purposes. Nowadays, almost everyone in touch with *computer technology* is somehow connected online. To serve this huge number of users, great volumes of data are stored in web application databases in different parts of the globe. From time to time, users need to interact with the backend databases via user interfaces for various tasks such as updating data, making queries, extracting data, and so forth. For all these operations, the design interface plays a crucial role, the quality of which has a great impact on the security of the stored data in the database. A less secure web application design may allow crafted injection and malicious update on the backend database. This trend can cause much damage and theft of trusted users' sensitive data by unauthorized users. In the worst case, the attacker may gain full control over the web application and totally destroy or damage the system. This is successfully achieved, in general, via Structured Query Language (SQL) injection attacks on the online web application database. In this chapter, we review most of the well-known and new SQL injection attacks (SQLIAs), vulnerabilities, and prevention techniques. We present this topic in a way that the work could be beneficial for both the general readers and the researchers in the area for their future research works.

SQL injection is a type of injection or attack in a web application, in which the attacker provides SQL code to a user input box of a web form to gain unauthorized and unlimited access. The attacker's input is transmitted into an SQL query in such a way that it forms an SQL code [1,2]. In fact, SQL injection is categorized as the top-10 2010 web application vulnerabilities experienced by web applications according to the Open Web Application Security Project (OWASP) [3].

SQL injection vulnerabilities (SQLIVs) are one of the open doors for hackers to explore. Hence, they constitute a severe threat for web application contents. The key root and basis of SQLIVs is quite simple and well understood: insufficient validation of user input [1]. To mitigate these vulnerabilities, many prevention techniques have been suggested, such as manual approach, automated approach, secure coding practices, static analysis, and use of prepared statements. Although the proposed approaches have achieved their goals to some extent, SQLIVs in web applications remain a major concern among application developers.

Relating to the above-mentioned texts, the key objective of this work is to present a detailed survey on various types of SQLIVs, attacks, and their prevention techniques. Alongside presenting our findings from the study, we also note future expectations and possible development of countermeasures against SQLIAs. The key purpose of this study is to address the issue from all necessary angles so that the work could be used as a reference by researchers and practitioners.

Although there are some previous works on SQL injections, they have mainly the following limitations:

1. *Not up-to-date:* The growth of e-commerce is almost parallel to the alarming threats targeting web applications using SQL injections. Hence, the relevance and accuracy of some previous publications are now questionable. The more time passes by, the more kinds of attacks evolve and put less confidence on the previously noted information. Hence, we believe up-to-date information with rigorous analysis should be presented to the research community.
2. *Lack of practice:* In almost all the previous works, there is a critical lack of a discussion about the web application security training tutorials used in practice. Sometimes, there is a huge gap between theory and practice. Hence, in our work, we mention the tools that should be known for practical use and tackling SQLIAs. The information about these tools is missing in most, if not all, of the previous works we have analyzed.

6.2 Next-Generation Networks and Security

Although there is considerable research effort underway to define the boundary and standards of next-generation networks (NGNs), a proper boundary is yet to be finalized. NGNs are used to label the architectural evolutions in telecommunications and access networks. The term is also used to depict the shift to higher network speeds using broadband, migration from the public switched telephone network to an Internet Protocol (IP) network, and greater integration of services on a single network, and often is representative of a vision and a market concept. NGNs are also defined as "broadband managed IP networks" [4]. The IP address sometimes used as NGNs are built around the IP.

From a more technical point of view, NGNs are defined by the International Telecommunication Union as a "packet based network able to provide services including telecommunication services and able to make use of multiple broadband, QoS-enabled transport technologies and in which service related functions are independent from underlying transport-related technologies." NGNs offer access by users to different service providers and support "generalized mobility which will allow consistent and ubiquitous provision of services to users [5]."

6.2.1 Security Concerns

Looking at NGNs from a security perspective, it is quite challenging to state any claim. However, there is no doubt that, within an integrated service network, there will always be major security concerns. In new IP-based NGNs, there is a high chance of exposure to different types of threats and attacks from both within and external than ever before. Consumers, being increasingly dependent on information, are vulnerable to attacks. Voice-over-IP (VoIP) services can be a specific example of possible security issues in an NGN environment. In fact, access can be gained through any access point to the voice network (particularly if there are wireless access points in the same network that supports the VoIP service). Once access (via SQLIAs) has been gained, network sniffer tools are commonly available to intercept IP-based traffic [4]. Adding to that, NGNs are, at an early stage, without any solid background; hence, we could expect more threats. Loopholes could be easily exploited. Having that in mind, as a precaution, a proper awareness campaign toward security should be launched. Policies should be in place in case of any data breach. NGNs' firewall is currently being explored.

6.2.2 SQLIAs in Recent Years

The media heavily publicize data breaches mainly when victims are high-profile industries. Information about victims can be found at Zone-H (www.zone-h.org), BBC, and through some other web sites and channels. Zone-H reported a huge number of cases in the year 2010, where the SQLIA method was used. Here is a list of some of those cases witnessed in the last 3 years where SQL injection was used:

- *Sites hit in massive web attack:* On April 1, 2011, it was reported that "Hundreds of thousands of web sites appear to have been compromised by a massive cyber attack." The redirection of these web sites was carried out by SQLIA. Experts acknowledge that it was the most successful SQLIA ever seen.
- *Cleanup begins after massive web site attack:* A massive attack against hundreds or thousands of web sites was reported on April 4, 2011. SQL injection was one of the attack method used.

Sony PlayStation hacked: By using simple SQL injection, 77 million users' details were compromised during the third major hack that hit the Sony PlayStation Network. This was reported in the news in June 2011. A LulzSec press release said, "SonyPictures.com was owned by a very simple SQL injection, one of the most primitive and common vulnerabilities, as we should all know by now. From a single injection, we accessed *everything.* Why do you put such faith in a company that allows itself to become open to these simple attacks?"

- *Lulz hackers group using SQL injection:* This famous hacker group used SQL injection in quite a number of attacks.
- *Nokia's developer network hacked (29 Aug 2011):* Members' details were stolen via SQL injection. It said, "A database table containing developer forum members' email addresses has been accessed, by exploiting vulnerability in the bulletin board software that allowed an SQL injection attack."
- *Turkish net hijack hits big name web sites (5 Sept 2011):* Based on an interview, Turkguvenligi revealed that it got access to the files using a well-established attack method known as SQL injection.
- *Royal Navy web site attacked by Romanian hacker (8 Nov 2010):* The hacker gained access to the web site via a common attack method known as SQL injection.
- *U.S. man "stole 130m card numbers" (18 Aug 2009):* Mr. Gonzalez, the attacker, used a technique known as an "SQL injection attack" to access the databases and steal information, the U.S. Department of Justice (DoJ) said.
- SQLIV in RockYou led to the disclosure of 32 million passwords in plain text.
- Recently, hackers gained access to the database of "PlentyOfFish.com," an online dating site, exposing nearly 30 million users.
- In 2005, CardSystems had 263,000 credit card numbers stolen from its database. Over 40 million credit card numbers were exposed. This happened through SQLIAs. The company assets were finally acquired by another company.

6.3 SQL Injection and New Internet Technologies

With the introduction of a number of new Internet technologies, we must be careful about the possible use of SQLIA in future networks. In fact, it is an alarming situation that the types of attacks are becoming more clandestine and blurring, but, at the same time, hackers' targets are more focused [6]. In this section, we will discuss some of those recent Internet technologies in relation to SQL injection.

6.3.1 Pervasive Computing

This postdesktop model of computing stems from human–computer interaction. It is used as a label of the usage of computing technologies and devices in daily human life [7]. More formally, this technology is defined as "machines that fit the human environment instead of forcing humans to enter theirs" [4]. Processing information via a wide range of devices becomes part of an ordinary life. This technology is also known as ubiquitous computing (Ubicomp), physical computing, or ambient intelligence based on the emphasis point.

With many devices being blended together, there is a high chance that, if one gets compromised or injected, others will be vulnerable as well. It should be mentioned that, in this particular environment, where technical devices and usual human life are consolidated, there will always be security concerns and data breach. The close interaction between objects within this paradigm makes it very difficult to apply some security concepts and restriction approaches. Due to the default built-in trust and familiarity, abuse of trust will be of ease. SQLIAs, reverse engineering, and the like will be part of common life, and attack types do not have to be that complicated or well prepared to do the job of the hackers. This is simply because the environment is supposed to be vulnerable from different angles.

Reputation-based systems have been identified as a way to build trust among electronic devices. However, its proper implementation is always challenged. Therefore, security concerns always remain a dilemma. There is still an urgent need to further invest in research to find an adequate solution to the threat aiming private data within this human–computer interaction arena.

6.3.2 Cloud Computing

John McCarthy, the Turing Award winning computer scientist, stated in 1906 that, "Computation may someday be organized as a public utility." That was a forecast of the cloud computing concept. Cloud computing is a recent technology aiming to deliver computing as a service, that is, "computing service" rather than a particular product. Customers are interacting with resources in the "cloud" and pay for the information technology (IT) service based on the resource usage. It also refers to the delivery of scalable IT resources over the Internet, as opposed to hosting and operating those resources locally, such as on a college or university network. Many giant industries began to explore this technology mainly because of its benefits (low cost, flexibility, bigger storage, etc.). Analysts expect cloud computing to see mainstream adoption in 2 to 5 years [8].

On the other hand, there are quite a number of security concerns and risks associated with this fast growing technology. The fact that someone else is in control of your data, someone else is managing your application, and the perimeter of the cloud is different, is a real concern that is worth being considered. Cloud computing raises significant concerns about data privacy, authenticity, security, data integrity, intellectual property management, audit trails, etc. [8]. SQLIAs and similar attacks against the cloud database can make it vulnerable and lead to cyber criminals harvesting one's data on someone else's watch (service providers). According to Bloomberg news [9], the Sony PlayStation Network was hacked via Amazon's web services cloud. Figure 6.1 shows possible vulnerability in the cloud environment: in fact, if the application is vulnerable

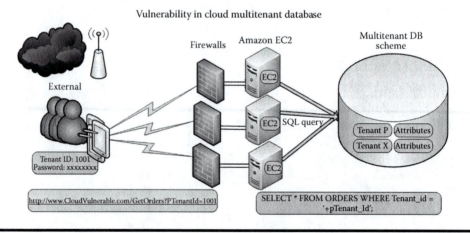

Figure 6.1 Vulnerability in cloud.

to SQL injection, then a particular tenant can easily access data belonging to another client through different means. In this scenario, we can simply identify a few possible data breaches and vulnerabilities:

■ Random manipulation of the ID: a tenant or hacker can simply try different IDs with this pattern to retrieve (select) the data of a different user from the database.
■ Malicious input codes at the input box from the user interface.

Most of the world's sensitive data are stored in database systems (Oracle, Microsoft SQL Server, IBM DB2, Sybase, etc.), making these storages an increasingly favorite target for criminals. This may explain why external attacks such as SQL injection increased 134% in 2008, jumping from an average of a few thousand per day to several hundred thousand per day, according to a report recently published by IBM [10].

6.3.3 Internet of Things

This terminology was first used by Kevin Ashton in 1999. It refers to the uniquely identifiable objects (or things) and their virtual representations in an Internet-like structure [7]. Pervasive computing (or Ubicomp) and Internet of things (IoT) considers the same environment. When viewed from a conceptual approach, this technology is noted as Ubicomp, but when the identifiable things are meant, then it is IoT. Similar concerns in terms of security are relevant in this environment as are those in pervasive computing.

6.3.4 What Do These Technologies Have in Common?

We are in an era of IT where almost everyone is connected online or accesses some kind of web technology. Concretely, we all have our data stored somewhere around the globe. Nearly every web application interacts with a database. Martin G. Nystrom, in his book *SQL Injection Defenses*, states, "In each place that the application accepts input from the user and does something with it, there's an opportunity for the attacker to supply malicious input." If this is the case, we are all vulnerable to SQLIAs, are we not?

After having all these discussions, let us thoroughly check what SQLIA is really about.

6.4 SQL Injection: The "Need-to-Know" Aspects

6.4.1 What Is SQL?

SQL (pronounced as "S-Q-L" or "sequel") stands for Structured Query Language [11]. It is the high-level language used in various relational database management systems (DBMSs) [11]. SQL was originally developed in the early 1970s by Edgar F. Codd at IBM. It was commercial and the most widely used language for all relational databases. This language is a declarative computer language that has elements that include clauses, expressions, predicates, queries, and statements. It allows the users to do mainly (1) data insertion, (2) data updating, (3) query, and (4) deletion, and many more features (thus gives the user the power of manipulating databases) [12,13].

6.4.2 SQLIV versus SQLIA

Vulnerability in any system is defined as a bug, loophole, weakness, or flaw existing in the system that can be exploited by an unauthorized user in order to gain unlimited access to the stored data. *Attack* generally means an illegal access, gained through well-crafted mechanisms, to an application or system. An SQLIA is a type of attack [14] where an attacker (a crafted user) adds malicious keywords or operators into an SQL query (e.g., SQL malicious code statements) and then injects it to a user input box of a web application. This allows the attacker to have illegal and unrestricted access to the data stored at the backend database. Figure 6.2 shows the normal user input process

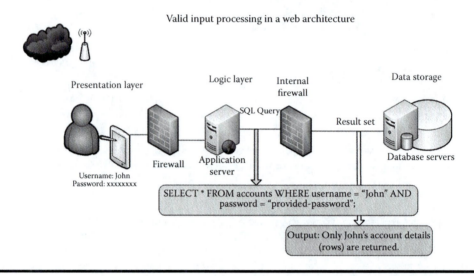

Figure 6.2 Normal user input process in a web application.

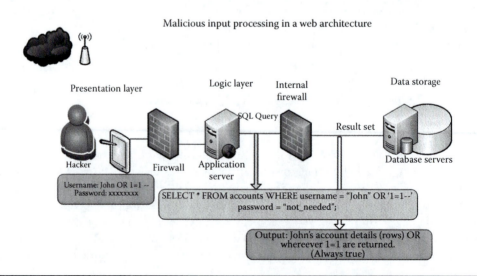

Figure 6.3 Malicious input process in a web application.

in a web application, which is self-explanatory. Figure 6.3 shows an example of how a malicious input could be processed in a web application. In this case, the malicious input is the carefully formulated SQL query that passes through the system's verification method. To explore this area more, in this chapter, we investigate both the SQLIVs and SQLIAs.

6.4.3 Why Is SQL Injection a Threat?

Injecting a web application is the synonym of having illegal access to the data stored in the database. The data sometimes could be confidential and of high value, like the financial secrets of a bank or list of financial transactions or secret information of some kind of information system. An unauthorized access to these data by a crafted user can pose a threat to their confidentiality, integrity, and authority. As a result, the system could bear a heavy loss in giving proper services to its users, or it may even face complete destruction. Sometimes such a type of collapse of a system can threaten the existence of a company, bank, or industry. If it happens against the information system of a hospital, the private information of the patients may be leaked out, which could threaten their reputation or may become a case of defamation. Attackers may even use such a type of attack to get confidential information that is related to the national security of a country. Hence, SQL injection could be very dangerous in many cases, depending on the platform where the attack is launched and where it gets success in injecting rogue users to the target system.

6.4.4 Types of Vulnerabilities in Web Programming Languages

There could be various types of vulnerabilities that could be exploited for SQL injection. In this section, we present the most common security vulnerabilities found in web programming languages [15] through which SQLIAs are usually launched. We show the major types of vulnerabilities at a glance in Table 6.1.

6.4.5 Types of SQLIAs: Past and Present

It is not an easy task to find out and categorize all types of SQLIAs. The same attack may be called with different names in different cases, depending on the system scenario. In this section,

Table 6.1 Types of Vulnerabilities at a Glance

Vulnerability Types	Basic Idea
Type I	Lack of clear distinction between data types accepted as input in the programming language used for the web application development.
Type II	Delay of operation analysis until the runtime phase where the current variables are considered rather than the source code expressions.
Type III	Weak concern of type specification in the design: a number can be used as a string, or vice versa.
Type IV	The validation of the user input is not well defined or sanitized. Inputs are not checked correctly.

Table 6.2 Types of SQLIAs at a Glance

Types of Attack	Working Method
Tautologies	SQL injection codes are injected into one or more conditional statements so that they are always evaluated to be true.
Logically incorrect queries	Using error messages rejected by the database to find useful data facilitating injection of the backend database.
Union query	Injected query is joined with a safe query using the keyword UNION in order to get information related to other tables from the application.
Stored procedure	Many databases have built-in stored procedures. The attacker executes these built-in functions using malicious SQL Injection codes.
Piggy-backed queries	Additional malicious queries are inserted into an original injected query.
Inference • Blind injection • Timing attacks	An attacker derives logical conclusions from the answer to a true/false question concerning the database. • Information is collected by inferring from the replies of the page after questioning the server true/false questions. • An attacker collects information by observing the response time (behavior) of the database.
Alternate encodings	It aims to avoid being identified by secure defensive coding and automated prevention mechanisms. It is usually combined with other attack techniques.

we present all the commonly known SQLIAs [1,16] that so far have been discovered along with newly invented innovative attacks. We use the terminologies as deemed to be appropriate. Table 6.2 shows the types of SQLIAs with brief descriptions.

6.5 An In-Depth Look at the Most Common SQLIAs

Among various types of SQLI attacks, some are frequently used by the attackers. It is imperative to know the commonly used major attacks among all available attacks. Hence, in this section, we present an in-depth look at some of the most common SQLIAs. We explain each of these major attacks with simple examples, wherever appropriate.

6.5.1 Tautology

SQL injection codes are injected into one or more conditional statements so that they are always evaluated to be true. Under this technique, we may have the following types and scenarios of attacks.

6.5.1.1 String SQL Injection

This type of injection is also referred to as an AND/OR attack [17,18]. The attacker inputs SQL tokens or strings to a conditional query statement that always evaluates to a true statement. The interesting issue with this type of attack is that, instead of returning only one row in a table, if it is successful, it causes all of the rows in the database table targeted by the query to be returned. The goal behind this type of attack may include the following: (1) bypassing authentication, (2) identifying parameters that can be injected, and (3) extraction of data [1].

Scenario

- Normal Statement: SELECT * FROM users WHERE name = 'Lucia01'
- Input: Lucia01 Output: Lucia's Rows only
- Injected Statement: SELECT * FROM users WHERE name = 'Lucia01' OR '1' = '1'

Input: 'Lucia01' OR '1' = '1'
Output: this will return rows for Lucia01 OR wherever one equals to one which is true for all rows. Hence, all rows will be returned.

6.5.1.2 Numeric SQL Injection

This type of injection is quasi-similar to that previously discussed. The main difference is that, here, numeric values are used instead of strings. Therefore, the attacker would input numeric values to a conditional query statement that would always evaluate to a true statement.

Scenario

- Normal Statement: SELECT * FROM users WHERE id = '101'
- Input: 101 Output: id '101's Rows only.
- Injected Statement: SELECT * FROM users WHERE name = '101' OR '1' = '1'.
- Input: '101' OR '1' = '1'
- Output: this will return rows for '101'id or wherever one equals to one (ALL ROWS)

Note: The crafted user can be more specific by adding ORDER BY clause to get exactly what he or she wants on time. The malicious input will look like: 101 OR 1 = 1 ORDER BY salary desc;

6.5.1.3 Comments Attack

This type of attack takes advantage of the inline commenting allowed by SQL [19]: the malicious code and comments whatever comes after the "—" in the WHERE clause. The point is that everything after the comment characters will be ignored. Comments attack can be combined with either string or numeric SQL injection so that it performs as a tautology, which always evaluates to a true statement.

Scenario

- User Input: 'user1 OR '1' = '1— '.
- Generated SQL Query: SELECT username, password FROM clients WHERE username = 'user1 OR '1' = '1— ' AND password = 'whatever'.

In this case, not only is the WHERE clause transformed into a tautology by the (OR 1 = 1) but the password part is also completely ignored; hence, only the username part will be checked [1,19].

6.5.2 Inference

An attacker derives logical conclusions from the answer to a true/false question concerning the database. Through a successful inference, crafted users change the behavior of the database.

6.5.2.1 Blind SQL Injection

In this type of attack, useful information for exploiting the backend database is collected by inferring from the replies of the page after questioning the server some true/false questions. It is very similar to a normal SQL injection [17,18]. However, when attackers attempt to exploit an application, rather than getting a useful error message, they get a generic page specified by the developer instead. This makes exploiting a potential SQLIA more difficult but not impossible. An attacker can still get access to sensitive data by asking a series of true and false questions through SQL statements.

Scenario

http://victim/listproducts.asp?cat = books
SELECT * from PRODUCTS WHERE category = 'books'
http://victim/listproducts.asp?cat = books' or '1' = '1.
SELECT * from PRODUCTS WHERE category = 'books' or '1' = '1'.

6.5.2.2 Timing Attacks

An attacker collects information by observing the response time (behavior) of the database. Here, the main concern is to observe the response time that will help the attacker to decide wisely on the appropriate injection approach.

6.5.2.3 Database Backdoors

Databases are used not only for data storage but also to keep malicious activity like a trigger. In this case, an attacker can set a trigger in order to get the user input and get it directed to his or her e-mail, for example.

Scenario

101; CREATE TRIGGER myBackDoor BEFORE INSERT ON employee FOR EACH ROW BEGIN
UPDATE employee SET email = 'hacker@me.com'WHERE userid = NEW.userid.

6.5.2.4 Command SQL Injection

The purpose of this injection is to inject and execute commands specified by the hacker in the vulnerable application. The application executing the unwanted system commands is like a pseudo-system shell controlled by the attacker. Lack of correct input data validation (forms, cookies, HTTP headers, etc.) is the main vulnerability exploited by attackers for a successful injection. It differs from code injection in the sense that the attacker adds his or her own code to the existing code. Hence, the default functionalities of the application are extended without executing system commands. An operating system (OS) command injection attack occurs when an attacker attempts to execute system level commands through a vulnerable application. Applications are considered vulnerable to OS command injection attack if they utilize user input in a system level command.

6.6 Web Application Security Training Tutorials Used

In this section, we discuss some existing web application security tutorials that we have used either online or offline for analyzing various mechanisms. These tutorials purposefully contain vulnerabilities for the user to discover and exploit.

OWASP is a 501c3 not-for-profit worldwide charitable organization focused on improving the security of application software [20]. Tutorials are written in Java. This tutorial covers the 10 most common web application vulnerabilities: (1) injection flaws, (2) cross-site scripting (XSS), (3) broken authentication and session management, (4) insecure direct object references, (5) cross-site request forgery (CSRF), (6) security misconfiguration, (7) insecure cryptographic storage, (8) failure to restrict URL access, (9) insufficient transport layer protection, and (10) invalidated redirects and forwards. In addition, they provide hints, prevention, solution, and Java options. Every year, they present the top ten web application vulnerabilities. The source code of the project and LiveCD are free of charge and accessible to almost every user. Although it provides comprehensive practice, the explanation of the topics is lacking and is left to the user to learn. It focuses more on the hands-on part rather than the teaching side. Because of being completely Java oriented, it is not concerned about applications built using other languages such as PHP or RoR (Ruby on Rails). Figure 6.4 shows an OWASP environment.

Figure 6.4   OWASP environment/interface.

Damn Vulnerable Web Application (*DVWA*) [21] is another practice tool built using PHP/ MySQL. It is an aid for security professionals and Web developers to test and try out their skills and tools in a legal practice environment. Besides that, it is a handy approach to train/teach users (i.e., students, teachers, researchers, and security professionals) on secure web development. The source code and LiveCD are made available for free. This tutorial covers the following topics: brute force, command execution, CSRF, file inclusion, SQL injection (blind), upload, XSS reflected, and XSS stored. Compared to OWASP, it is less comprehensive and covers only a few topics. In this tutorial, there is a lack of adequate information not only of direct topic-related discussions but also of guidelines, hints, and solutions. Users can only find information on topics through some provided Internet links/sources. Figure 6.5 shows the DVWA environment.

The *Web Security Dojo* is a "free open-source self-contained training environment for Web Application Security penetration testing. Tools + Targets = Dojo" [22]. The VmWare image is provided for free. Users can download and install it in a virtual machine at their own pace with full documentation. Some default targets provided are DVWA, REST Demos, and JSON demos. They also provide WebGoat, Hackme Casino vulnerable application, Insecure Web App, and some tools like burp suite. It is a very efficient environment for practice; however, it seems to be a bit of an advanced tool for a security-practice beginner. The environment of Web Security Dojo is shown in Figure 6.6.

Daffodil is also an open-source web application project designed for learning purposes [23]. It is similar to the OWASP and DVWA. It contains both exercises and solutions for the selected web application vulnerabilities. This tutorial also lacks proper topic discussions. The user has to look for other sources to find out more information on selected practices. It should be made more user-friendly so that beginners (i.e., the naive practitioners) can work without much hassle of finding information from here and there. Figure 6.7 shows a snapshot of Daffodil lessons.

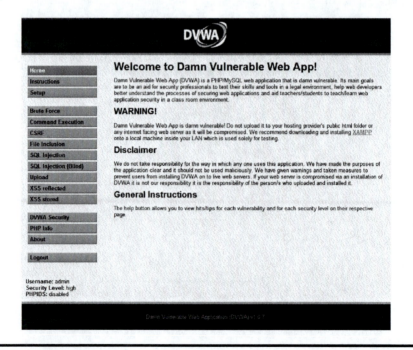

Figure 6.5　DVWA environment.

Figure 6.6 Web security dojo environment.

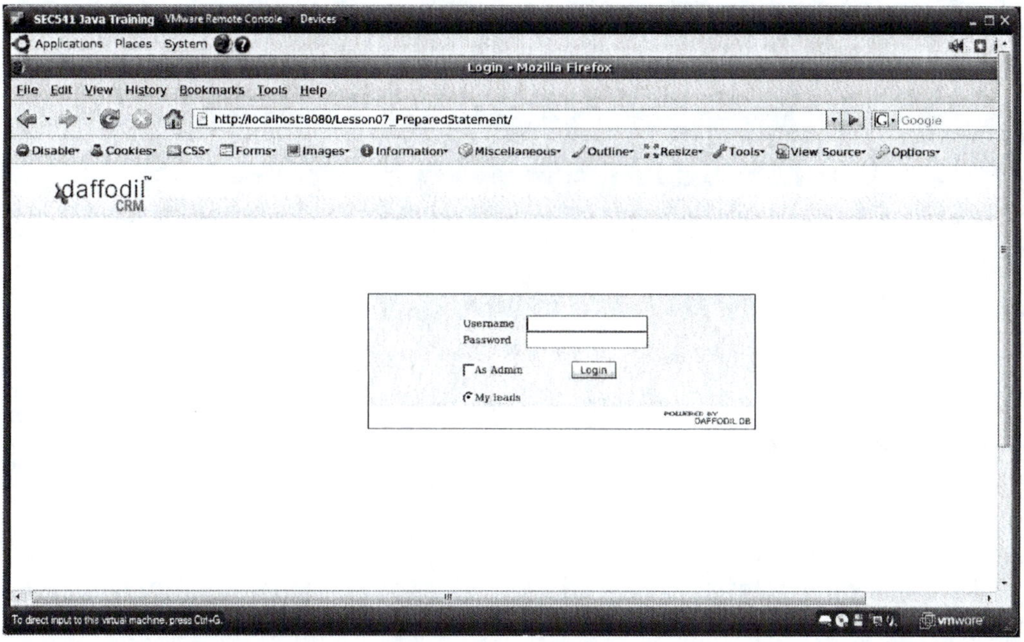

Figure 6.7 Daffodil lessons.

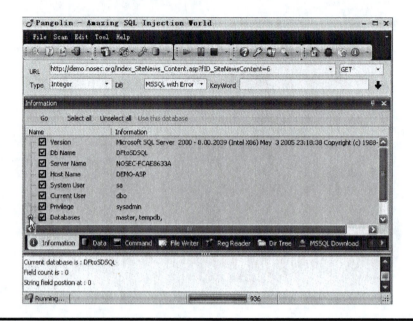

Figure 6.8   Pangolin screenshot.

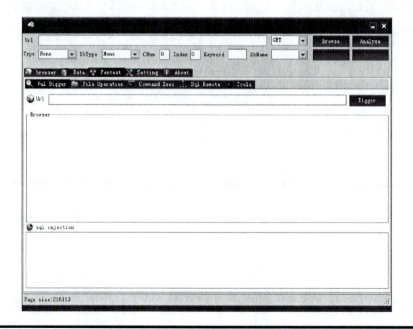

Figure 6.9   Safe3SI screenshot.

We have discussed all these training tutorials here for better understanding of the topic by the readers. In fact, for our work, we have used all these tutorials side by side with our literature survey, problem definitions, possible solutions, analysis, and comparisons of various approaches.

Pangolin (Automated SQL Injection Test Tool) is a penetration testing tool on database security developed by NOSEC. Its main aim is to detect and exploit existing SQLIVs on web applications. Once one or more vulnerabilities are found, the pen tester has the choice to perform an extensive backend database management system fingerprint; retrieve DBMS session user and database; enumerate users, password hashes, privileges, and databases; dump the entire or a user's specific DBMS tables/columns; run his or her own SQL statement; read specific files on the file system; and more [24]. Figure 6.8 shows a screenshot of Pangolin's working interface.

aidSQL is a PHP application used for testing vulnerabilities in web applications. It is a modular application, meaning that someone can develop his or her very own plugins for SQL injection detection and exploitation [25].

Safe3SI is a very efficient penetration testing tool that automates the vulnerability detection and exploitation process as well (Figure 6.9). It also allows taking over of database servers. This tool is equipped with a kick-ass detection engine, many additional niche features for the ultimate penetration tester, and a broad range of switches ranging from database fingerprinting over data fetching from the database to accessing the underlying file system and executing commands on the OS via out-of-band connections [26].

6.7 Detecting SQL Injection

In order to protect a web application from SQLIAs, there are two major concerns. First, there is a great need of a mechanism to detect and exactly identify SQLIAs. Second, knowledge of SQLIVs is a must for securing a web application. So far, many frameworks have been used and/or suggested to detect SQLIVs in web applications. Here, we mention the prominent solutions and their working methods in brief to let the readers know about the core ideas behind each work.

6.7.1 SAFELI

Fu et al. [27] proposed a static analysis framework in order to detect SQLIVs. The SAFELI framework aims at identifying the SQLIAs during the compile time. This static analysis tool has two main advantages. First, it does a white-box static analysis, and second, it uses a hybrid-constraint solver. For the white-box static analysis, the proposed approach considers the byte code and deals mainly with strings. For the hybrid-constraint solver, the method implements an efficient string analysis tool that is able to deal with Boolean, integer, and string variables. The implementation of this framework was done on ASP.NET web applications, and it was able to detect vulnerabilities that were ignored by the black-box vulnerability scanners. The methodology is an efficient approximation mechanism to deal with string constraints. However, the approach is only dedicated to ASP.NET vulnerabilities.

6.7.2 Thomas et al.'s Scheme

Thomas et al. [28] suggested an automated prepared statement generation algorithm to remove SQLIVs. They implemented their research work using four open-source projects: (1) Net-trust,

(2) ITrust, (3) WebGoat, and (4) Roller. Based on the experimental results, their prepared statement code was able to successfully replace 94% of the SQLIVs in four open-source projects. However, the experiment was conducted using only Java with a limited number of projects. Hence, the wide application of the same approach and tool for different settings still remains an open research issue to investigate.

6.7.3 Ruse et al.'s Approach

In [29], Ruse et al. proposed a technique that uses automatic test case generation to detect SQLIVs. The main idea behind this framework is based on creating a specific model that deals with SQL queries automatically. In addition, the approach identifies the relationship (dependency) between subqueries. Based on the results, the methodology is shown to be able to specifically identify the causal set and obtain 85% and 69% reduction, respectively, while experimenting on a few samples. Moreover, it does not produce any false positive or false negative, and it is able to detect the real cause of the injection. In spite of the claimed and apparent efficiency of the technique, the major drawback of the work is that it was not tested with real queries on a real-life existing database.

6.7.4 Haixia and Zhihong's Database Security Testing Scheme

In [13], Haixia and Zhihong proposed secure database testing design for web applications. They suggested a few things: first, detection of potential input points of SQL injection; second, generation of test cases automatically; and then finally, finding the database vulnerability by running the test cases to make a simulation attack to an application. The proposed methodology is shown to be efficient as it was able to detect the input points of SQL injection exactly and on time, as the authors expected. However, after analyzing the scheme, we find that the approach is not a complete solution, but rather, it needs additional improvements in two main aspects: the detection capability and the development of the attack rule library.

6.7.5 Roichman and Gudes's Fine-Grained Access Control Scheme

In [30], Roichman and Gudes, in order to secure web application databases, suggested using fine-grained access control to web databases. They developed a new method based on the fine-grained access control mechanism. Access to the database is supervised and monitored by built-in database access control. This approach is efficient in the fact that the security and access control of the database is transferred from the application layer to the database layer. This is a solution of the vulnerability of the SQL session traceability. Besides that, it is a framework that is applicable to almost all database applications. Therefore, it significantly decreases the risk of attacks at the backend of the database application.

6.7.6 Shin et al.'s Approach

In [31], Shin et al. suggested SQLUnitGen, which is a static-analysis-based tool that automates testing for identifying input manipulation vulnerabilities. They applied the SQLUnitGen tool, which is compared with FindBugs, a static analysis tool. The proposed mechanism is shown to be efficient (483 attack test cases) as regard to the fact that false positive was completely absent in the experiments. However, for different scenarios, false negatives at a small number were noticed. In addition to that, it was found that, due to some shortcomings, a more significant rate of false negatives may occur "for other applications." Hence, the authors talk about concentrating on getting rid of those significant false negatives and further improvement of the approach to cover input manipulation vulnerabilities as their future works.

6.7.7 SQL-IDS Approach

Kemalis and Tzouramanis [32] suggested using a novel specification-based methodology for the detection of exploitations of SQLIVs. The proposed query-specific detection allowed the system to perform focused analysis at negligible computational overhead without producing false positives or false negatives. This new approach is very efficient in practice; however, it requires more experiments and comparison with available detection methods under a shared and flexible benchmarking environment.

6.8 SQL Injection Countermeasures: Detection and Prevention Techniques

In the previous section, we have discussed various schemes that only deal with SQL injection detection. After having successfully detected any vulnerability or any kind of attack that exploits the vulnerability, other schemes could be applied to cure the system. In the usual case, there are mainly two types of schemes: some are for prevention, and others are for curing the system once it is under attack. In case of SQL injection, those schemes that work for preventing SQL injection also do the curing of the system (or application) in the early stage. Hence, in plain terms, we could call the schemes *countermeasures*. A strong countermeasure can remove or at least block all the available vulnerabilities in a system, and thus, it could protect it against various types of attacks that take advantage of the vulnerabilities. Once a system is under attack, the curing mechanisms include some other techniques like resetting the system, reorganizing the various elements in the system, etc., which are not the topics of our current study. As those mechanisms mainly deal with other aspects of network setting, database reshuffling, reorganizing, and utilizing the clean-slate approach of reinstalling the system (or application), the curing schemes are irrelevant for our survey. After our analysis of the available steps and guidelines (after-attack scenario), we found that they are more related to the managerial and administrative policies set for the system (or application) once the attacks are launched against it, and it suffers from damage.

In this section, we list a number of countermeasures that could be employed before and during running the system. It should be noted that these schemes not only detect SQL injection but also take necessary measures so that the vulnerabilities are not exploited by the rogue entities. Thus, these schemes defer from the schemes mentioned in the previous section in the point that they do more than just detection of SQL injection. Here, we also present brief descriptions and analyze each scheme from the critical *need-to-know* angles. Table 6.3 shows a summary of so far known countermeasures against SQL injection.

Now, let us see what these schemes are actually about. The remaining texts in this section will analyze the various aspects covered in the different types of countermeasures.

Table 6.3 SQL Injection Countermeasures

Countermeasure	Overview
SQL-IDS [32]	A specification-based approach to detect malicious intrusions.
Prepared statements [28]	It is a fixed query "template," which is predefined, providing type-specific placeholders for input data.
AMNESIA [12]	This scheme identifies illegal queries before their execution. Dynamically generated queries are compared with the statically built model using runtime monitoring.
SQLrand [33]	A strong random integer is inserted in the SQL keywords.
SQL DOM [17]	A set of classes that are strongly typed to a database schema are used to generate SQL statements instead of string manipulation.
SQLIA prevention using stored procedures [18,34]	Combination between static analysis and runtime monitoring.
SQLGuard [35]	The parse trees of the SQL statement before and after user input are compared at a runtime. The web script has to be modified.
CANDID [36]	Programmer-intended query structures are guessed based on evaluation runs over nonattacking candidate inputs.
SQLIPA [37]	Using user name and password hash values to improve the security of the authentication process.
SQLCHECK [38]	A key is inserted at both the beginning and end of the user's input. Invalid syntactic forms are the attacks. The key strength is a major issue.
DIWeDa [39]	To detect various types of intrusions in web database applications.
Manual approaches [12]	Defensive programming and code review mechanisms are applied.
Automated approaches [12]	Static analysis FindBugs and web vulnerability scanning frameworks are implemented.

6.8.1 AMNESIA

In [12], Junjin proposed the AMNESIA approach for tracing SQL input flow and generating attack input, JCrasher for generating test cases, and SQLInjectionGen for identifying hotspots. The experiment was conducted on two web applications running on MySQL1 1 v5.0.21. Based on three attempts on the two databases, SQLInjectionGen was found to give only two false negatives in one attempt. The proposed framework is efficient, considering the fact that it emphasizes on attack input precision. Besides that, the attack input is properly matched with method arguments. Better than all the previous advantages, the proposed approach has no false positives and counts a small number of false negatives. The only disadvantage of this approach is that it involves a number of steps using different tools.

6.8.2 SQLrand Scheme

In [33], the SQLrand approach (approach using randomized SQL query language, targeting a particular *Common Gateway Interface* application) was proposed by Boyd and Keromytis. For the implementation, they used a proof-of-concept proxy server in between the web server (client) and the SQL server; they derandomized queries received from the client and sent the request to the server. This derandomization framework has two main advantages, that is, portability (applied with wide range of DBMS) and security (database content highly protected). The proposed scheme has a good performance: 6.5 ms is the maximum latency overhead imposed on every query. Hence, it is efficient considering the performance obtained and defense against injected queries. However, this is a proof of concept; it still requires further testing and support from programmers in building tools using SQLrand targeting more DBMS backends.

6.8.3 SQL DOM Scheme

The SQL DOM (a set of classes that are strongly typed to a database schema) framework was suggested by McClure and Krüger [17]. They closely considered the existing flaws while accessing relational databases from the object-oriented programming language's point of view. They mainly focus on identifying the obstacles in the interaction with the database via call level interfaces. The SQL DOM object model is the proposed solution to tackle these issues through building a secure environment (i.e., creation of SQL statement through object manipulation) for communication. The qualitative evaluation of this approach has shown many advantages and benefits in terms of error detection during compile time, reliability, testability, and maintainability. Although this mechanism is efficient, it can be further improved with the more advanced and latest tool such as CodeSmith [40].

6.8.4 SQLIA Prevention Using Stored Procedures

Stored procedures are subroutines in the database to which the applications can make calls [18]. The prevention in these stored procedures is implemented by a combination of static analysis and runtime analysis. The static analysis used for command identification is achieved through a stored procedure parser and the runtime analysis by using an SQLChecker for input identification. Huang et al. proposed in [34] a combination of static analysis and runtime monitoring to fortify the security of potential vulnerabilities. Web application Security by Static Analysis and Runtime Inspection (WebSSARI) was used and implemented on 230 open-source applications on SourceForge.net. The approach was effective; however, it failed to remove the SQLIVs. It was only able to list the input in either white or black.

6.8.5 Parse Tree Validation Approach

Buehrer et al. [35] adopted the parse tree framework. They compared the parse tree of a particular statement at runtime and its original statement. They stopped the execution of statement unless there is a match. This method was tested on a student web application using SQLGuard. Although this approach is efficient, it has two major drawbacks, that is, additional overheard computation and listing of input only (black or white).

6.8.6 Dynamic Candidate Evaluation Approach

In [36], Bisht et al. proposed Candidate Evaluation for Discovering Intent Dynamically (CANDID). It is a dynamic candidate evaluation method for automatic prevention of SQLIAs. This framework dynamically extracts the query structures from every SQL query location, which are intended by the developer (programmer). Hence, it solves the issue of manually modifying the application to create the prepared statements. Although this tool is shown to be efficient for some cases, it fails in many other cases. For example, it is inefficient when dealing with external functions and when applied at a wrong level. Besides that, sometimes it also fails due to the limited capability of the scheme.

6.8.7 Ali et al.'s Scheme

Ali et al. [37] adopted the hash value approach to further improve the user authentication mechanism. They used the user name and password hash values. The SQL Injection Protector for Authentication (SQLIPA) prototype was developed to test the framework. The username and password hash values are created and calculated at runtime the first time a particular user account is created. Hash values are stored in the user account table. Although the proposed framework was tested on a few sample data and had an overhead of 1.3 ms, it requires further improvement to reduce the overhead time. It also requires to be tested with a larger amount of data.

6.8.8 SQLCHECKER Approach

Su and Wassermann [38] implemented their algorithm with SQLCHECK on a real-time environment. It checks whether the input queries conform to the expected ones defined by the programmer. A secret key is applied for the user input delimitation [1]. The analysis of SQLCHECK shows no false positives or false negatives. Also, the overhead runtime rate is very low and can be implemented directly in many other web applications using different languages. It is a very efficient approach; however, once an attacker discovers the key, it becomes vulnerable. Furthermore, it also needs to be tested with online web applications.

6.8.9 DIWeDa Approach

Roichman and Gudes [39] proposed intrusion detection systems for the backend databases. They used Detecting Intrusions in Web Databases (DIWeDa), which is a prototype that acts at the session level rather than the SQL statement or transaction stage, to detect intrusions in web applications. DIWeDa profiles the normal behavior of different roles in terms of the set of SQL queries issued in a session and then compares a session with the profile to identify intrusions [39]. The proposed framework is efficient and could identify SQL injections and business logic violations as well. However, with a threshold of 0.07, the true positive rate (TPR) was found to be 92.5%, and the false positive rate (FPR) was 5%. Hence, there is a great need of accuracy improvement (increase in TPR and decrease in FPR). It also needs to be tested against new types of web attacks.

6.8.10 Manual Approaches

Junjin [12] highlighted the use of manual approaches to prevent SQLI input manipulation flaws. In manual approaches, defensive programming and code review are applied. In defensive programming, an input filter is implemented to disallow users to input malicious keywords or characters. This is achieved by using white lists or black lists. As regards to the code review [41], it is a low-cost mechanism in detecting bugs; however, it requires deep knowledge on SQLIAs.

6.8.11 Automated Approaches

Besides using manual approaches, Junjin [12] also highlighted the use of automated approaches. The author noted that the two main schemes are Static analysis FindBugs and web vulnerability scanning. The Static analysis FindBugs approach detects bugs on SQLIAs and gives warning when an SQL query is made of a variable. However, for web vulnerability scanning, it uses software agents to crawl, scans web applications, and detects the vulnerabilities by observing their behavior to the attacks.

6.9 Comparative Analysis

It would be difficult to give a clear verdict as to which scheme or approach is the best as each one has some proven benefits for specific types of settings (i.e., systems). Hence, in this section, we note down how various schemes work against the identified SQLIAs. Table 6.4 shows a chart of the schemes and their defense capabilities against various SQLIAs. This table shows the comparative analysis of the SQL injection prevention techniques and the attack types. Although many approaches have been identified as detection or prevention techniques, only a few of them were implemented in practicality. Hence, this comparison is not based on empirical experience, but rather, it is an analytical evaluation.

In Table 6.5, we note the major approaches to deal with SQL injection and classify them based on their features.

6.10 Concluding Remarks

Although many approaches and frameworks have been identified and implemented in many interactive web applications, security still remains a major issue. SQL injection prevails as one of the top ten vulnerabilities and threats to online businesses targeting the backend databases. In this chapter, we have reviewed the most popular existing SQL injection-related issues. We have investigated the possibilities of SQLIAs for future networking technologies. To build NGNs, alongside many other issues, this particular area must also be given proper emphasis. Otherwise, a new system may come, but an old problem could still remain as a significant or even more complex threat.

Key findings of this study could be summarized as follows:

- Detailed survey report on various types of SQLIAs, vulnerabilities, detection, and prevention techniques
- Assessment of techniques based on their performance and practicality

Table 6.4 Various Schemes and SQLIAs

Schemes	Tautology	Logically Incorrect Queries	Union Query	Stored Procedure	Piggy-Backed Queries	Inference	Alternate Encodings
AMNESIA [12]	✓	✓	✓	×	✓	✓	✓
SQLrand [33]	✓	×	✓	×	✓	✓	×
SQLDOM [17]	✓	✓	✓	×	✓	✓	✓
WebSSARI [18,34]	✓	✓	✓	✓	✓	✓	✓
SQLGuard [35]	✓	✓	✓	×	✓	✓	✓
CANDID [36]	✓	×	×	×	×	×	×
SQLIPA [37]	✓	×	×	×	×	×	×
SQLCHECK [38]	✓	✓	✓	×	✓	✓	✓
DIWeDa [39]	×	×	×	×	×	✓	×
Automated approaches [12]	✓	✓	✓	×	✓	✓	×

Table 6.5 Various Approaches and Types of Tasks

Approaches	Goals	
	Detection	Prevention
SQL-IDS [32]	Yes	Yes
AMNESIA [12]	Yes	Yes
SQLrand [33]	Yes	Yes
SQL DOM [17]	Yes	Yes
WebSSARI [18,34]	Yes	Yes
SQLGuard [35]	Yes	No
CANDID [36]	Yes	No
SQLIPA [37]	Yes	No
SQLCHECK [38]	Yes	No
DIWeDa [39]	Yes	No

■ Awareness information of the threat of SQL injections by providing recent and updated cases and information
■ Exploration of "web application security training tutorials" to train security practitioners to deal with SQLIAs

The findings of this study could be used for penetration testing purposes to protect data either in academic or industrial fields. Our research outcomes help

■ To measure the security level of web applications using proposed tools
■ To find/detect vulnerabilities of online applications
■ To protect applications against using the proposed secure coding approaches
■ To train security practitioners on SQL injection using the proposed tutorials
■ To investigate the threat of SQLIA for future next-generation and emerging networks and concepts

We believe that the work would be useful both for the general readers of the topic as well as for the practitioners. The reality is that hackers are very innovative, and as time passes by, new attacks are being launched that may need new solutions.

References

1. Halfond, W.G., Viegas, J., and Orso, A., A Classification of SQL-Injection Attacks and Countermeasures. In *Proc. of the Intl. Symposium on Secure Software Engineering*, Mar. (2006).
2. Tajpour, A., Masrom, M., Heydari, M.Z., and Ibrahim, S., SQL injection detection and prevention tools assessment. *Proc. 3rd IEEE International Conference on Computer Science and Information Technology (ICCSIT'10)*, 9–11 July (2010), 518–522.

3. http://www.owasp.org/index.php/Top_10_2010-A1-Injection, retrieved on 13/01/2010.
4. http://en.wikipedia.org/wiki/Ubiquitous_computing#cite_note-0.
5. ITU-T Recommendation Y.2001, approved in December 2004, available at http://www.itu.int/rec/T-RECY. 2001-200412-I/en.
6. The Secure Online Business Handbook: a practical guide to risk management and business continuity (4th Edition), by Jonathan Reuvid, Kogan Page, United Kingdom, 2006.
7. Pathan, A.-S.K., "On the Boundaries of Trust and Security in Computing and Communications Systems" International Journal of Trust Management in Comp. and Commun., Vol. x, No. x, (To be published).
8. Seven things you should know about Cloud Computing, Educause 2009. Available at: http://www.educause.edu/Resources/7ThingsYouShouldKnowAboutCloud/176856.
9. Playstation hack came from Amazon EC2, by Carl, Available at: http://www.kitguru.net/channel/carl/playstation-hack-came-from-amazon-ec2/.
10. IBM Global Technology Services, "IBM Internet Security Systems X-Force® 2008 Trend and Risk Report," January 2009.
11. Foundations of security: what every programmer needs to know, by Neil Daswani, Christoph Kern, and Anita Kesavan, Apress, NY, USA, 2007.
12. Junjin, M., An Approach for SQL Injection Vulnerability Detection. *Proc. of the 6th International Conference on Information Technology:* New Generations, Las Vegas, Nevada, April (2009), 1411–1414.
13. Haixia, Y. and Zhihong, N., A database security testing scheme of web application. *Proc. of 4th International Conference on Computer Science and Education 2009* (ICCSE '09), 25–28 July (2009), 953–955.
14. Kindy, D.A. and Pathan, A.-S.K., A Survey on SQL Injection: Vulnerabilities, Attacks, and Prevention Techniques. (Poster) *Proceedings of The 15th IEEE Symposium on Consumer Electronics (IEEE ISCE 2011)*, June 14–17, Singapore (2011), 468–471.
15. Seixas, N., Fonseca, J., Vieira, M., and Madeira, H., Looking at Web Security Vulnerabilities from the Programming Language Perspective: A Field Study. *Proc. of 20th International Symposium on Software Reliability Engineering 2009 (ISSRE '09)*, 16–19 Nov. (2009) 129–135.
16. Tajpour, A., JorJor Zade Shooshtari, M., Evaluation of SQL Injection Detection and Prevention Techniques. *Proc. of 2010 Second International Conference on Computational Intelligence, Communication Systems and Networks (CICSyN'10)*, 28–30 July (2010), 216–221.
17. McClure, R.A. and Krüger, I.H., SQL DOM: Compile time checking of dynamic SQL statements. *27th International Conference on Software Engineering (ICSE 2005)*, 15–21 May (2005), 88–96.
18. Amirtahmasebi, K., Jalalinia, S.R., and Khadem, S., A survey of SQL injection defense mechanisms. *International Conference for Internet Technology and Secured Transactions (ICITST 2009)*, 9–12 Nov. (2009), 1–8.
19. Luong, V., Intrusion Detection And Prevention System: SQL-Injection Attacks. Master's Projects. Paper 16. (2010), available at: http://scholarworks.sjsu.edu/etd_projects/16.
20. http://www.owasp.org/index.php/Main_Page, retrieved on 31/01/2011.
21. http://www.dvwa.co.uk, retrieved on 31/01/2011.
22. http://www.mavensecurity.com/web_security_dojo/, retrieved on 31/01/2011.
23. http://crm.daffodilsw.com/article/call-centre-crm.html, retrieved on 04/06/2011.
24. http://www.nosec.org/en/productservice/pangolin.
25. http://code.google.com/p/aidsql.
26. http://code.google.com/p/safe3si.
27. Fu, X., Lu, X., Peltsverger, B., Chen, S., Qian, K., and Tao, L., A Static Analysis Framework for Detecting SQL Injection Vulnerabilities. *Proc. 31st Annual International Computer Software and Applications Conference 2007* (COMPSAC 2007), 24–27 July (2007), 87–96.
28. Thomas, S., Williams, L., and Xie, T., On automated prepared statement generation to remove SQL injection vulnerabilities. *Information and Software Technology*, Volume 51 Issue 3, March (2009), 589–598.

29. Ruse, M., Sarkar, T., and Basu. S., Analysis & Detection of SQL Injection Vulnerabilities via Automatic Test Case Generation of Programs. *Proc. 10th Annual International Symposium on Applications and the Internet* (2010), 31–37.
30. Roichman, A., Gudes, E., Fine-grained Access Control to Web Databases. *Proceedings of 12th SACMAT Symposium*, France (2007).
31. Shin, Y., Williams, L., and Xie, T., SQLUnitGen: Test Case Generation for SQL Injection Detection. North Carolina State University, Raleigh Technical report, NCSU CSC TR 2006-21 (2006).
32. Kemalis, K. and T. Tzouramanis. SQL-IDS: A Specification-based Approach for SQLinjection Detection. SAC'08. Fortaleza, Ceará, Brazil, ACM (2008), 2153–2158.
33. Boyd S.W. and Keromytis, A.D., SQLrand: Preventing SQL Injection Attacks. *Proceedings of the 2nd Applied Cryptography and Network Security (ACNS'04) Conference*, June (2004), 292–302.
34. Huang, Y.-W., Yu, F., Hang, C., Tsai, C.-H., Lee, D.-T., and Kuo, S.-Y., Securing Web Application Code by Static Analysis and Runtime Protection. *Proc. of 13th International Conference on World Wide Web*, New York, (2004), 40–52.
35. Buehrer, G., Weide, B.W., and Sivilotti, P.A.G., Using Parse Tree Validation to Prevent SQL Injection Attacks. *Proc. of 5th International Workshop on Software Engineering and Middleware*, Lisbon, Portugal (2005), 106–113.
36. Bisht, P., Madhusudan, P., and Venkatakrishnan, V.N., CANDID: Dynamic Candidate Evaluations for Automatic Prevention of SQL Injection Attacks. *ACM Transactions on Information and System Security*, Volume 13 Issue 2, (2010), doi.10.1145/1698750.1698754.
37. Ali, S., Shahzad, S.K., and Javed, H., SQLIPA: An Authentication Mechanism Against SQL Injection. *European Journal of Scientific Research*, Vol. 38, No. 4 (2009), 604–611.
38. Su, Z. and Wassermann, G., The essence of command injection attacks in web applications. In *ACM Symposium on Principles of Programming Languages (POPL'2006)*, January (2006).
39. Roichman, A., and Gudes, E., DIWeDa–Detecting Intrusions in Web Databases. Atluri, V. (ed.) DAS–2008. LNCS, vol. 5094, Springer, Heidelberg (2008), 313–329.
40. http://www.codesmithtools.com.
41. Baker, R.A., Code Reviews Enhance Software Quality. In *Proceedings of the 19th international conference on Software engineering (ICSE'97)*, Boston, MA, USA (1997), 570–571.

Chapter 7

Wireless Network Security: An Overview

Danda B. Rawat, Gongjun Yan, Bhed Bahadur Bista, and Vigyan "Vigs" Chandra

Contents

7.1 Introduction

Wireless communications is the fastest growing segment of the communication industry. Wireless technologies and applications have been widely deployed in various areas. Successful deployment of wireless local area networks (WLANs) in the unlicensed industrial, scientific, and medical (ISM) band and cellular wireless telephone networks in the licensed band during the past decades have shown the widespread use of wireless technologies and applications. Numerous wireless applications and technologies are under development and deployment. Wireless networks consist of various types of devices that communicate without a wired medium. Generally, wireless networks can be categorized into two different types based on the structure of the networks: infrastructure-based wireless networks and infrastructureless wireless networks [1].

An infrastructure-based wireless network has a central unit through which client stations communicate with each other. Cellular telephone systems such as Global System for Mobile Communications (GSM) or code-division multiple access (CDMA) and the IEEE 802.11 WLAN in access point (AP) mode and the IEEE 802.16 Worldwide Interoperability for Microwave Access (WiMAX) are some examples of infrastructure-based wireless networks. GSM, CDMA, and their variants are the most widely deployed cellular communication technologies that made mobile communications possible. GSM and CDMA use base stations through which mobile phones communicate with each other. Generally, cellular wireless networks cover a wide area and are known as wireless wide-area networks. Similarly, the WiMAX network also has centralized base stations used by wireless clients when they communicate with each other. The coverage area of WiMAX is closer to metropolitan areas and is known as a wireless metropolitan area network (WMAN). WLANs in infrastructure mode use centralized wireless APs through which wireless client stations communicate with each other. As the centralized base stations or APs in infrastructure-based wireless networks are mostly static and costly, such networks require serious and careful topology design for better performance and coverage.

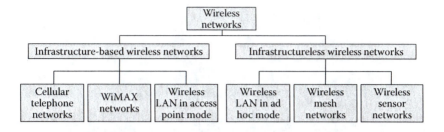

Figure 7.1   Classification of wireless networks.

An infrastructureless wireless network does not contain any centralized infrastructure, and thus, wireless client stations communicate with each other directly in a peer-to-peer manner. These types of networks are also known as wireless ad hoc networks. The network topology of the wireless ad hoc network is dynamic and changes constantly, and the participating wireless stations adapt to changes in topology on the fly [2,3].

Subcategories of wireless networks under centralized infrastructure-based and infrastructureless wireless networks are depicted in Figure 7.1. Cellular networks are for voice communications but also carry data. WiMAX, on the other hand, is for last-mile Internet delivery for a larger coverage area. WLANs are for data communication within smaller areas, typically for office and residential use. However, voice-over-Wi-Fi is also part of WLANs. Recent advancements have shown that infrastructure-based wireless networks support both voice and data communications.

Infrastructure-based wireless networks need fixed infrastructures such as base stations in cellular telephone networks and WiMAX networks, or WAPs in WLANs to facilitate communications among mobile users. The stationary equipment serves as a backbone for these kinds of wireless networks. Mobile users connect to this equipment through wireless links and can move anywhere within a coverage area of a base station. They can also move from one base station's coverage area to another by using handover features. For example, a cellular telephone system consists of a fixed base station for a region called a cell [1], and each cell can handle a number of mobile users. While communicating, mobile users can move within a coverage area of a base station and from one base station to another by using roaming features. To cover a large area and a large number of users, multiple base stations are needed, and base stations are connected with each other by a reliable wired or wireless link to provide seamless wireless service. Interconnecting links should be robust in terms of reliability, efficiency, fault tolerance, transmission range, etc., to provide uninterrupted service [4,7].

7.2 Cellular Telephone Networks

Cellular communication has become an important part of our daily life. At present, almost 2.3 billion users have subscribed for telephone services. It is predicted by Gartner that, by 2013, mobile devices such as personal digital assistants will surpass personal computers for Internet browsing as cellular telephone networks offer mobile communications. Cellular telephone communications uses a base station to cover a certain area known as *cell* [1]. Mobile users connect to their base station to communicate with each other. They can move within a cell during communications and can move from one cell to another using handover technique without breaking communications. Wireless systems are prone to interference from other users who share the same frequency for communications. To avoid interference between cells, adjacent cells use different frequencies, as shown in Figure 7.2.

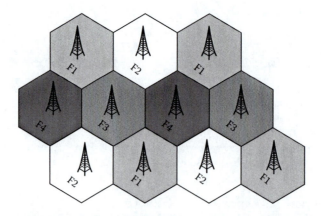

Figure 7.2 Cells with different frequencies in cellular telephone networks.

Cellular networks have been commercially available since the early 1980s. Japan implemented cellular telephone systems in 1979 and became the first country to deploy a cellular telephone network. European countries implemented Nordic Mobile Telephony in 1982. The United States deployed Advanced Mobile Phone System (AMPS) as the first cellular telephone network in 1983 [4].

There are different generations of cellular telephone systems [1,4]. First-generation (1G) wireless telephone networks were the first cellular networks that are commercially available. A 1G network was able to transmit voice with maximum speed of about 9.6 kb/s. 1G telecommunication networks used analog modulation to transmit voice and are regarded as analog telecommunication networks.

The 1G cellular system has some limitations such as poor voice quality, no support of encryption, inefficient use of frequency spectrum, and poor interference-handling techniques. Personal communication services introduced the concept of digital modulation, in which the voice was converted into digital code and became the second-regeneration (2G) cellular telephone system. 2G being digital addressed some of the limitations of 1G and was deployed using different signal representation and transmission techniques.

In the United States, CDMA, North American Time-Division Multiple Access, and Digital AMPS (D-AMPS) have been deployed as the 2G cellular network. In Europe, time-division-multiplexing-based GSM has been deployed, whereas in Japan, Personal Digital Cellular has been deployed. The GSM-based cellular system has become the most widely adopted 2G technology in the world.

2G's primary focus was voice communications, although it served as a remedy for the several limitations of 1G. Active research for data communications, along with voice communication service, resulted in data services over 2G being developed and became 2.5G. 1xEV-DO and 1xEV-DV have been deployed as 2.5G in the United States. 1xEV-DV uses a single radio-frequency channel for data and voice, whereas 1xEV-DO uses separate channels for data and voice.

High-speed circuit switched data (HSCSD), General Packet Radio Service (GPRS), and Enhanced Data Rate for GSM Evolution (EDGE) have been deployed in Europe. HSCSD was the first attempt at providing data at high-speed data communication over GSM with speeds of up to 115 kb/s.

However, this technique cannot support large bursts of data. The GPRS can support large burst data transfers, and it had a service GPRS support node (SGSN) for security mobility and access control and a Gateway GPRS support node (GGSN) to connect to external packet switched networks. EDGE provides data rates of up to 384 kb/s. Cellular digital packet data detect idle voice channels and use them to transmit data without disturbing voice communications.

The third-generation (3G) cellular system was developed with the goal of providing fast Internet connectivity, enhanced voice communication, video telephone, etc. CDMA2000 in the United States, Wideband CDMA (WCDMA) in Europe, and Time-Division Synchronous CDMA in China were deployed as 3G cellular networks. This process started in 1992 and resulted in a new network infrastructure called International Mobile Telecommunications 2000 (IMT-2000). IMT-2000 aimed the following [5,6]:

- To offer wide range of services over a wide coverage area
- To provide the best quality of service (QoS) possible
- To accommodate a variety of mobile users and stations
- To admit the provision of service among different networks
- To provide an open architecture and a modular structure

3G has been deployed in most countries and is being used in major communication networks. Service providers have already started deploying fourth-generation (4G) cellular communication systems, which offer data rates of up to 20 Mb/s and support mobile communication in moving vehicles with speed up to 250 km/h.

4G aims to incorporate high QoS and mobility in which a mobile user terminal will always select the best possible access available. 4G also aims to use mobile Internet Protocol (IP) with the IPv6 address scheme, in which each mobile device will have its own globally unique IP address.

It is important to understand the architecture of cellular networks to see its related security issues. A cellular network has two main parts [7]:

- Radio access network (RAN)
- Core network (CN)

Mobile users gain access wirelessly to the cellular network via the RAN, as shown in Figure 7.3. The RAN is connected to the CN. The CN is connected to the Internet via gateways through which mobile users can receive multimedia services. The CN is also connected to a public switched telephone network (PSTN). A PSTN is a circuit switched telephone public telephone network that is used to deliver calls to landline telephones. It uses a set of signaling protocols called Signaling No. 7 (SS7) that is defined by the International Telecommunication Union. SS7 provides telephony functions. The CN provides the interface for the communication among mobile users and landline telephone users.

The RAN consists of existing GPRS, GSM, or CDMA cellular telephone networks in which the radio network controller or base station connector is connected to packet switched CN to provide the interaction between the RAN and the CN.

The CN consists of circuit switch networks, packet switched networks, and IP multimedia networks. The high-end network servers facilitate the CN and provide several functions through the home location register to maintain subscriber information, the visitor location register to maintain temporary data of subscribers, the mobile switching center (MSC) to interface the RAN and the CN, and the gateway switching center to route the calls to the actual location of mobile users [8].

Every subscriber is permanently assigned to a home network and is also affiliated with a visiting network onto which it can roam. The home network is responsible in maintaining subscriber

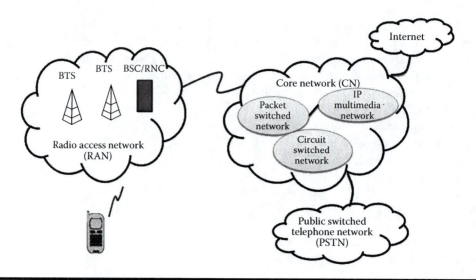

Figure 7.3   Cellular telephone network architecture.

profile and current location. The visiting network is the network where a mobile user is currently roaming. It is important to note that the visiting networks provide all the functionality to mobile users on behalf of the home network.

IP-based servers such as DNS, Dynamic Host Configuration Protocol (DHCP), and RADIUS servers interact with the gateways and provide control and management functions needed for mobile users while getting service from the Internet.

7.2.1 Security Issues in Cellular Networks

Multiple entities are incorporated in cellular telephone networks, and the infrastructure for supporting these services is massive and complex. IP multimedia Internet connection with the CN in a telephone network presents a big challenge for the network to provide security. Wireless networks, in general, have many limitations compared to wired networks [6,7]:

- Radio signal travels through an open wireless access medium such as air
- Limited bandwidth shared by many mobile users
- Mobility in wireless networks makes the system more complex
- Mobile stations run on limited time batteries resulting in power issue in wireless systems
- Small mobile devices have limited processing capabilities
- Unreliable network connection for mobile users

Apart from the above listed limitations, several security issues need to be considered when deploying a cellular network. There are varieties of attacks in the wireless cellular network:

1. Denial of service (DOS) caused by sending excessive data to the network so that the legitimate users are unable to access network resources.
2. Distributed DOS is a result of attack by multiple DoS attackers.
3. Channel jamming by sending a high power signal over the channel that denies access to the network.

4. Unauthorized access to the network by illegitimate users.
5. Eavesdropping in wireless communications.
6. Message replay: it can be done even if the transmission is encrypted by sending an encrypted message repeatedly.
7. Man-in-the-middle attack: attacker poses as a relay station between a sender and a receiver.
8. Session hijacking: hijack the established session and pretend as a legitimate user.

7.2.1.1 Security in RANs

In RANs, mobile users connect with each other wirelessly through a base station. A determined attacker with a radio transmitter/receiver can easily capture the radio signal transmitted on the air. In 1G and 2G systems, there was no encryption mechanism to hide voice from a malicious user and no guard mechanism against eavesdropping on conversations between the mobile user and the base station. Because of the lack of security provision in 1G and 2G cellular telephone systems, the attacker can not only enjoy wireless service without paying for any service usage fees but can also entice mobile users through a rouge or false base station to get secret information. The 3G cellular system has security provision to prevent these types of attacks. It has an encryption mechanism with integrity keys (IKs) to encrypt the conversation, and thus, the attacker cannot change the conversation between the mobile user and the base station. 3G has also improved radio network security. However, it still cannot prevent DOS attacks when a large number of requests are sent from the RAN to the visiting MSC, in which the MSC needs to verify every request through an authentication process. Because of excessive requests and authentication, the MSC may fail to serve legitimate users.

7.2.1.2 Security in CNs

CN security deals with security issues at the service node and wire-line-signaling message between service nodes. Protection is provided for the services that use the Mobile Application Part (MAP) protocol. Security for the MAP protocol is provided either through MAP security (MAPSec) when the MAP runs on the SS7 protocol stack or IPSec when the MAP runs on top of the IP. 3G also lacks in security for all types of signaling messages. However, the end-to-end security (EndSec) protocol proposed in [9] can prevent misrouting the signal.

Internet connectivity through a mobile device introduces the biggest threat to cellular network security. Any attacks that are possible on the Internet can now be entered into the CN via gateways located between the CN and the Internet. One example of this is the attack to the E-911 service [10]. Short message and voice conversation still use the same channel, resulting in contention and collision between them. Preclusion of the entire CN (servers for PSTN, circuit, and packet switched network services) from attacks that are coming through the Internet link is an important consideration. As the PSTN uses the SS7 protocol that does not have any authentication mechanism and transmits voice messages in plain text, the attacker can easily introduce fake messages or attack by DOS. Only a limited amount of research has been undertaken for securing PSTN [11].

As mentioned above, the cellular network has many new services, and the security architecture needs to provide security for all these services.

7.2.1.3 Cellular Network Security Architecture

A cellular network security architecture consists of five sets of features, as shown in Figure 7.4.

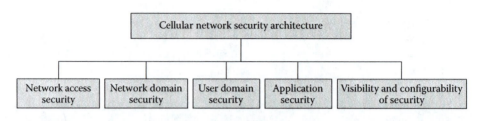

Figure 7.4   Cellular network security architecture.

Network access security is responsible for providing authentication of user and mobile device, confidentiality, and integrity. It enables mobile users to access cellular network services securely. International Mobile Equipment Identifier and secret cipher key (CK) are used to provide confidentiality of both the device and the user. The challenge response method using a secret key is used to achieve authentication. It is worth noting that the Authentication and Key Agreement provides mutual authentication for the user and the network. A CK and an IK for which the user and the network agree are used until their time expires. Integrity protection in a cellular network is necessary as control signaling communications between a mobile station and a network is sensitive. An integrity algorithm and IK provide the integrity service.

Network domain security enables nodes in the service provider to securely exchange the signaling data and prevent attacks on the wired networks.

User domain security enables mobile stations to securely connect to the base station and prevent external attacks.

Application security provides secure mechanisms to the exchange of messages between users of the user domain and services of the service provider domain for different applications.

The feature of *visibility and configurability of security* allows users to query what security features are available to them and what features they can use.

7.2.1.4 Wireless Application Protocol

Cellular networks are connected to the Internet through CNs to provide Internet access to mobile users using the Wireless Application Protocol (WAP) [12]. Thus, it is important to understand the security mechanisms of the protocol used to access the Internet via the CN. WAP is an open specification protocol, meaning that it is independent of the underlying networks. It is platform- and technology-independent and thus provides Internet access service to users who use WCDMA, CMDA 2000, UMTS, or any operating systems such as Windows CE, PALM OS, etc. The first version of WAP (WAP1) was released in 1998. WAP1 considers that a wireless mobile device has limited power and other resources and has limited security features and thus communicates through other gateways while communicating with the servers. The second version of WAP (WAP2) was released in 2002. It assumes that mobile devices are powerful. It has better security features and allows mobile users to directly communicate with the servers.

WAP2 protocol stack/layers shown in Figure 7.5 are briefly discussed as follows:

1. *Wireless application environment (WAE):* This layer is like an application layer in the OSI reference model and provides an environment for WAP applications such as web applications.
2. *Hypertext Transfer Protocol:* This layer deals with a platform-independent protocol that is used for transferring web content/pages.

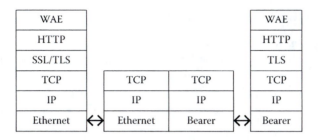

Figure 7.5 WAP2 protocol stack.

3. *Transport layer security (TLS):* This is the fourth-layer (from bottom) protocol that provides security features such as confidentiality, integrity, and authentication. TSL used in WAP2 is known as profiled TLS, which consists of a cipher and authentication suites, session resume from identification suites, and tunneling capability.
4. *Transport Control Protocol (TCP):* This is the third-layer (from bottom) protocol that is a standard reliable TCP.
5. IP: This is the second-layer (from bottom) protocol that is responsible for routing data in a network.
6. Bearer Protocol: This is the lowest level protocol that can be used by any wireless techniques (e.g., CDMA, GSM, WCDMA, etc.) used in cellular telephone networks.

Overall, multiple layers of the protocol stacked with multiple layers of encryption address the security issues in existing 3G wireless cellular networks, which consume more power and introduce high transmission delay. In 4G, only one layer is responsible for encrypting data using interlayer security [13], which reduces communications delay.

7.3 Worldwide Interoperability for Microwave Access

WiMAX [14] is a WMAN that can offer data transfer rates of up to 75 Mb/s or an area of radius of about 50 km (30 mi.) and is part of 4G wireless communication technology. WiMAX was released in December 2001 as an IEEE 802.16 standard. The IEEE 802.16 uses three major frequency bands: 10–66 GHz (licensed bands), 2–11 GHz (licensed bands), and 2–11 GHz (unlicensed bands).

WiMAX still has some shortcomings in terms of security as designers have incorporated the use of the preexisting standard Data over Cable Service Interface Specifications (DOCSIS) that was used in cable communication [15]. Among different IEEE 802.16 standards, 802.16a/d standards make use of public-key encryption keys (which are exchanged at connection setup time), and the base station authenticates the clients using 56-bit data encryption standard-based digital certificates [15]. However, it does not provide adequate protection against data forgery. IEEE 802.16e implements a 128-bit encryption key mode based on the Advanced Encryption Standard (AES) to remove the flaws that are present in 802.16a/d. The man-in-the-middle attacks launched using rouge base stations are mitigated by client-to-base-station and base-station-to-client authentication [15].

7.4 Wireless Local Area Networks

The successful deployment of WLANs in the past decade is due to their advantages such as flexibility, scalability, mobility, and freedom from wires, which wired networks lack [16]. Wireless networks are easy to install in rural areas, where wired network infrastructures are either difficult or impossible to create due to physical obstacles. Wireless networks are easily scalable, flexible, and esthetic since wireless devices communicate using mainly either radio frequency (RF) or infrared frequency.

The main standard in the WLAN is IEEE 802.11 and is also known as Wi-Fi, that is, IEEE-standardized WLAN in 1999. Wireless communications were tested in 1971 by a researcher at the University of Hawaii. The recent standard of the WLAN is IEEE 802.11-2007. IEEE 802.11 WLANs can be configured to operate in an infrastructure (AP) mode or in an ad hoc mode.

7.4.1 WLAN in AP Mode

WLANs in AP mode consist of wireless client stations (STAs) and an AP in which clients are equipped with wireless adapters that allow wireless communication among other wireless stations. In this case, the AP functions like a regular switch or router in a wired network for the wireless client stations. In AP-mode WLANs, all communications pass through an AP, meaning that wireless clients cannot communicate with each other directly.

The basic structure of a WLAN is called the basic service set (BSS), as shown in Figure 7.6, in which the network consists of an AP and several wireless devices. In order to form a wireless network, the AP continually broadcasts its service set identifier (SSID), which is the logical name of the wireless network. This allows wireless client stations to locate and join the wireless network. The area covered by a transmission range of an AP is called the basic service area.

A WLAN operating in AP mode is connected to a wired network through an AP. Thus, the AP is a gateway for wireless client stations to join a wired network. One example is shown in Figure 7.6 where the AP is connected to a wired network through a switch.

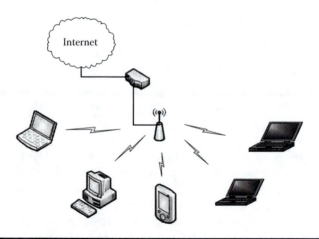

Figure 7.6   WLAN in AP mode (also known as BSS).

For roaming support, BSSs can be combined to form an extended service set (ESS). In ESSs, APs are connected to a single-backbone system to provide roaming (moving from one BSS to another BSS) for wireless client stations (STAs), as shown in Figure 7.7.

In order to avoid interference, WAPs should be configured in such a way that they transmit in nonoverlapping adjacent channels, as shown in Figures 7.7 and 7.8. If multiple APs overlap transmission ranges in the same channel, the performance of the WLAN will be significantly degraded [16].

Channel occupancy information along with the MAC address, received signal strength indication, vendor information, network types (infrastructure or ad hoc), privacy/security mode, scan time, etc., can be easily obtained using freely available tools such as inSSIDer [17], as shown in Figure 7.9. The inSSIDer is a freeware wireless auditing tool that is compatible with many vendors' wireless adaptors. It can be downloaded from the MetaGeek web site [18]. Using the result of the inSSIDer, the network administrator can change the orientation or position of a WAP or clients to increase the signal strength. Furthermore, one can change the security features to secure the wireless network and channel used for wireless transmission to have the least interference in a wireless network.

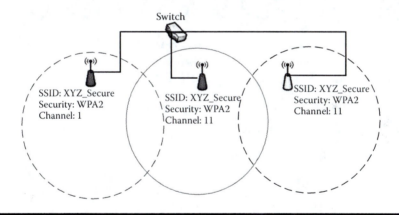

Figure 7.7   ESS.

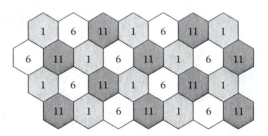

Figure 7.8   WLAN channel assignment for multiple APs.

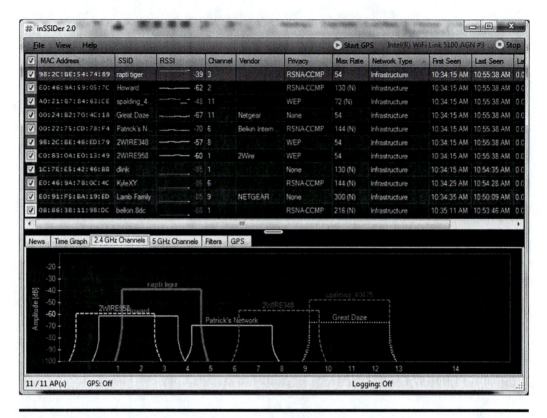

Figure 7.9   WLAN channel assignment for multiple APs.

7.4.2 WLANs in Ad Hoc Mode

When wireless devices communicate with each other directly without using centralized AP as shown in Figure 7.10, the WLAN configuration is called an independent BSS (IBSS).

One of the ad hoc wireless nodes (e.g., computer) should be configured to provide SSID for wireless ad hoc networking.

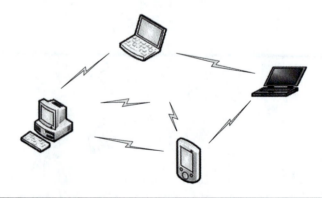

Figure 7.10   WLAN in ad hoc mode: IBSS.

7.4.3 Security Attacks in WLANs

As in other wireless networks, the medium used to transfer data from a source to a destination is the RF signal. The RF signal in a WLAN is also freely available and thus makes the WLAN susceptible to attack if it is not properly configured to secure transmission. Typical transmit power of APs lies in the range of 50–100 mW [maximum allowed range by the Federal Communications Commission (FCC) in the Unites States is 4 W], and the range of AP is about 300–1800 ft. [19].

After successful deployment of WLAN and handheld devices, wireless applications and devices increased exponentially, which has created major security-related issues in the network. The following is the list of most common attack types in wireless networks [16,17].

7.4.3.1 Network Traffic Analysis

To find information of the target network, the attacker uses the statistics of network connectivity, activity, AP location, SSID, etc.

7.4.3.2 Passive Eavesdropping

Attackers sniff the packet transmitted over the network and extract the network information. Networks with unencrypted setup are the victims of this type of attacks. Attackers use the extracted information to attack the network.

7.4.3.3 Active Eavesdropping

In this type of attack, the attacker tries to inject a complete packet in the data stream to change the data on the packet. Both unencrypted and encrypted types of networks can be victims of this type of attack.

7.4.3.4 Unauthorized Access or War-Xing

Unauthorized access attack can be just for free Internet access [20,21] using unauthorized login. Information about the wireless network can be obtained by War-Xing (wardriving, warwalking, warcycling, warflying, etc.) [20].

7.4.3.5 Man-in-the-Middle Attacks

In this type of attack, the attacker stays between the intended transmitter and receiver, and works as a relay station. The attacker (relay station) manipulates and pretends to be an intended sender.

7.4.3.6 Session Hijacking

In this type of attack, the attacker hijacks an authorized session and pretends to be an intended sender.

7.4.3.7 Replay Attacks and Rouge AP

In replay attacks, the attacker sends a legitimate packet several times or changes the content of the packet before transmitting it. In this type of attack, attackers set a wireless device as AP (called rouge AP) using a special type of software and entice the legitimate users to get secret information. By imposing mutual authentication between AP and network devices, rouge AP and reply attack can be solved.

7.4.3.8 DoS Attacks

In this type of attack, the attacker sends noise continually on a specific channel to ruin network performance. RF jamming is an example of DoS attack in the wireless network [16,22].

7.4.4 Security in WLAN 802.11

The IEEE 802.11 standard consists of three layers:

1. *Physical layer:* it is responsible for providing an interface to exchange frames with the upper MAC layer.
2. *MAC layer:* it provides the functionality needed to control media access and allow reliable transfer of frames to the upper layers.
3. *Logical link control (LLC) layer:* it provides connection-oriented service to the upper layers. It also provides addressing and data link control through the LLC.

7.4.4.1 802.11 Authentication

Wireless clients must be authenticated and associated before any data transmissions. In WLANs, there are two types of authentication, that is, open authentication and shared key authentication [16,23]. Open authentication is actually no authentication at all. Any clients can be authenticated and associated in the open authentication system. In shared key authentication, when the client wants to connect to the AP, it sends a request to the AP. Once the AP receives a request, it sends a packet in unencrypted text as a challenge message. The client then encrypts this message a pre-shared key and sends it back to the AP. The AP decrypts it and compares it with that sent previously as a challenge. If both texts match, the client will be authenticated; otherwise, connection will be denied. In actual data transmission, wired equivalent privacy (WEP) can be used in both preshared and open authentication. It is worth noting that open key authentication is more secure than the preshared key because the latter does not have a challenge response and does not expose the WEP key to traffic sniffers [24].

7.4.4.2 Wired Equivalent Privacy

WEP was designed to provide the security level that is available in wired networks. It has three goals to achieve for WLANs: confidentiality, availability, and integrity of information [16,23]. However, WEP was proved to be breakable and thus is now considered unsecure for many reasons; nonetheless, it is used to provide general security instead of leaving the network unsecure. WEP provides encryption only between the wireless client station and the AP. When data travel over the wired network, it is unencrypted.

As shown in Figure 7.11, WEP uses stream cipher RC4 (Ron's Code 4) for the encryption. RC4 needs an initial vector (IV) as a seed, which is used along with the shared WEP key to

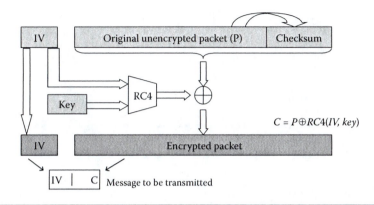

Figure 7.11   WEP packet encryption.

encrypt and decrypt the packets. From the packet to be transmitted, a checksum (cyclical redundancy checking) is calculated and attached with the payload. An exclusive OR (XOR) operation is performed between the payload and RC4 stream (generated from the shared key and the IV) to generate an encrypted packet. The unencrypted IV is appended with an encrypted packet, and the combined packet is transmitted over the wireless network. At the receiving end, reverse process takes place for decrypting the packet.

The IV is 40 bits long, and the key length is 40 bits in WEP and is 104 bits in WEP2. Using freely available tools, anyone can break WEP security used in a WLAN. After collecting a sufficient number of packets (20,000–100,000 packets), one can easily break the WEP key using freely available tools such as BackTrack, Russix, and Aircrack-ng [17].

When a WEP key is fixed, mathematically, if the same IV is used to encrypt two different packets, you can know P_2 when you have C_1, C_2, and P_1 [17,22,23], that is,

$$C_1 XOR\ C_1 = P_1 XOR\ P_1.$$

Because of many weaknesses in WEP, the WLAN was designed with Wi-Fi protected access (WPA) security modes.

7.4.4.3 IEEE 802.1x: Extensible Authentication Protocol over LAN

The IEEE 802.1x is port-based authentication to authenticate users in IEEE 802 networks. The Extensible Authentication Protocol (EAP) allows any of the encryption schemes to be implemented on top of it, adding flexibility to the security design module. The Remote Authentication Dial-In User Service (RADIUS) server is used for authentication in the 802.1x framework to provide authentication, authorization, and accounting (AAA) service for network clients, as shown in Figure 7.12 [17,22–25]. The 802.1x framework defines three entities/ports, that is, supplicant (client STA that wants to be authenticated), authenticator (AP that connects the supplicant to the wired network), and authentication server (performs the authentication process from the supplicant based on their credentials) [22,23].

Authentication   AP (Authenticator)   Supplicant (STA)
server (RADIUS)

Figure 7.12   802.1x authentication.

7.4.4.4 IEEE 802.11i Standard

The IEEE 802.11i, which was released in June 2004, improves authentication, integrity, and data transfer in WLANs. To get rid of WEP weaknesses, the Wi-Fi Alliance developed WPA, which was released in April 2003. Vendors or Wi-Fi Alliance implemented the full specifications under the name WPA2, that is, 802.11i [16,17,22,23].

Two methods of authentication are supported in IEEE 802.11i:

■ 802.1x and EAP to authenticate users: this is described above.
■ Per-session-key per-device authentication: this is an alternative method of authentication to the first method. Similar to WEP, the shared key called group master key, with pair transient key and pair session key, is used for authentication and data encryption.

Michael algorithm is used to solve the integrity problem with WEP, which protects both the header and data. The IEEE 802.11i specifies three protocols [16,23]:

■ Temporal Key Integrity Management: it provides a short-term solution that fixes all WEP weaknesses using per-packet key mixing, message integrity check, and a rekeying mechanism.
■ Wireless Robust Authenticated Protocol: it was introduced to get the benefits of AES in WLAN Offset Codebook mode of AES.
■ Counter with Cipher Block Chaining Message Authentication Code Protocol [26]: it uses AES for encryption and requires hardware upgrade to support the new encryption algorithm. It is considered to be the best solution to secure wireless data transfer under 802.11i.

Robust secure/security network (RSN) is part of the IEEE 802.11i standard that provides mechanism to create a secure communication channel between an AP and wireless clients by broadcasting an *RSN Information Element* message across the wireless network.

7.4.5 Best Practices

There is not a single solution that can completely secure a wireless network. Therefore, we need to follow best practices [16,17,22,23], which are given as follows:

■ Define, enforce, and monitor a wireless security policy: the policy should cover all wireless services and users such as Wi-Fi and Bluetooth services and users.
■ Always conduct a site survey to collect the information about all WAPs and Wi-Fi devices, which helps to eliminate rouge APs and unauthorized users.

- Configure APs and user stations for security:
  - Change the WEP key on a regular basis in home networks to weaken the chances of being attacked.
  - Configure the AP to stop broadcasting its SSID to hide your network.
  - Turn off "ad hoc" mode operation.
  - Implement layers of security schemes such as MAC address filtering and protocol filtering, along with WEP and SSID hiding.
  - Deploy a wireless intrusion detection system to identify or log threats and attacks. Analyze log and resolve incidents in a timely manner.
  - Define and develop institution-wide policies with detailed procedures regarding wireless devices and usage.
  - Conduct regular security awareness and training sessions for both system administrators and users to make them become aware of recent advances in computer network and wireless security. Train users not to respond to social engineering or phishing emails.
  - Define acceptable encryption and authentication protocols:
    - Implement WPA or WPA2 wherever possible.
    - Use strong encryption with at least 128-bit keys (WPA, AES recommended).
    - Turn off "open" authentication.
    - Deploy a layer-3 virtual private network for wireless communication.
  - DHCP: Use static IP addresses instead of DHCP. As DHCP automatically provides an IP address to anyone (authorized or not) and facilitates access to your wireless network, it creates a big threat to the network from unauthorized users.
  - Plan for AP coverage to radiate out toward windows but not beyond.
  - Use directional antennas for wireless devices to better contain and control the RF array and thus prevent unauthorized access.
  - Use remote authentication dial-in user service, which can be built into an AP or provided via a separate server. RADIUS is an additional authentication step. Interface this authentication server to a user database to ensure that the requesting user is authorized.
  - Force periodic (every 15 min or so) reauthentication for all wireless users.
  - Implement physical security controls: because of small size and portability of wireless devices, they are easy to steal or lose so it is recommended to implement strong physical security controls (such as guard, video camera, and locks) to prevent theft of equipment and unauthorized access.
  - To secure wireless network through lost or stolen devices, implement device-independent authentication.

7.4.6 Protocol for Carrying Authentication for Network Access

Protocol for Carrying Authentication for Network Access (PANA) is the recent proposal to enhance wireless security mechanisms through improved authorization between WLAN clients and AAA servers over IP-based networks [27]. In other words, PANA carries EAP to perform authentication between the access network and the wireless client. After successful PANA authentication, the client is authorized to receive an IP forwarding service from the network.

PANA is the network-layer protocol and is intended to authenticate PaC (PANA Client) with a PANA authentication agent (PAA) in situations where no prior trust between PAA and PaC exists. PANA consists of four parts, that is, a wireless client known as PaC (PANA Client); an

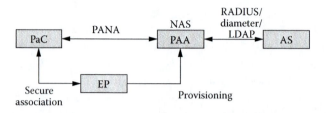

Figure 7.13 PANA framework.

enforcement point, which is the physical point where inbound and outbound traffic filters are applied; a PAA representing access authority on the network; and the AAA servers. Using an initial sequence number and cookie-based authentication between PAA and PaC, PANA can provide a mechanism to prevent DoS attacks [27,28]. The PANA framework is shown in Figure 7.13.

7.5 Wireless Personal Area Networks (PANs)

7.5.1 IEEE 802.15: PANs

Personal area networks (PANs) span a small area within personal premises such as a home or an office [29]. Mostly, they are formed by using peer-to-peer basis or master–slave basis. Bluetooth, ZigBee, and ultrawideband (UWB) networks are some examples of PANs.

7.5.2 Bluetooth Network Security

Bluetooth is an example of a wireless PAN in which clients use a *pairing* process to establish encryption and authentication between two devices. Bluetooth operates in an ISM radio band. The association process takes about up to 4 s. Bluetooth devices form a master/slave-like structure while pairing and use 48-bit hardware address of a master, shared 128-bit random number, and a user-specified personal identification number (PIN) of up to 128 bits. Some Bluetooth devices only allow 1–4-digit PINs. Hardware address and random number are exchanged using plain text, and a user-specified PIN is entered by users similar to the password. It is assumed that the Bluetooth network is secured; unfortunately, it is possible to break a Bluetooth network [30] by sniffing the packet for a PIN when a 1- to 4-digit PIN is used. Exploiting vendor-specific flaws such as default setting of allowing any pairing, attackers exploit Bluetooth devices. In order to protect the Bluetooth network, users need to change the default setting and choose strong PINs.

7.5.3 IEEE 802.15.4: ZigBee Security

To provide security in the ZigBee network [31], it is built on top of the IEEE 802.15.4AES-128 algorithm. ZigBee operates in the ISM radio bands, and its data transmission rates vary from 20 to 900 kb/s. Two devices take about 30 ms to get associated. To provide network security, ZigBee runs in two different security modes, that is, residential mode and commercial mode.

In residential mode, all users use a predeployed key for the entire PAN and for all applications. Residential mode security protects the PAN from external eavesdroppers; however, it does not provide security from the user within the same PAN. In commercial mode, the coordinator node in a trust center is used to preshare the two master keys that provide extra security on top of the residential mode. This method is costly since infrastructure is needed to have a centralized coordinator node for the trust center to store sessions for each link.

7.5.4 UWB Security

UWB radios use low transmit power; as a result, they have a low coverage area. To attack this type of networks, the attacker should be close enough to the UWB network. The FCC in the United States authorizes unlicensed use of UWB in the range of 3.1–10.6 GHz. There are no standard security modes in UWB networks. According to WiMedia [32], there are three levels of link-layer security: (1) *Security Level 0*, in which communication is fully unencrypted; (2) *Security Level 1*, which has both encrypted communications with AES-128 for encrypted links and unencrypted communications for unencrypted links; and (3) *Security Level 2*, in which all communications must be encrypted with AES-128.

7.6 Best Practices for Mobile Device Security

This section presents best practices for securing wireless or mobile devices in general. There is no perfect method to protect wireless networks and mobile devices/users, and thus, it is recommended to use multiple techniques and best practices.

7.6.1 Devices Choice

All devices are not designed equally when it comes to security. Wireless mobile devices for users should be chosen based on the security requirements. Wireless security configuration in mobile devices is highly dependent on the security features that are available on them. For example, iPods are not as secure as BlackBerry devices because iPods are built for general users who are not concerned with security, and BlackBerry devices are designed for enterprise users who need a higher level of security.

7.6.2 Enable Encryption

Encryption enables strong security features in mobile devices and mandates it for all users to provide security for the network. In general, many organizations do not enforce or mandate encryption through policies for mobile devices and users.

7.6.3 Configure Wireless Networks for Authentication

The best practice for mobile device security is to enable device authentication so that lost devices cannot be easily accessed by any person that finds or steals a device. The survey result published in September 2008 by Credent Technologies shows that, in a 6-month period, more than 31,000 passengers left their mobile devices in a taxicab. The fact of the matter is that these devices are too easy to lose, and they can be used to enter the network if authentication is not enabled.

7.6.4 Enable and Utilize Remote Wipe Capabilities

It is best practice to enable remote access to disable devices and wipe out data in the case of loss or theft. With the remote wiping capability, the user or network administrator would be able to delete data in the stolen or lost devices to protect these devices from malicious use. Additionally, the network administrator should be able and available to take necessary steps to wipe out the wireless/mobile device.

7.6.5 Limit Third-Party Apps

There are several applications available for smartphones. These apps provide many features but can also easily provide backdoors or security loopholes, which are the biggest threat to the privacy and security of the organization. There should be policy and recommendation to control the installation of unsigned third-party applications to prevent attackers from requisitioning control of wireless/mobile devices.

7.6.6 Implement Firewall Policies

It is recommended to set up firewall policies for traffic coming from smartphones to provide security to the network, as well as to the mobile devices.

7.6.7 Implement Intrusion Prevention Software

It is possible to run Metasploits on recent smartphones such as iPhones because smartphones are becoming powerful enough to run this. Smartphones can be exploited by hackers or attackers to attack the network system. Intrusion prevention systems can examine traffic coming through mobile devices and protect the system.

7.6.8 Bluetooth Policies

Bluetooth capabilities available on Wi-Fi devices and smartphones are easy to use for creating PANs. Hackers can take advantage of default always-on always-discoverable settings of Bluetooth to launch attacks. It is best practice to disable Bluetooth when it is not actively transmitting information and to switch Bluetooth devices to hidden mode. This type of configuration should be the part of the policy to limit the exposure of the wireless network and mobile devices within the organization.

7.7 Summary

This chapter provided an overview of concepts related to security features and issues in wireless voice and data communication networks. Discussions about why and how wireless networks are more vulnerable, as compared to wired networks, were presented. The combination of different systems within a wireless cellular network makes it complex and increases vulnerabilities and loopholes. Attackers can exploit the vulnerabilities available in any part of the network and can gain access to the network. The protocols and practices used to secure a wireless cellular network are presented. In order to secure a WiMAX network, the IEEE 802.16e standard that implements a 128-bit encryption key mode based on the AES is used to remove the flaws that are present in older WiMAX IEEE 802.16a/d standards. In IEEE 802.11, WEP is an old security mode used to protect WLANs. It is not secure but is still widely used since it provides at least one level of security to the network. Recent advances in WLANs have improved its security schemes. The IEEE 802.11i is assumed to be a secured solution to fix most of the security holes found in its predecessor WEP. A recently proposed PANA framework with different protocols is used as a secure messaging system between wireless clients and wireless network access authority. To protect the network, different security schemes can be implemented in PANs, including Bluetooth, ZigBee, and UWB networks. Furthermore, best practices and recommendations to secure different wireless networks and devices were presented.

Wherever wireless networks are deployed, security vulnerability will always exist. Security attacks and vulnerabilities can only be mitigated if best practices, as well as correct policies and standards, are used. We discussed some of the important and best practices that can be implemented for improving mobile and wireless security. Wireless security will continue to be a research topic as long as there are ways to attack or obtain unauthorized access to wireless networks.

References

1. Goldsmith, A. *Wireless Communications.* Cambridge University Press, New York, 2005.
2. Rawat, D. B., D. C. Popescu, G. Yan, and S. Olariu. "Enhancing VANET performance by joint adaptation of transmission power and contention window Size." *IEEE Transactions on Parallel and Distributed Systems*, vol. 22, no. 9, pp. 1528–1535, September 2011.
3. Rawat, D. B., B. B. Bista, G. Yan, and M. C. Weigle. "Securing Vehicular Ad-Hoc Networks Against Malicious Drivers: A Probabilistic Approach." *Proceedings of the International Conference on Complex, Intelligent, and Software Intensive Systems* (CISIS), pp. 146–151, Seoul, Korea, June 2011.
4. Lee, W. *Wireless and Cellular Telecommunications.* McGraw-Hill Press, New York, 2005.
5. Balderas-Contreras, T., and R. A. Cumplido-Parra. Security Architecture in UMTS Third Generation Cellular Networks, Coordinación de Ciencias Computacionales INAOE, Technical Report No. CCC-04-002 27, 2004.
6. Gardezi, A. I. *Security In Wireless Cellular Networks.* http://www.cs.wustl.edu/~jain/cse574-06/ftp/cellular_security/index.html, accessed December 10, 2011.
7. Yang, H., F. Ricciato, S. Lu, and L. Zhang. "Securing a wireless world." *Proceedings of the IEEE*, vol. 94, no. 2, 2006.
8. 3GPP, A guide to 3rd generation security. Technical Standard 3GPP TR 33.900 V1.2.0, 3G Partnership Project, January 2001.
9. Kotapati, K., P. Liu, and T. F. La Porta. "EndSec: An end-to-end message security protocol for mobile telecommunication networks." *Proceedings of the 2008 International Symposium on a World of Wireless, Mobile and Multimedia Networks*, 2008.
10. Moore, D., V. Paxson, S. Savage, C. Shannon, S. Staniford, and N. Weaver. "Inside the slammer worm." *IEEE Security and Privacy*, vol. 1, no. 4, pp. 33–39, 2003.
11. Moore, T., T. Kosloff, J. Keller, G. Manes, and S. Shenoi. "Signaling System 7 (SS7) Network Security." *Proceedings of the IEEE 45th Midwest Symposium on Circuits and Systems*, August 2002.
12. Mann, S., and S. Sbihli. *The Wireless Application Protocol (WAP): A Wiley Tech Brief.* John Wiley Press, Hoboken, NJ, 2002.
13. Carneiro, G. "Cross-layer design in 4G wireless terminals." *IEEE Wireless Communications*, vol. 11, issue 2, 2004.
14. Pareek, D. *WiMAX: Taking Wireless to the MAX.* John Wiley Press, Hoboken, NJ, 2006.
15. Johnston D., and J. Walker. "Overview of IEEE 802.16 security." *IEEE Security and Privacy Magazine*, vol. 02, issue 3, pp. 40–48, June 2004.

16. Roshan, P., and J. Leary. *802.11 Wireless LAN Fundamentals*, CISCO, 2009.
17. Rawat, D. B. et al. Comprehensive ComTIA Security+ Lab Manual, 2012, in preparation.
18. inSSIDer Software URL. http://www.metageek.net/products/inssider/, accessed December 2011.
19. Arbaugh, W. A. "Wireless security is different." *Computer*, vol. 36, issue 8, pp. 99–101, August 2003.
20. Hurley, C., and F. Thornton. *WarDriving: Drive, Detect, Defend: A Guide to Wireless Security*. Syngress Publishing Press, Rockland, MA, 2004.
21. Potter, B. C. "Wireless security's future." *IEEE Security and Privacy Magazine*, vol. 1, issue 4, pp. 68–72, Aug. 2003.
22. Welch, D., and S. Lathrop. "Wireless security threat taxonomy." *Proceedings of the IEEE Information Assurance Workshop 2003*, pp. 76–83, June 2003.
23. Earle, A. E. *Wireless Security Handbook*. Auerbach Publications, Boca Raton, FL, 2005.
24. http://www.cs.wustl.edu/~jain/cse574-06/ftp/wireless_security/index.html-startawisp
25. RFC for RADIUS server URL: http://www.ietf.org/rfc/rfc2865.txt.
26. RFC for CCMP, http://www.ietf.org/rfc/rfc3610.txt.
27. Protocol for Carrying Authentication for Network Access (PANA) RFCURL. http://tools.ietf.org/html/rfc5191, accessed December 2011.
28. RFC for PANA Threat Analysis and Security Requirements, URL http://www.armware.dk/RFC/rfc/rfc4016.html.
29. Surhone, L. M., M. T. Timpledon, and S. Marseken. *Personal Area Network*. Betascript Publishers, Beau Bassing, Mauritius, 2010.
30. Shaked, Y., and A. Wool. "Cracking the bluetooth PIN." *Proceedings of the 3rd International Conference on Mobile Systems, Applications, and Services*, pp. 39–50, 2005.
31. Elahi, A., and A. Gschwender. *ZigBee Wireless Sensor and Control Network*. Pearson Education, Boston, 2009.
32. ECMA International URL, http://www.ecma-international.org, accessed December 2011.

Chapter 8

Security and Access Control in Mobile ad hoc Networks

Soumya Maity and Soumya K. Ghosh

Contents

8.1 Introduction

The deployment of mobile ad hoc networks (MANETs) in several mission-critical organizations (such as military, banking, disaster management, and vehicular network) emphasizes the need of enforcing appropriate security policies to control the unauthorized access of network resources. This is becoming complex due to various security requirements. The security technologies (firewall, intrusion detection system, access control lists, etc.) that are used in traditional network are unable to meet the security requirements in MANETs. Security in MANETs is a major challenge due to the nature of the uncontrolled medium where MANETs operate, their dynamically changing topology, and the lack of centralized management. Moreover, different nodes connected in a MANET can have different roles; thus, role-based access control (RBAC) needs to be employed. Implementation of security and access control is required to protect the network resources from unauthorized access in MANETs. This is still in a premature stage.

This chapter primarily focuses on the following areas of MANETs:

■ Security threats
■ Security countermeasures
■ Access control

Advancement in wireless technologies, such as Bluetooth and IEEE 802.11, leads to the development of MANETs, wherein potential mobile users arrive within the common perimeter of radio link and participate in setting up the network topology for communication. Nodes within MANETs are mobile and communicate with each other within radio range through direct wireless links or multihop routing. A MANET is a network of wireless mobile nodes organized in arbitrary and temporary network topologies. Mobile nodes such as personal digital assistants, laptops, mobile phones, and handheld radio devices can thus be internetworked in areas without a preexisting communication infrastructure or when the use of such infrastructure requires wireless extension. In MANETs, nodes can directly communicate with all other nodes within their radio ranges, whereas nodes that are not in the direct communication range use intermediate node(s) to communicate with each other. A MANET has a number of security loopholes (which are discussed in the subsequent sections) that challenge researchers. This chapter mainly focuses on the security and access control in MANETs. As very few research works have been performed to ensure overall policy-based access control in MANETs, there is a need to pay attention. The different aspects of access control mechanisms are discussed in this chapter.

This chapter is organized as follows. In Section 8.1, the overview of MANETs and the basic routing protocols followed by the security needs of MANETs are discussed. Then, Section 8.2 is focused on MANET-specific security loopholes, threats, and different types of attacks. Security countermeasures and secured routing protocols are discussed in Section 8.3. This section also explains the defense strategies against specific attacks. Section 8.4 focuses on access control mechanisms over different resources in MANETs. Different access control frameworks proposed in the recent past are also discussed. Conclusions are presented in Section 8.5. In Section 8.6, the open challenges and future trends are discussed.

8.1.1 Overview of MANETs

A MANET is a wireless network that has two or more nodes connected in meshed network links. This network builds and configures itself automatically. Ad hoc networks (Figure 8.1) connect mobile devices (network nodes), such as mobile phones, personal digital assistants, laptops, and wireless access points, without any fixed infrastructure. Data are passed from node to node until they reach their destination, thus increasing the data load, which is more advantageous than in distributed networks with a central point of contact. In MANETs, network resources demand effective cooperation of network nodes to optimize time, energy, and data rate. Special routing protocols have been designed to ensure that the network adapts constantly with the mobility of the nodes. In other words, a MANET is a type of self-organizing multihop network. Figure 8.1 is an example of a typical MANET formed by different mobile devices. The dotted lines represent the wireless link between the devices that are in the radio range of each other.

Due to their wireless and distributed nature, MANETs pose a great challenge for system security designers. In the last few years, security problems in MANETs have attracted much attention. A number of research efforts focus on specific security areas, such as securing routing protocols,

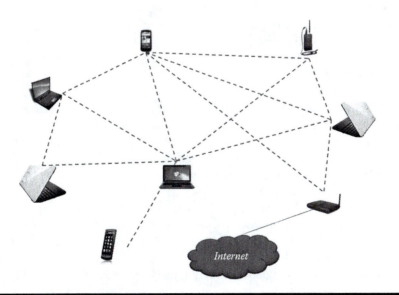

Figure 8.1 Example of a typical MANET.

establishing trust infrastructure, intrusion detection, and response. One of the main characteristics of MANETs with respect to security design point of view is the lack of a clear defense perimeter. In the case of wired networks, dedicated routers are used, which perform routing functionalities for devices; in MANETs, each mobile node acts as a router and forward packets for other nodes. It is also true that the wireless channel is accessible to both network users and attackers. There is no well-defined system by which traffic from different nodes should be monitored or access control mechanisms can be enforced. As a result, there is no defense perimeter that separates the inside network from the outside network. Thus, the existing ad hoc routing protocols, such as dynamic source routing (DSR)[1] and ad hoc on-demand distance vector(AODV) [2], and wireless medium access control protocols, such as IEEE 802.11, are typically assumed to be trusted. As a result, an attacker being a router can disrupt network operations.

8.1.2 Basic Routing Protocols of MANETs

To provide effective transmission of data in a MANET, special routing protocols are used, which determine a path from the source to the destination node. Another additional requirement for these protocols is a small routing table that must be constantly updated as nodes appear, disappear, or move. The time and the number of messages needed to locate a route should be minimal. Due to these constraints in MANETs, traditionally used routing algorithms are not used. The main constraints of routing in MANETs are as follows:

■ Nodes have no prior knowledge about the topology of the network.
■ No central instances for storing routing information.
■ Mobility of nodes and continuous topology change associated to mobility.
■ Changing metric of transmission lines (e.g., by interference).
■ Limited resources of nodes (e.g., system performance and energy consumption).

There are more than 70 competing designs for the routing of packets in a MANET. A classification of routing protocols can be done in four categories [3]: unicast routing (destination of the data transmission is a single node), multicast routing (destination are multiple nodes), geocast routing (target are all nodes in a given geographic area), and broadcast forwarding (target are all nodes in the range of the transmitter). Based on the technical approach of routing, protocols can be classified in the following categories:

■ *Position-based routing method.* Position-based routing techniques use geodetic information about the exact positions of the nodes. This information is used via a global positioning system receiver. Based on this location information, the shortest or the best paths between source and destination nodes are determined. An example of a position-based routing protocol is the location-aided routing protocol [4].
■ *Topology-based routing method.* Topology-based routing is based on logical information about the neighborhood of each node. The neighboring nodes can communicate with each other. The topological information is usually obtained by sending so-called HELLO packets. Depending on the timing of the construction of the topology data, proactive or reactive routing strategies are developed. Examples of routing protocols of this class are the neighborhood discovery protocol and optimized link state routing (OLSR) protocol [5].
■ *Proactive process.* Proactive routing process determines the paths between two nodes even before they are needed to transfer user data. User data will then be sent, so there is no need

to wait for the determination of the path to the destination node. However, paths between all pairs of nodes may not be required in the future. However, considerable bandwidth is wasted for the control packets that determine such paths. An example of a protocol of this class is the OLSR protocol [5].

■ *Reactive process.* In contrast to the proactive protocols, reactive routing techniques determine the path between two nodes only when data are to be transferred. The first packet is sent with a delay. The node waits for the control packets to be received, and then it determines the route. This suggests a positive impact in the energy consumption of nodes. The AODV routing protocol [2] is an example of a protocol in this category.

■ *Hybrid methods.* Hybrid methods are the mixture of proactive and reactive routing techniques by combining the advantages of both approaches in a new routing protocol. For example, a proactive process can be used locally in a restricted area, whereas further afield is used for a reactive process. This reduces the load on the network with control packets sent from a purely proactive process over the entire network. Zone routing protocol [6] is a routing protocol that implements this approach.

8.1.3 Security Needs in MANETs

MANETs are prone to suffer from malicious behaviors compared to the traditional wired networks. Therefore, the security in an ad hoc network is a major concern. In 1996, the Internet Engineering Task Force created a MANET Working Group with the goal of standardizing IP routing protocol functionality suitable for wireless routing applications within both static and dynamic topologies [7]. Possible applications of MANETs include soldiers relaying information for situational awareness on the battlefield, business associates sharing information during a meeting, attendees using laptop computers to participate in an interactive conference, and emergency disaster relief personnel coordinating efforts after a fire, hurricane, or earthquake. Other possible applications are personal area and home networking, location-based services, and sensor networks.

MANETs have the following typical features:

■ *Unreliability of wireless links between nodes.* Because of the limited energy supply for the wireless nodes and the mobility of the nodes, the wireless links between mobile nodes in the ad hoc network are not consistent for the communication participants.

■ *Constantly changing topology.* Due to the continuous motion of nodes, the topology of the MANET changes constantly: the nodes can continuously move into and out of the radio range of the other nodes in the ad hoc network, and the routing information will be changing all the time because of the movement of the nodes.

■ *Targeted attacks for MANETs.* Lack of incorporation of security features in a statically configured wireless routing protocol is not meant for ad hoc environments. Because the topology of the ad hoc networks is changing constantly, it is necessary for each pair of adjacent nodes to incorporate in the routing issue to prevent some kind of potential attacks that try to make use of vulnerabilities in the statically configured routing protocol.

Security is an essential service for wired and wireless network communications. The success of MANETs strongly depends on whether their security can be trusted. However, the characteristics of MANETs raise both challenges and opportunities in achieving the security goals, such as confidentiality, authentication, integrity, availability, access control, and nonrepudiation.

8.2 Security Loopholes in MANETs

The vulnerabilities of MANETs include the following:

- Nodes can be captured and compromised. The attacker may get the access to a legitimate node and uses the resources of the network.
- Algorithms are assumed to be cooperative, but some nodes may not respect the rules. A node may not forward the data packet to the next-hop node, or a node may hold the media access for an indefinite time.
- Routing mechanisms are susceptible to attacks. The malicious node can advertise nonexisting routes and create a false impression to the neighboring nodes.
- Public key can be maliciously replaced. As there is no trusted central authority, it is hard to manage the key.
- Some keys can be compromised. As MANET nodes cannot afford much computation power due to the limitation of the nodes and power constraints, the keys are simple and easy to compromise.
- The trusted server can fall under the control of a malicious party. That is why trust management is a crucial issue in MANETs.

An ad hoc wireless network does not have any predefined infrastructure (such as centralized servers), and all network services are configured and established dynamically in run time. Thus, it is obvious that, with the lack of infrastructural support, security in an ad hoc network becomes an inherent weakness. Achieving security within ad hoc networking is challenging due to the following reasons:

- Dynamic topologies.
- A network topology of an ad hoc network is very dynamic, as mobility of nodes or membership of nodes is very random and rapid. This emphasizes the need for secure solutions to be dynamic.
- Vulnerable wireless link.
- Passive/active link attacks such as eavesdropping, spoofing, denial of service (DoS), masquerading, and impersonation are possible.
- Roaming in dangerous environments.
- Any malicious or misbehaving node can create hostile attack or deprive all other nodes from providing any service.

Nodes within a dynamic environment with access to a common radio link can easily participate to set up ad hoc infrastructure. However, the secure communication among nodes requires the secure link to communicate. Before establishing a secure communication link, the node should be capable enough to identify another node. As a result, a node needs to provide its identity as well as associated credentials to another node. However, the delivered identity and credentials need to be authenticated and protected so that the authenticity and integrity of delivered identity and credentials cannot be questioned by the receiver node. Every node wants to be sure that delivered identity and credentials to the recipient nodes are not compromised. Therefore, it is essential to provide security architecture to secure ad hoc networking.

The abovementioned problem leads to a privacy problem. In general, a mobile node uses various types of identities that vary from link level to user/application level. Also, in a mobile environment, frequently a mobile node is not ready to reveal its identity or credentials to another mobile node from the privacy point of view. Any compromised identity leads an attacker to create privacy threat to a user's device. Unfortunately, the current mobile standards [8] do not provide any location privacy; in many cases, revealing identity is inevitable to generate communication links. Hence, a seamless privacy protection is required to harness the use of ad hoc networking.

The success of MANETs strongly depends on whether their security can be trusted. Achieving the security goals, such as confidentiality, authentication, integrity, availability, access control, and nonrepudiation, is still one of the challenges ahead. There is a wide variety of attacks that target the weakness of MANETs. For example, routing messages are an essential component of mobile network communications, as each packet needs to be passed quickly through intermediate nodes, wherein the packet must traverse from a source to the destination. Malicious routing attacks can target the routing discovery or maintenance phase by not following the specifications of the routing protocols. There are also attacks that target some particular routing protocols, such as DSR or AODV. Blackhole (or sinkhole) [3], byzantine [9], and wormhole [10] attacks have been recorded recently. Currently, security in routing is one of the major research areas in MANETs.

8.2.1 Potential Threats and Loopholes of Routing Protocols

There are two sources of threats related to nodes in MANETs. The first is an external attack, in which unauthenticated attackers can replay old routing information or inject false routing information to partition the network or increase the network load. The second is an internal attack, which comes from the compromised nodes inside the network. Because compromised nodes can be authenticated, internal attacks are usually much harder to detect and can create severe damage. Although passive (eavesdropping) attacks are also possible in MANETs, they can easily be controlled by using cryptographic mechanisms. Active attacks, which are more damaging, cannot be defended by only applying cryptography mechanisms. Exploits against MANET routing protocols can be classified into modification, fabrication, tunneling attack, DoS attack, invisible node attack, Sybil attack, rushing attack, and noncooperation.

8.2.1.1 Modification

By altering routing information, an attacker can cause network traffic to be dropped, to be redirected to a different destination, or to take a long route to the destination, increasing communication delays [3].

Using AODV as an example, a malicious node can either increase the `broadcast _ id` in a received route request packet (RREQ) to make the faked RREQ message acceptable or decrease

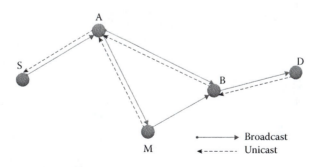

Figure 8.2   Redirection with modification.

the hop _ cnt to update other nodes' reverse routing tables. In the network illustrated in Figure 8.2, a malicious node M can increase the chances it is included on a newly created route from source node S to destination node D by consistently advertising to A a shorter route to D than that B advertises.

8.2.1.2 Fabrication

Fabrication refers to attacks performed by generating false routing messages. This attack is launched by sending a false route error message. This is an active attack where the data packet is fabricated or changed by the malicious node. In Figure 8.2, S has a route to D via nodes A and B. A malicious node M can launch a DoS attack by continually sending route error messages to A spoofing B, indicating a broken link between B and D. A receives the spoofed route error message thinking that it came from B. A deletes its routing table entry for D and forwards the route error message to the upstream node, which then also deletes its routing table entry. If M listens and broadcasts spoofed route error messages whenever a route is established from S to D, it can successfully prevent communications between S and D.

8.2.1.3 Tunneling Attack

Tunneling attack is also called wormhole attack. In a tunneling attack, an attacker receives packets at one point in the network, "tunnels" them to another point in the network, and then replays them into the network from that point. It is called tunneling attack because the colluding malicious nodes are linked through a private network connection that is invisible at higher layers. In Figure 8.3, M receives RREQ and tunnels it to N. When N receives the RREQ, it forwards the RREQ to D as if it had traveled S, M, and N. N also tunnels the RREQ back to M. By doing this, M and N falsely claim a path between them and fool S to choose the path through M and N.

8.2.1.4 DoS Attack

A DoS attack is an attack where a malicious node floods irrelevant data to consume network bandwidth or to consume the resources (e.g., power, storage capacity, or computation resource) of a particular node. With fixed infrastructure networks, DoS attack can be controlled by using round-robin scheduling; but with MANETs, this approach has to be extended to adapt to the lack of infrastructure, which requires the identification of neighbor nodes by using cryptographic tools, and the cost is very high.

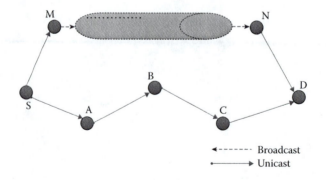

Figure 8.3   Tunneling attack.

8.2.1.5 Invisible Node Attack

This attack occurs when an intermediate node M does not append its IP address to the route record field of the secure routing protocol (SRP) header (see Section 8.3.1.1). In the SRP, the destination node uses the accumulated route record to establish a path between the source node and itself.

8.2.1.6 Sybil Attack

The Sybil attack refers to representing multiple identities for malicious intent [3]. This can be achieved if the malicious nodes collude and share their secret keys. As illustrated in Figure 8.4, A is connected with B, C, and the malicious node, M1. If M1 represents other nodes M2, M3, and M4 (e.g., by using their secret keys), this makes A believe it has six neighbors instead of three. In MANETs, where the functionality relies on the trust of each node, the Sybil attack is very harmful. By "being in more than one place at once," the Sybil attack disrupts geographic and multipath routing protocols. In a MANET that uses multipath routing, the possibility of choosing a path that contains a malicious node (e.g., M1) will be largely increased.

8.2.1.7 Rushing Attack

Generally, during the process of route discovery, only the first RREQ is processed. If the RREQ forwarded by an attacker is the first to reach the destination node, then the route discovered will include the hop through the attacker. Thus, an attacker that can forward route request packets more quickly than legitimate nodes can increase the probability of being included in the discovered route.

8.2.1.8 Noncooperation

In MANETs, the resource (e.g., power, storage capacity, and computation resource) of a mobile node is restricted. In order to get the most benefit, a mobile node may behave selfishly to save energy for itself; it may not participate in routing or may not forward packets for other nodes. This kind of node misbehavior caused by lack of cooperation is called node selfishness. A selfish node

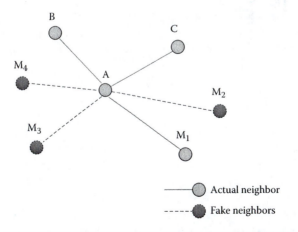

Figure 8.4　Sybil attack.

differs from a malicious node for it does not intend to damage other nodes with active attacks, but the damage selfish behaviors cause to a MANET cannot be underestimated.

8.2.2 Attacks on MANETs in Different Layers

Attacks in an ad hoc network can be categorized in many different ways.

■ *Active and passive attacks.* A passive attack obtains data communicated in the network without disrupting the communications, whereas an active attack involves information interruption, modification, or fabrication, thereby disrupting the normal functionality of a MANET. Eavesdropping, traffic analysis, and monitoring are passive attacks. Active attacks are jamming, spoofing, modification, replaying, and DoS.

■ *Internal and external attacks.* Internal attacks are performed by a node within the ad hoc network perimeter by compromising it. External attacks are performed by an agent external to the network.

■ *Stealthy versus nonstealthy attacks.* Some of the attackers try to hide their existence and action. These are stealthy attacks. The attacks where the action of the attacker is revealed are categorized as nonstealthy.

■ *Cryptographic versus noncryptography-related attacks.* The different nodes in an ad hoc network exchange messages in encrypted form, so attacks on a cryptographic area, such as RSA, ElGamal, and SHA, are cryptographic attacks. Attacks involving plain text are noncryptographic.

8.2.2.1 Attacks in Physical Layer

Ad hoc communication uses wireless broadcast for sending packets. It is easy to sniff packets from a wireless medium. Radio signal can easily be jammed or intercepted.

■ *Eavesdropping.* Eavesdropping is intercepting messages with unwanted nodes. The mobile hosts in MANETs share a wireless medium. The communications use radio frequency spectrum and broadcast by nature. The broadcasted signal can be easily intercepted using a receiver tuned in proper frequency.

■ *Interference and jamming.* The message can be tampered or blocked by using proper radio frequency signals to jam or interfere the signal. An attacker uses a powerful transmitter. The most common types of this form of signal jamming are random noise and pulse. Jamming equipment are commercially available.

8.2.2.2 Attacks in Link Layer

The MANET is an open multipoint peer-to-peer network architecture. Specifically, one-hop connectivity among neighbors is maintained by the link layer protocols, and the network layer protocols extend the connectivity to other nodes in the network. Attackers can exploit the cooperation of the layer's protocols. Wireless medium access control protocols have to coordinate over the common transmission medium. Because a token-passing bus medium access control protocol is not suitable for controlling a radio channel, IEEE 802.11 protocol is specifically devoted to wireless local area networks (LANs). By keeping the wireless media busy (by holding back the token) for an indefinite time, any node can make distributed DoS attacks.

8.2.2.3 Attacks in Network Layer

The connectivity between mobile hosts over a potentially multihop wireless link relies heavily on cooperative reactions among all network nodes. A variety of attacks targeting the network layer have been identified by contemporary research works. By attacking the routing protocols, attackers can intercept network traffic, inject themselves into the path between the source and the destination, and thus control the network traffic. The traffic packets follow a nonoptimal path, which may cause significant delay. In addition, the packets could be forwarded to a nonexistent path and get lost. The attackers can create routing loops, introduce severe network congestion, and channel contention into certain areas. Multiple colluding attackers may even prevent a source node from finding any route to the destination, causing the network to partition, which triggers excessive network control traffic and further intensifies network congestion and performance degradation. There are malicious routing attacks that target the routing discovery or maintenance phase by not following the specifications of the routing protocols. Routing message flooding attacks, such as HELLO flooding, RREQ flooding, acknowledgment flooding, routing table overflow, routing cache poisoning, and routing loop, are simple examples of routing attacks targeting the route discovery phase. Proactive routing algorithms, such as destination-sequenced distance vector protocol [11] and OLSR protocol [5], attempt to discover routing information before it is needed, whereas reactive algorithms [11], such as DSR and AODV, create routes only when they are needed. Thus, proactive algorithms perform worse than on-demand schemes because they do not accommodate the dynamic of MANETs, and clearly proactive algorithms require many costly broadcasts. Proactive algorithms are more vulnerable to routing table overflow attacks. Some of these attacks are listed below.

8.2.2.4 Attacks in Higher Layers

To provide end-to-end connectivity, the higher layers (transport layer and application layer) are presenting the ad hoc network. Very common attacks in the transport layer are SYN flooding and session hijacking. Applications in MANET nodes can attack on the network by malicious programs or repudiation. The SYN flooding attack is a DoS attack. The attacker initiates a large number of transmission control protocol (TCP) connections with a victim node but never completes the handshake to fully open the connection. For two nodes to communicate using TCP, they must first establish a TCP connection using a three-way handshake. The three messages exchanged during the handshake allow both nodes to learn that the other is ready to communicate and to agree on initial sequence numbers for the conversation.

8.3 Security Countermeasures

In MANETs, all networking functions, such as routing and packet forwarding, are performed by the nodes themselves in a self-organizing manner. For this reason, such networks have increased vulnerability and securing a MANET is very challenging. The following attributes are important issues related to MANETs, especially for those security-sensitive applications:

- Availability ensures the survivability of network services despite DoS attack.
- Confidentiality ensures that certain information is never disclosed to unauthorized entities.
- Integrity guarantees that a message being transferred is never corrupted.

- Authentication enables a node to ensure the identity of the peer node it is communicating with.
- Nonrepudiation ensures that the origin of a message cannot deny having sent the message.

Because of the nature of ad hoc, it is extremely difficult to achieve the above security goals in MANETs. Threats that MANETs have to face can be classified into two levels: attacks on the basic mechanism and attacks on the security mechanism.

8.3.1 Different Security Protocols for MANETs

There might be a number of attacks performed in routing protocol of ad hoc networks, which are discussed in an earlier section. Routing is an important resource of ad hoc networks. To secure routing, a number of modifications on the existing routing protocol have been proposed. Among them, SRP and authenticated routing for MANETs (ARAN) are discussed in detail. Few more protocols, such as ARIADNE and SEAD, are recently developed secured algorithms for ad hoc routing.

8.3.1.1 SRP

Papadimitratos and Haas [12] proposed the SRP as an extension of existing on-demand routing protocols. The SRP emphasizes the acquisition of correct topological information in a timely manner in the presence of malicious nodes. It introduces a set of features, such as the requirement that the query verifiable arrives at the destination, the consequent verifiable return of the query response over the reverse of the query propagation route, the query/reply identification by a dual identifier, the reply protection of the source and destination nodes, and the regulation of the query propagation. The only assumption of the proposed scheme is the existence of a security association between the node initiating the query and the destination. The trust relationship could be instantiated by the knowledge of the public key of the other communicating end. The two nodes can negotiate a shared secret key (KS,T) and then, using the secret key, verify that the nodes that participated in the communication are indeed the trusted node. The route request packet initiated by the source node S contains a pair of identifiers: a query sequence number and a random query identifier. The source and destination and the unique (with respect to the pair of end nodes) query identifiers are the input for calculating the message authentication code (MAC) along with KS,T. The identities of the traversed intermediate nodes are accumulated in the route query packet. The intermediate nodes relay route requests and maintain a limited amount of state information regarding the relayed queries so that previously seen route requests are discarded. When the route request reaches the destination T, T verifies the integrity and authenticity of the request by calculating MAC and comparing them with the MAC contained in the route request packet. If the route request is valid, T constructs the route replies, calculates a MAC covering the route reply contents, and returns the packet to S over the reverse of the route accumulated in the respective request packet. The destination response is to one or more request packets of the same query, so that it provides the source with as diverse topology as possible. The querying node will validate the replies and update its topology. The SRP copes with noncolluding malicious nodes that are able to modify, replay, spoof, and fabricate routing packets. However, the SRP suffers from the lack of a validation of route maintenance messages: route error packets are not verified. However, by source-routing error packets along the prefix of the route reported as broken, the source node can verify that the provided route error feedback refers to the actual route and is not generated by a node that is not even part of the route; that is, a malicious node can harm only the route in which it belongs. The SRP is also not immune to the wormhole attack: two colluding malicious nodes can misroute the routing packets on a private network connection and alter the network topology vision a benign node can collect.

8.3.1.2 ARAN

Sanzgiri et al. [10] proposed a secure MANET routing protocol (ARAN) that detects and protects against malicious actions by third parties and peers in one particular MANET environment. ARAN introduces authentication, message integrity, and nonrepudiation. It makes use of cryptographic certificates and requires the use of a trusted certificated server, whose public key is known to all valid nodes.

The route discovery packet (RDP) includes the packet type identifier (RDP), the IP address of D (IP_D), S's certificate ($Cert_S$), a nonce (N_S), and the current time (t), all signed with S's private key (K_S-). Each time S performs route discovery, it monotonically increases the nonce.

When S's neighbor B received the packet, it validates the signature, sets up a reverse path back to the source, and forward broadcasts the message:

$$[[RDP, IP_D, Cert_S, N_S, t] K_S-, Cert_S]K_B-, Cert_B.$$

The signature of B prevents spoofing attacks that may alter the route or form loops. B's neighbor C received the packet, validates the signature, sets up a reverse path by recording the neighbor from which it received the RDP, and forward broadcasts the message:

$$[[RDP, IP_D, Cert_S, N_S, t] K_S-, Cert_S]K_C-, Cert_C.$$

Each node along the path validates the previous node's signature, removes the previous node's certificate and signature, records the previous node's IP address, signs the original contents of the message, appends its own certificate, and forward broadcasts the message. Eventually, the message is received by the destination, D, which replies to the first RDP that it receives for a source and a given nonce. There is no guarantee that the first RDP received traveled along the shortest path from the source. The destination unicasts a route reply (REP) packet back along the reverse path to the source. Let the first node that receives the REP sent by D be node C. D will send to C the following message:

$$[REP, IP_S, Cert_D, N_S, t]K_D-, Cert_D.$$

The REP includes a packet type identifier (REP), the IP address of S (IP_S), the certificate belonging to D ($Cert_D$), the nonce (N_S), and the associated timestamp (t) sent by S. D also signs the REP using its private key (K_D-).

Nodes that receive the REP forward the packet back to the predecessor from which they received the original RDP. Each node along the reverse path back to the source signs the REP and appends its own certificate before forwarding the REP. Let C's next hop to the source be node B. C will send to B the following message:

$$[[REP, IP_S, Cert_D, N_S, t] K_D-, Cert_D]K_C-, Cert_C.$$

B validates C's signature, removes the signature, and then signs the contents of the message before unicasting the following RDP message to S:

$$[[REP, IP_S, Cert_D, N_S, t] K_D-, Cert_D]K_B-, Cert_B.$$

Each node checks the nonce and signature of the previous hop as the REP is returned to the source. This avoids attack where malicious nodes instantiate routes by impersonation and replay of D's message. When the source receives the REP, it verifies the destination's signature and the nonce returned by the destination.

8.3.2 Defense against Specific Attacks

Different defense mechanisms have been proposed for different attacks. As the different attacks have been discussed in an earlier section, the security strategies are discussed for those attacks. The security measures are discussed according to the different network layer they fit.

8.3.2.1 Security in Physical Layer

Spread spectrum technology, such as frequency hopping (FHSS) or direct sequence (DSSS), can make difficulties to the attackers to detect or jam signals. It changes frequency in a random fashion to make signal capture difficult. Directional antennas can also be deployed due to the fact that the communication techniques can be designed to spread the signal energy in space.

In FHSS, the signal is modulated by a seemingly random series of radiofrequencies that jump from one frequency at fixed intervals. The receiver uses the same spreading code that is synchronized with the trans-issuer, wherein the spread signals in their original form recombine. When the transmitter and the receiver are correctly synchronized, data are transmitted in one lane. However, the signal appears to be the duration incomprehensible impulse noise eavesdroppers. Meanwhile, interference is minimized when the signal is spread over several frequencies.

Then again, each data bit in the original signal is represented by a number of bits in the transmitted signal using a spreading code. The spreading code spreads the signal over a wider frequency band in direct proportion to the number of bits used. The recipient can use the spreading code signal to recover original data.

For a four-bit in the transmitted signal, the first bit of 0110 is sent as a 0 first four-bit of the spreading code. The second bit 1 is transmitted as 0110. This is the bit-wise complement of 4 s of the spreading code of the music. In turn, each input bit is connected to the exclusive or with four pieces of the diffusion code.

Both FHSS and DSSS are difficult for foreigners to try to intercept radio signals. The spy must know the frequency, application techniques, and code modulation to read the signals correctly. The property that the spread spectrum technology does not work with other difficulties added to the spy. The spread spectrum technology minimizes the risk of interference with radio and other electromagnetic devices. Despite the ability of technology to spread spectrum, it is certain that when the hopping pattern or spreading code is unknown these are eavesdroppers.

8.3.2.2 Security in Link Layer

Neighbors should monitor the misbehaviors of each node to identify whether it is disturbing the cooperative nature of the network. Although it is still an open challenge to prevent selfishness, some schemes have been proposed, such as ERA-802.11, where detection algorithms are proposed. Traffic analysis is prevented by encryption at the data link layer. The wired equivalent privacy (WEP) encryption scheme defined in the IEEE 802.11 wireless LAN standard uses link encryption to hide the end-to-end traffic flow information. However, WEP has been widely criticized for its weaknesses. Some secure link layer protocols have been proposed in recent research, such as the link-layer security protocol.

8.3.2.3 Security in Network Layer

In general, a kind of authentication and integrity mechanism, either the hop-by-hop or the end-to-end approach, is used to ensure the correctness of routing information. For instance, digital signature, one-way hash function, hash chain, MAC, and hashed MAC are widely used for this purpose. IPsec and ESP are standards of security protocols on the network layer used in the Internet that could also be used in MANETs, in certain circumstances, to provide network layer data packet authentication and a certain level of confidentiality; in addition, some protocols are designed to defend against selfish nodes, which intend to save resources and avoid network cooperation. Some SRPs have been proposed in MANETs in recent papers. Those defense techniques are outlined in the next sections. Sanzgiri et al. [10] described an SRP. In this regard, Ning et al. [22] described attacks and countermeasures of the AODV routing protocol.

8.3.2.4 Security in Higher Layers

Session hijacking takes advantage of the fact that most communications are protected (by providing credentials) at session setup but not thereafter. In the TCP session hijacking attack, the attacker spoofs the victim's IP address, determines the correct sequence number that is expected by the target, and then performs a DoS attack on the victim. Thus, the attacker impersonates the victim node and continues the session with the target. Sundaresan et al. [23] showed a number of protocols such as TCP feedback, TCP explicit failure notification, ad hoc TCP, and ad hoc transport protocol for transport layer in an ad hoc network, but none of these protocols are designed with security in mind. Extension of secure socket layer and transport layer security are the solutions proposed against SYN flooding and session hijacking. Viruses, trojans, and exploits can keep on executing in the nodes. A selfish node can deny any operation afterward. Therefore, to solve these kinds of attacks, intrusion detection system for MANETs is developed. Recently, Cheng and Tseng [13] have been working on a more advanced intrusion detection system for ad hoc networks.

8.4 Access Control in MANETs

Enforcing security policies in MANETs is in its premature stage. A framework to manage ad hoc networks with a policy-based administration was proposed by Chadhaet al. [14] in 2004. The concept of trust in MANET communication has been proposed to ensure collaboration and cooperation. The use of RBAC models in ad hoc networks is a recent trend [15]. In 2009, Alicherry et al. [16] proposed a framework to implement access control mechanism on the network resources such as a service. Maity et al. [17] proposed a framework for access control in MANETs.

8.4.1 Access Control Challenges in MANETs

The security technologies (firewall, intrusion detection system, access control lists, etc.) that are used in traditional structured networks are unable to meet the security requirements. For more than a decade, various security issues in MANETs are being addressed by researchers. Avramovic [18] first introduced the concept of policy-based routing.

Jin and Ahn [19] showed the use of RBAC models in ad hoc networks. However, this work did not focus on how the role information would propagate through the whole network. Alicherry et al. [16] proposed a framework to implement access control mechanism on the network resources such as a service. However, they did not consider the effect of access control mechanism on routing schemes.

The different resources in an ad hoc network, which are the concerns of researchers, are the physical media, group management, cooperation, and role.

8.4.1.1 Access Control on Physical Media

In ad hoc networks, coordination of the access from active nodes is controlled by the medium access control protocols. These protocols deserve significant attention because the wireless communication channel is inherently prone to errors and unique problems such as the hidden terminal problem, the exposed terminal problem, and signal fading effects.

8.4.1.2 Access Control on Group Membership and Admission

Group membership is an important resource in any network. In ad hoc networks, under constant change in topology, maintaining group membership and proper access control on group membership is a major challenge. Group membership is controlled by two techniques that are complementary to each other. One is admission control and the other is secure group communication. Admission control in an ad hoc network focuses on cryptographic techniques to perform secure group admission. The purpose is for a certain threshold of group members to make collaborative decisions regarding the admission of a prospective member and provide it with a signed group membership certificate. Among these signature schemes plain (RSA or DSA) signatures, accountable subgroup multi signature scan be mentioned. Then again, authentication in MANETs can be centralized or decentralized. There is another attempt for admission control in MANETs by Chen et al. [20]. They have introduced a public key infrastructure-based key management protocol. The key management protocol ensures secure admission control in MANET environments. By assigning the responsibilities of the authenticator to multiple certificate authorities (CAs), which are selected from a pool of users with the highest trust levels, security is achieved. In this approach, manual selection of CAs is not required. A certificate graph to represent the friendship among the participants is maintained. This approach is similar to human social networks in which good (i.e., nonmalicious) users are expected to have more friends than bad (i.e., malicious) ones. The most trustworthy subset of these good users in a MANET is represented by the maximum clique and is selected as the authenticator of this group.

8.4.1.3 Access Control on Cooperation

In MANETs, basic networking functions such as packet forwarding and routing are carried out by all available nodes in the network. There is no reason to assume that the nodes will cooperate with one another because network operation consumes energy, which is a particularly scarce resource in MANETs. A new type of node misbehavior is caused by lack of cooperation and is called node selfishness. A selfish node differs from a malicious node, as it does not intend to damage other nodes with active attacks but simply does not cooperate to the network operation, saving battery life for its own communication. However, damages caused by selfish behavior cannot be underestimated. Packet forwarding functions, trust, and group key management are the three important resources for cooperation.

■ Mechanisms to enforce a node to cooperate by forwarding the packets in a MANET can be divided into two categories: one is currency based (Nuglets and Sprite) and the other uses a local monitoring technique (Watchdog, Confidant, and CORE). Currency-based systems are simple to implement but may rely in a tamperproof hardware, and it is difficult to establish a way to exchange the virtual currency, making their use not realistic in a practical system. Cooperative security schemes based on a local monitoring offer a more suitable solution to the selfish problem. Every node monitors its local neighbors, evaluating for each of them a metric that is directly related to the node's behavior. The main drawback is related to the absence of a mechanism that securely identifies the nodes of the network: any selfish node could elude the cooperation enforcement mechanism and get rid of its bad reputation by changing its identity.

■ The trust establishment schemes are grouped into two groups based on the type of evidence that trust evaluation uses. The two groups investigated are certificate-based models and reputation-based trust models.

■ Group key management is another important aspect for cooperation. The most commonly accepted taxonomy of group key management protocols divides it into three approaches: centralized, decentralized, and distributed. The centralized approach uses only one server. This server is responsible for the generation, the distribution, and the renewal of the group key. This approach is clearly not scalable. Decentralized approach divides the multicast group into a prefixed number of subgroups. Each of them shares a local session key managed by a local controller. When a member joins or leaves the group, only the concerned subgroup will renew its local key. In distributed approach (also called key agreement approach), all group members cooperate and generate the traffic encryption key to establish secure communications between them. The key agreement approach eliminates the bottleneck in the network compared with the centralized approach but is less scalable because the traffic encryption key is composed of the contributions of all group members and needs more computation processing.

8.4.1.4 RBAC

RBAC was designed to provide a way for assigning and managing permissions to users. However, its centralized architecture and preconfiguration requirement made it unfit in MANETs. Barka and Gadallah [15] proposed a role-based protocol for MANETs. This framework is focused on control access to MANET multicast groups based on node credentials and control access to information exchanged within the group based on the access level of the different nodes. For this purpose, they utilized the features of the RBAC model to develop a secure multicast protocol. This protocol is based on the multicast features of the AODV protocol [2], with enhancements aiming to control access memberships to the multicast groups as well as data within these groups. The framework tried to ensure secure communication between members of different groups in an ad hoc mobile environment. Their approach was focused on the hierarchical organizational structure (e.g., in the military environment) where senior ranks inherit the permissions assigned to their juniors. Therefore, it can both control access to a certain multicast group and ensure that only the right member of the group gets access to the right data based on membership role privileges.

8.4.2 Access Control Frameworks Proposed for MANETs

There are a number of access control frameworks proposed by researchers to ensure the security of MANETs. DIPLOMA by Alicherry and Keromytis [21] is a novel framework for access control implementation in MANETs that uses deny-by-default paradigm. Maity et al. [17] also proposed another such framework. A generalized framework for enforcing access control in MANETs is discussed in this section.

These frameworks ensure network access control of each node in MANETs. The main reason that motivates researchers to ensure access control in MANETs is the reliability of the intermediate nodes. To ensure the packet sent by a particular node reaches its destination, the intermediate nodes should be trusted and reliable. To enforce policy and security access control mechanism, several challenges should be overcome.

Such access control architecture (Figure 8.5) implements the global policy over the distributed nodes in MANETs. A central authority configures the nodes before they come into the network. Unconfigured nodes are treated as guest nodes. The global policy is set by the central authority. These frameworks are designed to reflect the global policy on the ad hoc network by implementing access control mechanism in a distributed fashion. This motivates to develop a policy-based security management framework over the organizational ad hoc network. This framework considers two distinct entities: offline central authority and unstructured ad hoc network. Central authority is responsible for the authentication and policy administration, whereas ad hoc network is a collection of collaborative nodes. One or more nodes are chosen as a policy enforcing node or group head. A group head is responsible in enforcing the policy rules to the MANET nodes in a distributed fashion. Therefore, it is a novel approach to enforce security access control in MANETs. RBAC-supported and trust-based mechanism to secure MANETs with policies set by the administrator is implemented in the framework. A number of research challenges need to be overcome. These research challenges in different spectra are mapped into three broad categories: access control modeling, policy implementation, and trust enforcing. The proposed access control in MANETs, with minimal power consumption and trusted communication ensures that security of MANETs minimizes the effect of active and passive attacks by identifying and avoiding the selfish and malicious nodes.

Another example of such a framework is DIPLOMA proposed by Alicherry and Keromytis [21]. In this framework, group heads are predefined and may be one or more. They are called group

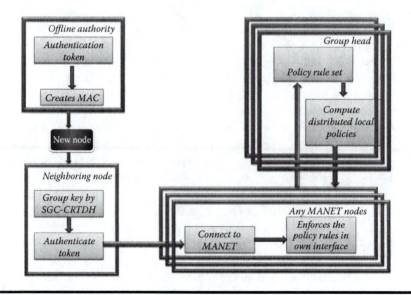

Figure 8.5   Framework for access control in MANET.

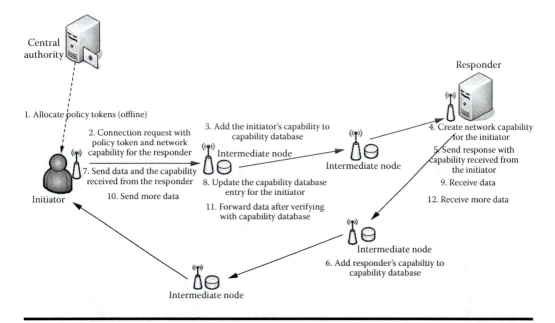

Figure 8.6 DIPLOMA framework. (From M. Alicherry and A.D. Keromytis. DIPLOMA: Distributed Policy Enforcement Architecture for MANETs, in International Conference on Network and System Security, Citeseer, 2010. With permission.)

controller (GC), which is trusted by all the group nodes. A GC has authority to assign resources to the nodes in MANETs. This resource allocation is represented as a credential (capability) called policy token, and it can be used to express the services and the bandwidth a node is allowed to access. They are cryptographically signed by the GC, which can be verified by any node in MANETs. When a node (initiator) requests a service from another MANET node (responder) using the policy token assigned to the initiator, the responder can provide a capability back to the initiator. This is called a network capability, and it is generated based on the resource policy assigned to the responder and its dynamic conditions (e.g., level of utilization). Figure 8.6 gives a brief overview of the system.

All nodes in the path between an initiator and a responder (i.e., nodes relaying the packets) enforce and abide by the resource allocation encoded by the GC in the policy token and the responder in the network capability.

8.5 Conclusion

The chapter covers the security mechanisms for different resources in ad hoc networks. There are still plenty of research scopes in the analysis of cooperation, trust, policy-based routing, routing strategies, coalition formation algorithms, RBAC, and firewall distribution. However, using a more realistic representation of the mechanisms underlying the management of the network still needs to be achieved. Indeed, several limitations of these implementations are due to initial hypotheses such as homogeneous networks, although it would be more realistic to consider heterogeneous networks. Game theoretical approach for determining strategies and cooperation is widely used in recent works. Regarding the final goal of designing distributed access control policy implementation from scratch, most solutions proposed are not capable to completely ensure access control. This chapter emphasized the need for a security architecture that implements access control mechanism in an ad hoc network.

8.6 Open Challenges and Future Trends

The research on MANETs is still at a premature stage. Existing research works are generally based on a specific attack. They could function well in the presence of the designated attack. Research is still ongoing and will lead to the discovery of new threats as well as creation of new counter-measures. The research interest is growing in the areas such as robust key management protocols based on trust, delivery of an integrated safety, and security of data on different layers. The future direction of research might be cryptographic solutions. Cryptography is the fundamental security technology that has been used in almost all aspects of security. The strength of a cryptographic system depends on good management of keys. The approach to public key cryptography is based on the CA central body that is not always available for pure MANETs. Some research papers focus on distributing the keys to several or all entities of the network via a centralized system, whereas some suggest a fully distributed trust model—in the style of pretty good privacy (PGP). Symmetric cryptography has computational efficiency but suffers from potential attacks on key agreement or key distribution. Many complex interactions or key distribution protocols are designed, but for MANETs, they are restricted by the resources of a node, the network topology, and the limited dynamic bandwidth. Efficient key agreement and distribution in MANETs is an area of ongoing research. Many recent researches focus on methods of prevention intrusion detection. An interesting problem is building a system based on trust, in which the level of implementation of the security is dependent on the level of trust. Implementing a secure system based on trust and its integration with traditional methods should be done in future research. Because most attacks are unpredictable, a resiliency-oriented security solution will be more useful, which depends on a multifence security solution. Cryptography-based methods offer a subset of solutions. Other solutions are the future scope of research. Trust-based security, policy-based security, secured routing, and RBAC are the noncryptographic approach toward the MANET security. To enforce policy and security access control mechanism, the following challenges are still the concerns of researchers.

■ *Finding a secure route in an ad hoc network.* In an ad hoc network, a good number of routing protocols have been proposed. DSR, OLSR protocol, and temporally ordered routing algorithm are the widely used protocols for routing in ad hoc networks. However, these routing protocols hardly consider the security or reliability of the channel and the intermediate nodes. The intermediate nodes can be malicious or selfish. Therefore, the routing protocol should be compatible with security access control of an ad hoc network. The route should be selected in such a way that the scope of access control violation will not occur under any circumstances. There is a scope to verify the routing protocols by formal methods or mathematical graph theory-based methods.
■ *Finding the priority of the nodes in terms of security.* In general, an ad hoc network gives the same importance to all nodes. There is no option to set priority based on security. Trust, role, connectivity, and behavior should be the major concerns for setting priority for the nodes. According to the priority of the nodes, the different roles can be assigned in the network.

- *Controlling admission in a network.* There is no definite and standard admission control mechanism for ad hoc networks. WEP and Wi-Fi protected access-based authentication are there. However, the point of contact is only one single node in the network. If that node becomes malicious or compromised by the attacker, then the whole scheme of authentication will fail. Offline certification-based authentication using a secure group key is a smarter choice. However, that area is still in an immature state.
- *Modeling RBAC in an ad hoc network.* Modeling RBAC is a research issue. User nodes and their nodes vary from network to network. However, the temporal and spatial aspects of the role are the major challenge to researchers.
- *Modeling policy specification for an ad hoc network.* Access control policies should be specified for an ad hoc network. As the topology is constantly changing, and the trust values of nodes are often changing with respect to any other nodes, the simple policies will not fit. Specifying policy and modeling them for an ad hoc network is thus another area of research.
- *Policy algebra and policy distribution.* The set of rules combines to overall access control. However, those rules should be complete, sound, conflict-free, and unambiguous. The collection of rules must conform to the global policy. Therefore, the proper distribution of policy rules leads to an immense research interest.
- *Finding security metric for access control in MANETs.* Measuring security for an ad hoc network is still a tough job. Very few works target to frame a proper evaluation of security. Therefore, finding a good security metric considering all the aspects in an ad hoc network will be a major contribution.

References

1. D.B. Johnson, D.A. Maltz, and J. Broch. DSR: The dynamic source routing protocol for multi-hop wireless ad hoc networks, *Ad hoc Networking*, 5:139–172, 2001.
2. C.E. Perkins, E.M. Belding-Royer, and S. Das. Ad hoc on demand distance vector (AODV) routing (RFC 3561), IETF MANET Working Group, 2003.
3. Y.C. Hu and A. Perrig. A survey of secure wireless ad hoc routing, *IEEE Security and Privacy Magazine*, 2:28–39, 2004.
4. Y.B. Ko and N.H. Vaidya. Location-aided routing (LAR) in mobile ad hoc networks, *Wireless Networks*, 6(4):307–321, 2000.
5. T. Clausen and P. Jacquet. *RFC3626: Optimized Link State Routing Protocol (OLSR)*. RFC Editor United States, 2003.
6. Z.J. Haas, M.R. Pearlman, and P. Samar. The zone routing protocol (ZRP) for ad hoc networks, INTERNET-DRAFT, IETF, 2002.
7. Mobile ad-hoc networks (MANET), 2012, [Online], available at http://datatracker.ietf.org/wg/manet/charter/.
8. 3G Security: Security Architecture. 3GPP TS, 33.102, V3.6.0, October 2000.
9. B. Awerbuch, D. Holmer, C. Nita-Rotaru, and H. Rubens. An on-demand secure routing protocol resilient to byzantine failures, in *Proceedings of the 1st ACM Workshop on Wireless Security*, ACM, 2002, pp. 21–30.
10. K. Sanzgiri, B. Dahill, B.N. Levine, C. Shields, and E.M. Belding-Royer. A secure routing protocol for ad hoc networks, 2002.
11. C.E. Perkins. Ad hoc networking: an introduction. Ad hoc Networking, pp. 1–28, Addison-Wesley Longman Publishing Co. Inc., 2008.
12. P. Papadimitratos and Z.J. Haas. Secure routing for mobile ad hoc networks. In SCS Communication Networks and Distributed Systems Modeling and Simulation Conference (CNDS 2002), volume 31, pp. 193–204. San Antonio, TX, 2002.

13. B.C. Cheng and R.Y. Tseng. A context adaptive intrusion detection system for MANET. Computer Communications, 2010.
14. R. Chadha, H. Cheng, Y.H. Cheng, J. Chiang, A. Ghetie, G. Levin, and H. Tanna. Policy-based mobile ad hoc network management, 2004.
15. E.E. Barka and Y. Gadallah. A role-based protocol for secure multicast communications in mobile ad hoc networks, in *Proceedings of the 6th International Wireless Communications and Mobile Computing Conference on ZZZ, ACM*, 2010, pp. 701–705.
16. M. Alicherry, A.D. Keromytis, and A. Stavrou. Deny-by-default distributed security policy enforcement in mobile ad hoc networks, *Security and Privacy in Communication Networks*, 2009, pp. 41–50.
17. S. Maity, P. Bera, and S. K. Ghosh. An access control framework for semi-infrastructured ad hoc networks, in Computer Technology and Development (ICCTD), 2010 2nd International Conference, IEEE, 2010, pp. 708–712.
18. Z. Avramovic. Policy based routing in the defense information system network, in Military Communications Conference, 1992, MILCOM '92, Conference Record, Communications—Fusing Command, Control and Intelligence, IEEE, 3, 11–14, 1992, pp. 1210–1214.
19. J. Jin and G.J. Ahn. Role-based access management for ad-hoc collaborative sharing, in *Proceedings of the Eleventh ACM Symposium on Access Control Models and Technologies, ACM*, 2006, p. 209.
20. Q. Chen, Z.M. Fadlullah, X. Lin, and N. Kato. A clique-based secure admission control scheme for mobile ad hoc networks (MANETs), *Journal of Network and Computer Applications*, 34(6):1827–1835, 2011.
21. M. Alicherry and A.D. Keromytis. DIPLOMA: Distributed Policy Enforcement Architecture for MANETs, in International Conference on Network and System Security, Citeseer, 2010.
22. P. Ning and K. Sun. How to misuse AODV: A case study of insider attacks against mobile ad-hoc routing protocols. *Ad Hoc Networks*, 3(6):795–819, 2005.
23. K. Sundaresan, V. Anantharaman, H.Y. Hsieh, and R. Sivakumar. ATP: A reliable transport protocol for ad hoc networks, *IEEE transactions on mobile computing*, 588–603, 2005.

Chapter 9

Design of Framework for Safety and Resource Management of Converged Networks for Seaport Applications

Luca Caviglione, Cristiano Cervellera, and Roberto Marcialis

Contents

9.1 Introduction

As a consequence of the fast-growing world economy, both the organization and the exploitation of many complex operations of fundamental importance for enterprises and individuals have experienced a revolution ignited by the pervasive nature of network services (NSs). Besides, rich sets of novel services are now accessible through the Internet, thus enabling new business and organizational paradigms, including utilization of software as a service, data storage and synchronization through cloud computing facilities, or complex real-time interactions with third parties such as national or international authorities, public safety bodies, and other entrepreneurs [1].

In this perspective, a seaport is one of the most composite and distributed industrial infrastructures. Accordingly, it can heavily improve its internal setup by exploiting advancements of the most cutting-edge network and telecommunication technologies [2]. Thus, the enhancement of standard duties (e.g., the management of an incoming fleet or the coordination of operations for moving goods) can be heavily simplified or optimized through communication services. In fact, data communications are a relevant component for seaport-related operations as they play an essential role also in the perspective of assuring a safe work environment. To summarize, information exchanges within a seaport are performed for the following purposes: (1) to enable audio/video interaction among employees, especially while working outdoors; (2) to assure the correct delivery of a huge amount of data (e.g., to capture a real-time snapshot of the business in terms of both the status of operations and the placement of goods); (3) to provide the access to, and control of, remote applications and machineries as well as satellite offices or assets; (4) to strengthen the collaboration with third-party entities (e.g., customers or business partners) via Internet connectivity, also in the perspective of delivering added value services; (5) to perform actions aimed at flattening operational costs (i.e., by harnessing the convergence of services over full IP platforms); and (6) to exploit distributed safety frameworks [e.g., via wireless sensor networks (WSNs) or radiofrequency identification (RFID) or by using network's stimuli as an intrinsic probe of the surrounding environment].

However, each one of the aforementioned functionalities needs specific features available in the underlying network infrastructure. For example, Internet telephony needs some real-time guarantees, which are usually implemented within the data bearer through some quality of service (QoS) mechanism. At the same time, specific machineries for handling the required signaling [e.g., the session initiation protocol (SIP)] should be in place [3]. Also, additional technological constraints are due to the need of making services available also while moving. Specifically, this reflects in the adoption of wireless accesses that behave differently from wired ones. To sum up, the network infrastructure must be carefully designed to deliver a given service without neglecting the essential requisites of security and service availability (sometimes defined as the carrier-grade requirement).

This could bring to a highly balkanized (as used in the RFC1726 to express the risk of having an Internet composed by different technologically split portions) network scenario, where different ad hoc infrastructures have to be adopted and properly unified to avoid isolation. Pushing this approach to the limit, each specific service could be delivered through a different network, possibly managed (or owned) by a given provider. In many seaports, it is not unlikely to have the cellular access network provided by a mobile carrier operator, the wide area network (WAN) connectivity by an Internet service provider (e.g., via dark fiber), and satellite backup by another different actor. Even if specialization can be beneficial, such a fragmented scenario has several drawbacks; for example, capital expenditures and operating expense costs can become unsustainable, different technologies could require a variety of specific skills, and service integration could need additional machineries (such as application layer gateways). Besides, as a consequence of the more complex design needed to merge different technological solutions, achieving a satisfactory degree of scalability, fault tolerance, or quality of experience (QoE) could be impossible.

However, researchers and industrial developers have been studying, for years, various mechanisms to integrate heterogeneous networks or to avoid performance degradations. In a nutshell, relevant effort has been put to merge communication systems composed by different technologies, which can also support different service models or protocol architectures. A paradigmatic example is the scenario composed by wired trunks connected via wireless loops (e.g., IEEE 802.11 or satellite links), whereas popular solutions are those employing proxying or mediating entities, such as performance-enhancing proxies and middle boxes. Yet, the increasing diffusion of devices or appliances operated through the network, jointly with the need of ensuring end-to-end transparency (e.g., by avoiding the break of L4 semantic or pushing the adoption of IPv6), makes mandatory to pursue a next step into the design of solutions for making different services to coexist.

To this aim, converged networks (CNs) [4–6] are one of the most promising engineering paradigms to effectively support next-generation services, which also demand for broadband, mobility, and multimedia support. Put briefly, in CNs, the overall deployment is "flexible" enough to support a variety of services with different requirements, especially in terms of bandwidth resources and real-time constraints. Therefore, a seaport is an excellent playground for CNs because their characteristics are very challenging: (1) a seaport's communications must be scalable and configurable in an on-the-fly basis (e.g., when new ships or trucks arrive at a seaport, the resource provisioning should be adjusted at run time), and (2) the overall infrastructure is intrinsically mixed both in terms of applications (e.g., sensors-driven commands and alarms, data, multimedia content delivery, and traffic with real-time requirements [7]) and physical technologies (e.g., wired, General Packet Radio Service (GPRS)/Universal Mobile Telecommunication System (UMTS)/ Long Term Evolution (LTE), IEEE 802.11, satellite backup links, and ad hoc wireless networks). Nevertheless, seaport-related operations are often performed through mobile devices or resource-constrained terminals. Finally, to provide a suitable overlay to such a scenario [8], the adoption of a full IP Core can speed up the integration with different subsystems and service providers as well as their merging with other networks (e.g., at a national or continental level).

In this perspective, this chapter presents the design of a novel framework for resource and safety management of networking infrastructure employed in seaport-related environments. Such a framework has been developed within an Italian National Project aimed at advancing the state-of-the-art information and communication technologies applied to seaport infrastructures.

The contribution of the chapter is twofold: (1) to introduce a unified layered model to perform different duties in network infrastructures deployed within seaports and (2) to show the design of two entities for the management of resources and safety purposes. At the best of the authors' knowledge, this is the first work dealing with CNs and seaport-oriented applications.

The remainder of this chapter is structured as follows: Section 9.2 introduces the reference scenario and the layered system architecture of the overall framework by emphasizing the design choices. Section 9.3 analyzes the mathematical settings adopted to model the system to determine the proper decisions relevant to exploit the safety and resource management. Section 9.4 portrays a performance evaluation to prove the effectiveness of the proposed approach. Section 9.5 concludes the chapter.

9.2 Design of Framework for CNs in Seaport Environments

In order to describe the development of the proposed framework, we introduce a reference scenario general enough to characterize the most critical services for the successful exploitation of seaport operations. Figure 9.1 depicts the typical interconnection of different access networks employed for ad hoc duties.

IP Core network. It is responsible for the data transport among all the peripheral access networks. Also, it is assumed to be full IP (i.e., it only implements IPv4 or IPv6 at the network layer) as already happens in about the totality of transport and core installations [9]. However, other protocols could be used via proper tunneling mechanisms, such as those based on the generic route encapsulation.

Sensors. It is the network in charge of gathering information from the sensors distributed through the seaport. In other words, it represents the portion of the infrastructure gathering and dispatching data belonging to safety operations. It can also be a complex merge of WSNs and wired trunks (e.g., sinks collecting data from fixed RFID gates). Nevertheless, sensors or sinks can be connected to the IP Core through a dedicated access network or implement such a service itself (i.e., via ad hoc and cooperative mechanisms). Then, collected values can be routed through the IP Core to the remote facilities (e.g., through the Internet) or to local entities (e.g., for mining and processing operations).

Voice. It provides the infrastructure for the delivery of real-time communication services among the personnel of the seaport. We assume that voice-based conversation is the most adopted one; that is, it provides some voice over IP (VoIP) support. Similarly to the sensors portion of the infrastructure, it can also be the result of a complex interconnection among wired and wireless networks (e.g., to assure VoIP coverage outdoor to mobile endpoints). The IP Core enables peripheral nodes to reach other peers and assure routability to the Internet and the exchange of the needed signaling flow.

Data. It implements a bulk data transfer service, for example, for synchronizing databases or for providing Internet connectivity (e.g., web browsing and file transfer operations). Also, it can act as a generic service provider by allowing other infrastructures to reach the IP Core or access to other portions of the system. For instance, this is the case when new ships arrive in the seaport and need network connectivity. We point out that such third-party entities can be also routed to specific service networks (e.g., internal sensors can be merged within the sensor portion in order to have a unique network devoted to exploit safety operations).

Internet (WAN). It represents a communication path to the Internet or to a WAN. For example, a seaport can be connected to other seaports by means of national dedicated

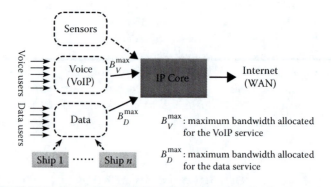

Figure 9.1 Reference network architecture employed to develop the proposed convergence framework for safety and resource management purposes.

infrastructures. Additionally, it is common to have a specific portion of the seaport remotely operated or under a 356/24 monitoring regime. This could be the case of a ship requiring connectivity to sync data with the ship owner's headquarter or the constant supervision of some hazardous goods. According to its critical importance, the WAN connectivity, at least for mission-critical communications, could be further strengthened by using proper back-ups (e.g., via a geostationary satellite link) [10]. As a consequence, the heterogeneity of the overall technological pool is further increased.

For the sake of simplicity, we assume the different networks as disjointed or at least grouped to form a coherent set in terms of provided functionalities. Yet, this cannot be possible in real or very complex deployments. For instance, the different services can be implemented through generic IP networks and then provided via suitable overlays superimposed over a multipurpose bearer. For example, a wireless network could be partitioned to serve as a VoIP attachment point and for generic data transfers. Moreover, with the increasing convergence of services also in mobile plat-forms, this can be a reasonable setup for forthcoming Beyond 3rd Generation telephony frame-works [11]. When services are merged over the same physical network, proper mechanisms to assure the correct requirements, such as bandwidth and jitter, must be present. In this case, the adopted infrastructure must support technologies such as multiprotocol label switching (MPLS), resource reservation protocol, traffic engineering mechanisms, or application layer resource man-agement techniques [12]. We point out that the introduced services are the most critical and adopted within a CN deployed in a seaport. Yet, other possible networks could be in place, also in the perspective of increasing the redundancy of the infrastructure, to physically decouple services to conform to some security and privacy regulations.

9.2.1 Design of Framework: Reference Layered Architecture

As discussed, a seaport environment presents a two-dimensional heterogeneity space: (1) tech-nological, which is mainly due to the different media used to assure the necessary physical coverage, and (2) functional, which is the result of the plethora of services needed to support specific operations. Additionally, services can coexist by exploiting different overlay architec-tures. Therefore, to better organize the framework and to have a unique and coherent abstraction of the overall CN deployment, a proper protocol functional architecture has been introduced. Figure 9.2 showcases the reference layered model, emphasizing the two main functionalities of the framework (i.e., resource management and safety mechanisms), which are the major contri-bution of this chapter.

Specifically, to handle the multifaceted nature of seaport requirements, the framework is sup-posed to act over a proper abstraction of the overall network deployment. This is indispensable in the perspective of having a network-independent method to manage resources and recognize misbehaviors for safety purposes. To this aim, we split the layered architecture within two main domains, belonging to different spaces, which we define as physical and logical, respectively. The physical space represents all the protocols and entities bounded to a specific technology. Therefore, all the components belonging to this area have to be considered as highly technology dependent. As a consequence, they reduce the reusability of software components and also prevent algorithms to be general enough to be simple and adaptable for other seaports. The last two properties are "good practices" when engineering CNs, because complexity should be handled as close as pos-sible to specific devices and machineries [6]. For such reasons, we introduce a higher set of com-ponents to abstract functionalities. In other words, we exploit convergence through abstraction.

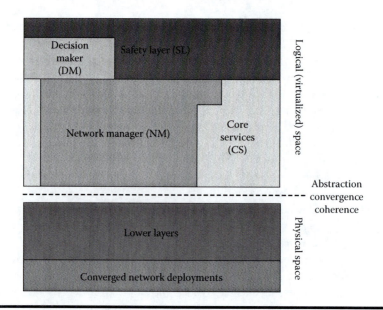

Figure 9.2   Reference layered architecture of the proposed framework (interactions among different entities are ruled according to standard data percolation within a protocol stack).

Even if this choice facilitates the development of resource management and safety algorithms, also by making the physical space easier to model, this requires that suitable software components and protocol entities are developed. For instance, "virtualization" could be achieved by mapping high-level descriptive quantities (e.g., bandwidth) into effective reservation or resource allocation within the underlying network [13].

Defining the precise engineering of such a virtualization layer is out of the scope of this chapter. Rather, we aim at showcasing how a CN can be enhanced with cutting-edge functionalities by proper modeling and imposing very loose functional requirements. In this perspective, we group all the needed abstraction features within a macrolayer defined as core services (CS). The latter is in charge of mapping stimuli produced by higher layers (i.e., those acting in the abstracted network model) to the network-dependent portion of the stack. In more detail, this layer must provide functionalities such as authorization authentication and accounting, QoS, management, and creation of service-oriented architecture like environments or development of proper network overlays, for instance, by using the peer-to-peer communication paradigm. CS should also interact with a proper database allowing one to capture configurations and policies of the managed network. This can be done through proper eXtensible Markup Language (XML) schema and data exchange via ad hoc mechanisms such as web services, simple object access protocol, and discovery methodologies like universal description discovery and integration and universal plug and play [14].

On the top of the stack, the safety layer (SL) is placed. Basically, it is the functional entity responsible for evaluating stimuli gathered by the CS and by the network manager (NM) in order to reveal possible behaviors that can lead to safety risks. For example, by observing descriptive parameters (e.g., the number of users and deviations from standard traffic profiles), it could be able to raise alarms, for instance, to support or start an operation aimed at public protection and disaster recovery (PPDR) [15]. Additionally, this layer can also directly interact with the NM and the CS to trigger specific actions on the network to support the aforementioned PPDR operations. Possible automatically handled duties enabled by the framework could be aimed at the following:

1. Reserving bandwidth for VoIP communications to assure communications during rescue operations. We point out that this can be done in different parts of the seaport infrastructure (e.g., in the IP Core, the data access network, or in a route to the public Internet). This also highlights why having a proper unified abstraction is beneficial.
2. Guaranteeing a proper data syncing for the evaluation of the status of some assets on receiving alarms from remote sensors.

Besides, it could be also possible to reserve additional bandwidth paths among sensors to avoid bottlenecks or excessive delays due to malfunctioning or high traffic loads triggered by users' reaction to safety critical phenomena.

In the perspective of joining the CN within the high-layer logic implemented by the framework, a very critical component is the NM. Put briefly, it is responsible for providing an alternative (and, possibly, thinner) abstraction layer over the physical network. It can interact directly with the CS or export an additional set of application program interfaces (APIs) for directly managing the network (see, e.g., the case of performing a reservation of VoIP channels or a bandwidth path for database synchronization). Nevertheless, it can be used to compute proper metrics useful in feeding the logic in charge of evaluating performance optimization or spawning alarms. As a paradigmatic example, the NM can estimate the network status by interacting with a set of probes (e.g., by using NetFlow, the simple network management protocol, or Open Source tools such as nTop) [16]. Besides, by interacting with SIP nodes (or proprietary VoIP call managers), it could estimate the number of active call as well as other parameters such as their duration. The NM can also inquire the dynamic host configuration protocol server (if any) to estimate the number of active network nodes and their evolution within a given time frame. Such operations are example of network-dependent quantities that can be "ported" into an abstract domain by means of proper software agents or modules. This also explains why CNs are needed because they serve many services concurrently; thus, the needed information is less diffuse or not spread over many operators.

Such values can be further elaborated to provide the SL with the needed information to perform computation. Conversely, if the SL needs to change something in the underlying network, it can use the API provided by the NM, which can exploit the proper conversions from the abstraction layer to the lower ones, also by relying on CS.

Finally, the decision maker (DM) is a software module containing all the needed logic to perform optimization of resources or to assist safety operations by using suitable algorithms. It can be invoked by the SL (e.g., to react against anomalies), or it can interact with the NM to adjust policies to maintain a proper degree of QoS or QoE [17]. As a possible use case, we mention the computation of the needed resource allocation strategies to guarantee voice call continuity. To this aim, it can directly interact with the CS, as depicted. As a final remark, we point out that the DM is conceptually a stand-alone entity, but it can be merged within the other proposed modules to simplify the overall implementation.

9.3 Development of Safety Agent and NM

In this section, we discuss the development of the two layers comprising the integrated framework for adding functionalities to a CN adopted in a seaport-based environment. In detail, we decided to engineer the functionalities of the SL and the DM by means of nonlinear models capable of predicting the behavior of the relevant parameters of the network. Such quantities are assumed as available by virtue of proper components, as explained in Section 9.2.1.

Put briefly, such models allow, starting from a set of real data collected from past observation of a real system, one to obtain a realistic prediction in real time of the future behavior of the system itself. The possibility of predicting future values of some interesting parameters can serve different purposes. From the point of view of the SL, it allows recognizing in advance the possible risks and unwanted or dangerous behaviors of the system. From the point of view of the DM, it enables the adoption of suitable policies to avoid potential disasters before they actually happen; in general, it allows allocating the shared resources in order to satisfy demands, taking into account trends of the near future.

Section 9.3.1 presents the basic mathematical settings about nonlinear regression algorithms that will be used by the SL. Section 9.3.2 discusses the possible bandwidth allocation strategies implemented within the DM.

9.3.1 Predictive Models for Safety Management of CNs Deployed in Seaports

In order to introduce the models that are implemented in the SL, let us consider a generic discrete-time dynamic system, that is, a system whose evolution is observed (and, possibly, controlled) over a finite or infinite number of temporal stages (in the following denoted by the indexes $t = 0, 1,...$) of fixed length Δt.

In a very general formulation of the problem of predicting a discrete-time dynamic system based on data, a set of measures is available corresponding to variables $x_t \in \mathfrak{R}^n$, for $t = 0, 1,...$, describing the state of the system at temporal stage t, together with a set of corresponding "output" values $s_t \in \mathfrak{R}$ representing the quantities that we want to predict. In many cases, such quantities are just the values of the state variables at the temporal stage next to the current one (i.e., their future value).

In a very general way, we can consider the output of the system at temporal stage t as a random variable with unknown probability distribution. Then, at temporal stage t, define:

- The vector of past state observations: $x^{(t-1)} = [x_0, x_1,..., x_{t-1}]$.
- The vector of past outputs: $s^{(t-1)} = [s_0, s_1,..., s_{t-1}]$.

As said, the aim is to find a model capable of capturing the functional relationship between past observations and future outputs of the system. Define $\tilde{s}_t$ as the value of the output at stage t estimated by the model, and consider the latter chosen within a class of functions γ having a prefixed structure.

This allows one to write:

$$\tilde{s}_t = \gamma\left(x^{(t-1)}, s^{(t-1)}\right). \tag{9.1}$$

Then, the aim of a "good" model γ becomes making $\tilde{s}_t$ as close as possible, in correspondence to past observations $x^{(t-1)}$ and $s^{(t-1)}$, to the "true" output s_t of the actual system at stage t. An issue with this approach is that the model needs, in general, a finite-length input vector, whereas the vectors $x^{(t-1)} = [x_0, x_1,..., x_{t-1}]$ and $s^{(t-1)} = [s_0, s_1,..., s_{t-1}]$, by definition, grow with t.

Then, we can obtain a finite-dimensional input for our model by defining a function φ mapping past observations into a vector of fixed dimension, commonly named regressor.

Another hypothesis that we need to introduce, in general, is that the output of the real system depends only on a limited number of past values. Such hypothesis, commonly named fading memory assumption, is not an actual limitation, because it realistically states that only the recent past of the system affects the future output values. This assumption allows one to define the regressor as a function of a finite number q of past observations.

There are various possibilities for the choice of a good regressor function. In particular, a very straightforward way is to consider the collection of past output observations themselves; for example, we can define a regressor in this way:

$$\varphi(x^{(t-1)}, s^{(t-1)}) = [s_{(t-q)}, s_{(t-q+1)}, \ldots, s_{t-1}]. \tag{9.2}$$

Another possibility is also to employ the measured past values of the state variables in the informative set:

$$\varphi(x^{(t-1)}, s^{(t-1)}) = [x_{(t-q)}, x_{(t-q+1)}, \ldots, x_{t-1}, s_{(t-q)}, s_{(t-q+1)}, \ldots, s_{t-1}]. \tag{9.3}$$

Both approaches have pros and cons. For instance, in the case depicted in Equation 9.2, the resulting model is lower-dimensional and thus simpler and more manageable from a computational point of view.

However, the resulting regressor may be not rich enough in case of a system characterized by very complex dynamics. In this case, the approach presented in Equation 9.3 would be preferable due to the larger amount of information from which the prediction can be based.

In general, once the regressor has been defined, we can write the system output predicted at stage t as:

$$\tilde{s}_t = \gamma(\varphi_t). \tag{9.4}$$

Notice that the definition of the regressor has the effect of making the input–output mapping formally independent from the actual temporal stage t. This means that all the measurements, even if made at different times, can be added to the same informative set made of input–output pairs, all contributing to obtaining a satisfactory prediction. Obviously, we must assume that the system is sufficiently time-invariant, that is, its behavior does not change completely over time (in that case, a new model would have to be built to adapt to the changed setting).

Concerning the class of functions where we look for a suitable model γ employed for the output prediction of temporal series and complex systems, nonlinear models have gained popularity in the literature, turning out nowadays as a well-established method from both the practical and theoretical points of view.

In fact, many examples of successful application to important problems from different fields such as engineering, artificial intelligence, and optimization (for a good survey on neural networks, see, e.g., reference [18]), as well as a large number of theoretical results, have proven that this kind of models, of which neural networks and radial basis function networks are popular instances, is more computationally manageable with respect to classic linear ones for the approximation of complex classes of nonlinear functions, especially in contexts characterized by high-dimensional input variables (see, e.g., reference [19] and references therein).

This is due to the higher flexibility given by parameters that act nonlinearly on the output, which allow obtaining better approximation with respect to linear structures having the same number of parameters.

A quite general way of defining nonlinear architectures of the above-mentioned kind is to consider a family of functions that depend on finite set α of p parameters ($\alpha \in \Lambda \subset \mathfrak{R}^p$) that can be "tuned" to the purpose of obtaining the best model within the family itself.

In particular, a standard possibility for the construction of this kind of architecture is to exploit linear combinations of K fixed basis functions:

$$\gamma(\varphi,\alpha) = \sum_{i=1}^{K} c_i \psi\left(\varphi, w_i\right) + c_0, \quad c_0, c_i \in \mathfrak{R}, \quad w_i \in \mathfrak{R}^r \tag{9.5}$$

where $\alpha = [c_0, c_1,\ldots, c_K, w_1,\ldots, w_K]^T$ and $p = K(r + 1)$.

As said, radial basis function networks and one-hidden-layer feedforward neural networks are popular examples of nonlinear structures that can be obtained by the aforementioned construction. In particular, the latter correspond to the class of approximating functions actually chosen to implement the functionalities of the SL described in Section 9.2.1.

Such very popular architectures, which have been applied to many different contexts, are defined as nonlinear mappings having the following structure:

$$\gamma(\varphi,\alpha) = \sum_{i=1}^{K} c_i \sigma\left(\sum_{j=1}^{d} a_{ij}\varphi_j + b_i \right) + c_0 \tag{9.6}$$

where σ is typically called the "activation function."

The kind employed for the tests in this work has the well-known hyperbolic tangent form:

$$\sigma(z) = \frac{e^z - e^{-z}}{e^z + e^{-z}}. \tag{9.7}$$

The term c_0 is commonly denoted as bias, and it ensures that the output is not necessarily 0 when the input is equal to 0. This class of nonlinear models is proven to be endowed by the universal approximation property, that is, the property of being able to approximate any sufficiently "well-behaved" function within arbitrary accuracy. Furthermore, the complexity of the network, measured in terms of the number of parameters to be tuned, in many cases, does not grow exponentially with the input vector dimension, which is not true, in general, for linear structures [19]. This makes feedforward neural networks with one hidden layer excellent candidates to model the evolution of the complex dynamics that characterizes a CN adopted for seaport-related operations.

Once the class of approximating functions to be used for the prediction of the system output is chosen, there arises the issue of finding, within such class, the best element (i.e., the corresponding value of the parameter α) in the sense that it approximates in the most accurate way the behavior of the system.

This is typically obtained through an optimization procedure for the vector of parameters α that, in the neural networks parlance, is called training.

Specifically, a cost is defined based on available data that measures, for a given value of the parameters, the difference between the network output and the true observed output values (targets, in neural network terminology).

Typically, such cost has the form of a mean square error (MSE). Consider the available informative set of L input–output pairs $[\varphi_i, s_i]$, $i = 1,...,L$, where the index i is assigned according to some arbitrary criterion (i.e., ignoring, as said, the particular temporal stage at which a pair has actually been measured). Then, the cost to be minimized to train the network assumes this form:

$$\text{MSE}(\alpha) = \frac{1}{L} \sum_{i=1}^{L} \left[s_i - \gamma\left(\varphi_i, \alpha\right) \right]^2 \tag{9.8}$$

The parameter vector that minimizes the mean square error (MSE) corresponds to the neural network that will be actually employed for the output prediction in the real-time phase. To obtain such an optimal vector, it is possible to solve the optimization problem by means of any nonlinear programming technique. However, some of these techniques have been specifically tailored for neural network training. This is the case, for instance, of back-propagation or the Levenberg–Marquardt algorithm [20]. Classic results from learning theory, guarantee that this procedure is consistent in the sense that the obtained neural network approximates the system behavior better over the whole input space as the number L of input–output pairs grows to infinity.

9.3.2 Exploiting DM for Bandwidth Allocation Policies in CNs for Seaports

As said, the DM does not have only the role of setting prescriptive policies in case of emergency. In fact, even if it can be assumed as a part of the NM, conceptually, it is an "intelligent" entity whose functionalities can be shared, depending on convenience, between the NM and the SL.

Then, under noncritical conditions, the DM can tune dynamically the network parameters to optimize its performance. For the sake of clarity, we define an example scenario to explain how the DM can be used. Additionally, Section 9.4 presents some numerical results produced via a simulation campaign. The reference scenario is the one depicted in Figure 9.1. Specifically, let us assume we want to reserve the bandwidths $B_V^{\max}(t)$ and $B_D^{\max}(t)$, which are the resource shares allocated within the CN for the voice and data services, respectively, at stage t.

The setting is performed adaptively by guaranteeing that the total resource, defined as B^{tot}, to be reserved within the IP Core network for enabling the two aforementioned services is shared efficiently, aiming at satisfying the bandwidth request of both and ensuring that possible priorities due to safety reasons are complied with. In case of a more structured scenario, where the range of actors and their relevant parameters are more stable, neural networks could be employed to find management policies able to optimize performance in advance over a long horizon of stages (see, e.g., reference [21]). In our seaport scenario, we can still take advantage of the neural models in the SL described in Section 9.3.1, which allow one to predict the requests of voice and data bandwidths in the near future time stages, to implement effective policies through the DM.

Specifically, define $B_V^*(t)$ and $B_D^*(t)$ as the predictions on the bandwidth requests for voice and data, respectively, at temporal stage t. Furthermore, define p_V and p_D as a default standard sharing of the available bandwidth between voice and data. Notice that p_V and p_D are between 0 and 1, and such that $p_V + p_D = 1$.

A possible strategy for bandwidth allocation for voice and data at the beginning of temporal stage t can be described by the following algorithm:

■ If the predicted requests for stage t (i.e., for the time interval of length Δt between stages t and $t + 1$) exceed the total available bandwidth (i.e., if $B_V^*(t) + B_D^*(t) > B^{tot}$), the standard default sharing is set for the bandwidths:

$$B_V^{max}(t) = p_V B^{tot}, \quad B_D^{max}(t) = p_D B^{tot}.$$

■ If $B_V^*(t) + B_D^*(t) \leq B^{tot}$, the strategy is to share the bandwidth in a way that is proportionally inverse to the relative distance between the value of the bandwidth requested by a given kind of service at stage $t - 1$ and the one predicted for stage t.
More formally, we define

$$\Delta B_V(t) = \frac{\left| B_V(t-1) - B_V^*(t) \right|}{B_V(t-1)} \tag{9.9}$$

$$\Delta B_D(t) = \frac{\left| B_D(t-1) - B_D^*(t) \right|}{B_D(t-1)}. \tag{9.10}$$

On the basis of these quantities, the following partitioning is adopted:

$$p_V'(t) = \frac{\Delta B_V(t)}{\Delta B_V(t) + \Delta B_D(t)} \quad \text{and} \quad p_D'(t) = \frac{\Delta B_D(t)}{\Delta B_V(t) + \Delta B_D(t)}.$$

Notice that the constraint $p_V'(t) + p_D'(t) = 1$ is respected.
This sharing of the bandwidth is then modified to take into account the standard default sharing, thus obtaining the definitive bandwidth sharing in the following way:

$$B_V^{max}(t) = \frac{\left(p_V'(t) + p_V \right)}{2} B^{tot} \tag{9.11}$$

$$B_D^{max}(t) = \frac{\left(p_D'(t) + p_D \right)}{2} B^{tot}. \tag{9.12}$$

This strategy aims at making sure that the bandwidth devoted to a given service is always sufficient to satisfy the predicted requests, taking into account at the same time the standard priorities corresponding to the parameters p_V and p_D. If necessary, it is possible to change such parameters to enforce priorities in a "soft" way following alerts from the SL together with the policies set by the NM. Table 9.1 presents a possible scheme for policies of such kind.

Table 9.1 Example of Interactions between NS and SL to Perform Safety-Oriented Operations via CN Deployed in Seaport

Type of Detected Anomaly	SL Actions	NS Actions
Single alarm sensed in the container yard.	Alarm revealed.	Alarm forwarded through the network.
Multiple alarms sensed in the container yard.	Alarms revealed. Additional safety risk could happen in the yard.	Alarms forwarded through the network. Bandwidth for VoIP communication is reserved to support PPDR operation (if needed).
Sudden increase in VoIP calls.	Possible safety hazard could be ongoing.	Alarms are forwarded through the network. Bandwidth for VoIP communication is further reserved to deal with possible emergency communication.
Sudden traffic reduction within the network (both for voice and data).	Some access network as well as the IP Core could be malfunctioning.	Diagnostic tests are performed over the network.

9.4 Performance Evaluation

In order to test the effectiveness of the proposed approach, a simulation campaign was conducted by implementing an ad hoc software simulator based on the converged scenario depicted in Figure 9.1. The simulator takes as inputs some parameters of the scenario, such as available total bandwidth, number of users populating the networks, mean interarrival times and length of calls, data connections, and ship stays in the terminal, and, over a desired horizon of time stages, generates ship arrivals, calls, and data connections. At the same time, it implements the predictive algorithms at the core of the SL and the policies computed by the NS, changing the bandwidth allocation accordingly and dynamically. Besides, the entire infrastructure is assumed to be a full IP deployment (i.e., there is a unique network layer for all the underlying deployments). The software implementation assumes both the NS and the SL component directly interacting with the IP Core portion of the CN, which sends back the needed descriptive parameters through the discussed layered architecture. However, such a requirement could be too tight. In fact, a seaport deployment is typically highly mixed with a huge variety of access networks, which cannot have the same requirements (e.g., in terms of QoS support, access control functionalities, and, more importantly, availability of fine-grained performance indicators). For instance, a cost-effective, commercial off-the-shelf IEEE 802.11 access point seldom offers built-in traffic management services or fine-grained traffic statistics (e.g., implemented within the medium access control layer). Even if IEEE 802.11e can cope with this, it is very unlikely to have it deployed due to both scarce diffusion and a relevant population of legacy services. Also, to flatten costs, it could send back to a requestor only coarse-grained traffic specifications possibly in a non-real-time manner. Therefore, the model assumes that proper probes can be placed in the proximity of interested access networks to suitably gather information used by the proposed framework by proper collection and processing by the CS. Anyhow, the optimum would be having such a monitoring facility natively incorporated within the NM, which is a reasonably midterm requirement for quite complex CNs. Additionally, we assume that proper overlays can be deployed to overcome the lack of QoS support in the lower layers of specific network trunks. This is a reasonable requirement, which can be exploited by using well-established algorithms for the management of resources at the application level (see, e.g., [22] and references therein).

Concerning the use case, we focused on two of the most used services (i.e., VoIP communications and bulk data transfers [23,24]). For the sake of investigating the correct behavior of the proposed scheme, we imposed that the framework can manage a fixed amount of bandwidth within the IP Core. In our scenario, B^{tot} has been set to 100 Mb/s. Then, such a resource is shared among the voice and data service, that is, the following constraint holds:

$$B_V^{max} + B_D^{max} = B^{tot} \qquad (9.13)$$

where B_V^{max} and B_D^{max} are the total amount of resource assigned to the voice and data service, respectively. We point out that, in this scenario, we do not assume data belonging to sensors competing for the resources described in Equation 9.8, that is, a proper reservation for such mission-critical transmissions within the IP Core has been computed offline (e.g., via proper MPLS overlays or other traffic engineering mechanisms).

As regards the mutability of the scenario, especially in terms of new entities connecting to the seaport network, we assumed that arriving ships dynamically require joining the CN. In this perspective, each ship accounts for an increased usage of resources within the IP Core, which ends when the ship leaves. To represent a realistic behavior, we modeled the traffic of new ships with a sudden increase in data bursts upon joining the seaport environment to reflect the need of syncing databases, for example, for uploading/downloading procedures and to exchange proper documentation, such as the ships' manifests. The time horizon has been discretized in equal intervals of fixed size (i.e., $\Delta t = 10$ s).

To generate the load of VoIP traffic, we adopted a Poisson distribution with an average of 120 calls/h [25]. We assumed that 50% of calls happen within two seaport's endpoints, whereas the other 50% comprises conversation with a remote peer (therefore accounting for a utilization also within the WAN link). The conversation length is randomly selected by using a gamma distribution with an average duration of 5 min. To model the bandwidth occupation, a value randomly sampled with a uniform distribution ranging from 56 to 256 kb/s is employed. Such a choice has been performed to better capture the mixed set of features and codecs adopted nowadays in Internet telephony client interfaces and appliances. Also, it can help in taking into account the high variability of codecs, which can scale according to the available bandwidth [26]. Specifically, with 56 kb/s, we represent a high-quality adaptive voice coding [26], whereas, with 256 kb/s, we also want to model low-definition video calls or the usage of additional features offered by modern services, such as desktop or slides sharing.

Regarding the characterization of the data traffic, it has been modeled with a similar approach. We used a transport-layer connection degree of granularity. The number of new requests is generated according to a Poisson distribution having an average of 240 connections/h. The duration is randomly selected by using a gamma distribution with a mean of 15 min. The bandwidth occupation of each connection is randomly extracted from a uniform distribution ranging from 56 to 512 kb/s.

Finally, a Poisson process, resulting in an average of a new ship requesting to join the hosting CN every 5 h, also regulates new ships' arrival and departure. We point out how this is different from the actual amount of ships populating the seaport. The time spent connected to the seaport's communication infrastructure is randomly selected from a uniform distribution ranging from 12 to 18 h.

Figure 9.3 depicts the evolution trends over a 12-h-long horizon (equal to 4320 time steps of Δt) of the parameters adopted by the NS to reconstruct the status of the CN and provide a suitable knowledge of the environment to the SL and the DM. Specifically

- $N_V(t)$ is the total number of VoIP calls at t.
- $B_V(t)$ is the total bandwidth need by the above voice calls.
- $N_D(t)$ is the total number of transport layer data connections.
- $B_D(t)$ is the total bandwidth used for transporting data generated by the aforementioned transport connections.

Referring to Figure 9.3, we underline that parameters are computed by the NS every Δt. Then, such knowledge is exploited by the predictive models implemented within the framework (as explained in Section 9.3). In particular, interesting parameters are used to train the neural components (i.e., they compose the regressor). To this aim, we used values collected in the 30 previous time steps, amounting to a regressor spanning a total length of 5 min. In order to train the neural network, we adopted data collected from a daylong observation of the CN, that is, accounting for $L = 8610$ input–output pairs. After the MSE has been minimized with such a data set, the resulting neural model has been employed by the DM to predict the evolution of the aforementioned

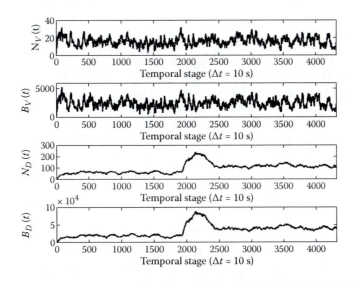

Figure 9.3 Evolution of parameters adopted by the NS component.

parameters. Sections 9.4.1 and 9.4.2 introduce how to use such model for safety purposes and resource allocation, respectively.

9.4.1 Detection of Anomalies

As said, the SL is used to enhance the overall safety level of the seaport by using the CN itself as an intrinsic sensors network. Obviously, the SL can use the same techniques presented in Section 9.3.1 for evaluating data gathered by a standard WSN or via a sink collecting data from surrounding RFID-tagged environments. However, such a case is out of the scope of this work. Rather, we are interested in exploiting the rich set of services offered by a CN in the perspective of revealing potentially risky behaviors. Besides, this is another benefit of using a CN, which allows one to concentrate all the knowledge within a unique underlying physical deployment, thus increasing the chance of having descriptive quantities or control data flows available (which is often unfeasible because different service providers usually do not disclose diagnostic data or usage statistics of their infrastructures).

As an example to prove the efficiency of the proposed mechanism, Figure 9.4 deals with a possible anomalous situation revealed by the SL. Specifically, the SL exploits an unobserved behavior of the total amount of voice communications [i.e., $N_V(t)$] jointly with a reduction of their average duration. This is a typical behavior in case of disasters (e.g., earthquakes or wreckage of machineries), where many users concurrently perform calls to broadcast a hazard or ask for help/intervention.

The figure clearly shows that, until the temporal stage 75 occurs, the prediction is quite accurate and very close to the real (in the sense of normal) behavior. However, when the "anomaly" happens, the deviation increases. As a consequence, we can use a threshold between the actual value and the predicted value to reveal anomalies. Such a value should be carefully selected in order to reduce possible false alarms. For instance, the SL can become aware of the anomalous situation starting from $t = 80$ (i.e., in a time frame less than 1 min). However, it could be better to wait until $t = 90$ to avoid possible misinterpretation due to circumscribed deviation from the standard behavior. Still, the time needed by the system to react also varies according to the size of the sampling interval Δt. Notice that an excessive reduction of Δt accounts for an increased size of the regressor, which reflects into a model with more computationally demanding requirements, possibly leading to a reduction of the model accuracy. In any case, the correct time interval Δt should be chosen in such a way that the anomaly is detected in a time that can be considered acceptable depending on the actual conditions of the port and the level of safety-critical activities carried on in the environment.

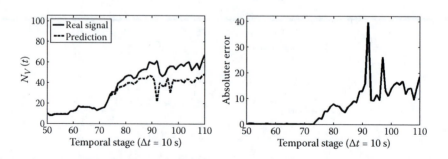

Figure 9.4   Anomaly detected on inspecting the behavior of the number of VoIP communications.

9.4.2 Role of DM

As explained in Section 9.3.2, the DM can be also used jointly with the NM to exploit optimal bandwidth allocation within the CN. In this section, we introduce a simulative analysis showcasing the usefulness of using such autonomous procedures within a CN. The reference scenario is the same as depicted in Figure 9.1 and explained in Section 9.4. In this simulated scenario, the overall manageable resource B^{tot} has been set equal to 25 Mb/s. Also, the evolution of the system is sampled with time steps of length $\Delta t = 300$ s. As a consequence, the DM computes the bandwidth shares every 5 min. The number of past observations used to construct the regressor has been selected equal to 6, thus corresponding to 1-h-long observation. To better evaluate long-term behaviors, the simulation has been done over a 10-day-long horizon, which amounts to 2880 temporal stages. The algorithm utilized to partition the resource is the one presented in Section 9.3.2, with $p_V = p_D = 0.5$.

Figure 9.5 portraits the results over a window of 1500 stages. It illustrates, in particular, the trends of resource requests [i.e., $B_V(t)$ and $B_D(t)$] as well as the maximum quantities assigned by the DM [i.e., $B_V^{max}(t)$ and $B_D^{max}(t)$]. We point out that the DM will be able to effectively interact with the underlying CN by exploiting high-level services such as those offered by the NM and the CS.

As depicted in Figure 9.5, the evolution of the maximum assigned bandwidths follows in an efficient manner the requests needed for the exploitation of the two services. In fact, both $B_V^{max}(t)$ and $B_D^{max}(t)$ are always sufficient to fulfill peak of requests but promptly decrease when needed to avoid resource trashing or the endangering of other competing services.

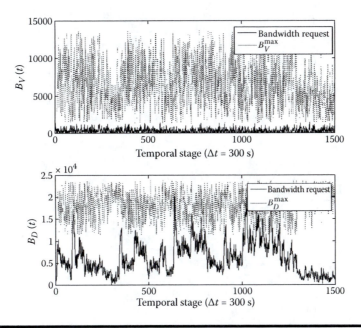

Figure 9.5 Dynamic bandwidth allocation between voice and data traffic performed by the DM according to specific requests of the SL and NM.

9.5 Conclusions and Future Work

In this chapter, a novel framework for safety and resource management has been proposed in CNs deployed to support seaport-oriented applications. Specifically, we introduced a layered architecture that, on one hand, helps in augmenting the abstraction level of NSs and, on the other hand, offers a suitable playground to develop nonlinear regression algorithms for the diagnosis of possible faults or hazardous events. Also, we have showcased how it can be used to perform resource optimizations for supporting PPDR operations as well. To prove the effectiveness of the proposed approach as well as its applicability within a typical CN, we presented numerical results obtained through an ad hoc software simulator.

Future works aim at porting this framework, at least at a prototypal level, over a real CN deployment, also in the perspective of quantifying its degree of fault tolerance and hardware requirements. Nevertheless, the adoption of such an integrated approach can better clear the functional requirements of CNs when injected in a complex organization, thus reflecting in a real advancement to their adoption and representing an input to potential standardization efforts, like, the needed management and control interfaces to develop the next generation of services.

Acknowledgments

This work has supported by the Project Industria 2015—SlimPort and funded by the Italian Ministry for the Industrial Development.

References

1. L. Caviglione, R. Podestà, "Evolution of peer-to-peer and cloud architectures to support next-generation services," in A. Prasad, J. Buford, V. Gurbani, Eds., *Advances in Next Generation Services and Service Architectures*, River Publishers, Denmark, 2010, ISBN: 978-87-92329-55-4.
2. O. Podevins, "Sea port system and the inland terminals network in the enlarged European Union," International Symposium on Logistics and Industrial Informatics (LINDI2007), pp. 151–155, Wildau, Germany, September 2007.
3. L. Caviglione, "Enabling cooperation of consumer devices through peer-to-peer overlays," *IEEE Transactions on Consumer Electronics, IEEE*, 55(2), pp. 414–421, May 2009.
4. J. Yelmo, R. Trapero, J. del Alamo, "Identity management and web services as service ecosystem drivers in converged networks," *IEEE Communications Magazine*, 47(3), pp. 174–180, March 2009.
5. S. Chatterjee, S., B. Tulu, T. Abhichandani, L. Haiqing, "SIP-based enterprise converged networks for voice/video-over-IP: Implementation and evaluation of components," *IEEE Journal on Selected Areas in Communications*, 23(10), pp. 1921–1933, October 2005.
6. S. Ou, K. Yang, H.-H. Chen, "Integrated dynamic bandwidth allocation in converged passive optical networks and IEEE 802.16 networks," *IEEE Systems Journal*, 4(4), pp. 467–476, December 2010.
7. A. Alessandri, C. Cervellera, M. Cuneo, M. Gaggero, G. Soncin, "Modeling and feedback control for resource allocation and performance analysis in container terminals," *IEEE Transactions on Intelligent Transportation Systems*, 9(4), pp. 601–614, 2008.
8. L. Caviglione, F. Davoli, "Using P2P overlays to provide QoS in service-oriented wireless networks," *IEEE Wireless Communications Magazine*, Special Issue on Service-Oriented Broadband Wireless Network Architecture, IEEE, 16(4), pp. 32–38, August 2009.
9. M. Tatipamula, F. Le Faucheur, T. Otani, H. Esaki, "Implementation of IPv6 services over a GMPLS-based IP/optical network," *IEEE Communications Magazine*, 43(5), pp. 114–122, May 2005.

10. L. Franck, R. Suffritti, "Multiple alert message encapsulation over satellite," Wireless Communication, Vehicular Technology, 1st International Conference on Information Theory and Aerospace and Electronic Systems Technology, (Wireless VITAE 2009). pp. 540–543, Princeton, New Jersey, USA, May 2009.

11. L. Caviglione, S. Oechsner, T. Hoßfeld, K. Tutschku, F.-U. Andersen, "Using Kademlia for the configuration of B3G radio access nodes," 3rd IEEE International Workshop on Mobile Peer-to-Peer Computing (MP2P'06), Pisa, Italy, pp. 141–145, March 2006.

12. L. A. DaSilva, "Pricing for QoS-enabled networks: A survey," *IEEE Communications Surveys and Tutorials*, 3(2), pp. 2–8, second quarter 2000.

13. K. Shiomoto, T. Miyamura, A. Masuda, "Resource management of multi-layer networks for network virtualization," Telecommunications: The Infrastructure for the 21st Century (WTC), pp. 1–3, September 2010.

14. E. Thomas, *Service-Oriented Architecture: Concepts, Technology and Design*, Prentice Hall, 2005, ISBN 0-131-85858-0, Upper Saddler River, NJ, USA.

15. L. Caviglione, "Introducing emergent technologies in tactical and disaster recovery networks," *International Journal of Communication Systems*, Wiley, 19(9), pp. 1045–1062, November 2006.

16. L. Deri, F. Fusco, J. Gasparakis, "Towards monitoring programmability in future Internet: Challenges and solutions," in L. Salgarelli, G. Bianchi, N. Blefari-Melazzi, Eds., *Trustworthy Internet*, Springer, Milan, pp. 249–259, 2011.

17. S. Jelassi, G. Rubino, H. Melvin, H. Youssef, G. Pujolle, "Quality of experience of VoIP Service: A survey of assessment approaches and open issues," *IEEE Communications Surveys and Tutorials*, (99), pp. 491–513, Vol. 14, No. 2, 2012.

18. S. Haykin, *Neural Networks: A Comprehensive Foundation*, Prentice Hall, 1999, Upper Saddler River, NJ, USA.

19. R. Zoppoli, T. Parisini, M. Sanguineti, "Approximating networks and extended Ritz method for the solution of functional optimization problems," *Journal of Optimization Theory and Applications*, 112, pp. 403–439, 2002.

20. M.T. Hagan, M. Menhaj, "Training feedforward networks with the Marquardt algorithm," *IEEE Transactions on Neural Networks*, 5, pp. 989–993, 1994.

21. C. Cervellera, M. Muselli, "Efficient sampling in approximate dynamic programming algorithms," *Computational Optimization and Applications*, 38(3), pp. 417–443, 2007.

22. L. Caviglione, C. Cervellera, "Design of a peer-to-peer system for optimized content replication," *Computer Communications Journal*, Elsevier, 30(16), pp. 3107–3116, November 2007.

23. K. Forward, *Recent Developments in Port Information Technology*, Digital Ship Ltd., London, 2003.

24. L. Faulkner, B. Kritzstein, P. Brian, J. Zimmerman, "Security infrastructure for commercial and military ports", IEEE OCEANS 2011, pp. 1–6, Santander, Spain, September 2011.

25. K. Singh, H. Schulzrinne, "Failover, load sharing and server architecture in SIP telephony," Computer Communications, Elsevier, 5(30), pp. 927–942, March 2007.

26. S. Baset, H. Schulzrinne, "An analysis of the Skype peer-to-peer Internet telephony protocol," in *Proceedings of the 25th IEEE International Conference on Computer Communications* (INFOCOM 2006), pp. 1–11, April 2006.

NETWORK MANAGEMENT AND TRAFFIC ENGINEERING

Chapter 10

Wavelet *q*-Fisher Information for Scale-Invariant Network Traffic

Julio Ramírez-Pacheco, Deni Torres-Román,
Homero Toral-Cruz, and Pablo Velarde-Alvarado

Contents

10.1 Introduction

The study of the properties of computer network traffic is important for many aspects of computer network design, performance evaluation, network simulation, capacity planning, and network algorithmic design, among others. In the very beginning of computer network traffic modeling, the traffic itself was considered Markovian because older telephone network traffic was suitably described by this model; thus, it was unsurprising to consider the characteristics of network traffic similar to those of the telephone network. Markovian models permitted straightforward computations of performance issues due to its being short-range dependent (SRD); moreover, because of the ease of computation and lack of memory, they became very popular. The modeling of computer network traffic with Markovian models ended when Leland et al. [1], based on detailed studies of high-resolution network measurements, discovered that network traffic did not follow the Markovian model but instead is more appropriately modeled by self-similar or fractal stochastic processes. Subsequent studies not only validated this finding but also found self-similar features in additional network configurations [2,3]. The self-similar nature of network traffic indicated that computer network traffic behaves "statistically" similar at different scales of observation. In fact, persistence behavior was observed in local area network traffic at small as well as high levels of observation. This finding was contrary to commonly observed features of Markovian models where, for large scale, the traffic appeared to reduce to white noise. The self-similar nature of network traffic implied that numerous results based on the Markovian model needed to be thoroughly revised. Later, many authors reported that, when considering traffic as a self-similar process, many Internet quality of service (QoS) metrics such as delay, packet-loss rate, and jitter increased. Because of this, it was obvious that a characterization of the traffic flowing through a network was necessary; based on the observed characteristics, actions specifically designed to maintain the QoS of the network under acceptable levels were required. The characterization of traffic was in principle performed by estimating the parameters that determine its behavior. The Hurst parameter (or the self-similarity parameter) provided a complete characterization of self-similar processes; however, due to the complex characteristics of observed traffic, nowadays, it is clear that complementary techniques are required [4–6]. Self-similar processes are related to long-memory, fractal, and multifractal processes, and it is common to find in the scientific literature claims that traffic is self-similar, fractal, or multifractal. Self-similar processes along with long-memory, fractal, $1/f$, and multifractal processes belong to the class of scaling or scale-invariant signals. The theory of scaling signals has been relevant for the study of many phenomena occurring in diverse fields of science and technology. Some aspects of physiology such as heart rate variability [7] are suitably modeled by scaling signals, and the parameter of scaling signals determines much of the properties of the heart and the individual under study [7]. Electroencephalogram (EEG) signals obtained in humans and animals are also appropriately described by scaling signals [8], but they also model the traffic flowing through computer communications [2,9,10], the turbulence in physics, the noise observed in electronic devices [11], and the time series obtained in economy [3] and finance, among others. Many techniques and methodologies have been proposed to analyze these processes [12–14]; however, they have shown to be limited for the rich set of complexities observed in the data [5,15]. In addition, many articles have concluded that no single technique of analysis is sufficient for providing efficient and robust estimation of the scaling parameter [12]. Because of this, current works concentrate in developing cutting-edge techniques that are robust to trends, level-shifts, and missing values embedded in the data under study. The presence of these phenomenologies significantly impacts the estimation process and can lead to misinterpretation of the phenomena [7,12]. In this context, recent results that attempt to study the complexities of the underlying process using wavelet-based entropies provide interesting alternatives. In fact, it has been demonstrated, for example, that wavelet Tsallis q-entropies behave as a sum-cosh window [6] and that this behavior can be used to detect multiple mean level shifts embedded in the scaling signal under study and for the classification of scaling signals as stationary or nonstationary as well [15]. This chapter presents novel techniques based on wavelet information tools for the important problem of detecting level shifts embedded in scaling signals. This problem has been recognized as of sufficient importance because it impacts the estimation of the scaling parameter α [16,17]. The chapter therefore defines the concept of wavelet q-Fisher information and provides a thorough study of its properties for scaling signal analysis. Information planes that attempt to describe the complexities of scaling signals are constructed for these processes. In fact, this chapter shows that wavelet q-Fisher information provides plausible explanations of the complexities associated to scaling signals; based on this, level-shift detection capabilities can be attached to it. Extensive experimental studies validate the theoretical findings and allow one to study the effect of the parameter q on wavelet Fisher's information's behavior and the level-shift detection capabilities within scaling signals. The parameter q allows further flexibility and can be adapted to the characteristics of the data under study. In the limit of $q \to 1$, it reduces to the standard wavelet Fisher information as defined in the work of Ramírez-Pacheco et al. [18].

The rest of the chapter is organized as follows: In Section 10.2, the properties and definitions of scaling signals are studied with sufficient detail, and their wavelet analysis is explored. Also, some important results are reviewed for fractional Brownian motion (fBm), fractional Gaussian noise (fGn), and discrete pure power-law (PPL) signals. Section 10.3 derives the wavelet q-Fisher information for scaling signals and studies its properties. In this section, generalizations of the wavelet q-Fisher information are presented in terms of the q-analysis. Section 10.4 details the level-shift detection problem and presents some results in which the wavelet q-Fisher information is applied for this problem. Section 10.6 draws the conclusions of the chapter.

10.2 Wavelet Analysis of Scaling Processes

10.2.1 Scaling Processes

Scaling processes of parameter α, also called $1/f^\alpha$ or power-law processes, have been extensively applied and studied in the scientific literature because they model diverse phenomena [9,10] within these fields. These processes are sufficiently characterized by the parameter α, called the scaling index, which determines many of their properties. Various definitions have been proposed in the scientific literature; some are based on their characteristics such as self-similarity or long memory, and others are based on the behavior of their power spectral density (PSD). In this section, a scaling process is a random process for which the associated PSD behaves as a power-law in a range of frequencies [2,19], that is,

$$S(f) \sim c_f |f|^{-\alpha}, f \in (f_a, f_b) \tag{10.1}$$

where c_f is a constant, $\alpha \in R$ is the scaling index, and f_a, f_b represent the lower and upper bound frequencies on which the power-law scaling holds. Depending on f_a, f_b and α, several particular scaling processes and behaviors can be identified. Independently of α, local regularity and band-pass power-law behavior is observed whenever $f_a \to \infty$ and $f_a > f_b \gg 0$, respectively. When the scaling index α is taken into consideration, long-memory behavior is observed when both $0 < \alpha < 1$ and $f_a > f_b \to 0$. Self-similar features (in terms of distributional invariance under dilations) are observed in all the scaling index range for all f. Scaling index α determines not only the stationary and nonstationary conditions of the scaling process but also the smoothness of their sample path realizations. The greater the scaling index α, the smoother their sample paths. In fact, as long as $\alpha \in (-1,1)$, the scaling process is stationary [or stationary with long memory for small f and $\alpha \in (0,1)$] and nonstationary when $\alpha \in (1,3)$. Some transformations can make a stationary process appear nonstationary and vice versa. Outside the range $\alpha \in (-1,3)$, several other processes can be identified; for example, the so-called extended fBm and fGn defined in the work of Serinaldi [12] provide generalizations to the standard fBm and fGn signals. The persistence of scaling processes can also be quantified by the index α, and within this framework, scaling processes possess negative persistence as long as $\alpha < 0$, positive weak long persistence when $0 < \alpha < 1$, and positive strong long persistence whenever $\alpha > 1$. Scaling signals encompass a large family of well-known random signals, such as fBm and fGn [20], PPL processes [19], and multifractal processes [2]. fBm, $B_H(t)$, comprises a family of Gaussian, self-similar processes with stationary increments; because of the Gaussianity, it is completely characterized by its autocovariance sequence (ACVS), which is given by

$$\mathbb{E}B_H(t)B_H(s) = R_{BH} = \frac{\sigma^2}{2} \left\{ |t|^{2H} + |s|^{2H} - |t-s|^{2H} \right\}, \tag{10.2}$$

where $0 < H < 1$ is the Hurst index. fBm is nonstationary, and as such, no spectrum can be defined on it; however, fBm possesses an average spectrum of the form $S_{\text{fBm}}(f) \sim c|f|^{-(2H+1)}$ as $f \to 0$, which implies that $\alpha = 2H + 1$ [21]. fBm has been applied very often in the literature; however, it is its related process, fGn, that has gained widespread prominence because of the stationarity of its realizations. fGn, $G_{H,\delta}(t)$, obtained by sampling an fBm process and computing increments of the form $G_{H,\delta}(t) = 1/\delta \{B_H(t+\delta) - B_H(t), \delta \in Z_+\}$ (i.e., by differentiating fBm), is a well-known Gaussian process. The ACVS of this process is given by

$$\mathbb{E}G_{H,\delta}(t)G_{H,\delta}(t+\tau) = \frac{\sigma^2}{2} \left\{ |\tau+\delta|^{2H} + |\tau-\delta|^{2H} - 2|\tau|^{2H} \right\} \tag{10.3}$$

where $H \in (0,1)$ is the Hurst index. The associated PSD of fGn is given by [19]

$$S_{\text{fGn}}(f) = 4\sigma_X^2 C_H \sin^2(\pi f) \sum_{j=-\infty}^{\infty} \frac{1}{|f+j|^{2H+1}} \quad |f| \le \frac{1}{2}, \tag{10.4}$$

where σ_x is the process variance and C_H is a constant. fGn is stationary and, for large enough τ and under the restriction of $1/2 < H < 1$ possesses long-memory or long-range dependence (LRD). The scaling index α associated to fGn signals is given by $\alpha = 2H - 1$ as its PSD, given by Equation 10.4, behaves asymptotically as $S_{\text{fGn}}(f) \sim c|f|^{-(2H+1)}$ for $f \to 0$. Another scaling process of interest is the family of discrete PPL processes, which are defined as processes for which their PSD behaves as $S_x(f) = C_s |f|^{-\alpha}$ for $|f| \le 1$, where $\alpha \in \mathbb{R}$ and C_s represents a constant. PPL signals are stationary when the power-law parameter $\alpha < 1$ and nonstationary whenever $\alpha > 1$. As stated in the work of Percival [19], the characteristics of these processes and those of fBm/fGn are similar; however, the differences between fBm and PPLs with $\alpha > 1$ are more evident. In fact, differentiation of stationarity/nonstationarity is far more difficult for PPL than for fBm/fGn. Figure 10.1 displays some realizations of fGn, fBm, and PPL processes. The scaling index α of the PPL signals are identical to the scaling index of the associated fGn and fBm. Note that the characteristics of the sample paths of fGn are fairly different from those of fBm. In the case of PPL processes, this differentiation is not so evident; in fact, when the scaling indexes approach the boundary $\alpha = 1$,

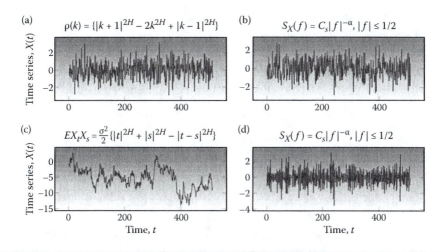

Figure 10.1 **Sample path realizations of some scaling processes. (a) fGn with α = –0.1; (b) PPL process with α = –0.1; (c) fBm signal with α = 1.9; (d) PPL process with α = 1.9.**

classification becomes complex. For further information on the properties, estimators, and analysis techniques of scaling processes, please refer to references [2,9,10,12,13,19,22].

10.2.2 Wavelet Analysis of Scaling Signals

Wavelets and wavelet transforms have been applied for the analysis of deterministic and random signals in almost every field of science [5,23,24]. The advantages of wavelet analysis over standard techniques of signal analysis have been widely reported, and its potential for nonstationary signal analysis is proven. Wavelet analysis represents a signal X_t in the time-scale domain by the use of an analyzing or mother wavelet, $\psi_o(t)$ [25]. For our purposes, $\psi_o(t) \in L_1 \cap L_2$ and the family of shifted and dilated $\psi_o(t)$ form an orthonormal basis of $L_2(R)$. In addition, the finiteness of the mean average energy $\left(\mathbb{E} \int |X(u)|^2 \, du < \infty \right)$ on the scaling process allows one to represent it as a linear combination of the form

$$X_t = \sum_{j=1}^{L} \sum_{k=-\infty}^{\infty} d_x(j,k)\psi_{j,k}(t), \tag{10.5}$$

where $d_x(j,k)$ represents the discrete wavelet transform (DWT) of X_t, and $\{\psi_{j,k}(t) = 2^{-j/2}\psi_o(2^{-j}t - k), j,k \in \mathbb{Z}\}$ is a family of dilated (of order j) and shifted (of order k) versions of $\psi_o(t)$. The coefficients $d_x(j,k)$ in Equation 10.5, obtained by DWT, represent a random process for every j and a random variable for fixed j and k, and as such, many statistical analyses can be performed on them. Equation 10.5 represents signal X_t as a linear combination of L detail signals, obtained by means of the DWT. DWT is related to the theory of multiresolution signal representation, in which signals (or processes) can be represented at different resolutions based on the number of detail signals added to the low-frequency approximation signal. Detail random signals ($d_x(j,k)$) are obtained by projections of signal X_t into wavelet spaces W_j, and approximation

Table 10.1   Wavelet Spectrum or Wavelet Variance Associated to Different Types of Scaling Processes

Type of Scaling Process	Associated Wavelet Spectrum or Variance
Long-memory process	$\mathbb{E}d_X^2(j,k) \sim 2^{j\alpha}C(\psi,\alpha), \quad C(\psi,\alpha) = c_\gamma \int |f|^{-\alpha}|\Psi(f)|^2\, df$
Self-similar process	$\mathbb{E}d_X^2(j,k) = 2^{j(2H+1)}\mathbb{E}d_X^2(0,k)$
Hsssi process	$\mathrm{Var}d_X^2(j,k) = 2^{j(2H+1)}\mathrm{Var}d_X(0,0)$
Discrete PPL process	$\mathbb{E}d_X^2(j,k) = C2^{j\alpha}$

Note: $\mathbb{E}(.)$, Var(.), and $\psi(.)$ represent expectation, variance, and Fourier integral operators, respectively.

coefficients $(a_x(j,k))$ are obtained by projections of X_t into related approximation spaces V_j. In the study of scaling processes, wavelet analysis has been primarily applied in the estimation of the wavelet variance [4,5]. Wavelet variance or spectrum of a random process accounts for computing variances of wavelet coefficients at each scale. Wavelet variance not only has permitted to propose estimation procedures for the scaling index α but also to compute entropies associated to the scaling signals. Wavelet spectrum has also been used for detecting nonstationarities embedded in Internet traffic [5]. For stationary zero-mean processes, wavelet spectrum is given by

$$\mathbb{E}d_X^2(j,k) = \int_{-\infty}^{\infty} S_X\left(2^{-j}f\right)\left|\psi\left(f\right)\right|^2 df, \tag{10.6}$$

where $\psi(f) = \psi(t)e^{-j2\pi ft}\, dt$ is the Fourier integral of $\psi_o(t)$ and $S_X(.)$ represents the PSD of X_t. Table 10.1 summarizes the wavelet spectrum for some standard scaling processes. For further details on the analysis, estimation, and synthesis of scaling processes, please refer to the works of Abry and Veitch [25] and Bardet [26] and references therein.

10.3   Wavelet *q*-Fisher Information of 1/*f*^α Signals

10.3.1   Time-Domain Fisher's Information Measure

Fisher's information measure (FIM) has recently been applied in the analysis and processing of complex signals [27–29]. In the work of Martin et al. [27], FIM was applied to detect epileptic seizures in EEG signals recorded in human and turtles; later, Martin et al. [28] reported that FIM can be used to detect dynamical changes in many nonlinear models such as the logistic map and Lorenz model, among others. The work of Telesca et al. [29] reported on the application of FIM for the analysis of geoelectrical signals. Recently, Fisher information has been extensively applied in quantum mechanical systems for the study of single particle systems [30] and also in the context of atomic and molecular systems [31]. FIM has also been used in combination with Shannon entropy power to construct the so-called Fisher–Shannon information plane/product (FSIP) [32]. The FSIP was recognized in that work to be a plausible method for nonstationary signal analysis. Let X_t be a signal with associated probability density $f_X(x)$. Fisher's information (in time domain) of signal X_t is defined as

$$I_X = \int \left(\frac{\partial}{\partial_x} f_X(x) \right)^2 \frac{dx}{f_X(x)}. \tag{10.7}$$

Fisher's information I_X is a nonnegative quantity that yields large (possibly infinite) values for smooth signals and small values for random disordered data. Accordingly, Fisher's information is large for narrow probability densities and small for wide (flat) ones [33]. Fisher information is also a measure of the oscillatory degree of a waveform; highly oscillatory functions have large Fisher information [30]. Fisher's information has mostly been applied in the context of stationary signals using a discretized version of Equation 10.7:

$$I_X = \sum_{k=1}^{L} \left\{ \frac{\left(p_{k+1} - p_k \right)^2}{p_k} \right\}, \tag{10.8}$$

for some probability mass function (pmf) $\left\{ p_k \right\}_{k=0}^{L}$. Equation 10.8 can be computed in sliding windows resembling a real-time computation. In this case, Fisher's information is often called FIM. Generalizations of Fisher's information have been defined in the literature. In fact, Plastino et al. [8] defined the q-Fisher information of a pmf as

$$I_q = \sum_j \left\{ p_{j+1} - p_j \right\}^2 p_j^{q-2}. \tag{10.9}$$

The parameter q provides further analysis flexibility and can highlight nonstationarities embedded in the signal under study. In this context, q-Fisher information is again a descriptor of the complexities associated to random signals and can attain high values.

10.3.2 Wavelet q-Fisher Information

This section defines a generalized version of Fisher information in the wavelet domain, derives a closed-form expression for this quantifier, and explores the possibility of using wavelet Fisher information for the analysis of scaling signals. Let $\left\{ X_t, t \in \mathbb{R} \right\}$ be a real-valued scaling process satisfying Equation 10.1, with DWT $\{d_x(j,k), (j,k) \in Z^2\}$ and associated wavelet spectrum $\mathbb{E}|d_X(j,k)|^2 \sim c_{X_t} 2^{j\alpha}$ (c_{X_t} is a constant and $\mathbb{E}$ is the expectation operator) [5]. A pmf obtained from the wavelet spectrum of scaling signals is given by the expression [6]

$$p_j \equiv \frac{1/N_j \sum_k \mathbb{E}d_X^2(j,k)}{\sum_{i=1}^{M} \left\{ 1/N_i \sum_k \mathbb{E}d_X^2(j,k) \right\}} = 2^{(j-1)\alpha} \frac{1-2^\alpha}{1-2^{\alpha M}}, \tag{10.10}$$

where N_j (N_i) represents the number of wavelet coefficients at scale j (i), $M = \log_2(N)$ with $N \in \mathbb{Z}_+$ the length of the data, and $j = 1,2,\ldots, M$. Substituting Equation 10.10 into Equation 10.9 results in the wavelet q-Fisher information of a scaling signal, which is given by

$$I_q = \left(1 - 2^\alpha\right)^2 \left\{\frac{1 - 2^\alpha}{1 - 2^{\alpha M}}\right\}^q \left\{\frac{1 - 2^{\alpha q(M-1)}}{1 - 2^{\alpha q}}\right\}$$

$$= 2^{\alpha\left(1-\frac{q}{2}\right)+2} \left\{\sinh^2_{1-v_1}\left(u_2\right)\right\} \left[\frac{\sinh^q_{1-\frac{v_2}{M-1}}\left(u_2\right)}{\sinh_{1-v1}\left(u_1\right)}\right] \tag{10.11}$$

$$\times \left\{\frac{\sinh_{1-v_1}\left(u_1\right)}{\sinh^q_{1-v_2}\left(u_2\right)}\right\} \left\{\frac{P_{\text{num}}}{P_{\text{den}}}\right\}, \tag{10.12}$$

where P_{num} and P_{den} are given by the following polynomial expressions:

$$P_{\text{num}} = 2\cosh_{1-\frac{v_1}{(M-2)}}\left(u_1(M-2)\right) + 2\cosh_{1-\frac{v_1}{(M-4)}}\left(u_1(M-4)\right)$$

$$+ 2\cosh_{1-\frac{v_1}{(M-6)}}\left(u_1(M-6)\right) + \ldots \tag{10.13}$$

$$P_{\text{den}} = 2\cosh_{1-\frac{v_2}{(M-1)}}\left(u_2(M-1)\right) + 2\cosh_{1-\frac{v_2}{(M-3)}}\left(u_2(M-3)\right)$$

$$+ 2\cosh_{1-\frac{v_2}{(M-5)}}\left(u_2(M-5)\right) + \ldots \tag{10.14}$$

with $u_1 = \alpha q ln_q(2)/2$, $u_2 = qu_1$, $v_1 = 2(1-q)/(\alpha q)$, and $v_2 = qv_1$. Equations 10.12 through 10.18 involve the use of the q-analysis [34], where $\sinh_q(x) \equiv \left\{e^x_q - e^{\ominus_q x}_q\right\}/2$ and $\cosh_q(x) \equiv \left\{e^x_q + e^{\ominus_q x}_q\right\}/2$ denote the q-sinh and q-cosh functions, respectively. $e^x_q \equiv \left\{1 + (1-q)x\right\}^{1(1-q)}$ and $\ominus_q x \equiv (-x)/\left\{1 + (1-q)x\right\}$ denote the q-exponential and q-difference functions, respectively. Equation 10.12 allows one to relate the results of wavelet q-Fisher information with the ones of the standard wavelet FIM. In fact, in the $q \to 1$ limiting case, wavelet q-Fisher information turns out to be the standard wavelet Fisher information for which the following holds:

$$I_1 = \frac{\left(2^\alpha - 1\right)^2 \left(1 - 2^{\alpha(M-1)}\right)}{1 - 2^{\alpha M}} \tag{10.15}$$

$$= 2^{\frac{\alpha}{2}+2} \sinh^2\left(\frac{\alpha \ln 2}{2}\right) \cdot \left\{\frac{P^M_{\text{num}}\left(2\cosh(\alpha \ln 2/2)\right)}{P^{M+1}_{\text{den}}\left(2\cosh(\alpha \ln 2/2)\right)}\right\}, \tag{10.16}$$

where $P^M_{\text{num}}(.)$ and $P^{M+1}_{\text{den}}(.)$ denote polynomials of argument $2\cosh(\alpha \ln 2/2)$ that are given by

$$P_{\text{num}}^{M}(.) = (2\cosh u)^{M} - \frac{2(M-3)}{2!} 2\cosh u^{M-2}$$
$$+ \frac{3(M-4)(M-5)}{3!}(2\cosh u)^{M-5} - \dots \tag{10.17}$$

$$P_{\text{den}}^{M+1}(.) = (2\cosh u)^{M+1} - \frac{(M-2)}{1!} 2\cosh u^{M-1}$$
$$+ \frac{(M-3)(M-4)}{2!}(2\cosh u)^{M-3} - \dots \tag{10.18}$$

where $u = \alpha \ln 2/2$. An interesting question is how the behavior of wavelet q-Fisher information is affected by q. To answer this question, Figure 10.2 displays the wavelet q-Fisher information for $q \in (0,1)$. Note that, as q approaches 1, wavelet q-Fisher information attains higher values for nonstationary signals ($\alpha > 1$). Therefore, if the signal is smooth or has a narrow probability density, then it is more likely to have a large wavelet q-Fisher value. Note also that, as long as $q \in (0,1)$, the form of the Fisher information is similar to that of the standard wavelet Fisher information [18] (high for highly oscillatory data and low for smooth signals). For the case where $q \in (0,1)$, wavelet q-Fisher has a behavior that is similar to that of the top left plot of Figure 10.3. In fact, when $q = 2$, wavelet q-Fisher information is symmetric with respect to $\alpha = 0$. Wavelet q-Fisher information reverses its behavior as long as $q > 2$, that is, in this case, highly oscillatory functions or functions with narrow probability densities display low values of their Fisher information, whereas smooth and flat probability densities display high values. Unlike the case $q > 1$, the $q > 2$ case decreases the range of variation of the wavelet q-Fisher information; thus, the detection of nonstationarities

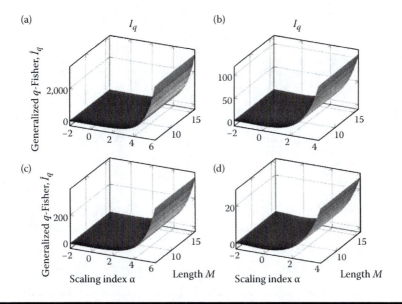

Figure 10.2 Wavelet q-Fisher information for $1/f^{\alpha}$ signals. (a) Fisher information with $q = 0.2$; (b) $q = 0.4$; (c) Fisher information for $q = 0.6$; (d) $q = 0.8$.

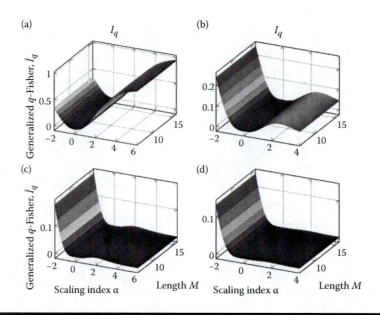

Figure 10.3 Wavelet q-Fisher information for $1/f^\alpha$ signals. (a) Fisher information with $q = 2.5$; (b) $q = 3$; (c) Fisher information for $q = 3.5$; (d) $q = 4$.

embedded in a signal is more difficult. Based on this behavior of the wavelet q-Fisher information, it is clear that values of $q \in (0,1)$ are suitable for detecting nonstationarities embedded in the data.

In this case, the value of the q-Fisher information is significantly higher for nonstationary signals. Figure 10.4 presents the theoretical wavelet q-Fisher information for scaling signals with $\alpha \in (-4,4)$ and fixed-length $M = 16$. According to Figure 10.4, for $q < 1$, wavelet q-Fisher information is high for nonstationary signals ($\alpha \geq 1$) and low for stationary ones ($\alpha < 1$). Fisher information

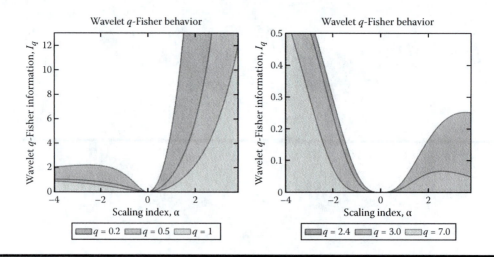

Figure 10.4 Wavelet q-Fisher information of scaling signals. Wavelet q-Fisher information is exponentially increasing (or decreasing) for signals with $\alpha > 0$ (or $-\infty < \alpha < 0$) and minimum for scaling signals with $\alpha = 0$.

consequently is high for correlated scaling signals and low for anticorrelated ones [18]. Fisher information is minimum ($I_q = 0$) for completely random signals ($\alpha = 0$).

10.3.3 Applications of Wavelet FIM

Because wavelet q-Fisher information describes properly the characteristics and complexities of fractal $1/f^\alpha$ signals, many applications can be identified using this complexity-based framework. In fact, based on the fact that wavelet q-Fisher information achieves large values for nonstationary signals and small values for stationary ones (for the case $q \in (0,1)$), a potential application area of wavelet q-Fisher information is in the classification of fractal signals as fractional noises and motions. Classification of $1/f^\alpha$ signals as motions or noises remains as an important, attractive, and unresolved problem in scaling signal analysis [7,35,36] because the nature of the signal governs the selection of estimators, the shape of quantifiers such as qth-order moments, and the nature of correlation functions [37]. Another important potential application of wavelet q-Fisher information, related to the classification of signals, is in the blind estimation of scaling parameters [38]. Blind estimation refers to estimating α independently of signal type (stationary or nonstationary). Wavelet q-Fisher information can also be utilized for the detection of structural breaks in the mean embedded in $1/f^\alpha$ signals. Structural breaks in the mean affect significantly the estimation of scaling parameters leading to biased estimates of α and consequently in misinterpretation of the phenomena. In fact, in the work of Stoev et al. [5], it was demonstrated that the well-known Abry–Veitch estimator overestimates the scaling index α in the presence of a single level shift leading to values of $H = (\alpha + 1)/2 > 1$, which, in principle, is not permissible in theory. In the following, the section concentrates on the detection of structural breaks in the mean embedded in synthesized stationary fGn signals by the use of wavelet q-Fisher information. The section studies anticorrelated and correlated versions of fGn and the power of wavelet q-Fisher information in detecting single structural breaks in the mean in these signals.

10.4 Level-Shift Detection Using Wavelet q-Fisher Information

10.4.1 Problem of Level-Shift Detection

Detection and location of structural breaks in the mean (level shifts) have been recognized as an important research problem in many areas of science and engineering [16,17]. In the Internet traffic analysis framework, detection, location, and mitigation of level shifts significantly improve on the estimation process. In fact, the presence of a single level shift embedded in a stationary fGn results in an estimated $H > 1$ [5]. This, in turn, results in misinterpretation of the phenomena under study and also in inadequate construction of qth-order moments. Let $B(t), t \in \mathbb{R}$, be a $1/f$ signal with level shifts at time instants $\{t_1, t_{1+L}, \ldots, t_i, t_{i+L}\}$. $B(t)$ can be represented as

$$B(t) = X(t) + \sum_{j=1}^{j=\infty} \mu_j 1_{\left[t_j, t_{j+L}\right]}(t), \tag{10.19}$$

where $X(t)$ is a signal satisfying Equation 10.1 and $\mu_j 1_{[a,b]}(t)$ represents the indicator function of amplitude μ_j in the interval $[a,b]$. The problem of level-shift detection reduces to identify the points $\{t_j, t_{j+L}\}_{j \in J}$, where a change in behavior occurs. Often, the change is perceptible by eye, but frequently, this is not the case and alternative quantitative methods are preferred. In what follows, a description of the procedure for detecting level shifts in $1/f$ signals by wavelet q-Fisher information is described; later, results on simulated fGn signals are presented.

10.4.2 Level-Shift Detection Using Wavelet q-Fisher Information

To detect the presence of level shifts in fractal $1/f$ signals, wavelet q-Fisher information is computed in sliding windows. A window of length w, located in the interval $m\Delta \leq t_k < m\Delta + w$ applied to signal $\{X(t_k), k = 1,2,\ldots, N\}$, is

$$X\left(m; w, \Delta\right) = X\left(t_k\right) \prod \left(\frac{t - m\Delta}{w} - \frac{1}{2}\right) \tag{10.20}$$

where $m = 0,1,2,\ldots, m_{max}$, Δ is the sliding factor, and $\Pi(.)$ is the well-known rectangular function. Note that Equation 10.20 represents a subset of $X(t_k)$; thus, by varying m from 0 to m_{max} and computing wavelet q-Fisher information on every window, the temporal evolution of wavelet FIM is followed. Suppose the wavelet q-Fisher information at time m (for sliding factor Δ) is denoted as $I_x(m)$. Then a plot of the points

$$\left\{\left(w + m\Delta, I_x(m)\right)\right\}_{m=0}^{m_{max}} := I_X \tag{10.21}$$

represents such time evolution. In the work of Stoev et al. [5], it was demonstrated that the presence of a sudden jump in a stationary fractal signal will cause the estimated $\hat{H} > 1$. The level shift thus causes the signal under observation to become nonstationary. In the wavelet q-Fisher information framework, this sudden jump will cause its value to increase suddenly [according to its studied behavior for $q \in (0,1)$]. Therefore, a sudden jump increase in the plot of Equation 10.21 can be considered as an indicator of the occurrence of a single level shift in the signal. These theoretical findings are experimentally tested by the use of a synthesized scaling signal with level shifts. The synthesized signals correspond to fGn signals generated using the circular embedding algorithm [39,40] (also known as the Davies and Harte algorithm).

10.5 Results and Discussion

Figure 10.5 displays the level-shift detection capabilities of wavelet q-Fisher information for a correlated fGn signal with Hurst exponent $H = 0.7$ and a single structural break located at $t_b = 8192$. The length of the signal is $N = 2^{14}$ points with the break located in the middle and amplitude $\sqrt{\sigma_X^2}$, where σ_X^2 is the fGn variance. The top plot displays the signal and also the level shift (in white) added to its structure for illustration purposes only. It is important to note that the amplitude of the considered and studied level shifts are weak; however, wavelet q-Fisher information detects appropriately its location. The presence of a single level shift embedded in the fGn signal is therefore detected by a sudden increase (in the form of an impulse) in the wavelet q-Fisher information value when computed in sliding windows.

In Figure 10.5, wavelet q-Fisher information was computed with $q = 0.6$. The Hurst exponent $H = 0.7$ means that the signal under study is stationary with LRD. Figure 10.6 displays the level-shift detection capabilities of wavelet q-Fisher information when considering anticorrelated fGn signals. Anticorrelated signals have the property that high values are likely to be followed by low values and vice versa. Note that, for these types of signals, wavelet q-Fisher information effectively detects and also locates level shift embedded in the signal structure. Therefore, independently of the type of anticorrelated fGn signal ($H < 0.5$), wavelet q-Fisher information appropriately detects weak level shift with amplitudes higher than $\sqrt{\sigma_X^2/2}$. The analysis performed in Figure 10.6 was performed in sliding windows of length $W = 2^{11}$ at steps of $\Delta = 90$.

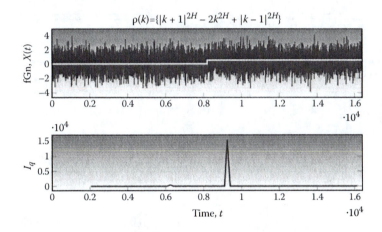

Figure 10.5 Detection of a single structural break at $t_b = 8192$ embedded in an fGn signal with parameter $H = 0.7$.

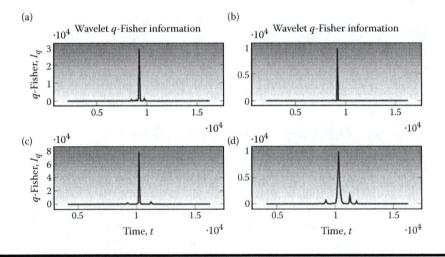

Figure 10.6 Wavelet q-Fisher information for anticorrelated fGn signals. (a) wavelet q-Fisher information for an fGn signal with $H = 0.1$; (b) $H = 0.2$; (c) $H = 0.3$; (d) $H = 0.4$. The amplitude of the level shifts was set to $\sqrt{\sigma_X^2/2}$.

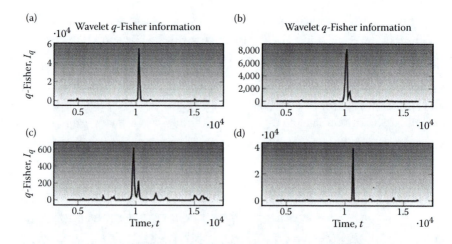

Figure 10.7   Wavelet *q*-Fisher information for Gaussian white noise and correlated fGn signals. (a) wavelet *q*-Fisher information for a Gaussian white noise signal (*H* = 0.5); (b) an fGn signal with *H* = 0.6; (c) an fGn signal with *H* = 0.8; (d) an fGn with *H* = 0.9. The amplitude of the level shifts was set to $\sqrt{\sigma_X^2/2}$.

The parameter *q* of Fisher was set to *q* = 0.6, and the length of the considered signal was $N = 2^{14}$ points. Similar results were obtained when increasing $q \to 1$. In fact, by increasing the amplitudes of the level shifts, better detection capabilities can be observed in wavelet *q*-Fisher information; however, a higher level shift can also be detected by eye. Figure 10.7 presents the level-shift detection capabilities of wavelet *q*-Fisher information for correlated fGn signals with long-memory and Gaussian white noise. The top left plot displays the wavelet *q*-Fisher information for a totally disordered Gaussian white noise signal (*H* = 0.5). Note that, wavelet *q*-Fisher information effectively detects the level shift within this signal. For correlated signals, wavelet *q*-Fisher information performs well and appropriately detects the presence of the level shifts.

Note, however, that, in some cases, wavelet *q*-Fisher information values decrease in amplitude but are sufficiently high to be considered as level shifts. Based on these results, wavelet *q*-Fisher information therefore detects appropriately level shifts embedded in stationary fGn signals. The detection is accomplished independently of the range of the Hurst parameter and in consequence of the correlation structure in the signal. Wavelet *q*-Fisher information therefore provides an interesting alternative to the problem of level-shift detection in fGn signals.

10.5.1 Application to Variable Bit Rate Video Traces

Variable bit rate (VBR) video is expected to account for a large amount of the traffic flowing through next-generation converged networks. The study of the properties of VBR video traffic is therefore important because novel algorithms can be designed to the characteristics observed within this traffic. VBR video traffic is long-range dependent and the long-memory parameter is in many cases *H* > 1, which, in theory, is not permissible. The *H* > 1 case suggests that the VBR video traffic may be subjected to level shifts or nonstationarities embedded within the signal. In this context, we apply the wavelet *q*-Fisher information to a large set of VBR video traces and found that many traces display many impulse-shaped peaks in their wavelet *q*-Fisher information. Figure 10.8 displays the wavelet *q*-Fisher information for an H.263 encoded video signal. Note

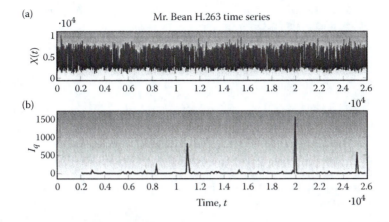

Figure 10.8 **Wavelet *q*-Fisher information for an H.263 encoded video signal. (a) time series (frame size) in bits of Mr. Bean movie; (b) corresponding wavelet *q*-Fisher information of Mr. Bean movie with *q* = 0.8.**

that, this VBR video signal presents many impulse-shaped peaks, which indicate the presence of level shifts embedded within the VBR video signal. This result also explains why wavelet-based estimator displays a long-memory index estimation of $H > 1$.

10.6 Conclusions

In this chapter, the notion of wavelet *q*-Fisher information was introduced. A closed-form expression for this quantifier was developed for scaling signals, and its properties and behavior, in a range of the parameter α and for various *q*, were studied. It was demonstrated through experimental studies with simulated fGn signals that wavelet *q*-Fisher information not only provides appropriate descriptions of the complexities of these signals but also allows one to detect structural breaks in the mean embedded in their structure. In fact, wavelet *q*-Fisher information allows to effectively and timely detect structural breaks embedded in anticorrelated and correlated fGn signals.

References

1. Leland, W. E., Taqqu, M. S., Willinger, W., and Wilson, D. V.: "On the self-similar nature of Ethernet traffic (extended version)", *IEEE/ACM Transactions on Networking*, 1994, **2**, (1), pp. 1–15.
2. Lee, I. W. C., and Fapojuwo, A. O.: "Stochastic processes for computer network traffic modelling", *Computer Communication*, 2005, **29**, pp. 1–23.
3. Beran, J.: "Statistical methods for data with long-range dependence", *Statistical Science*, 1992, **7**, pp. 404–416.
4. Shen, H., Zhu, Z., and Lee, T. M. C.: "Robust estimation of the self-similarity parameter in network traffic using wavelet transform", *Signal Processing*, 2007, **87**, (9), pp. 2111–2124.
5. Stoev, S., Taqqu, M. S., Park, C., and Marron, J. S.: "On the wavelet spectrum diagnostic for Hurst parameter estimation in the analysis of Internet traffic", *Computer Networks*, 2005, **48**, (3), pp. 423–445.
6. Ramírez-Pacheco, J., and Torres-Román, D.: "Cosh window behaviour of wavelet Tsallis *q*-entropies in $1/f^\alpha$ signals", *Electronics Letters*, 2011, **47**, (3), pp. 186–187.

7. Eke, A., Hermán, P., Bassingthwaighte, J. B., Raymond, G., Percival, D. B., Cannon, M., Balla, I., and Ikrényi, C.: "Physiological time series: distinguishing fractal noises and motions", *Pflugers Archiv*, 2000, **439**, (4), pp. 403–415.

8. Plastino, A., Plastino, A. R., and Miller, H. G.: "Tsallis nonextensive thermostatistics and Fisher's information measure", *Physica A*, 1997, **235**, pp. 557–588.

9. Samorodnitsky, G., and Taqqu, M. S.: *Stable Non-Gaussian Random Processes*, Chapman & Hall, New, York, USA, 1994.

10. Beran, J.: *Statistics for Long-Memory Processes*, Chapman & Hall, New, York, USA, 1994.

11. Mandal, S., Arfin, S. K., and Sarpeshkar, R.: "Sub-pHz MOSFET 1/f noise measurements," *Electronics Letters*, 2009, **45**, (1), pp. 81–82.

12. Serinaldi, F.: "Use and misuse of some Hurst Parameter estimators applied to stationary and nonstationary financial time series", *Physica A*, 2010, **389**, pp. 2770–2781.

13. Malamud, B. D., and Turcotte, D. L.: "Self-affine time series: measures of weak and strong persistence", *Journal of Statistical Planning and Inference*, 1999, **80**, (1–2), pp. 173–196.

14. Gallant, J. C., Moore, I. D., Hutchinson, M. F., and Gessler, P.: "Estimating the fractal dimension of profiles: a comparison of methods", *Mathematical Geology*, 1994, **26**, (4), pp. 455–481.

15. Ramírez Pacheco, J., Torres Román, D., and Toral Cruz, H.: "Distinguishing stationary/nonstationary scaling processes using wavelet Tsallis q-entropies", *Mathematical Problems in Engineering*, 2012, **2012**, pp. 1–18.

16. Rea, W., Reale, M., Brown, J., and Oxley, L.: "Long-memory or shifting means in geophysical time series", *Mathematics and Computers in Simulation*, 2011, **81**, (7), pp. 1441–1453.

17. Cappelli, C., Penny, R. N., Rea, W. S., and Reale, M.: "Detecting multiple mean breaks at unknown points in official time series", *Mathematics and Computers in Simulation*, 2008, **78**, (2–3), pp. 351–356.

18. Ramírez-Pacheco, J., Torres-Román, D., Rizo-Dominguez, L., Trejo- Sanchez, J., and Manzano-Pinzon, F.: "Wavelet Fisher's information measure of $1/f^{\alpha}$ signals", *Entropy*, 2011, **13**, pp. 1648–1663.

19. Percival, D. B.: "Stochastic models and statistical analysis of clock noise", *Metrologia*, 2003, **40**, (3), pp. S289–S304.

20. Mandelbrot, B. B., and Van Ness, J. W.: "Fractional Brownian motions, fractional noises and applications", *SIAM Review*, 1968, **10**, pp. 422–437.

21. Flandrin, P.: "Wavelet analysis and synthesis of fractional Brownian motion", *IEEE Transactions on Information Theory*, 1992, **38**, (2), pp. 910–917.

22. Lowen, S. B., and Teich, M. C.: "Estimation and simulation of fractal stochastic point processes", *Fractals*, 1995, **3**, (1), pp. 183–210.

23. Hudgins, L., Friehe, C. A., and Mayer, M. E.: "Wavelet transforms and atmospheric turbulence", *Physical Review Letters*, 1993, **71**, (20), pp. 3279–3282.

24. Cohen, A., and Kovacevic, A. J.: "Wavelets: the mathematical background", *Proceedings of the IEEE*, 1996, **84**, (4), pp. 514–522.

25. Abry, P., and Veitch, D.: "Wavelet analysis of long-range dependent traffic", *IEEE Transactions on Information Theory*, 1998, **44**, (1), pp. 2–15.

26. Bardet, J. M.: "Statistical study of the wavelet analysis of fractional Brownian motion", *IEEE Transactions on Information Theory*, 2002, **48**, (4), pp. 991–999.

27. Martin, M. T., Pennini, F., and Plastino, A.: "Fisher's information and the analysis of complex signals", *Physical Letters A*, 1999, **256**, pp. 173–180.

28. Martin, M. T., Perez, J., and Plastino, A.: "Fisher information and non-linear dynamics", *Physica A*, 2001, **291**, pp. 523–532.

29. Telesca, L., Lapenna, V., and Lovallo, M.: "Fisher information measure of geoelectrical signals", *Physica A*, 2005, **351**, pp. 637–644.

30. Romera, E., Sánchez-Moreno, P., and Dehesa, J. S.: "The Fisher information of single particle systems with central potential", *Chemical Physics Letters*, 2005, **414**, pp. 468–472.

31. Luo, S.: "Quantum Fisher information and uncertainty relation", *Letters in Mathematical Physics*, 2000, **53**, pp. 243–251.

32. Vignat, C., and Bercher, J. F.: "Analysis of signals in the Fisher–Shannon information plane", *Physical Letters A*, 2003, **312**, pp. 27–33.

33. Frieden, B. R., and Hughes, R. J.: "1/f noise derived from extremized physical information", *Physical Review E*, 1994, **49**, pp. 2644–2649.
34. Borges, E. P.: "A possible deformed algebra and calculus inspired in nonextensive thermostatistics", *Physica A*, 2004, **340**, pp. 95–101.
35. Eke, A., Hermán, P., Kocsis, L., and Kozak, L. R.: "Fractal characterization of complexity in temporal physiological signals", *Physiological Measurement*, **23**, 2002, R1–38.
36. Deligneres, D., Ramdani, S., Lemoine, L., Torre, K., Fortes, M., and Ninot, G.: "Fractal analyses of short time series: A re-assessment of classical methods", *Journal of Mathematical Psychology*, 2006, **50**, pp. 525–544.
37. Castiglioni, P., Parato, G., Civijian, A., Quintin, L., and Di Rienzo, M.: "Local scale exponents of blood pressure and heart rate variability by detrended fluctuation analysis: Effects of posture, exercise and aging", *IEEE Transactions on Biomedical Engineering*, 2009, **56**, (3), pp. 675–684.
38. Esposti, F., Ferrario, M., and Signorini, M. G.: "A blind method for the estimation of the Hurst exponent in time series: Theory and methods", *Chaos*, 2008, **18**, (3), 033126-033126-8.
39. Davies, R. B., and Harte, R. S.: "Tests for Hurst effect", *Biometrika*, 1987, **74**, pp. 95–101.
40. Cannon, M. J., Percival, D. B., Caccia, D. C., Raymond, G. M., and Bassingthwaighte, J. B.: "Evaluating scaled windowed variance for estimating the Hurst coefficient of time series", *Physica A*, 1996, **241**, pp. 606–626.

Chapter 11

Characterizing Flow-Level Traffic Behavior with Entropy Spaces for Anomaly Detection

Pablo Velarde-Alvarado, Cesar Vargas-Rosales, Homero Toral-Cruz, Julio Ramirez-Pacheco, and Raul Hernandez-Aquino

Contents

11.1 Introduction

Cyber-attacks carried out directly against networking infrastructure are becoming more and more prevalent. Both the number and the complexity of attacks have increased dramatically in the last years. At the same time, the surges of network security threats have the potential to significantly impede productivity, disrupt business and operations, and result in information and economic losses.

Typical perimeter defenses, such as firewalls, conventional network intrusion detection systems (NIDS), application proxies, and virtual private network servers, have become important components of a security infrastructure. However, they do not provide a sufficient level of protection against sophisticated attacks. Besides, the network has a dynamic behavior for which more elaborate and complicated security procedures need to be derived; this is emphasized when different network segments work with different protocols and services. In addition, when a new service is introduced, or the number of users changes, network dynamics is altered, and this produces modifications to which the security tasks need to adapt in order to protect the network. To overcome these limitations, one promising approach makes use of entropy to obtain knowledge of the structure and composition of traffic, which is summarized by behavioral traffic profiles. This approach is being proposed as a good candidate in traffic characterization for the development of a new generation of NIDS. Some of today's major challenges in NIDS design are to achieve sensitivity in the detection of sophisticated attacks, to successfully obtain an early detection, and to minimize false-positive and false-negative rates, among others.

In this chapter, we present methodologies based on information theory and present entropy spaces and pattern recognition (PR) techniques for decision-making processes in order to detect anomalies in traffic traces. We introduce the entropy space technique together with the excess point methodology, and both help us characterize special features by using probability density function (PDF) estimations. In some cases, a Gaussian distribution is used to show that it is close in the description of distances to a central reference point in the entropy spaces that allows us to classify anomalies for different time slots in a traffic trace.

11.2 Background

In this section, we review the basic concepts of intrusion detection systems (IDSs), entropy, principal component analysis (PCA), and PR that are relevant to the approach we are proposing.

11.2.1 Intrusion Detection Systems

The attacks launched against hosts or networks are identified as intrusions. An intrusion to a host or network is defined in [1] as an unauthorized attempt or achievement to access, alter, render unavailable, or destroy information on a system or the system itself. Individuals regarded as intruders include both those without the proper authorization for the use of resources and those who abuse their authority (insiders) [2]. Heady et al. [3] also define intrusion as any set of actions that attempt to compromise the confidentiality, integrity, or availability of a computational resource. Throughout this chapter, we will use the terms intrusion and attack interchangeably. The definitions for intrusion detection and IDS used in this chapter are as follows: Intrusion detection is the problem of identifying an attempt to compromise the confidentiality, integrity, or availability of a computational resource or information. IDS is a computer system (possibly a combination of hardware and software) that automates the intrusion detection process. Note that, intrusion detection is generally decoupled from the response to the intrusion. Typically, an IDS only alerts the information technology (IT) manager when an attack or a potential attack condition is recognized. The IT manager then takes necessary and appropriate actions to minimize (preferably avoid) any actual damage. Basically, IDSs can be classified into two types: host IDS (HIDS) and NIDS. A HIDS operates on individual devices or hosts in the system to monitor and analyze all the incoming and outgoing traffic on the device only; such systems are not addressed in this chapter. On the other hand, NIDS sensors are typically connected to a network through the use of switched port analyzer connections or taps placed at strategic locations having the ability to monitor all traffic on a network, and its position allows them to detect attacks on any host in the network they protect. When intrusive activity occurs, the NIDS generates an alarm to let the network administrator know that the network is possibly under attack. A passive NIDS is not proactive in that it does not take any action by itself to prevent an intrusion from happening once detected. A reactive system, also known as an intrusion prevention system (IPS), autoresponds to threats, for example, reconfigures access control list on routers and firewalls dynamically, closes network connections, kills processes, and resets transmission control protocol (TCP) connections. Intrusion prevention technology is considered by some to be a logical extension of intrusion detection technology [4,5]. NIDSs play a fundamental role in any enterprise's multilayered defense-in-depth strategy as they constitute, besides firewalls and cryptography, the third security layer. NIDSs can be subdivided into two categories with respect to the implemented detection technique, namely, misuse-based NIDS, also sometimes referred to as signature-based NIDS (S-NIDS), and behavior-based NIDS, also known as anomaly-based NIDS (A-NIDS). S-NIDSs are relying on pattern matching techniques; they monitor packets and compare with preconfigured and predetermined attack patterns known as signatures [6]. According to [7], these signatures can be classified into two main categories based on the information they examine to detect subversive incidents: content-based signatures and context-based signatures. Content-based signatures are triggered by data contained in packet payloads, and context-based signatures are triggered by data contained in packet headers. These signatures are based on known vulnerabilities such as those published by Common Vulnerabilities and Exposures [8]. In terms of advantages, these systems are very accurate at detecting known attacks and tend to produce few false positives (i.e., incorrectly flagging an event as malicious when it is legitimate). There are two important drawbacks to signature-based IDS solutions. First, there will be a lag between the new threat discovered and the signature being applied in NIDS for detecting the threat. During this lag time, the NIDS will be unable to identify the threat. Second, on large-scale networks, where the amount of data passing through the network is extensive, S-NIDSs have performance and scalability problems. Most widely deployed NIDS are signature based; an example is Snort [9], an open-source NIDS or IPS capable of performing real-time traffic analysis to detect possible attacks, instances of buffer overflows, stealth port scans, etc. With respect to anomaly-based systems, they must create a profile that describes the normal behavior of certain traffic features based on information gathered during a training period. These profiles are then used as a baseline to define normal activity. If any network activity deviates too far from this baseline, for example, when an attack occurs, then the activity generates an alarm. Anything that does not fall within the boundaries of this normal use is flagged as an anomaly that may be related to a possible intrusion attempt. Anomaly is a pattern in the data that does not conform to the expected behavior, also referred to as outliers, exceptions, peculiarities, and surprises [10]. Anomaly-based systems have the advantage over signature-based systems in the ability to detect never-seen attacks (zero-day attacks) from the moment they appear. Nevertheless, if the profile is not defined carefully, there will be lots of false alarms, and the detection system will suffer from degraded performance. Another limitation that is well known in the research community is the difficulty of obtaining enough high-quality training data. Traditional anomaly-based systems use a volumetric approach to detect anomalies by monitoring the amount of traffic in the network. However, volumetric intrusion detection has become ineffective because the behavior patterns of the attacks have evolved; these attacks attempt to avoid detection by low-rate infection techniques. This issue has been addressed with the introduction of solutions based on robust traffic features that are sensible to low-rate attacks [11]. The traffic features are extracted from the packet headers, payload, or a combination of both. Lakhina et al. [12] applied an information-theoretic analysis on feature distributions of IP addresses and service ports using entropy to quantify the changes in those traffic distributions. Some examples of anomaly-based systems include network traffic anomaly detector (NETAD) [13], learning rules for anomaly detection (LERAD) [14], and payload-based anomaly detector (PAYL) [15].

Network anomaly detection has been approached using various techniques such as artificial intelligence [16], machine learning [17], statistical signal processing [18], and information theory [19]. A malicious activity has behavior patterns that can be distinguished from a legitimate one. For example, scanning is a typical process that an intruder often uses to gather knowledge from the target. These probes are atypical of a legitimate user and alter the natural diversity of network traffic. Information theory metrics have been used to detect specific types of malicious traffic such as port scanning attacks, distributed denial-of-service (DDoS) attacks, and network worms.

11.2.1.1 Data Preparation for Traffic Profiling

Supervised detection builds specifications of normal behavior from training data, and it is clear that the effectiveness of the detection critically relies on the quality and completeness of the training data. Therefore, one requirement imposed on a training data set is that it should be attack-free, that is, it should not contain malicious activity that would induce the resulting models to consider malicious behavior as normal. The cleaning process of the training data can be performed manually; however, this time-consuming alternative may be prohibitive for a large amount of raw traffic data. In addition, this cleaning process has to be performed periodically because the system needs to be updated regularly to adapt to network changes. Manual inspection can be assisted by an S-NIDS, which can preprocess the training set to discover known attacks (e.g., web scanners and old exploits).

11.2.2 Entropy

Entropy has been used by various disciplines, one of which is physics. Mathematically, entropy measures the concentration of a PDF, so that the entropy is large when the PDF is not concentrated. The entropy therefore has a similar role as the variance. However, in contrast to the variance, the entropy is completely independent of the expected value and of any norm structure on the sample space [20]. Commonly, it is necessary to describe events in terms of its complexity, disorder, randomness, diversity, or uncertainty; it is in these cases where the entropy becomes a fundamental tool to quantify these types of properties in the data. Shannon [21], in a well-known publication of 1948, proposed a mathematical tool to measure the entropy and it was the origin of information theory. In information theory, entropy is a function that measures the information content of a data set or, equivalently, measures the uncertainty of a random variable.

Consider a discrete random variable X that takes values from an alphabet $\chi = \{x_1, \cdots, x_M\}$ with cardinality M, where a realization is denoted by x_k. Let $p_x(x_k)$ denote the discrete PDF or probability mass function of X so that $p_x(x_k)$ is the probability that realization x_k occurs. Shannon entropy for a random variable X is defined by its expected information content:

$$H(X) = E\left[I_X\right] = -\sum_{k=1}^{M} p_X(x_k)\log_2 p_X(x_k),$$ (11.1)

where $I_X = -\log_2 p_x(x_k)$ is the information content of a particular realization x_k. Because we use the base 2 logarithm, entropy has units of bits (natural logarithms give nats, and base 10 logs give bans). In this chapter, base 2 logarithms are used and $0 \log_2 0$ is defined to be 0. The entropy attains its minimum value $H = 0$, when all the realizations in a sample space are the same, that is, the state is pure, which is when $p_X(x_k) = 1$ for some $k = 1, \ldots, M$. On the other hand, the maximum value $H = \log_2 M$ is related to realizations that are equally probable corresponding to a totally random state, that is, $p_X(x_k) = 1/M, \forall k$.

Probability distributions $p_X(x_k) = p_k$ can be approximated by relative frequencies from a training data set of size N. These are the naive empirical-frequency estimates of the probabilities denoted by $\hat{p}_k = n_k/N$, where n_k counts the number of times each x_k appears in the data set. Replacing the probabilities p_k by $\hat{p}_k$ in Equation 11.1, the Shannon entropy of a data set can be estimated as follows:

$$\hat{H}(X) = -\sum_{k=1}^{M} \hat{p}_k \log_2 \hat{p}_k = -\sum_{k=1}^{M} \frac{n_k}{N} \log_2 \left(\frac{n_k}{N}\right).$$ (11.2)

The entropy estimator underestimates the actual entropy for small data sets [22], but the data sets used in traffic analysis are large enough to make a small bias that vanishes in the limit.

Entropy has been extensively studied for anomaly detection and prevention [12,23–25]. Researchers have suggested the use of entropy as a succinct means of summarizing traffic distributions for different applications, in particular, in anomaly detection and in fine-grained traffic analysis and classification. With respect to anomaly detection, the use of entropy for tracking changes in traffic distributions provides two significant benefits. First, the use of entropy can increase the sensitivity of detection to uncover anomalous incidents that may not manifest as volume anomalies. Second, using such traffic features provides additional diagnostic information into the nature of the anomalous incidents (e.g., making distinction among worms, DDoS attacks, and scans), which is not available from just volume-based anomaly detection [26]. When an attack takes place, it is likely to produce some kind of unusual effect on the network traffic (illegitimate traffic); on the other hand, if the data flow is licit, there will be no unusual effect on the traffic (legitimate traffic). Entropy measures allow identifying changes in distance and divergence between the probability densities related to these two types of traffic. Entropy measure can be used to detect such anomalies in the traffic monitored; for example, in Figures 11.1 and 11.2, entropy for the traffic features of IP source address (srcIP), IP destination address (dstIP), source port (srcPrt), and destination port (dstPrt) is shown. These entropy values are obtained from the packets being monitored as a function of time. In both figures, the Blaster worm attack was introduced, and those entropy values are the result of the Blaster propagation. This analysis is at the packet level, that is, a promiscuous method where the traffic features are extracted from each and every packet that traverses the router where monitoring takes place. The monitoring is carried out at fixed time slot with durations of 60 s each, or each time slot corresponds to 1 min of traffic monitored. In Figure 11.1, one can see that, in the first 70 min, the entropy values vary within normal or typical levels and have normal randomness; after the 70th minute, when the Blaster worm was introduced, the entropy levels start incrementing significantly in a very short time for the srcIP and the dstIP features. During the Blaster worm propagation experiment, there can be seen two points in the 38 min that last such attack. First, the entropy value according to the specific traffic feature is stabilized to a high or low value with almost no randomness in such values; second, the entropy for the features corresponding to the sources (either port or address) and those of the destinations has a negative linear dependency that is different in sign from that of normal traffic. These effects are also caused by the worm dynamics, that is, variation in srcIP is small, but a high variation of dstIP is attempting to infect computers.

Figure 11.2 shows the entropy values for the four features of the traffic that correspond to the time slots 1 through 10 of Figure 11.1. This close view of Figure 11.1 permits the identification of another interesting property of the entropy-based traffic analysis that is not seen easily. In time slots 3 and 4, an anomalous behavior can be detected in the features corresponding to the srcIP and the dstIP because there is negative linear dependence; that is, when one increases in value, the other decreases in value and vice versa. A closer view at the traffic packets in such time slots allowed seeing that there was a scan attack directed toward the ports of the proxy server monitoring the network. The interesting point of this scan attack is that there was an attack that is within normal entropy levels and that the entropy levels should not be considered the only procedure to detect anomalies.

Figure 11.3 shows traffic as a function of time. Traffic is measured in units of packets per time slot, that is, number of packets per minute. From the point of view of traffic volume, it can be

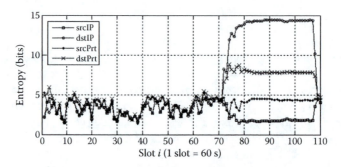

Figure 11.1   Entropy for the traffic features for a Blaster worm attack.

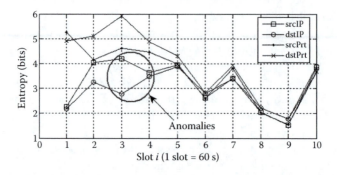

Figure 11.2   Entropy for the traffic features: a close view to slots 1 through 10 of Figure 11.1.

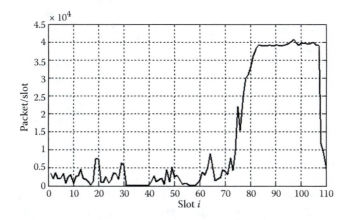

Figure 11.3   Traffic volume for the network with the Blaster attack experiments.

seen that the Blaster worm attack can be detected, but the attack detected by the negative linear dependency of the entropy values, as shown in Figure 11.2, cannot be detected (time slots 3 and 4) because the traffic volume in such time slots is within normal levels.

11.2.3 Pattern Recognition

Network intrusion detection (NID) is essentially a PR problem in which network traffic patterns are classified as either normal or abnormal. This section presents a brief introduction to the basic concepts of PR. For a more extensive treatment, see, for example [27–29].

PR is the scientific discipline of machine learning (or artificial intelligence) that aims at classifying data (patterns) into a number of categories or classes [30]. To understand how PR works, it is needed to describe some important concepts.

Pattern: It is a representation of a given object, observation, or data item. Examples of patterns are measurements of a voltage signal in electrocardiogram, a pixel in a two-dimensional (2D) image, fingerprints of different persons, and a cluster of traffic flows.

Features: It is a set of data or attributes obtained from measurements of patterns. These features are useful for their further characterization.

Feature selection: It is a process to determine which set of features is the most appropriate to describe a set of patterns according to their relevance. Feature selection reduces dimensionality by selecting a subset of original variables.

Feature extraction: It is a process to reduce the amount of redundant data given in a set of features; such a set should be transformed to obtain its reduced characteristic representation. This is done by dimensionality reduction applying (linear or nonlinear) projection of p-dimensional vector onto d-dimensional vector ($d < p$).

Feature vector: It is the reduced characteristic representation of a given set of features. It stores the relevant data of such features, and it is helpful for its further classification. For its easy management, these relevant data are represented in a vector form as shown in Equation 11.3:

$$x = [x_1, x_2, \cdots, x_p]^{\mathrm{T}} \tag{11.3}$$

where ^{T} denotes transposition and x_i are measurements of the features of an object. Each of the feature vectors identifies uniquely a single pattern.

Feature space: It is a given space where each pattern is a point in the space. Each feature vector is represented as a coordinate on such space. Thus, each axis represents each feature, and the dimension p should be equal to the number of used features.

Class label or class: A class can be viewed as a source of patterns whose distribution in the feature space is governed by a PDF specific to the class. For example $\{w_1, w_2, \cdots, w_c\}$ represents the finite set of c classes of nature.

Training set is a set of patterns already having been classified.

A PR system is an automatic system that aims at classifying the input pattern into a specific class. It proceeds into two successive tasks: (1) the analysis (or description) that extracts the characteristics from the pattern being studied and (2) the classification (or recognition) that enables us to recognize an object (or a pattern) by using some characteristics derived from the first task [30].

A PR system's main task is to classify patterns in a set of classes, where these classes are defined by a human who designs a system (supervised learning) or are learnt and assigned by using pattern similarity (unsupervised learning).

In this work, in particular, a statistical PR approach is implemented. By using this approach, each pattern is represented by a set of features or measurements in a p-space. An adequate feature selection allows establishing disjoint regions. Thus, each class may be distinguished accurately as a different class. Such disjoint regions are obtained by using training sets, and each disjoint region represents each class. By using the statistical approach, each decision region is determined by the PDF of the patterns belonging to each class.

11.2.4 Principal Component Analysis

PCA is an eigenvector method designed to model linear variation in high-dimensional data. PCA performs dimensionality reduction by projecting the original p-dimensional data onto the d-dimensional linear subspace ($d < p$) spanned by performing a covariance analysis between factors [31,32]. The goal of PCA is to reduce the dimensionality of the data while retaining as much as possible the variation present in the original data set. In pattern classification tasks, significant performance improvements can be achieved by a mapping between the original feature space to a lower-dimensional feature space according to some optimality criteria. In our proposed architecture, classification of traffic patterns for intrusion detection is performed through the features measured or extracted from flow clusters; this information is passed to a stage of feature selection implemented by PCA. Based on the above, it is possible to build more effective data analyses on the reduced-dimensional space: classification, clustering, and PR.

PCA problem can be described as follows: let $X \in \mathbb{R}^{p \times N}$ be the original data of N observations on p correlated variables, that is, $X = [x_1, x_2, \cdots, x_p]$, where $x_i = [x_{1i}, x_{2i}, \cdots, x_{pi}]^T$, $i = 1, 2, \ldots, N$, should be transformed into $Y = [y_1, y_2, \cdots, y_p] \in \mathbb{R}^{p \times N}$ of p uncorrelated variables and organized in decreasing order according to its variance. The covariance matrix of the random vector X is formally defined by $\text{cov}[X] = \sum_X = E\left[(X - \mu)(X - \mu)^T\right] \in \mathbb{R}^{p \times p}$, where $\mu = E[X]$. Let $\{v_i \in \mathbb{R}^{p \times 1} : i = 1, 2, \cdots, p\}$ denote the eigenvectors of $\sum_X$, and let $\{\lambda_i \in \mathbb{R} : i = 1, 2, \cdots, p\}$ denote the corresponding eigenvalues arranged in descending order. Furthermore, let $V = [v_1, v_2, \cdots, v_p]$ be the $p \times p$ orthogonal matrix constructed from the eigenvectors and $\Lambda = \text{diag}\{\lambda_1, \lambda_2, \cdots, \lambda_p\}$ be the $p \times p$ diagonal matrix constructed from the eigenvalues. The principal components are the eigenvectors with the d largest eigenvalues, that is, $\left[\sum_X \boldsymbol{v}_i = \lambda_i \boldsymbol{v}_i\right] \in \mathbb{R}^{p \times d}$ for $i = 1, 2, \cdots, d$. Finally, $\boldsymbol{Y}$ is given by the linear combination of optimally weighted variables $[x_{1i}, x_{2i}, \cdots, x_{pi}]$:

$$Y = \sum_{i=1}^{N}\left(x_{1i}v_1 + x_{2i}v_2 + \cdots + x_{pi}v_p\right) = VX. \tag{11.4}$$

11.3 Method of Entropy Spaces

In order to make use of the method of entropy spaces (MES), one needs to generate as input the flow-level network traffic and perform an abstraction of the data to obtain a three-dimensional (3D) representation by point cloud data. The coordinates of each point in the 3D space generated represent the naive entropy estimates of three features of a cluster of flows for a given cluster key. Such generated 3D spaces under typical traffic are used to define a behavior profile of network traffic.

The input that needs to be generated for the application of MES starts with the definition of a trace χ. A traffic trace is obtained where packets are captured from the network where anomaly detection wants to be applied, and then such a trace is divided into m nonoverlapping traffic slots with maximum time duration of t_d seconds, which is chosen empirically according to sensitivity analysis. In any slot i, K_i flows of packets are generated. The flows to be generated are defined under a five-tuple and interflow gap of 60 s. A five-tuple is a stream of packets having the same four flow fields and protocol, that is, the same source and destination IP addresses, same source and destination port numbers, and same protocol number [33]. The four flow r-fields are labeled as follows: $r = 1$ for source IP address (srcIP), $r = 2$ for destination IP (dstIP), $r = 3$ for source port (srcPrt), and $r = 4$ for destination port (dstPrt). After that, the set of flows are clustered for each slot i; in other words, one needs to fix an r-field, now named cluster key (CK-r). For a given i-slot, each cluster is formed by containing flows under the same r-field but leaving free the rest of the r-fields. As an example, for CK-$r = 1$ or CK srcIP, each cluster is formed with all flows that possess the same source IP address regardless of the value of the rest of the r-fields ($r = 2, 3, 4$). These fields are denoted as free dimensions. It is clear in this example that the complete number of flow clusters depends on the cardinality of the alphabet $\left|\mathbb{A}_i^{r=1}\right|$, where $\mathbb{A}_i^{r=1}$ is the set of IP source addresses shown in the traffic trace within time slot i.

Taking the aforementioned example, entropy estimation for each j-cluster flow that belongs to the CK srcIP is represented as a 3D Euclidean point, denoted as $(\hat{H}_{\text{srcPrt}}, \hat{H}_{\text{dstPrt}}, \hat{H}_{\text{dstIP}})j$, where $\hat{H}_{\text{srcPrt}}$, $\hat{H}_{\text{dstPrt}}$, and $\hat{H}_{\text{dstIP}}$ are the respective entropy estimates of the free dimensions of the cluster flows calculated using Equation 11.1 or 11.2 and $j = 1, 2, \cdots, \mathbb{A}_i^{r=1}$. The number of points generated in the 3D space for each slot i depends on the number of packets captured in such slot and hence on the cardinality of the set $\left|\mathbb{A}_i^{r=1}\right|$. Thus, the set of data points in a slot i is given by $(\hat{H}_{\text{srcPrt}}, \hat{H}_{\text{dstPrt}}, \hat{H}_{\text{dstIP}})_1, (\hat{H}_{\text{srcPrt}}, \hat{H}_{\text{dstPrt}}, \hat{H}_{\text{dstIP}})_2, \ldots, \left(\hat{H}_{\text{srcPrt}}, \hat{H}_{\text{dstPrt}}, \hat{H}_{\text{dstIP}}\right)\left|\mathbb{A}_i^{r=1}\right|$. One can repeat this procedure of obtaining the entropy points of the free dimensions for each time slot m of the traffic trace to generate the entropy space of CK srcIP. Afterward, one can plot those points using a scatter plot to form a point cloud data. A similar procedure is repeated in order to obtain three entropy spaces of the three remaining CKs. In this chapter, the analysis will be focused on the study of the CK srcIP.

Figure 11.1 shows the entropy spaces of the CK srcIP for three traffic traces. The traces were captured from a campus LAN. Figure 11.4a shows typical TCP traffic of 8 h during typical

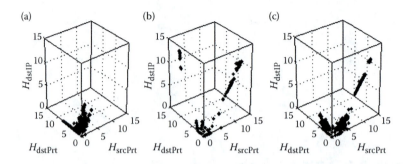

Figure 11.4   Entropy spaces for CK srcIP. (a) Typical traffic; (b) Blaster worm propagation; (c) Sasser worm propagation.

working hours; similar patterns were obtained for the rest of the weekdays. Figure 11.4b and c shows the behavior of Blaster and Sasser worms during 30 min of traffic capture, respectively. Figure 11.4a shows that, although typical traffic tends to be concentrated, the behaviors of Blaster and Sasser traffic tend to increase their entropy value and spread the points on the entropy space. In addition, there are patterns of positive correlation between the entropies of the free dimensions, specifically H_{dstIP} versus H_{srcPrt} for Blaster worm and H_{dstIP} versus H_{dstPrt} for Sasser worm.

In Figures 11.1 and 11.2, the negative linear dependency was discussed because it has sensitivity to some kind of attacks such as the scan attack. This anomalous behavior is observed and captured clearly through the use of the entropy spaces as it can be seen in Figure 11.5, with the point pattern organization in such spaces that is structured differently from that of a normal traffic entropy space. It can also be seen in Figure 11.5 that the values of entropy in such spaces are large. The entropy space shown in Figure 11.5 corresponds to a different data set from those of the Blaster and Sasser attacks shown in previous figures. The data set is from the CAIDA project and contains only the traffic generated by the propagation of the Witty worm that was captured in the backbones of the Internet. In Figure 11.5b, such linear dependency can be seen through the correlation coefficients of the srcPrt and dstIP features. We found that entropy estimates of the cluster flows exhibit linear relationships that can be seen through the scatter plots and that are dependent on the network activity. This linearized behavior has been detected in other traffic data sets, confirming its sensibility to different attacks. Hence, a detection tool based on the correlation

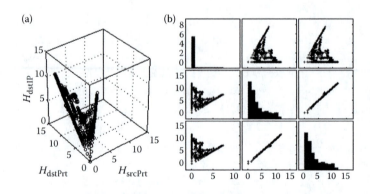

Figure 11.5   Entropy space for CK srcIP. (a) Witty worm traffic; (b) plot matrix for Witty worm propagation.

coefficients and the covariances can be applied to the traffic data sets at the time-slot level and detect such trends.

Once one has the entropy spaces, one can see that these are represented by the vector $X^p \in \mathbb{R}^3$. Because the amount of information is large, a dimension reduction procedure needs to be applied. In this case, PCA, as introduced in Section 11.2, is applied to this vector to reduce the dimensionality of it and to generate a new vector z-score $Z^r \in \mathbb{R}^k, k \leq 3$. To obtain accurate analysis estimation, tools such as kernel density estimation (KDE) are useful. This analysis was applied on data points at the slot level on the PCA-1 by using a Gaussian kernel of 200 points, with a bandwidth of $h = 1.06\sigma J^{1/5}$, the Silverman's criteria, where J is the number of observations and σ is the standard deviation of the set of observations. KDE shows that the traffic slots have Gaussian bimodality behavior in its PDF. Each mode was labeled as principal mode and far mode, respectively, as shown in Figure 11.6.

For PCA-1, the empirical obtained mean values were among 4.3 (positive far mode) and −4.2 (negative far mode) units of PCA-1. Negative far mode was the most frequent. In the case of the studied attacks, far mode presented anomalous values on slots with anomalous traffic. For instance, in the case of Blaster worm, the three first slots show values of −9, −11, and −13 PCA-1 units, respectively (Figure 11.7a). These values were classified as an anomaly

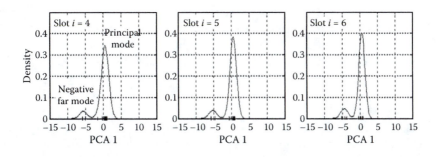

Figure 11.6   Density estimation of PCA-1 presents bimodal behavior on traffic slots.

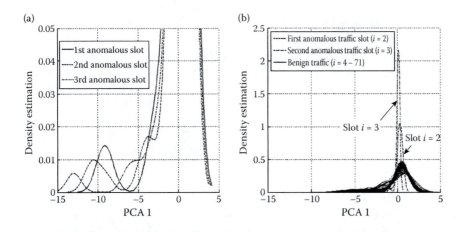

Figure 11.7   Cluster flow anomalies. (a) First three anomalous slots related to the Blaster worm traffic. (b) Two anomalous slots related to port scan attack.

because each value is lower than the mean and the threshold value (−4.2). Thus, this anomalous behavior can be detected easily in an early stage. A second feature that was used because of its sensitive behavior is the standard deviation of the principal mode. The standard deviation empirically obtained in typical conditions was 1.5 units. Then, the standard deviation of the principal mode was taken to show anomalous behavior. For instance, in the case of Sasser worm, the standard deviation of the principal mode on anomalous slots decreased to 0.4 unit in average. In other tests, this time detecting a port scanning attack, the standard deviation shows 0.7 and 0.4 unit on two anomalous traffic slots. In this case, Figure 11.7b shows an outlier-like density in slots 2 and 3.

Nevertheless, KDE only provides the density distribution shape but not its parameters. Therefore, a technique to extract these parameters (mean of the far mode and standard deviation of the principal mode) is needed. It is important because these parameters can be utilized in A-NIDS. The Gaussian mixture model (GMM) is the approach used to extract the parameters of the distribution [34]. A GMM implementation can be found in Statistic Toolbox of MATLAB®. gmdistribution.fit forms groups of data and then fits them on a GMM of $\mathbb{K}$ components. This function performs the maximum likelihood estimation of the GMM by using the expectation maximization algorithm. The algorithm returns the parameters $\theta_k = \left[\pi_k, \mu_k, \sigma_k\right]_{k=1}^{\mathbb{K}}$, where $\mathbb{K} = 3$ is the number of modes of the estimated PDF under KDE for PCA-1, π_k are the proportions of the mixture, and μ_k and σ_k are the mean and standard deviation, respectively, that are related to the required information about principal and far modes.

In the PCA-1 for CK srcIP analysis, we obtain the mixture vector θ_k. From θ_k, principal and far modes can be identified on an i slot. The principal mode corresponds to k mixture, which has max (π_k) (maximum proportion). The far mode corresponds to mixture with max $(|\mu_k|)$ (maximum mean value).

According to these empirical results, feature selection chooses the far mode mean value, from now on denoted as r_1, and the standard deviation of the principal mode, denoted as r_2, as the most appropriate set of features. The behavior of the aforementioned features allows one to establish that they are good parameters to identify anomalous and typical traffic. Experiments described throughout this document support the aforementioned assumption.

Once the results of the PCA are processed by using KDE, one needs to do a traffic analysis under typical conditions, which generally shows that the behavior of feature r_1 follows a trimodal distribution, with more prevalent trends to negative values with respect to positive values and a residual mode centered at zero, as shown in Figure 11.8a. The model that describes this behavior was first observed by KDE; subsequently, its distribution parameters were obtained by GMM.

Such a probabilistic model for feature r_1 based on GMM is as follows:

$$f_{GM}\left(r_1; \theta_k\right) = \sum_{k=1}^{3} \pi_k \mathcal{N}\left(r_1 \middle| \mu_k, \sigma_k^2\right) = \sum_{k=1}^{3} \frac{\pi_k}{\sqrt{2\pi}\sigma_k} e^{-\frac{\left(r_1 - \mu_k\right)^2}{2\sigma_k^2}} \tag{11.5}$$

where μ_k and σ_k are the mean and standard deviation of the kth component, respectively, and π_k are the mixture proportions. The three components form a three-component vector $\theta_k = \left[\pi_k, \mu_k, \sigma_k\right]_{k=1}^{3}$. $\mathcal{N}\left(r_1 \middle| \mu_k, \sigma_k^2\right)$ represents a Gaussian multivariate distribution. The fit values of this mixture are shown in Table 11.1.

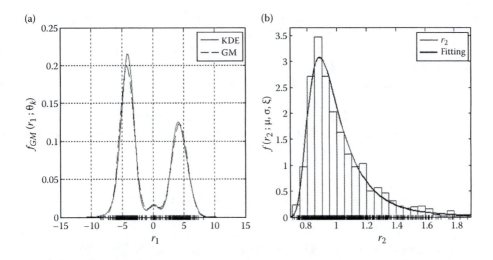

Figure 11.8 **Fitting distribution to selected features in cluster flows. (a) Fitting feature by using GMM. (b) Fitting curve of feature by using a GEV distribution.**

Table 11.1 **Fit Parameters for Mixture**

Parameter	$k = 1$	$k = 2$	$k = 3$
Mixing (π_k)	0.3946	0.5740	0.0314
Mean (μ_k)	4.3239	−4.1988	0.0980
Variance (σ_k^2)	1.6378	1.3208	0.7546

In the case of feature r_2, analysis on the set of typical traffic traces showed that their behavior followed a generalized extreme value (GEV) distribution, with the following profile behavior model:

$$f\left(r_2;\mu,\sigma,\xi\right)=\frac{1}{\sigma}\left[1+\xi*\left(\frac{r_2-\mu}{\sigma}\right)\right]^{-1/\xi-1}*\exp\left\{-\left[1+\xi*\left(\frac{r_2-\mu}{\sigma}\right)\right]^{-1/\xi}\right\} \quad (11.6)$$

for $1 + \xi*(x_2 - \mu/\sigma) > 0$, where $\mu \in \mathbb{R}$ is the localization parameter, $\sigma > 0$ is the scale parameter, and $\xi \in \mathbb{R}$ is the shape parameter.

Fitting parameters for Equation 11.6 that describe the behavior of r_2 are $\xi = 0.2228$, $\sigma = 0.1221$, and $\mu = 0.9079$. The obtained shape with these values is shown in Figure 11.8b.

11.3.1 Excess Point Method

In this section, we describe the method of excess points, where the 3D entropy spaces created are used to determine the presence of anomalies by detecting the region where such points in the 3D space are located. This method is an alternative method to that just described. It starts with the generation of the entropy spaces as described in the previous section for the application of MES. The excess point method consists of finding a centroid for the benign traffic and then characterizing it by the distances of the normal traffic points to that benign centroid. After the normal traffic characterization, the anomaly traffic is detected by taking the anomaly points' distances to the benign centroid previously found that exceed a maximum distance found for the benign traffic (excess points). A PDF can then be found using KDE for both the benign and the anomalous traffic distances to the benign centroid with which one can obtain a decision region that will determine if a given window can be declared as having a normal or an anomalous behavior. The steps that describe the excess point method are the following:

■ One must choose a number of days D in which Internet traces in the network segment to be analyzed will be gathered to have reliable information about the normal behavior or the network. There is a need to make a compromise in this matter, as having lots of information (many days of collected data) will have the effect of finding a more accurate overall behavior of the network, but it will be time consuming and of heavy processing to analyze the traffic. Then again, having few training days for the data results in a less accurate modeling of the network but will consume less memory, which helps in processing speed. The proposed period to train the data is 1 to 2 weeks.

■ A window time of t_w seconds is declared. This period will be the basic period used to capture packets in every trace. Just as in the number of days, there must be a compromise in choosing the window size, as it must be sufficiently large to gather enough information about the trace, and it should be sufficiently small to have a faster response when an anomaly appears in the network segment being analyzed. The data captured in this time window will constitute the actual training data for the IDS. In this method, a window size of 60 s was used.

■ Traffic from the D days is captured in time windows, each one with duration of t_w seconds. A total of Q time windows containing traffic over the D days will be obtained. Then, a data mining process must be applied to extract the information about the packet headers in each window time. Although there are other parameters that could be exploited to have a different analysis, in this system, the features that will be used will be those regarding computer information (IPs) and service information (ports).

■ Using the method described in this section, the entropy spaces must be found for each CK and for every window obtained in the D days from which the data were gathered, such as those depicted in Figure 11.9 for $D = 5$ days.

■ The next step consists of finding the benign or reference centroid $c^r = (x_{cn}, y_{cn}, z_{cn})$ from the captured data. The benign centroid constitutes the representative point that helps in differentiating a benign from a malicious traffic. To obtain the centroid of the captured data, a method based on "bins" b_j is proposed.

■ The method divides the whole entropy space H_i^r of each cluster key into m smaller cubes (bins) of side length s entropy units (according to the logarithm base, the units may be different). The value of s must be chosen between certain range, as it must be small enough to preserve the behavioral characteristics of the traces, and it must be big enough to reach an appropriate centroid. Once a side length s is chosen, the number of bins in which each entropy space will be divided to is given by

$$m = \left(\left\lceil \frac{\max\{x_c\}}{s} \right\rceil \right) \cdot \left(\left\lceil \frac{\max\{y_c\}}{s} \right\rceil \right) \cdot \left(\left\lceil \frac{\max\{x_c\}}{s} \right\rceil \right). \tag{11.7}$$

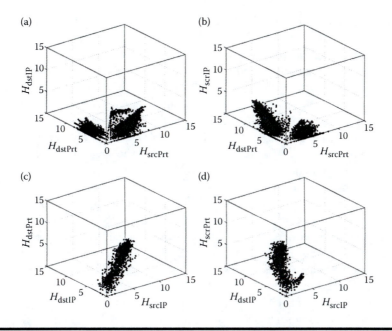

Figure 11.9 **Entropy spaces for the four cluster keys. (a) Source IP entropy space; (b) destination IP entropy space; (c) source TCP port entropy space; (d) destination TCP port entropy space.**

■ After choosing s, an average is obtained from all the points that fall into the volume of each cubic bin found. The average of each bin will function as a representative point for that bin, that is,

$$\left(x_b, y_b, z_b\right) = \frac{1}{J}\sum_{u=1}^{J}\left(x_u, y_u, z_u\right), \quad b = 1, 2, \ldots, m, \tag{11.8}$$

where J is the total number of points that fall within the volume.

■ Once the representative points for all the bins in which the space is divided (m) have been obtained, an average of all these representative points must be found, and it will become the centroid (c_i^r) for each ith window:

$$c_i^r = \left(x_i, y_i, z_i\right) = \frac{1}{m}\sum_{b=1}^{m}\left(x_b, y_b, z_b\right), \quad i = 1, 2, \ldots, Q. \tag{11.9}$$

Finally, the benign centroid is found by calculating an average of all the windows' Q centroids:

$$c^r = \left(x_{cn}, y_{cn}, z_{cn}\right) = \frac{1}{Q}\sum_{i=1}^{Q}c_i^r. \tag{11.10}$$

The centroids must be found for each cluster key. In Figure 11.10a and b, it can be seen that the entropy space resulting from a Blaster worm attack is easily distinguished from the benign traffic entropy space as there are many excess points.

The reason for using this procedure (the use of bins) to find the benign centroid follows the fact that the normal behavior of the traffic inside the network analyzed reflects many entropy points that are very near the space borders. Having that many points gathered into specific areas has the effect of pulling the centroid near that area, and as a result, the centroid found will not be a good representation of the whole extension of the entropy points. With this procedure, a centroid closer to the middle point of the range of the space points in every dimension can be found.

■ Once the benign centroids are found for each cluster key, the Euclidian distances δ_p of all the points $p = 1, 2, \ldots, P$, in each of the entropy spaces to the benign centroid can be found. The set of benign points' distances to the benign centroid $D_b = \{\delta_{b1}, \delta_{b2}, \ldots, \delta_{bP}\}$ is then obtained.

■ The next step in the excess point method consists of obtaining the anomalous traffic distances from the anomalous points to the benign centroid. The entropy points are obtained for traffic traces known to be anomalous using the same method used for the benign traffic. Then, the malicious distances set $D_m = \{\delta_{m1}, \delta_{m2}, \ldots, \delta_{mT}\}$ corresponding to the T anomaly points distances to the benign centroid.

■ Having found the distance sets for the benign and anomalous traffic entropy points, we use elements of Bayes decision theory to provide a Bayesian classifier with which a traffic trace can be classified as having normal or unusual behavior. The rules of the Bayes decision theory, explained in more detail in [29–31], are then used.

■ In this methodology, there are two classes defined as C_i with $i = 1, 2$, corresponding to benign and anomalous classes, respectively. With the event of an input entropy point (x_p, y_p, z_p), its distance to the benign centroid x and the a priori probabilities of the incoming point $p(x)$ and $p(C_i)$, the Bayes classifier rule is applied as follows:

$$\text{if } P(C_1|x) > P(C_2|x), \quad x \rightarrow C_1, \tag{11.11}$$

$$\text{if } P(C_1|x) < P(C_2|x), \quad x \rightarrow C_2. \tag{11.12}$$

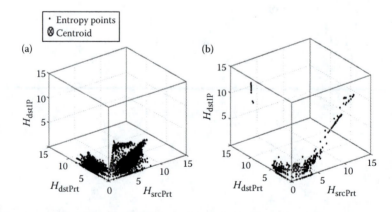

Figure 11.10 Entropy space identification for Blaster worm attack. (a) Benign traffic entropy space for srcIP; (b) Blaster worm attack traffic entropy space for srcIP.

By using Bayes rule, we can obtain the a posteriori probabilities and decisions as follows:

$$\text{if } P(x|C_1) \, P(C_1) > P(x|C_2) \, P(C_2), \quad x \rightarrow C_1, \tag{11.13}$$

$$\text{if } P(x|C_1) \, P(C_1) < P(x|C_2) \, P(C_2), \quad x \rightarrow C_2. \tag{11.14}$$

Then, probabilities $P(C_i|x)$ can be associated with PDFs obtained for the benign and anomalous entropy point distances. The excess points, as stated in the initial description of the method, are those points from the malicious distances set that exceed a maximum benign distance threshold t_h. The PDF of the benign traffic distances is obtained. The method used for the PDF estimation was the kernel smoothing algorithm. After applying the kernel smoothing algorithm, the PDF describing the distances from the benign points to the centroid is found. Figure 11.11a shows a histogram of the distances of the normal points to the srcIP benign centroid. Figure 11.11b shows a PDF found with the kernel estimation for the destination IP and also a normal PDF with the mean and variance of the distance set. Then, for the anomaly traffic, the excess points distance set is found by taking all the distances from the malicious set that exceed the maximum benign distance threshold. The threshold is obtained by the likelihood ratio method, where the total number of malicious points whose distance exceeds the maximum benign distance is divided by the total number of malicious points. An empirical analysis of training data sets helps in determining the value this likelihood ratio must have.

The excess point methodology can be summarized as follows:

■ Captured traces are gathered just as in the training period in windows of t_w seconds. For every window, there will be a captured trace Y.
■ Information about the trace packet headers' features is retrieved.
■ The entropy spaces for the current window are obtained for every cluster key using the same methodology described.

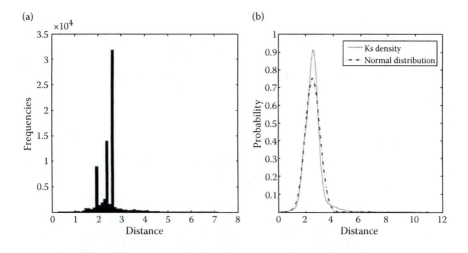

Figure 11.11 **PDF estimation for srcIP benign traffic. (a) Histogram of the distances to the benign centroid for srcIP; (b) PDF for the distances to the benign centroid for srcIP.**

- A new set of distances from each of the entropy space points to the benign centroid is found for every cluster key.
- Every point in the set Y can be classified as benign or anomalous by using the Bayes decision rule considering the PDFs found in the training stage.

11.4 Architecture for A-NIDS Based on MES

In this section, we describe the overall design of our anomaly-based NID architecture. Also, some test results are shown where Blaster and Sasser attacks are detected using the methodology described in Section 11.3. This framework (Figure 11.12) implements behavioral profiles based on entropy spaces. This architecture can be divided into two layers. The first layer does training functions based on the selected features r_1 and r_2, and its respective parameters of the behavior profiles are obtained by using probabilistic techniques from transformed data by PCA, whose models are specified in Equations 11.5 and 11.6. The second layer performs actual measurements on the cluster flows in an i traffic slot. This procedure converts cluster flows to a 2D point in the feature space. If the feature measurements are deviated from the typical traffic profile, then this entire traffic slot is labeled anomalous. Each block that forms the framework for the training layer is described as follows:

- *Preprocessing:* In this stage, each typical traffic trace is sanitized. Such process is done to remove duplicated or miscaptured packets. The trace is filtered according to the protocol (TCP, user datagram protocol, and Internet control message protocol).
- *Flow generator:* On each training traffic slot, flows are generated according to the five-tuple definition.

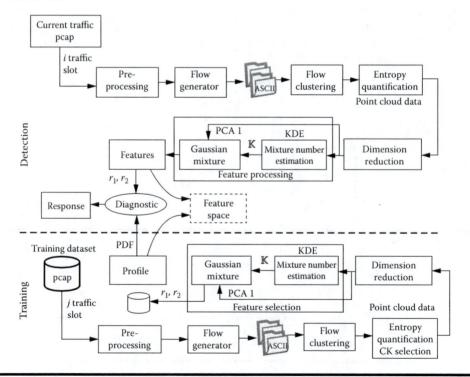

Figure 11.12 Block diagram architecture of the A-NIDS proposal.

- *Flow clustering:* For a given CK-*r*, the generated flows are clustered. The number of cluster flows will depend on the cardinality of the alphabet of the fixed *r*-field, that is, A_i^r.
- *Entropy quantification:* For each cluster flow, the naive entropy of the free dimensions is estimated. The result is represented as a point cloud in a 3D space.
- *Dimension reduction:* The set of entropy points in a traffic slot is transformed by using PCA. The first principal component behavior has properties that capture the changes in traffic patterns caused by illegitimate activities.
- *Feature selection:* The two selected features of the first principal component are obtained by a process that identifies the $\mathbb{K}$ number of mixtures in the KDE, and their respective density parameters are estimated using GMM.
- *Profile:* The analysis of the training data set under this procedure generates new data with the values of the features; these data are fitted with the models described in Equations 11.5 and 11.6. With this, the typical behavior of the traffic is characterized.

In the detection layer, the process is similar to that of the training layer; in other words, it does perform the functions as in the training layer from preprocessing up to the feature selection, and the only difference is that a profile is not generated. Instead of this, from each traffic slot being monitored, and after passing the aforementioned listed stages, r_1 and r_2 features are extracted and compared to the reference profile provided by the training layer. In case a difference is found between such features and the obtained profile, that is, upon finding feature deviations, an anomaly is flagged to the network administrator so that it can be further analyzed and decisions can be done regarding possible intrusions. The excess point method can be incorporated within the A-NIDS architecture as an alternate or a backup diagnostic tool that provides the response in the detection layer. The use of excess point methodology can help reduce information processing because the point cloud data are the ones used to perform such methodology avoiding the use of dimension reduction.

Figure 11.13a shows the two types of traffic, the normal and anomalous traffic, produced by Blaster. The former is grouped into three different regions, and the latter is grouped only into two regions. It can help to apply supervised classification techniques by dividing their corresponding

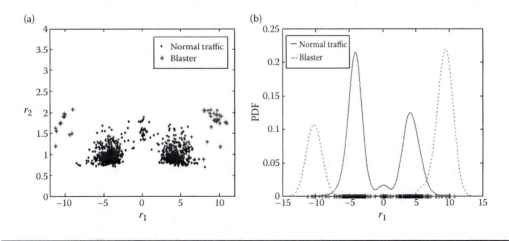

Figure 11.13 **Comparison between typical and anomalous traffic slots caused by Blaster worm. (a) Feature space. (b) PDFs.**

classes. In Figure 11.13b, traffic distribution is shown for the feature r_1 on both cases of normal and anomalous traffic produced by Blaster worm.

11.4.1 Results of Tests

In this section, we present results that show how, with the help of the PCA methodology, we can reduce the dimensionality of the entropy spaces and still detect anomalies based on a new 2D feature space. We also show that the characterization of the points in such feature space can be characterized by a mixture of Gaussian PDFs.

Figure 11.13a shows the points in the feature space after applying the PCA for the two types of traffic: normal and anomalous. Each time slot produces several of the points shown. The points that are not concentrated in the 2D graph are those points produced by the Blaster worm attack. The normal traffic is grouped into three different regions, and the anomalous traffic is grouped only into two regions. Both regions are clearly separated from one another; it helps to apply supervised classification techniques by dividing their corresponding classes.

As seen in Figure 11.13b, the model for r_1 shows a marked difference in the properties of traffic caused by the Blaster worm with respect to typical traffic. The extreme points (vertices) of the convex hull of the data points in the feature space contain the class of normal traffic. Then, a classifier can be designed in terms of the data instance that belongs to the outside of the boundaries of the class of typical traffic.

In the case of the Sasser worm, traffic slots are grouped in one region as shown in Figure 11.14a. In Figure 11.14b, it shows the difference between normal traffic and anomalous traffic produced by the Sasser worm as a decrease in the variance in the first principal component, which is captured by feature r_2.

11.4.2 Excess Point Results

The benign centroid for the excess point method was found using the bins method with a bin's side length of $s = 1.5$ bits, which showed good results for the centroid's location in the entropy space. The likelihood ratios for the Blaster and Sasser worms where excess points were found are presented in Table 11.2, where a value of zero means that there were no excess points found for a given cluster key.

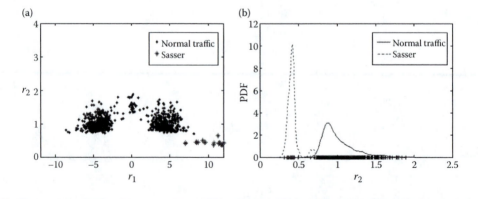

Figure 11.14 Comparison between typical and anomalous traffic slots caused by Sasser worm. (a) Feature space. (b) Probability densities.

Table 11.2   Likelihood Ratios of Blaster and Sasser Worms

Attack	srcIP	srcPrt	dstIP	dstPrt
Blaster	0.01	0.00046	0	0.000911
Sasser	0.0034	0.000325	0.0018	0.000454

In Figure 11.15, the results of the PDFs found for all the parameters of a Blaster worm attack are shown, whereas, in Figure 11.16, the results for all the parameters of a Sasser worm attack are shown. As seen in Figure 11.15, for three of the four cluster keys (having dstIP as an exception, i.e., there were no excess points for this parameter), there is a clear detection of a Blaster attack, as the PDFs are easily distinguished. In Figure 11.16, for the four cluster keys, there is a clear distinction between both PDFs; therefore, the attack is detected. The PDF fit found with the kernel smoothing method could be replaced by a normal distribution for both benign and anomalous traffic to simplify the analysis, as has been long time studied, and the probabilities of error are well established for that type of distribution. The normal distribution can be used given that the goal of effectively identifying a distinction between two classes of traffic behaviors is accomplished.

The excess point method performs well if attacks have the same characteristics as those of the Blaster and Sasser worms, where IP source and destination addresses produce high entropy variability in the entropy spaces. The reason for which the methodology of the excess points method was inefficient for other types of attacks is because some of them exhibit a behavior that derives into entropy points very near the benign centroid; therefore, the amount of anomaly points that

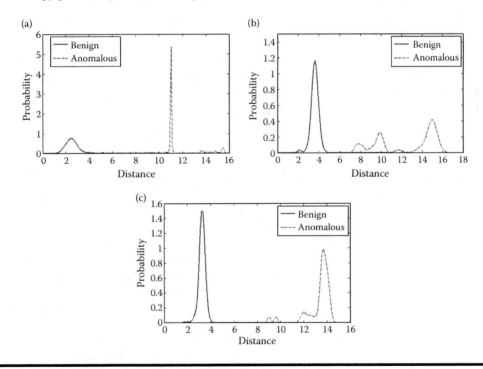

Figure 11.15   Probability comparison for a Blaster worm attack for the excess point method. (a) PDF comparison for CK srcIP; (b) PDF comparison for CK srcPrt; (c) PDF comparison for CK dstPrt.

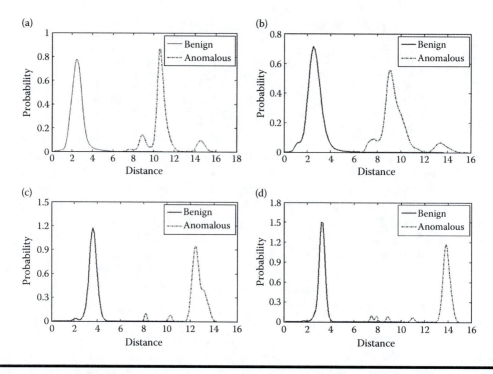

Figure 11.16 Probability comparison for a Sasser worm attack for the excess point method. (a) PDF comparison for CK srcIP; (b) PDF comparison for CK dstIP; (c) PDF comparison for CK srcPrt; (d) PDF comparison for CK dstPrt.

are located outside the boundary provided by the benign behavior (i.e., the excess points) is very small or zero in some cases, and the profiling for that traffic becomes impossible with this method. For other attacks, the centroid obtained from the traffic data sets can be close to the benign centroid obtained by the training data set, confirming the need to use a combination of detection tools in parallel to make a decision about the anomaly.

11.5 Experimental Platform, Data Set, and Tools

The worm propagation has been tested in a class C IP network subdivided into four subnets. There are 100 hosts running Windows XP SP2 mainly. Two routers connect the subnets with 10 Ethernet switches and 18 IEEE 802.11b/g wireless access points. The data rate of the core network is 100 Mbps. A sector of the network is left vulnerable on purpose, with 10 not patched Windows XP stations. In the experiments, Blaster and Sasser worms were released in the vulnerable sector.

The data set was collected by a network sniffer tool based on libpcap library used by tcpdump. This data set contains traces corresponding to a 6-day period of standard traffic in a user's typical work hours and one standard traffic trace combined with anomalous traffic of the day of the attacks. All traces were sanitized to remove spurious data using plab, a platform for packet capture and analysis. Traces were split in segments using tracesplit, which is a tool that belongs to Libtrace. The traffic files in ASCII format suitable for MATLAB processing were created with ipsumdump. Entropy estimation was implemented by a Perl script, which uses hash functions proving efficiency and speed in the analysis of large data sets.

11.6 Conclusions

We have proposed an architecture for anomaly detection based on information theory concepts such as the entropy. We have also shown that the methodology proposed is successful by detecting different attacks using different methodologies. The proposed method has been tested under load on real-time network traffic data and deploying real attacks. Empirical results have shown that the methods introduced in this chapter are able to detect traffic anomalies. Techniques such as entropy spaces, excess point method, PR, and fitting models were used. Their use allows identifying sensible features (mean of the far mode and standard deviation of the principal mode) that help in detecting anomalous traffic in specific slots and comparing them with normal traffic. We also introduced the sensibility of the entropy to attacks of different characteristics, where negative linear dependency captured the anomalies; this was shown by the use of correlation coefficients between the four features that were analyzed from the packets in the data sets. By using these results, a framework architecture is proposed to analyze network traffic anomalies. Deviations from a given reference obtained from typical traffic are considered anomalous. Future work will consider scenarios with high volume of traffic, the evaluation of more attacks, and adding a preprocessing stage by using an S-NIDS like Snort. The preprocessing implies filtering well-known attacks to obtain features that represent normal traffic in a better way.

11.7 Future Research

The results presented in this chapter direct us to consider several actions to pursue as future research areas related to the area of anomaly detection in networks. For example, regarding the entropy spaces, we can see that these need to be defined formally through the use of vector spaces, and once having this base, the entropy space can be considered as composed of two regions: the benign and anomaly regions. Maximum likelihood and maximum a posteriori probability theories from information and communication theory could be applied in order to have a decision process.

Another area of interest would be the anomaly detection through the characterization of the traffic interarrivals of packets. From the networking point of view, traffic characteristics such as heavy tails, self-similarity, long-range dependence, and alpha stability are changed when an anomaly is present, and metrics from information theory can be used to track the amount of change and determine the presence or absence of anomalies.

Another important point of view is that of adaptability and dynamism that the references must have in order to cope with the changing characteristics of the benign traffic because applications and new users are added to the network on a daily basis. Some of the results, although iterative as introduced in this book, can be seen as a product of a set of discrete-time sequences that need to be processed by digital filters to produce a response sensible to anomalies. These filters must adapt to the network conditions and need to be changing with time by dynamic procedures that instruct how parameters must be set at different times.

Acknowledgments

We would like to thank the sponsorship of CONACyT through the project Investigación Básica SEP-CONACyT CB-2011-01 "Modelos y algoritmos basados en Entropía para la Detección y Prevención de Intrusiones de Red." This chapter and the results presented here would not have been possible without such support.

References

1. Canavan, J. E., *Fundamentals of Network Security*, Artech House, Norwood, Massachusetts, USA, 2001, 194.
2. Mukherjee, B., Heberlein, T. L. and Levitt, K. N., Network intrusion detection, *IEEE Network*, 8 (3), 26–41, May/June 1994.
3. Heady, R., Luger, G., Maccabe, A. and Servilla, M., The Architecture of a Network Level Intrusion Detection System, Technical Report CS90-20, Department of Computer Science, University of New Mexico, August 1990.
4. Furnell, S. M., Katsikas, S., Lopez, J. and Patel, A., *Securing Information and Communications Systems: Principles, Technologies, and Applications (Information Security & Privacy)*, Artech House, Norwood, Massachusetts, USA, 2008, 166.
5. Solomon, G. and Chapple, M., *Information Security Illuminated*, Jones and Bartlett Learning, Sudbury, Massachusetts, USA, 2005, 327.
6. Whitman, M. E. and Mattord, H. J., *Principles of Information Security*, Fourth Edition, Course Technology, Cengage Learning, Boston, Massachusetts, USA, 2012, 305.
7. Malik, S., *Network Security Principles and Practices*, Cisco Press, Indianapolis, Indiana, USA, 2002, 420.
8. Common Vulnerabilities and Exposures. CVE is a dictionary of publicly known information security vulnerabilities and exposures. Available at http://cve.mitre.org/.
9. Snort: The De Facto Standard for Intrusion Detection and Prevention. Available at http://www.source fire.com/security-technologies/open-source/snort.
10. Chandola, V., Banerjee, A. and Kumar, V., Anomaly detection: A survey, *ACM Computing Surveys*, 41 (3), 2009, article 15.
11. Thatte, G., Mitra, U. and Heidemann, J., Detection of low-rate attacks in computer networks, in Proceedings of the 11th IEEE Global Internet Symposium, Phoenix, AZ, USA, IEEE, April 2008, 1–6.
12. Lakhina, A., Crovella, M. and Diot, C., Mining anomalies using traffic feature distributions, in ACM SIGCOMM, 2005, 217–228.
13. Mahoney, M., Network traffic anomaly detection based on packet bytes, in *Proceedings of the ACM SIGSAC*, 2003.
14. Mahoney, M. and Chan, P. K., Learning nonstationary models of normal network traffic for detecting novel attacks, in Proceedings of the SIGKDD, 2002.
15. Wang, K. and Stolfo, S., Anomalous payload-based network intrusion detection, *Recent Advances in Intrusion Detection. Lecture Notes in Computer Science*, Jonsson, E., Valdes, A. and Almgren, M. (Springer Eds.), Berlin, Germany, 2004, 203–222.
16. Karim, A., Computational intelligence for network intrusion detection: Recent contributions, in CIS 2005, Part I, LNAI 3801, Hao, Y. et al. (Eds.), Springer-Verlag, Berlin, 2005, 170–175.
17. Tsai, C., Hsu, Y., Lin, C., and Lin, W., Intrusion detection by machine learning: A review, *Expert Systems with Applications, Elsevier*, 36 (10), 2009, 11994–12000.
18. Thottan, M. and Ji, C., Anomaly detection in IP networks, *IEEE Transactions on Signal Processing*, 51 (5), 2003, 2191–2204.
19. Celenk, M., Conley, T., Willis, J. and Graham, J., Predictive network detection and visualization, *IEEE Transactions on Information Forensics and Security*, 5 (2), 2010, 287–299.
20. Toft, J., Minimization under entropy conditions, with applications in lower bound problems, *Journal of Mathematical Physics*, 45 (8), August 2004.
21. Shannon, C., A mathematical theory of communication. *Bell System Technical Journal*, 27, 1948, 379–423 and 623–656.
22. Paninski, L., Estimation of entropy and mutual information, *Neural Computation*, 15, 2003, 1191–1253.
23. Gu, Y., McCallum, A., and Towsley, D., Detecting anomalies in network traffic using maximum entropy estimation, in Proceedings of the 5th ACM SIGCOMM Conference on Internet Measurement (IMC'05), ACM, New York, 2005, 1–6.

24. Wagner, A. and Plattner, B., Entropy based worm and anomaly detection in fast IP networks, in Proceedings of the 14th IEEE International Workshop on Enabling Tech.: Infrastructure for Collaborative Enterprise, 2005, 172–177.
25. Xu, K., Zhang, Z. and Bhattacharyya, S., Internet traffic behavior profiling for network security monitoring. *Transactions on Networking, IEEE/ACM*, 16 (3), 2008, 1241–1252.
26. Lall, A., Sekar, V., Ogihara, M., Xu, J. and Zhangz, H., Data streaming algorithms for estimating entropy of network traffic, in International Conference on Measurement and Modeling of Computer Systems, Saint Malo, France, 2006.
27. Marques de Sa, J. P., *Pattern Recognition: Concepts, Methods and Application*, Springer, Berlin, Germany, 2001.
28. Bishop, C. M., *Pattern Recognition and Machine Learning*, Springer Science + Business Media, Berlin, Germany, LLC, 2006.
29. Duda, R. O., Hart, P. E. and Stork, D. G., *Pattern Classification*, 2nd edition, Wiley, New York, 2001.
30. Kpalma, K. and Ronsin, J., An overview of advances of pattern recognition systems in computer vision. *Vision Systems: Segmentation and Pattern Recognition*, Obinata, G. and Dutta, A. (Eds.), I-Tech, Vienna, Austria, June 2007, 546.
31. He, X., Yan, S., Hu, Y., Niyogi, P. and Jiang Zhang, H., Face recognition using Laplacianfaces. *IEEE Transactions on Pattern Analysis and Machine Intelligence*, 27 (3), 2005, 328–340.
32. Härdle, W. and Hlávka, Z., *Multivariate Statistics: Exercises and Solutions*, Springer Science + Business Media, LLC, Berlin, Germany, 2007, ISBN 978-0-387-70784-6.
33. Barakat, C., Thiran, P., Iannaccone G., Diot, C. and Owezarski, P., Modeling Internet backbone traffic at the flow level. *IEEE Transactions on Signal Processing, Special Issue on Networking*, 51 (8), August 2003.
34. McLachlan, G. J. and Peel, D., *Finite Mixture Models*, Wiley Interscience Publication, USA, 2000.

Chapter 12

Network Management SYSTEMS: Advances, Trends, and Systems Future

Sumit Goswami, Sudip Misra, and Chaynika Taneja

Contents

12.1 Introduction

12.1.1 Overview

Network management is one of the most crucial tasks that network administrators and managers face today. It ensures effective utilization of the network resources as well as smooth operation of the services running on it. With networks growing exponentially, network management has become indispensable for detecting faults in the network, maintaining service-level agreements (SLAs) and ensuring the reliability of the services. "Network management thus encompasses all the activities and tools dealing with the operation, maintenance, and administration of a network" [1].

12.1.2 NMS Protocols

Network management protocols are the languages that managers and agents use to communicate with each other. They define the formats and message structures for requests and response messages between the managing and the managed entities.

12.1.2.1 SNMP

Simple Network Management Protocol (SNMP) is one of the most commonly used protocols for network management. The protocol was designed and standardized by the Internet Engineering Task Force (IETF). SNMP is a management model consisting of managers and agents. The network device to be monitored and managed is called the *managed object*. The device may be a router, switch, hub, or even a desktop. The agent is a small software program that resides on the managed object. Agents provide an interface between managers and managed objects [2]. The agent stores the managing information in a virtual information database called the Managed Information Base (MIB). The manager is the component that collects the information from the agents and processes it. The manager is sometimes referred to as the network management system itself because it issues requests to the devices, returns messages, and, hence, manages the flow.

SNMP is the protocol that governs the communication between managers and agents. Each characteristic of a managed device is called a MIB object. The object, in turn, comprises one or more variables. SNMP uses the following set of messages for communication [3]:

- "Get" is used by the manager to retrieve the value of a specific variable.
- "GetNext" is used to retrieve the value of the next variable on the list from the agent.
- "Set" is used by the network management system to change the value of a particular variable.
- "GetResponse" is the message sent by the agent to the manager in response to a "Get" or "GetNext" request. It returns the value of the required variable or an error message if the requested variable is not manageable.
- Trap messages are sent by the agents to managers to notify them about any error or fault in the device [3]. Traps are generated by the agent and not requested by the manager. These are essential as with an absence of immediate notification, the fault would remain undiagnosed until the next polling by the manager.

The flow of the different messages is illustrated in Figure 12.1.

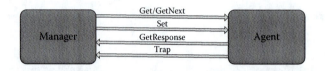

Figure 12.1 SNMP messages.

12.1.2.1.1 Message Structure

The SNMP message consists of three parts [4]:

SNMP version: This specifies the version of the SNMP being used. The field is 4 bytes long and helps to maintain compatibility between the different versions.

Community string: The community string is used to protect the information residing on the managed device. The string in the SNMP message should be identical to the community string configured on the managed device.

SNMP protocol data unit (PDU): This section contains the SNMP encoded information. The PDU consists of the following fields: PDU type, request ID, error status, error index, name, and value pairs for object instances.

PDU type indicates the type of the operation, that is, Get/Set, etc. Required parameters are specified in the following fields including the names and values of the variables. Request ID is used to link the request messages with the corresponding responses by using an identifier. Error status and error ID fields are applicable only for response messages. Error status gives an error type if an error occurs, and error ID specifies the corresponding message ID. The name and value fields that follow give the names of the variable instances and their corresponding values shown in Figure 12.2.

12.1.2.1.2 SNMP Communication

Figure 12.3 shows a typical SNMP communication model [4]. When the SNMP manager wants to know the state of the device on which the agent resides, it prepares an SNMP packet with the (OID) of the required variable. The message is then passed on to the UDP layer. The UDP layer, in turn, adds a UDP header to the message and passes the resulting datagram to the IP layer. Here again, an IP header containing the IP and MAC information from the manager is added, and the packet is transferred to the network interface layer. The packet travels through routers and media and reaches the agent. Here again, it passes through the four layers in the opposite order. The response follows the reverse path to reach the manager [5].

Version	Community	PDU

Figure 12.2 SNMP message structure.

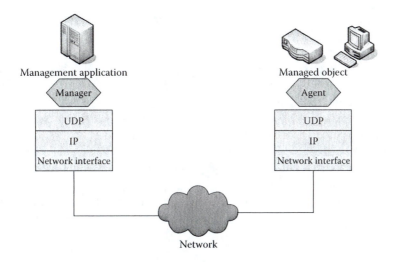

Figure 12.3 SNMP communication model.

12.1.2.1.3 SNMP Versions

SNMP is available in three versions: SNMPv1, v2, and v3. Each version built upon the functionality of the previous version. All the versions coexist today in different implementations.

12.1.2.1.3.1 SNMPv1

SNMP in its earliest form was referred to as SNMPv1. It was the simplest and most primitive form of the protocol, and as such, it had certain inherent limitations. SNMPv1 used sequential retrieval; hence, it was considered inefficient when dealing with large amounts of management information. It was vulnerable to security threats and hence was used only for monitoring purposes. Its use for configuration management was limited because of inadequate security features [3].

12.1.2.1.3.2 SNMPv2

SNMPv2 added more functionality to the protocol. A "Get" bulk operation was introduced for gathering large chunks of management information. In addition, PDU formats were also redefined, so the same PDU may be used for different operations. Traps were simplified by making its PDU the same as that of get and set PDUs [4].

12.1.2.1.3.3 SNMPv3

This version of the protocol addressed the security deficiencies in the earlier versions. It introduced the encryption of SNMP messages as well as the authentication of the SNMP manager [2]. As a result of these measures, SNMPv3 was less prone to security threats.

12.1.2.2 Netconf

Netconf is a management protocol that uses extensible markup language (XML) for configuration of managed devices. It provides operations to manage the files and databases storing the configuration details of the managed devices. It organizes the information in a hierarchical fashion [6]. Netconf uses XML for encoding the management operations. The following operations are provided by Netconf [6]:

- "Get-config" is used to retrieve a config file from the managed device. The source of the configuration file has to be provided as a parameter. If no parameter is specified, running config is retrieved as a default option.
- "Get" is a general operation to obtain the value of any state variable of the managed object.
- "Edit-config" is used to modify the configuration of the device.
- "Copy-config" is used to copy the configurations in a file to some target location.
- "Delete-config" is used to remove a configuration from a managed device. However, a running configuration is not removed.
- "Lock" and "unlock" are used by the administrators to control access to the device configuration. Once locked, the configuration cannot be edited until unlocked by the administrator.

12.1.2.3 Syslog

Syslog is the protocol that is used for generating interpretable logs. System messages are generated in log files, which are interpreted by management applications. Syslog messages have two parts: message header and message body. The body contains the message content, which is mostly in English text. The header contains information about the message, such as the time when the message was generated, ID and name of the generating host, message severity, etc. [6].

12.1.2.4 NetFlow

NetFlow is a special-purpose protocol used for accounting and performance applications; it collects statistical data about network traffic [6]. A "flow" refers to all the traffic that is a part of the same connection. The parameters that characterize a flow are the source address, source port, destination address, destination port, protocol type, and service. A flow record refers to the data collected for a flow. The flow records from the router are enclosed into NetFlow packets and sent to a NetFlow collector, which stores them for further processing and analysis [6].

Figure 12.4 shows the structure of the NetFlow packet [6].

A header consists of the following information:

- NetFlow version
- Sequence number
- Number of records

The analysis of flow records provides useful information about the usage of network resources by different users, helps in network planning and traffic analysis, and may also be used to identify potential threats.

NetFlow also comes in different versions. NetFlow5 is the most commonly used. NetFlow9 is the latest available version.

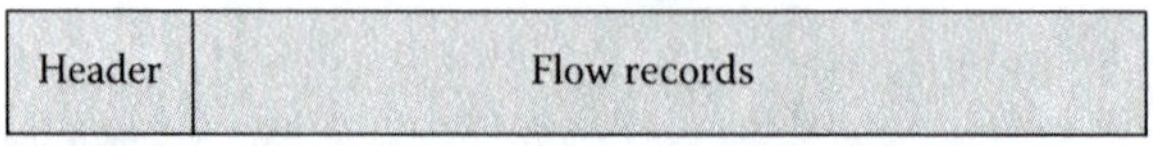

Header	Flow records

Figure 12.4   NetFlow packet structure.

12.2 Architecture of Network Management System

The major components of any network management system are the network element, which is the device to be monitored; the agent; and the manager. The agent is a small program that runs as a daemon on the network element. It acts as an intermediary between the managed device and the managing system or the manager. An agent consists of a conceptual database called the management information base to store the management information. The manager or the management system runs applications that collect information from the agent and processes them. The manager uses a management protocol to periodically poll the agents in the network and request required management information. The information collected from the agents is in raw form. The manager performs computational operations on it to generate meaningful reports.

The relationships among the various components are represented in Figure 12.5.

12.2.1 Software Probes/RMON

Remote monitoring (RMON) devices or probes are monitoring instruments that collect data for network management. RMON is a special SNMP MIB that is used to delegate some management tasks to the RMON probes [6]. RMON can be a standalone device or software. Many network elements, such as routers, may also have RMON capabilities shown in Figure 12.6.

Probes collect and provide statistics about traffic analysis, users, and applications. This information is valuable for network management and for traffic engineering.

RMON was designed for flow-based monitoring, packet capture, and management functions; such as data analysis, along with data collection and storage.

Both active as well as passive probes can be used. In active probing, functions of some network elements are used to mirror the flowing traffic to a port. These data are then passed on to the management applications. In passive probes, the probe is placed between two network elements using an electrical or optical splitter. This is analogous to the tapping of the line.

RMON has two versions: RMON-1 and RMON-2. RMON-1 is used for monitoring traffic up to the medium access control (MAC) layer of the OSI Model. It is composed of 10 MIB groups, which are essential for network monitoring. The MIB groups in RMON-1 are Ethernet Statistics,

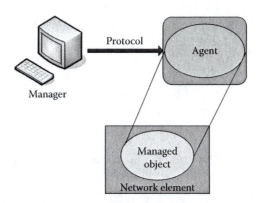

Figure 12.5 Network management system architecture.

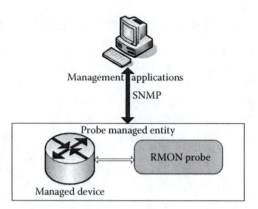

Figure 12.6 RMON probes.

History Control, Ethernet History, Alarm, Host, HostTopN, Matrix, Filter, Packet Capture, and Event [6].

RMON-2 is an extension of RMON-1 and aims at monitoring traffic in network and application layers.

12.2.2 Collector

The collector gathers and stores the flow of data packets and analyzes it. The data may be collected passively or by using NetFlow. For NetFlow, the router is configured to export the data in the form of flow records. The flow records are collected by the NetFlow collector and stored for further analysis and applications. For passive collection, all the data are sent through an interface to the collector. A special port is configured for it. This mode of data collection is promiscuous and more secure [6].

The collector may also perform other functions, such as the following:

- Flow analysis
- Mediation and storage
- Speech analysis
- Network behavioral analysis

12.2.3 Hardware Acceleration

Hardware acceleration is the process of using hardware to perform operations faster than similar software.

The need for hardware acceleration in modern network management systems was felt because of the following reasons [7]:

- The volume of data in the present high-speed next-generation networks (NGNs) is quite high. General-purpose computers are often incapable of processing such high volumes of data.
- Software-based applications are insufficient to meet the requirements of multi gigabit networks.

■ Data received from network devices needs postprocessing to enable analysis for network management, thus requiring high computational power.

The performance limitations of general-purpose processors were proving to be a bottleneck for network management applications. New hardware design technologies, such as application-specific integrated circuits (ASICs) and systems on chip (SoCs), offered solutions to this problem [7].

Field programmable gate arrays (FPGAs) are one of the most commonly used technologies used to implement hardware acceleration. FPGAs are reconfigurable hardware chips that can be programmed to implement logic. They can thus be used to develop prototype systems and implement algorithms. Hardware description languages, such as VHDL and Verilog, can be used to program the hardware. The code is then compiled and written to the FPGA.

Hardware acceleration speeds up the management systems and improves their performance as the high computation processing functions are assigned to the coprocessors while the processor can concentrate on the core functionality [8].

12.2.3.1 Networking Cards with Programmable Processing Chip

Various vendors [9,10] have come up with networking cards with FPGAs. The presence of a programmable processing chip enables these cards to be used to accelerate solutions related to network security and for packet capture. The cards can also be used for prototyping of algorithms. Standard server network interface cards (NICs) are not suitable for network analysis and security applications because they cannot handle high volumes of data and tend to get overloaded, leading to packet drops. Hence, NICs specially designed for high-speed networks are used to handle networks operating at gigabit speeds. The difference in performance is a result of the approach that NICs use for handling data. Standard NICs process the data frame by frame, using the NIC driver and the operating system. The CPU consumption in this approach increases drastically at high data rates, and thus throughput drops. Specially designed NICs, which are also called real-time NICs, overcome this limitation by bypassing the NIC driver and the OS and transmitting data directly. These adapters are capable of decoding the frames and defining flows [9].

12.2.3.2 Development Platforms and Hardware-Accelerated Appliances

A development platform is required to speed up the development process of hardware-accelerated solutions. These platforms help develop scalable FPGA-based applications [10]. The programmable cards and development platform help in the development of a hardware-accelerated appliance that can be used for speedy processing tasks for network-monitoring applications [11]. Such appliances can capture packets at high speed with almost 0% packet loss. Such accelerated solutions are capable of processing data at any frame size and any line rate. At the same time, they are completely transparent to the network users as the latency introduced is minimal. These appliances equipped with specially designed NICs perform tasks such as packet filtering, packet classification, tagging, and flow distribution at high speeds [9].

12.3 NMS Functions and Reference Models

The functionality of network management systems is characterized by various reference models. These models classify the management functions into functional areas.

12.3.1 FCAPS Reference Model

The FCAPS reference model was proposed by the IETF and is also known as the OSI Open System Architecture Reference Model. FCAPS is an abbreviation for fault, configuration, accounting, performance, and security, which are the management functions covered under the model.

12.3.1.1 Fault Management

This management function involves detecting, isolating, and resolving faults and anomalies in the network [6]. This is the umbrella term covering several tasks including but not limited to the following [1]:

■ Monitoring the network resources and services on a real-time basis
■ Fault diagnosis, root cause analysis, log generation, and proactive measures to prevent faults in the network
■ Maintaining a database of network faults and events
■ Proactively monitoring the network to prevent faults

Fault management increases the reliability of the network by helping the network administrators to detect the faults in a timely manner and to take corrective measures. To detect the faults, occasional polling of the network elements is carried out. The frequency of the poll, also called the polling interval, is decided by the network administrator. There is a tradeoff involved as specifying the parameter as a low value of the polling interval ensures a timely detection of the faults and a higher value conserves the bandwidth [1]. In addition to the normal polls, unexpected events are notified to the manager by the generation of traps by the agents.

The reporting of the faults, once detected, is also a crucial factor in network management. Some common forms of fault reporting include text messages, email alerts, SMS alerts, audio alarms, pop-ups, etc. The graphical user interfaces of various tools also use color coding to indicate the alarms. A green means the device is in upstate, red indicates the device is down, and yellow may indicate some error.

12.3.1.2 Configuration Management

Configuration management is the function associated with monitoring the configuration of the network's components as well as updating it. The configurations are also stored for further analysis and correlation to network faults [12].

Configuration management increases the network administrator's control over the network as any changes in any configurations may be tracked. It also helps to keep the configurations updated. Inventory management of the various components is also simpler with configuration management.

The configurations are stored at a central location in either ASCI files or in a database management system (DBMS) [1].

12.3.1.3 Accounting Management

Accounting management involves measuring and regulating network utilization. It also includes setting user quotas and generating billing information as per usage. It helps the network administrator in regulating the network usage and ensuring the optimal use of resources. The steps involved are as follows [1]:

- Measure utilization of resources
- Analyze usage patterns and set quotas
- Measure usage and generate billing information

12.3.1.4 Performance Management

Performance management involves measuring and maintaining the performance levels of the network [6]. For this, various parameters or variables are monitored continuously and any degradation is reported. Some of the network performance variables include network throughput, link capacity utilization, and response time.

The process involves collecting information about the performance variables, analyzing the data, and deciding the threshold levels. If the variables exceed the defined thresholds, alerts are generated.

Performance management also involves taking proactive measures, such as observing the present bandwidth utilization to predict future usage for timely planning and augmentation of the resources [1].

12.3.1.5 Security Management

Security management involves controlling access to sensitive network information and detecting unauthorized access. The process involves identifying secure resources and mapping the access points or user sets that can access the sensitive information. Access to these resources is also monitored, and unauthorized attempts are logged.

Various security techniques are used to secure the network [1]:

- Encryption of data may be done at the datalink layer.
- Packet filters are used at the network layer.
- User authentication and key authentication may be used at the host level.

12.3.2 OAM&P Model

An alternative to the FCAPS model is the operations, administration, maintenance, and provisioning (OAM&P) model [6].

- *Operations* covers the activities required for the normal day-to-day running of the network. Coordinating with the other three functions of the OAM&P model is also a part of this function.
- *Administration* includes the various support activities required to operate the network, such as tracking usage, maintaining inventory records, billing the users, etc.
- *Maintenance* is the function that covers detecting faults in the network and rectifying them so that the network is fully functional.
- *Provisioning* is the process of configuring the various network components as well as services.

12.3.3 Performance Metrics

The effectiveness of network management systems may be evaluated by taking into account certain metrics. These are the performance metrics that assess the reliability of the network [6]:

■ *Availability:* This measures the percentage of time during which the network is functioning properly. A modified calculation of availability is also used, and it takes the planned shutdowns into account.
■ *MTBF:* Mean time between failures is another performance measure commonly used. It is measured as the average time that will pass between random network failures. It indicates how frequently the network faces faults and is independent of the duration for which the faults persist.
■ *MTTR:* Mean time to repair indicates how long it takes for the services to be restored after a disruption or outage. It is measured as the average time elapsed between the occurrence of a fault and restoration of services.

The three metrics are interrelated by the following relation:

$$\text{Availability} = \text{MTBF}/(\text{MTBF} + \text{MTTR}).$$

12.3.4 Network Management Tools

Technology has added several tools to the arsenal of the network manager, ranging from simple device managers and network analyzers to application-level network management systems that assist the network operators in managing the network efficiently. The network management systems widely used in managing organizational networks include open-source solutions, such as OpenNMS, as well as proprietary ones, such as CiscoWorks. The choice of a particular system depends on the nature of the network to be managed and the services running on it.

12.4 NGN Management (NGNM)

With the evolution of networks into the next generations, network management has also attained new dimensions. NGNs involve the convergence of voice, data, video, and enhanced services. So there is a need to have an integrated platform to manage all these services.

12.4.1 Challenges

In NGNs, existing and new networks coexist, thereby making the network composition diverse. NGNs are often composed of heterogeneous network components from several service providers across the globe. These features combine to make management of NGNs a challenging task.

12.4.2 NGN Network Scenario

NGNs are characterized by a separation of network and services, treating each as a separate entity [13]. The architecture of a typical NGN is illustrated in Figure 12.7.

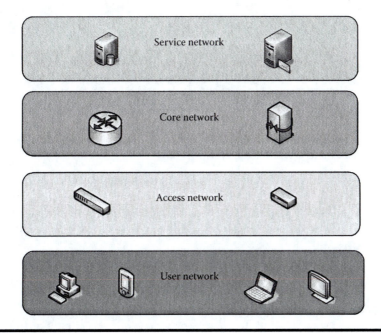

Figure 12.7　NGN architecture.

The service network is responsible for providing services to the users in a NGN. It comprises service servers, such as the web server, mail server, proxy server, etc. The servers are connected to each other through a high-speed network. A core network has the responsibility of transportation of data and interfacing with the Internet. It forms the backbone of the network and comprises routers and gateways. An access network acts as an interface between the core network and the user network. Different access media that characterize an NGN, including satellite, optical fiber, ADSL, LAN, etc., are a part of the access network. The user network is a web of end-user equipment, including desktops, laptops, mobile phones, personal digital assistants (PDAs), etc.

12.4.3 NGNM Features

Management strategies for NGNs are different from those used for conventional public switched telephone network (PSTN) and data networks [14]. Some of the features of NGN management are covered in the following sections.

12.4.3.1 Integrated Management

NGNs involve the amalgamation of several new technologies, such as multiprotocol label switching (MPLS), automatically switched optical networks (ASONs), etc. It is a common practice to involve more than one service provider to avail these services. This translates into heterogeneous components and platforms in the network. Management of such heterogeneous entities is a challenging task for network management systems. An integrated solution that interoperates between these diverse entities is required [13].

12.4.3.2 Quality of Service

In NGN, voice, data, video, and Internet services all interoperate on the same media. Hence, maintaining the quality of service (QoS) for different services is a challenge as the traffic volumes are quite high. It is necessary to monitor the traffic on a real-time basis and, in the event of congestion, transmit the high-priority packets first. It is also imperative to monitor performance parameters, such as bandwidth usage, packet loss, jitter, etc., and keep them below the permissible thresholds [13].

12.4.3.3 Topology Creation

NGNs are often characterized by changing topologies and distributed nodes. As new elements join the network, the topology may keep varying. This requires that the network management system reflects the changes in the management database. In addition, the configurations of the network elements may also be variable; hence, the configuration management should be dynamic and free of human intervention [14].

12.4.3.4 Dedicated Service Monitoring

In NGNs, services and customer experience are given utmost priority. Service providers and network managers are expected to maintain the SLAs and keep the downtime to a minimum. It is therefore necessary that traffic and service monitoring should be completely automatic in NGNM systems. Traffic management and control should also be automated and should involve minimum operator control [13].

12.4.3.5 Policy-Based Management

Because NGN architectures are often distributed, the management framework is also required to be decentralized. The management centers, to monitor the services and apply policies are also distributed over the network. These centers are interconnected to coordinate the management activities. A new approach that is being adopted in view of the distributed architectures and management is the policy-based network management (PBNM). This approach provides distributed control and management of the network. The architecture of a PBNM system is illustrated in Figure 12.8.

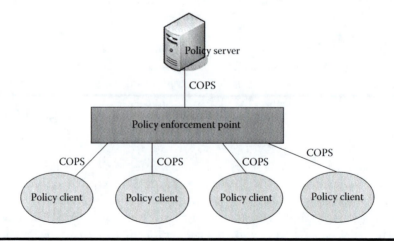

Figure 12.8 PBNM architecture.

The components of a PBNM system are the following [13]:

(1) Policy decision point: This is a policy server that makes policy implementation decisions.
(2) Policy enforcement point: This component interacts with the policy clients in the network.
(3) Policy clients: These receive decisions to implement a policy through the policy enforcement point.

A policy database is also used, which is a repository of predefined policies, such as bandwidth allocation, performance parameters, etc. Communication between the various components is through a policy protocol, such as common open policy service (COPS) protocol. PBNM enables the administrator to automate and simplify the management process by distributing the control.

12.5 Cognitive Behavioral Analysis

Network management systems today have acquired cognitive capabilities. They analyze the behavioral characteristics of the network using statistical methods and intrusion-detection techniques. This enables network administrators to deploy mechanisms to safeguard the network resources against any impending attacks and threats. Some of the key areas that future network management systems address are covered in the following sections.

12.5.1 *Vulnerability Management*

Vulnerability may be defined as an exposure or flaw in a system, operating system, or application that makes it susceptible to attacks. With the expansion of the networks and the Internet technologies, the vulnerabilities have increased at a tremendous rate. The attackers and hackers are always on the lookout for such vulnerabilities, which serve as easy targets for cyber attacks. Exploitation of vulnerabilities may lead to data theft, denial of service, OS damage, and many other problems. Hence, it becomes essential for the network administrator to manage and patch the vulnerabilities in their systems.

Vulnerability management (VM) refers to the process of detecting and fixing the vulnerabilities in the network and the system. It also includes improving the communication between the various security components of the network to guard against attacks that may exploit the vulnerabilities [15].

VM systems work at a level above intrusion detection systems (IDSs), firewalls, and antivirus systems. IDSs collect information from various sources and analyze it to detect attacks and misuse. However, they often give false positive errors, which may not be actual attacks. Antivirus systems offer protection against known viruses, but because they are dependent on the signatures released by the vendors, they often offer no protection against threats for which no update has been released as yet. Such vulnerabilities that have been just detected and for which no signatures and protection mechanism is available are called "zero day vulnerabilities" [15].

Common vulnerabilities and exposures (CVE) is a naming convention that has been developed to standardize the names of vulnerabilities. It lists the names of publicly known vulnerabilities and other information security exposures. Another standard in the field of VM is the open vulnerability assessment language (OVAL). It is the common language developed to check for the existence of vulnerabilities on a system. The mechanism involves OVAL queries written in structured query language (SQL), which look for the vulnerabilities listed in the CVE.

12.5.2 Automated Fuzz Testing

Fuzz testing or fuzzing is the process of supplying invalid inputs to the network application and monitoring them for faults. This helps in discovering vulnerabilities and hence protecting the network from attacks by hackers. In network-based fuzz testing, the input stream is mutated before being sent to the network component. Unexpected behavior, such as crashes, indicates a vulnerability that has to be patched. Fuzzing techniques are being used by hackers to scan the network for any flaws and launch attacks. Hackers use black box fuzz testing, that is, they randomly generate the various combinations of the input.

In the past, fuzz testing was primarily manual and ad hoc. However, today, the most prevalent mode of fuzz testing is automated fuzz testing or white box testing. This is based on systematic test case generation. Starting with a fixed input, constraints are then applied to the input; a constraint solver is used to generate new inputs. This is done to maximize the code coverage.

Many tools are available to perform fuzz testing of networks. Some of them are the following:

- *Protosis*: a Java-based network fuzzer that supports protocols such as HTTP, SNMP, DNS, etc.
- *Scapy*: a Python-based fuzz testing tool

12.5.3 Advanced Persistent Threats

Advanced persistent threats (APTs) are sophisticated and organized attacks in which unauthorized access is gained to the network resources. The attacks are targeted against organizations and aimed at stealing valuable information on the organization's network.

12.5.3.1 Attacks Based on Buffer Overflows and File Format Vulnerability

Buffer overflow attacks are designed to inject a malicious code into the program by assigning data more than the buffer size. The buffer is a fixed size, continuous chunk of memory used to store user input. When the user input's size exceeds the size of the buffer, the extra data is written to an adjacent memory location, which is often the return address of the code. Attackers can deliberately overwrite the return address to point to the injected malicious code. Programming languages such as C/C++ that do not perform automatic bound checking on user input often result in such errors.

Buffer overflow is the most common form of vulnerability used to launch network penetration attacks. Some common strategies to control such errors include the following [16]:

- Writing safe code by performing boundary checks and using safe library functions, such as "strncpy" instead of "strcpy"
- Performing array bound checking and integrity checking on code pointers by using compilers
- Using languages, such as Java, to write a sensitive code

Format string vulnerabilities are another class of vulnerabilities that are caused by improper use of functions that use format strings, such as the printf functions in C. The attackers use the format specifiers to read areas of memory that they do not have access to or to overwrite the instruction pointer to execute malicious shell code.

12.5.3.2 Digital Masquerade and Attribution Problems

Masquerade attacks are those in which the attackers launch attacks and protect themselves by hiding in the network as legitimate nodes. Such attacks are often a result of security thefts and are difficult to detect. The hacker often poses as a legitimate user on the network. Such attacks may cause huge losses, especially if they are carried out by an inside attacker.

Such attacks may be caused when hackers steal passwords and gain access to genuine users' accounts or through key loggers. It may also occur if users leave their systems' account open or logged in.

Masquerade detection is the process of collecting information about each user and modeling the user's profile [17]. The profile contains login time, location, session time, commands, etc. Once the profile is modeled, the user logs are compared to the developed profiles. Anomalies in the log patterns indicate a masquerade attack [18].

Another detection technique is to model the users' behavioral profile on the basis of the search commands. Abnormal patterns in the search profiles indicate that an impersonator has gained access to the user credentials or resources. The assumption used is that a genuine user knows his resources and file systems and searches in a targeted and well-defined manner. On the other hand, a hacker or impersonator searches randomly. The difference is detected by masquerade detection tools [17].

12.5.3.3 Remote Access Tools/Trojans

Remote access tools or Trojans (RATs) are malwares that provide the attacker administrative access to the computer by opening backdoors in the affected system. The infected host spreads the RAT to other systems on the network, thereby establishing a botnet. The RATs usually come camouflaged as harmless programs, such as email attachments, games, etc. Once the attacker gains access to the network resources, he or she may attempt malicious activities, such as accessing confidential information, data destruction, or intercepting live video by activating a system webcam. The detection of RATs is a challenging task as they do not appear in the running tasks or services.

Protection

Administrative steps that may be taken to prevent the network resources from RATs include the following:

- Blocking unused ports on network elements
- Turning off services that are not required
- Monitoring outgoing data to detect unexpected patterns
- Keeping antivirus and the operating system updated

12.5.3.4 Digital Intruders

Digital intruders gain access to the network through well-known vulnerabilities. They stay in the network and conceal their presence by deleting logs, altering time stamps, and other techniques.

Intruders can cause data pilferage, data destruction, and losses to the file system and the operating system. They use sophisticated encryption techniques to steal and exfiltrate data. The data are sent over covert channels inside legitimate traffic streams, such as HTML. Some common digital intruders that exploit vulnerabilities are the following:

- *Poison ivy:* This digital intruder belongs to a family of malware that comes in the form of a kit. It operates on a client server model and converts the affected machine into a RAT-generating and -distributing server. The malicious binaries that it creates are then distributed to other systems on the network through exploits and vulnerabilities.
- *Spear phish attacks:* In such attacks, customized emails from trusted sources are used by attackers to lure the users into disclosing confidential information. Alternately, Trojans may be sent through attachments in such mails that take over the system and enable control and command for the hacker to launch further attacks. Because these attacks are targeted, the volume of information in such emails is often small, thereby making the detection difficult. Email authentication technologies and regular updates of the mail systems, along with user awareness, can help prevent loss from such attacks.
- *Rootkits:* A rootkit is a stealth malware that hides processes and files from the user. It gives privileged access to the computer and hides it from the system owner. Rootkits are available in various modes. User-mode rootkits affect a single-user process. Kernel-mode rootkits operate like dynamic kernel modules in the memory of the affected system. On loading, it can directly make changes to operating system components, such as the data structures, and to the file system. Detection techniques include inspecting the file system for patterns of known rootkits, keeping track of the number of instructions executed by processes, etc. The presence of rootkits increases the number of executed instructions [19].

12.5.3.5 Mitigation Techniques

Various techniques are being adopted by network managers to protect the network from attacks. The mitigation strategies differ with the type of attack. However, some common techniques are the following [20]:

- Tools to extract the information from physical memory dumps of operating systems are used by managers to bypass the rootkits and observe hidden processes and files.
- Studying the logs of network devices, as well as applications, can help the administrator detect patterns that may be an indication of a network attack. Unusual traffic patterns, IP addresses, and nonstandard ports are signs that should be looked into.
- Analysis of the virtual memory of compromised hosts provides useful data about data exported by digital intruders through backdoors.
- Intruders and malware often leave behind files. The timestamps and MD5 hash values of such files help the administrator detect their presence.

12.6 Lawful Interception

Lawful interception (LI) is the process of lawfully intercepting and monitoring the communication details of a particular user. The user being intercepted is called the LI subject [21]. The information extracted from the communication may be about the location, content, or service associated and is referred to as the intercept-related information (IRI). The agency that has the authority to request the intercept information is called the law enforcement agency (LEA). The key requirements of LI are that it should be completely transparent to the subject under interception, the IRI should be kept separate from the data traffic, and the service provider should deliver only the authorized information to the LEA. Interception may be of two types: actual content of communication or of related information, that is, IRI.

12.6.1 Architecture

Figure 12.9 illustrates the generalized architecture for LI. The architecture clearly demarcates and separates the interception functions from the administration functions and the intercepted information. This is essential to prevent any interference with the user communication. The major components of a LI system are the following:

1. *Content IAP:* The communication content Intercept Access Point is the device that intercepts the user content and passes it to the central mediation device [21]. The device may be an edge router or a gateway.
2. *IRI IAP:* The IRI Intercept Access Point is the device that provides information about the intercepted content to the mediation device. The information varies with the mode of communication. For data communication, the IRI may be IP addresses and ports of the source and the destination. For voice communication, IRI often comprises the call timing or called numbers. IRI for email communications usually consists of the header of the message being sent [22].
3. *Central mediation device:* This component is the nerve center of the LI process. It controls and coordinates the different components. The major functions of the mediation device are the following [21]:
   a. Sending configuration commands to the intercept access points
   b. Collecting intercepted information from content IAP and IRI IAP
   c. Perform postprocessing on the received information to convert it into the format interpretable by the LEA

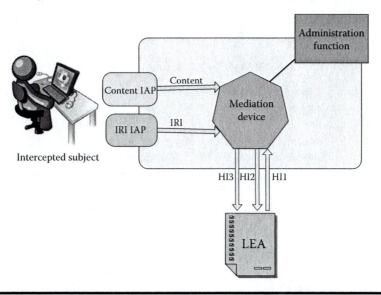

Figure 12.9   LI architecture.

d. Filtering the information to extract the information relevant to the LEA
e. Protects the intercept data from unauthorized access
4. *Administration function:* This function is used by the service provider to keep control over the various activities of the LI process. It is used to track and monitor the intercepts, maintain logs, etc.
5. *Law enforcement agency:* The LEA has a collection function or device that receives the intercepts from the mediation agency, stores them, and performs operations, such as sorting to study the information on a case-by-case basis.

The LEA thus is a separate entity that interacts with the service provider's network only through a set of instructions called handover interfaces (HIs) [22]. These HIs are the following:

■ HI1 is used to deliver the interception order from the LEA to the mediation device through the administration function.
■ HI2 is used to deliver the IRI from the mediation device to the LEA.
■ HI3 is the interface that helps to deliver the intercepted content to the LEA.

The process of LI consists of capturing the data, that is, content and IRI, and then filtering only the information about the required target [23]. The information is then formatted into a predefined format and given to the LEA. Several speech technologies, such as speaker identification, gender identification, and keyword scanning, are being utilized by service providers to analyze the intercepted information and provide valuable results to the LEA [22]. In order to ensure that the LI process does not interfere with the normal traffic, the LI components are kept completely separate from the normal architecture, and the interaction between the components is only through some predefined interfaces. With network forensics playing a crucial role in managing networks today, LI has become an activity that is now being associated with network management.

12.6.2 Interception Modes

LI may be carried out in two different modes.

12.6.2.1 Internal Interception

This mode of interception uses the network equipment of the service provider to capture the data and send them to the mediation device, which, in turn, sends them to the LEA. The data may be collected directly from the application servers (email, chat), routers, switches, or network access servers, such as RADIUS [23]. When the capabilities of the network equipment are sufficient to fulfill the LI requirements, the internal interception function (IIF) of the equipment is used. IIF utilizes the collection capabilities of the network equipment. However, if the volume of the data to be intercepted is large or the IIF of the device is not available, the external interception function (EIF) is used. In EIF, interception is carried out by connecting external network probes to the equipment [22].

12.6.2.2 External Interception

In case the access to the target's service provider's network is not available, external interception is used. This is also used by LEAs to carry out interception secretly. Here, the interception is carried out on adjacent networks or at the point where the network meets the public network. The equipment used is an edge router with interception capabilities. Probes may also be utilized. For low-volume interceptions, open-source programs may be used to monitor data paths and ports used [23].

12.6.3 Speech Processing

Speech processing techniques are now increasingly being used by interception agencies to analyze and process the intercepted voice traffic. The process involves coding of data that converts the speech signal into a bit stream, which is then transmitted over a digital channel [24]. The next step is recognition, which uses advanced techniques to extract intelligent information that may be of interest to the LEA. Some of the techniques used in the recognition process are the following.

12.6.3.1 Speaker Recognition

This technique is used to recognize the speaker from an archive of records. Using this method, both the verification of the speaker as well as identification may be achieved. Interception agencies normally use this to identify a target user in a large database of speech records [25]. The result of such systems is often a list of audio records arranged in order of similarity to the target speaker [24].

One approach used in the process is to take the long-term averages of acoustic features, such as pitch, which represent the speaker's vocal identity [25].

Another approach using pattern recognition uses speech patterns to train the system. Once the training process is accomplished, the algorithm can characterize the acoustic properties of any incoming pattern [25].

12.6.3.2 Gender Identification

This technique helps the agencies predict the gender of the speaker accurately. This reduces the search space by almost half and hence saves on the computational efforts substantially. Because there are physiological differences in the vocal tract between males and females, the acoustic features of the speech signal differ with gender. For the purpose of identification, the frequency spectrum of the speech is analyzed using statistical methods [24].

12.6.3.3 Language Identification

This technique identifies the language of a speech record so that it may be routed to an operator who knows the identified language. The technique has other potential applications, such as monitoring multilingual speech sources, identifying the nationalities that are using the network infrastructure, etc. This technique uses a combination of both phonetic and acoustic features of the speech signal. First, language models are created using training data, which are recordings in different languages. When acoustic features are used, the target speech signal is converted to frequency spectra, and statistical methods are used with channel compensation techniques to extract the information. In the other approach, a phoneme recognizer is used to convert the speech signal to phoneme strings, which are then statistically processed [24].

12.6.3.4 Keyword Spotting

This technique helps the agencies in spotting crucial words in speech to help detect relevant information. The user specifies a list of keywords that are used as an input to generate a list of records in which the detection equals a predefined threshold [24]. The system takes into consideration the variation in the pronunciation of the words. The agency may implement a policy of generating alarms when keyword detection occurs in the intercepted call, or it may store the calls for analysis at a later stage. The technology used in this technique is artificial neural networks.

12.6.3.5 Speech Transcription

This technique converts the intercepted speech signal to text so that it may be processed as text files. The generated text files may be analyzed using text-mining tools or archived for later use. Because text is easier to process than speech, this is used to utilize the captured information with search applications and other text-based processing systems. The technologies used for this include discriminative training, neural networks, channel adaptation techniques, etc. [24].

12.7 Management of Wireless Sensor Networks

Wireless sensor networks (WSNs) are being increasingly used today in a wide variety of applications, ranging from environmental monitoring to home automation and traffic control. The diversity in applications and the ad hoc nature of WSNs make their management a challenging task.

12.7.1 WSN Network Architecture

WSNs are wireless networks consisting of small autonomous devices equipped with sensors. The node also has a microcontroller, a transceiver, and a battery. The nodes in the WSNs are categorized as follows:

- Normal nodes have the function of collecting sensor data. Their processing capabilities are limited only to coordinating with neighboring nodes. They do not have surplus storage as they transmit the collected data to sink nodes after small intervals.
- Sink nodes receive data from the normal nodes and store them for further processing. They also interact with applications and the outside-world components.

With advances in very large scale integration (VLSI) and other technologies, sensor devices are being produced in huge quantities and are available in very small sizes and at very low costs. Thus, the management of WSNs is an area that is of interest to many researchers today.

12.7.2 Management Requirements

The management of WSNs is different from legacy networks because of inherent differences in the architectures and because of the deployment environments. Most sensor networks are application specific, and each sensor node has a built-in functionality. The management solution should help the WSNs in the achievement of their functions and goals. The management framework should fulfill certain basic requirements. Some of these are discussed in the following [26].

12.7.2.1 Scalability

WSNs exhibit dynamic topology; hence, it is important for the management structure to be scalable. Because all the nodes are not active at the same time, the network resides in different states at different time instants. In addition, it should be possible to add sensor nodes to the network, even in large numbers, without affecting the performance [26].

12.7.2.2 Light Weight

Sensor nodes are characterized by limited resources, including battery lifetime, memory, bandwidth, etc. The management operations and protocol should therefore be lightweight. The overhead involved should be minimal to ensure that the management tasks do not deplete the limited resources and hence decrease the lifetime of the sensor nodes.

This may be represented as

$$E_{\text{Transmission}} >>>> E_{\text{Processing}}$$

where $E_{\text{Transmission}}$ is the energy required for the transmission process, and $E_{\text{Processing}}$ is the energy required for the processing overheads.

12.7.2.3 Security

Sensor networks are often used for military applications and may be deployed in hostile environments. In such environments, the possibility of the attacker overtaking the sensor node physically is very high. The security of the entire network should be defended in such a scenario. It is therefore essential that additional security is embedded in the management structure. It is therefore essential to ensure that the communication between the base station and the sensor nodes should be secure and should have a robust authentication mechanism.

12.7.3 Management Framework

The management framework for WSNs is different from traditional management systems because of the enhanced needs for security and scalability. The architecture is more evolved and may be centralized, distributed, or hierarchical. The functional areas for WSN management include some additional dimensions, such as topology and power management, apart from the standard areas of fault, configuration, performance, and security management.

12.7.3.1 Architecture

Sensor management systems may be classified into the following types on the basis of the architecture [27]:

- *Centralized architecture.* In such a system, the base station performs the functions of the manager. It collects information from all the nodes and processes it to manage the network. This method is the least preferred for WSNs as it consumes a high percentage of both the energy as well as the bandwidth of the nodes for management purposes. The system is also not suitable for scaling the network up by adding more nodes. In case the base station fails, the entire system is affected as all the management information is located only at the base station.
- *Distributed architecture.* In this kind of system, there are several managers spread across the network, which communicate directly with each other. Each manager node controls the

nodes falling under a subsection of the network. This approach overcomes the drawbacks of the centralized approach. However, it is complex and hence computationally expensive.

■ *Hierarchical architecture.* This is the most commonly adopted management model as it combines the benefits of both the centralized and the distributed approaches. Here, the manager of a subnetwork controls all the nodes in it. The manager, in turn, receives management instructions from a higher-level manager and also passes the collected information to it. The higher-level manager consolidates the information received from several managers and passes it to the manager a level above it, thereby forming a hierarchical tree structure.

The generic architecture of a WSN management system is illustrated in Figure 12.10 using a single level of hierarchy. The central manager is usually the base station or the sink node. Each cluster has a submanager, which is usually the cluster head. The central manager delegates the management functions of a particular cluster to its submanager. Hence, a submanager of a particular cluster manages the nodes only in that particular cluster [27].

12.7.3.2 Functional Areas

The functional areas in management of sensor networks include some additional areas in addition to the conventional domains discussed in the FCAPS model in Section 12.3.1. The existing functions of fault, configuration, accounting, and performance also include some more activities that are relevant for WSNs.

■ *Topology management.* An important function of management systems in WSNs is the management of the topology and network configuration. The nodes have to be organized into clusters; a cluster head is chosen, and it communicates with the sink node or the manager on behalf of the entire cluster. Clustering in WSNs helps in conservation of energy and bandwidth as only a few nodes now correspond with the sink node. The management system also has to keep a record of the network states, so reconfiguration may be performed when new nodes join the cluster or when nodes move from one cluster to the other. A subset of this management function is mobility management, which involves planning and registering the movement of nodes from one cluster to another [27].

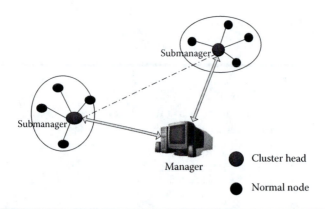

Figure 12.10 WSN management architecture.

■ *Fault management.* Sensor networks are usually deployed in inhospitable conditions with limited battery life. Hence, fault management is a crucial management function because faults occur frequently. Fault detection in sensor networks is different from traditional networks where periodic polling is used for detecting the wellbeing of a node. Such an approach will create a high overhead and is not desirable. The faults in a normal node are detected by the cluster head or by its neighbors. If the node sends data to the cluster head periodically, any deviation in the traffic pattern indicates a fault. Each node maintains a timer for its neighbors. If no data is received from a neighbor for long, the timer expires, and an alarm is sent to the cluster head about the failure of the node.

Faults in cluster heads are detected by the central manager by analyzing the communication between the cluster heads and the sink. A timer in the manager is reset when a packet is received from the cluster head to the sink. If failure occurs, the timer expires. The manager then sends a query message to the node and waits for a fixed period. If no reply is received, it is assumed the node has failed [27].

The fault management in sensor nodes may also be self-diagnostic and self-healing, that is, the faults are detected by themselves and the network recovers without any manual intervention.

■ *Security management.* Because sensor networks are deployed in open environments and in battlefields, they are easily accessible to outsiders and hence vulnerable to attacks. Security management thus attains utmost importance in WSN management. Because the messages from the base station to the nodes are often broadcast, it is essential for the sensor node to authenticate the message to verify its credibility. pTESLA is a authentication protocol that uses cryptographic keys and a random function to authenticate the broadcast messages [27].

12.8 Mobile Agent Approach

The traditional SNMP approach to network management is based on the client–server model, in which the agent acts as a server and the manager acts as a client requesting information from the server. However, this centralized approach has limitations when applied to large-scale and future networks. Because data collection is achieved through polling, latency may creep in when large-scale networks are involved [28]. The use of mobile agents (MAs) is a more efficient approach for managing large networks. It saves on the bandwidth involved in transmitting management information between the manager and the static agents.

12.8.1 Introduction

A static agent resides on a single network device and acts as an interface between the manager and the MIB. A MA, on the other hand, can move from one network element to the other, along with its code, and start executing on the new device. By moving closer to the network device from which the data are to be collected, the bandwidth involved in transmitting the management information is reduced. This increases the performance as the processing of management information may be done locally, and only the relevant information is passed on to the manager.

12.8.2 Architecture

The network management architecture using MAs is illustrated in Figure 12.11.

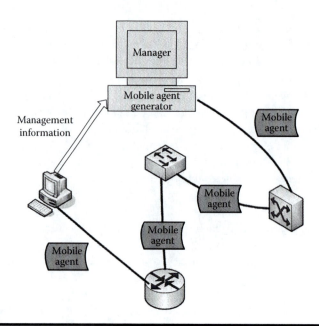

Figure 12.11 MA management architecture.

The management system consists of a management agent generator, which creates MAs with customized code. The MA code is generated so that it is capable of migrating from one device to the other on the basis of predefined policies [28]. The agent moves to a network device, collects information, processes it locally, and transmits the result to a local manager or to the central manager. It may stay at the same device if monitoring of the device is required, or it may move on to another device. The manager has the additional task of hosting and launching the MA generator in addition to obtaining results from the agents and displaying them. A domain manager in the management application decides the travel plan of the MA and its security policies [29].

12.8.3 Challenges

The MA technology faces several challenges, which should be addressed to harness its full potential and make it the de facto approach for distributed network management [29]. Some of the challenges in the implementation of the MA approach are the following:

- *Security.* The foremost challenge faced by the technology today is the need for enhanced security. Because the MAs traverse the network, they are susceptible to attacks and capture by malicious users who may modify the code and cause disruption in the network activity. The agents contain sensitive information, such as state and environment variables, which can cause serious security issues if compromised. Hence, the biggest challenge faced by researchers today is to develop a robust security framework to guard the MAs against threats. To address the issue of security, the network elements contain an agent server, which authenticates the MA before allowing it to execute. It also verifies that the MA code does not perform any unauthorized operation [29].
- *Coordination.* The coordination between agents, network elements, and the network manager is a challenging task. The mobility of the agents poses problems in the coordination

activity. Because the agents are mobile, more than one agent may try to reach the same network element at the same instant of time. However, as the management variables are allowed only a single point of access at a time, it may create a conflict. Hence, it is important to coordinate the movement of agents across the network.

■ *Resource utilization*. Because MAs operate autonomously in the network, it is essential to manage the allocation of resources to avoid conflicts and deadlocks between the agents. This helps prevent congestion and network breakdowns.

12.9 Conclusion

Network management has achieved new dimensions today as a result of the growing complexity of networks, evolution of new technologies, and enhanced security requirements. The classic manager–agent model has given way to distributed and mobile management frameworks. Introduction of NGNs and increasing applications of WSNs have added new functional areas and paradigms to network management. Passive network monitoring has now been replaced by proactive management to safeguard the network against impending threats and vulnerabilities. Advanced approaches such as policy-based management and MAs have been incorporated to meet the increasing demands of managing critical networks and related services. Future network management systems will see incorporation of new technologies and models to keep pace with ever-increasing demands for enhanced service quality and seamless connectivity.

References

1. A. Leinwand and K. Fang Conroy, *Network Management: A Practical Perspective*. Addison-Wesley Publishing Company, Inc., Boston, Massachusetts, 1996.
2. Cisco. "Cisco CPT Configuration Guide." *CTC and Documentation Release 9.3 and Cisco IOS Release 15.1(01)SA, SNMP, pp 655–665*. July 19, 2011. http://www.cisco.com/en/US/docs/optical/cpt/r9_3/configuration/guide/cpt93_configuration.pdf.
3. Simple Network Management Protocol (SNMP). *Internetworking Technology Overview*. June 1999. http://www.pulsewan.com/data101/pdfs/snmp.pdf.
4. J. Case, M. Fedor, M. Schoffstall, and J. Davin, "A Simple Network Management Protocol." Request for Comments: 1157, Network Working Group, IETF, 1990.
5. D. Mauro and K. Schmidt, *Essential SNMP*, 2nd ed. O'Reilly Media, Inc., California, USA, 2005.
6. A. Clemm, *Network Management Fundamentals*. Cisco Press, USA, 2007.
7. Accelerating UTM with Specialized Hardware. Fortinet. http://www.fortinet.com/sites/default/files/whitepapers/Accelerating_UTM_Specialized_Hardware.pdf.
8. J. Novotny, P. Celeda, T. Dedek, and R. Krejci, *Hardware Accleration for Cyber Security*. IST-091—Information Assurance and Cyber Defence, Estonia, November 2010.
9. Napatech. Napatech White Papers. http://www.napatech.com/resources/white_papers.html.
10. Invea-Tech, NetCOPE FPGA Platform. http://www.invea-tech.com/products-and-services/netcope-fpga-platform.
11. Invea-Tech, NIC Appliance. http://www.invea-tech.com/products-and-services/nic-appliance.
12. Cisco. "Network Management System: Best Practices White Paper." http://www.cisco.com/en/US/tech/tk869/tk769/technologies_white_paper09186a00800aea9c.shtml.
13. G. Yu and D. Cao, "The Changing Faces of Network Management for Next Generation Networks." *Proceedings of the 1st International Conference on Next Generation Network*, Korea, 192–197, 2006.
14. M. Li and K. Sandrasegaran, "Network Management Challenges for Next Generation Networks." *Proceedings of the IEEE Conference on Local Computer Networks 30th Anniversary* (LCN'05), 598, 2005.

15. W. Wu, F. Yip, E. Yiu, and P. Ray, "Integrated Vulnerability Management System for Enterprise Networks." *Proceedings of The 2005 IEEE International Conference on e-Technology, e-Commerce and e-Service*, EEE '05, 698–703, 2005.

16. C. Cowan, F. Wagle, P. Calton, S. Beattie, and J. Walpole. "Buffer Overflows: Attacks and Defenses for the Vulnerability of the Decade." *Proceedings of the DARPA Information Survivability Conference and Exposition, 2000* (DISCEX'00), 119–129, 2000.

17. M. Ben Salem and S. J. Stolfo, "Masquerade Attack Detection Using a Search-Behavior Modeling Approach." Tech report CUCS-027-09, Dept. of Computer Science, Columbia Univ., 2009.

18. R.F. Erbacher, S. Prakash, C.L. Claar, and J. Couraud, "Intrusion Detection: Detecting Masquerade Attacks Using UNIX Command Lines." Tech Rep, Utah State University.

19. P. Bravo and D.F. Garcia, "Proactive Detection of Kernel-Mode Rootkits." *Sixth International Conference on Availability, Reliability, and Security (ARES)*, 515–520, 2011.

20. E. Casey, "Investigating Sophisticated Security Breaches." *Communications of the ACM* 49, no. 2 (2006).

21. F. Baker and B. Foster, "Cisco Architecture for Lawful Intercept in IP Networks." RFC-3924, Network Working Group, Internet Society, 2004.

22. International Telecommunication Union. "Technical Aspects of Lawful Interception." ITU-T Technology Watch Report 6, 2008.

23. Aqsacomna. "Lawful Interception for IP Networks." White Paper, http://www.aqsacomna.com/us/articles/LIIPWhitePaperv21.pdf.

24. Phonexia. Phonexia White Papers. http://www.phonexia.com/download/.

25. R.V. Pawar, P.P. Kajave, and S.N. Mali, "Speaker Identification Using Neural Networks." *Proceeding of World Academy of Science, Engineering and Technology* 7, 31–35, 2005.

26. S. Duan and X. Yuan, "Exploring Hierarchy Architecture for Wireless Sensor." *Proceedings of the IFIP International Conference on Wireless and Optical Communications Networks*, IFIP 6, 6, 2006.

27. W.-B. Zhang H.-F. Xu, and P.-G. Sun, "A Network Management Architecture in Wireless Sensor Network." *Proceedings of the International Conference on Communications and Mobile Computing (CMC)*, 401–404, 2010.

28. F.L. Guo, B. Zeng, and L. Zhong Cui, "A Distributed Network Management Framework Based on Mobile Agents." *Proceedings of Third International Conference on Multimedia and Ubiquitous Engineering (MUE'09)*. 511–515, 2009.

29. M.A.M. Ibrahim, "Distributed Network Management with Secured Mobile Agent Support." *Proceedings of International Conference on Hybrid Information Technology (ICHIT)*, 244–251, 2006.

Chapter 13

VoIP in Next-Generation Converged Networks

Homero Toral-Cruz, Julio Ramirez-Pacheco,
Pablo Velarde-Alvarado, and Al-Sakib Khan Pathan

Contents

13.1 Introduction

In telecommunications, there are many networks, such as the public switched telephone network (PSTN), signaling system 7 (SS7), the integrated services digital network (ISDN), the internet protocol network (IP network), the wireless local area network (WLAN), the global system for mobile communication (GSM), general packet radio service (GPRS), the universal mobile telecommunications system (UMTS), etc. These telecommunications networks were designed to provide specific services, which are initially distinct and then converge into a single network [1]. Voice-over IP is a clear example of such convergent services [2]. To visualize this convergence, it is necessary to understand the historical development, which is summarized in the following events [3]: the introduction of automatic telephone exchange, the digitalization of telecommunications systems, the integration of circuit-switched technology and packet-switched technology (VoIP emergence), and the evolution of the mobile systems (1G, 2G, 3G, 4G, and beyond).

The development of technology and the enhanced services that can be offered to the end users have been captured in the concept of next-generation networks (NGNs). This concept encapsulates the convergent network process, where the convergent network is called a multiservice transport network and is based on IP technology. In the multiservice transport network, packets should be transported transparently between endpoints without excessive protocol conversion through an IP network core [4].

However, with this convergence, a new technical challenge has emerged. The IP network provides best-effort services in most of the cases and cannot guarantee the quality of service (QoS) of real-time multimedia applications, such as VoIP [5]. VoIP has emerged as an important service, poised to replace the circuit-switched telephony service in the future and carry voice packets transparently through the multiservice transport network. To achieve a satisfactory level of voice quality, the VoIP applications must be designed with a reconfiguration capability of some parameters [voice data length, Coder–Decoder (CODEC) type, size of forward error correction (FEC) redundancy, de-jitter buffer size, etc.] [6]. On the other hand, in the convergent network there must be consideration of the implementation of robust QoS mechanisms in order to guarantee acceptable QoS levels.

13.2 Telecommunication Networks

A communications network is a collection of terminals, links, and nodes, which connect together to enable communication between users via their terminals. The network sets up a connection between two or more terminals by making use of their source and destination addresses [1]. Behind this very general connection concept, there are a number of very different realities, bearing in mind the great variety of telecommunications networks, such as fixed telephone networks (PSTN, SS7, ISDN), IP networks, wireless networks (WLAN), mobile networks (GSM, GPRS, UMTS), etc. In the fixed telephone network, the term used is "connection," a direct relationship established at a physical level. In an IP network, the term used is usually "session," as there is normally no physical connection. On the other hand, the terminals connected will not only be fixed subscribers, but also mobile subscribers, giving rise to mobile networks.

13.2.1 Fixed Telephone Network

The telephone network was designed to carry voice, and the terminal main device is a simple telephone set. The network is more complex and is provided with intelligence necessary for providing various types of voice services [7]. The role of the telephone network is to connect two fixed terminals by means of the circuit-switched technology.

The communication via circuit-switched technology implies that there is a dedicated communication path between two or more terminals all through the communication session. Therefore, the resources (links and nodes) are reserved exclusively for information exchanges between origin and destination terminals. Before communication can occur between the terminals, a circuit is established between them. Thus, link capacity must be reserved between each pair of nodes in the path, and each node must have available internal switching capacity to handle the requested connection. The nodes must have the intelligence to make these allocations and to devise a route through the network.

In this switching technology, the nodes do not examine the contents of the information transmitted; the decision on where to send the information received is made just once at the beginning of the connection and remains during the connection. Thus, the delay introduced by a node is almost negligible. After the circuit has been established, the transmission delay is small and is kept constant throughout the duration of the connection. However, this is rather inefficient in terms of bandwidth utilization. Once a circuit is established, the resources associated to it cannot be used for another connection until the circuit is disconnected. Therefore, even if, at some point, both terminals stop transmitting, the resources allocated to the connection remain in use.

The most common examples of circuit-switched networks are the PSTN and ISDN. In the beginnings of the PSTN, a physical cable was needed between each terminal (telephone). Therefore, a large number of links is needed for this topology, that is, the number of relations grows proportionately to the square of the total number of terminals n [there are $n(n - 1)/2$ links], for example, Figure 13.1 shows a basic, four-telephone network.

Because of the cost concerns and the impossibility of running a physical cable between every terminal, another mechanism was developed, and it consists of a centralized operator (switch). By means of this centralized operator, the terminals needed only one cable to the centralized switch office instead of $n - 1$ links. At first, a human switch was used (see Figure 13.2).

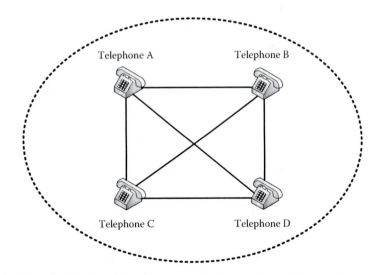

Figure 13.1   Basic, four-telephone network.

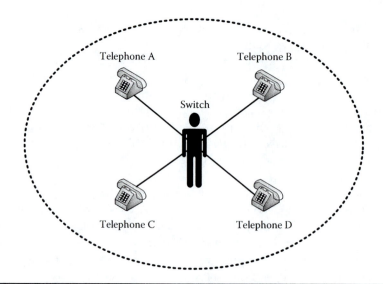

Figure 13.2   Basic, four-telephone network with a human switch.

Nowadays, the human switches have been replaced by electronic switches, and the telephone infrastructure starts with a simple pair of copper wires (local loop), which have the function of physically connecting the home telephone to the central office switch (class 5 switches). The communication path between the central office switch and your home is known as the phone line, and it normally runs over the local loop. The communication path between several central office switches is known as a trunk. Just as it is not cost-effective to place a physical wire between every terminal, it is also not cost-effective to place a physical wire between every central office switch. Therefore, the switches are currently deployed in hierarchies. The central office switches interconnect through trunks to tandem switches (class 4 switches). Higher-layer tandem switches connect local tandem switches. Figure 13.3 shows a typical model of switching hierarchy.

In discussing the PSTN network and the evolution of networks, reference must be made to SS7. The SS7 network is used to carry control information between the different elements of the network (telephone switches, databases, servers, etc.). The SS7 enhances the PSTN by handling call establishment, exchange of information, routing, operations, billing, and support for intelligent network (IN) services. The SS7 network consists of three signaling elements [8], the service switching point (SSP), signal transfer point (STP), and service control point (SCP), and several link types as shown in Figure 13.4.

- *SSPs* are end office or tandem switches that connect voice circuits and perform the necessary signaling functions to originate and terminate calls.
- *STPs* route all the signaling messages in the SS7 network.
- *SCPs* provide access to databases for additional routing information used in call processing. Also, the SCP is the key element for delivering IN applications on the telephony network.

All the signaling elements (SSP, STP, and SCP) form a dedicated network, which is completely separate from the voice transmission network, that is, all signaling information is carried on a common signaling plane. The signaling planes and the voice circuit planes are separated in logical terms at the link level because it makes use of the same physical resources. It is important to note

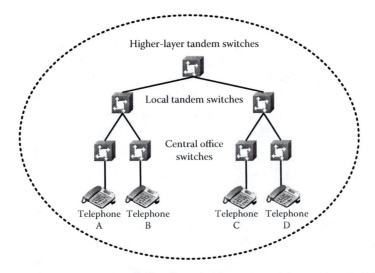

Figure 13.3   Circuit-switched network: hierarchy of switches.

that with this network the exchange of signaling is independent of the actual setup of a switched circuit. This independence makes the network well suited for evolution to the new generation of networks such as the NGN [8].

SS7 was initially designed for telephony call control; however, today's system includes database queries, transactions, network operations, and ISDN. ISDN is comprised of a set of communications standards for the transport of digital telephony, video, and data over the traditional circuit-switched network. The emergence of ISDN represents an effort to standardize subscriber services, user/network interfaces, and network and internetwork capabilities [8].

The SS7 and ISDN play an important role in the migration from traditional voice traffic to IP networks (VoIP), pushing toward the convergence between the voice and data networks.

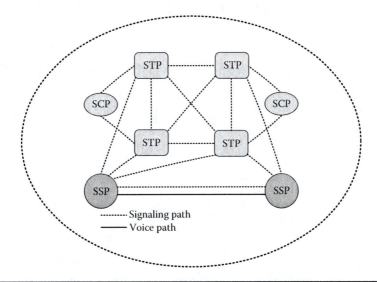

Figure 13.4   SS7 network.

13.2.2 IP Network

The IP network was designed to carry traffic in bursts presented by data transmission because the terminals do not transmit continuously, that is, they are idle most of the time and experience bursts at certain times. Data rates are not kept constant through the duration of the connection, but they vary dynamically [7]. Therefore, employing dedicated circuits (circuit-switched technology) to transmit traffic with these characteristics is a waste of resources. This network is based on the packet-switched technology and the most intelligence was placed in the terminal device, which is typically a computer, and the network only offers the best effort service [9].

The IP network contains a collection of machines (hosts or end systems) intended for running user programs (applications). These hosts were primarily traditional desktop PCs, Linux workstations, and servers that store and transmit information, such as web pages and email messages. The hosts are connected together by a communication subnet (or just a subnet). The job of the subnet is to carry messages from host to host, just as the telephone network carries voice from speaker to listener. The subnet consists of two distinct components (see Figure 13.5) [10]: communication links (coaxial cable, copper wire, fiber optics, radio spectrum, etc.) and switching elements.

The switching elements (routers) are specialized computers that connect three or more communication links. The sequence of communication links and routers traversed by a message from the sending host to the receiving host is known as a route or path through the network.

In the IP networks, the information is split up by the terminal into blocks of moderate size, called packets. These packets can be autonomous, that is, they are capable of moving on the network thanks to a header that contains the source and destination addresses [1].

The packet is sent to the first router, and the router receives the packet. It examines the header and forwards the packet to the next appropriate router. This technique of inspection and retransmission is called "store and forward," and it is accomplished in all the routers on the path until the packet reaches its destination unless the packet is lost. After reaching the destination, the destination terminal strips off the header of the packet to obtain the actual data that originated at the source.

In the communications process based on the packet-switched technology, the source sends packets and the network multiplexes the packets from various origins in the same resources to

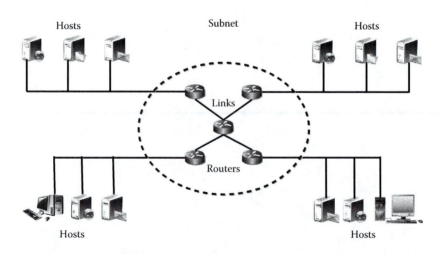

Figure 13.5 IP network components.

optimize their use. In this way, several communications can share the same resources. The packet-switched technology enables a better use of the transmission resource than circuit switched because the resources are shared. However, the multiplexing of different connections on the same resources causes delays and packet losses, which do not happen with circuit-switched technology [10].

Finally, it must be noted that in packet-switched technology a distinction is drawn between two modes of operation: connection-oriented mode and connectionless mode [11]. In connection-oriented mode, a path is established; this path is called a virtual circuit. There is a prior exchange of initial signaling packets to reserve resources and establish the path. The user establishes a connection, uses the connection, and then releases the connection. In most cases, the order of the packets is preserved, so the packets arrive in the order they were sent.

In connectionless mode, each packet is treated independently with no reference to packets that have gone before, and the routing decisions are taken on at each node. Each packet carries the full destination address, and each one is routed through the system independent of all the others. Normally, when two packets are sent to the same destination, the first one sent will be the first one to arrive. However, it is possible that the first one sent can be delayed so that the second one arrives first. The connectionless mode has been popularized mainly by Internet protocol. The IP networks have progressed to the point that it is now possible to support voice and multimedia applications but do not guarantee QoS because they are based on best-effort services.

13.2.3 Wireless and Mobile Networks

Wireless does not imply mobility. There are wireless networks in which both ends of the communications are fixed, such as in wireless local loops. In the historical development of telecommunications networks, wireless networks with mobility provide the biggest challenge to the network designer [12]. In this section, the most important wireless and mobile technologies are studied.

13.2.3.1 Wireless Local Area Network

As the number of mobile communication devices grows, so does the demand to connect them to the outside world. Even the very first mobile telephones had the ability to connect to other telephones. The first portable computers did not have this capability, but soon afterward, modems became commonplace on notebook computers. To be connected to the outside world, these computers had to be plugged into a telephone line. By means of this wired connection to the fixed network, the computers were portable but not mobile. In order to achieve true mobility, notebook computers need to use radio (or infrared) signals for communication. In this manner, dedicated users can read and send email while hiking or boating. A system of notebook computers that communicate by radio can be regarded as a WLAN [11].

Currently, there are many technologies and standards for WLANs, but one particular class of standards that is enjoying the most widespread deployment is the IEEE 802.11 WLAN, also known as Wi-Fi. There are several 802.11 standards for WLAN technology, including 802.11b, 802.11a, 802.11g, and 802.11n.

The IEEE 802.11 standards use infrared and the unlicensed spectra. These spectra are allocated in many countries for research and development in industry (I), science (S), and medicine (M) (called the ISM band). The IEEE standard PHY (physical layer) provides several mechanisms for the use of the ISM band designed to combat interference from other sources on the same bands. This is necessary because the use of such a system does not require a license from the government, which could result in numerous sources of interference. The infrared band specifies only one type of radiation, that is, indirect radiation reflected from a coarse surface (called diffused infrared) [12].

The IEEE 802.11 defines several device types, such as a wireless station (STA), which is a user terminal; a central base station, known as an access point (AP); and basic service set (Bss). A Bss contains one or more wireless stations and an AP. The wireless stations may be either fixed or mobile. This gives rise to two configurations of WLANs, the infrastructure WLAN (see Figure 13.6) and ad hoc WLAN (see Figure 13.7).

In an infrastructure WLAN, two wireless stations exchanging data can communicate only through an AP. Figure 13.6 shows an AP connected to a router, which, in turn, leads to the Internet. Multiple APs may be connected together to form a so-called distribution system (DS).

In an ad hoc WLAN, two stations communicate directly with each other without an AP and no connection to the outside world (see Figure 13.7). Wireless stations for such networks may require the capability to forward a packet, thus acting as a repeater. With this relaying capability, two wireless stations can exchange data packets even if they are unable to receive signals directly from each other.

The main advantages of WLAN are its simplicity, flexibility, and cost-effectiveness. In the past several years, WLAN has become a ubiquitous networking technology and has been widely deployed around the world. Although most existing WLAN applications are data centric, such as web browsing, file transfer, and electronic mail, there is a growing demand for multimedia services over WLANs. Recently, VoIP over WLAN (VoWLAN) has been emerging as an infrastructure to provide wireless voice service with cost-efficiency. However, supporting voice traffic over WLANs poses significant challenges because the performance characteristics of the physical and MAC layers are much worse than that of their wire line counterparts. Therefore, the applications of VoWLAN raise several deployment issues concerning the system architecture, network capacity, and admission control, QoS provisioning, etc. [13].

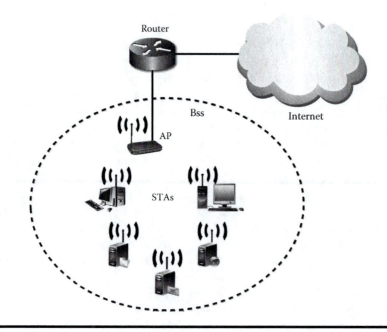

Figure 13.6 WLAN components.

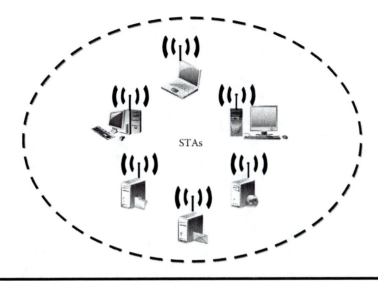

Figure 13.7   Ad hoc WLAN.

13.2.3.2 Global System for Mobile Communication

This is the basic mobile telephone network. The functions are the same as the fixed telephone network; however, the terminals are mobile. The GSM system consists of the following subsystems (see Figure 13.8) [3,14,15]:

- *Mobile station subsystem (MSS).* An MSS is basically a human–machine interface performing functions to connect the user and the public LAN mobile network (PLMN). These functions include voice and data transmission, synchronization, monitoring of signal quality, equalization, display of short messages, location updates, and others. To carry out all its functions, an MSS includes the terminal equipment (TE), the terminal adapter (TA), the mobile termination (MT), and a subscriber identity module (SIM).
- *Base station subsystem (BSS).* A BSS is composed of one or more base transceiver stations (BTSs) and one or more base station controllers (BSCs). The role of the BSS is to provide transmission paths between the MSS and the NSS. The MSSs are connected to the network via a radio link with the pilot station of the cell in which they are located, the BTS. The BTS is the radio AP, which has one or more transceivers. The BSC monitors and controls several base stations. The main functions of the BSC are cell management, control of a BTS, and exchange functions.
- *Network and switching subsystem (NSS).* An NSS includes switching and location management functions. It consists of the mobile switching center (MSC); databases for location management, which include the home location register (HLR) and visitor location register (VLR); the gateway MSC (GMSC); as well as the authentication center (AuC) and equipment identity register (EIR). The MSC is the core switching entity in the network and carries out the functions of control and connection of the subscribers located in its geographical zone. It also acts as a gateway (GW) between the fixed network and the mobile network, or between mobile networks, for incoming calls for which the location of the called party is not known. An MSC that receives a call from another network (PSTN, ISDN, etc.) and that routes this call toward the MSC, where the called subscriber is in fact located, is called the GMSC. To do this, it consults the location database, the HLR. Subscribers may move

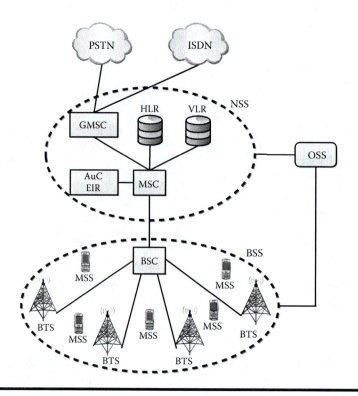

Figure 13.8   GSM network.

from one cell to another, even during a call (referred to as a handover), because the radio system continuously tracks their location. A GSM subscriber is normally associated with one particular HLR. The HLR is the system that stores subscriber data (identity, number, etc.) of all subscribers belonging to the mobile operator, no matter if they are currently located in the network or abroad (i.e., roaming). These data are permanent, such as the unique implicit international mobile subscriber identity (IMSI) number. The IMSI is embedded on the SIM card and is used to identify a subscriber. In addition to these static data in the HLR are added dynamic data, such as the last known location of the subscriber, which enables routing to the MSC where the subscriber is, in fact, located. Finally, it is the VLR that updates the data relating to subscribers visiting its zone and that notifies the HLR. The AuC is related to HLR and contains sets of parameters needed for authentication procedures for the mobile stations. EIR is an optional database that is supposed to contain the unique international mobile equipment identity (IMEI), which is the number of the mobile phone equipment. The EIR is specified to prevent usage of stolen mobile stations or to bar malfunctioning equipment.

■ *Operation and support subsystem (OSS).* An OSS performs the operation and maintenance functions through two entities, namely, the operation and maintenance center (OMC) and the network management center (NMC). The GSM standards do not fully specify these elements. Therefore, different manufacturers may have different implementations, which may be a problem for interoperability between different GSM systems.

GSM is a system created mainly for telephony service, but it also supports low data rate modem connections up to 9600 bps. For support of higher data rates in the radio access network (which are demanded by some multimedia services, such as Internet applications), GSM, on its way toward the third-generation mobile systems, is extended to GPRS.

13.2.3.3 General Packet Radio Service

The GPRS network should be seen essentially as an evolution of the existing GSM network. It basically adds to GSM the possibility of sending data in packet mode [1]. GPRS is the first step toward integration of the Internet and mobile cellular networks. GPRS uses the same radio access network as GSM but a different core network (CN) infrastructure [14]. In order to integrate GPRS into existing GSM architecture, two new network nodes should be added as shown in Figure 13.9 [3]:

- *Serving GPRS support node (SGSN):* An SGNS transmits data between the mobile terminals and the mobile network. Its main tasks are mobility management, packet routing, logical link management, authentication, and charging functions.
- *GW GPRS support node (GGSN):* A GGSN is an interface between the mobile network and the data network (e.g., the Internet). It converts protocol data packet (PDP) addresses from the external packet-based networks to the GSM address of the specified user and vice versa.

All GPRS support nodes are connected via an IP-based GPRS backbone network. In the case of GPRS, HLR stores the user profile, the current SGSN address, and the PDP address for each user. MSC/VLR is extended with additional functions that allow coordination between GSM circuit-switched services and GPRS packet-switched services [3]. Essentially, the IP-based GPRS backbone network has a core packet network that is overlaid with call-control functions and GW functions to support VoIP and other multimedia services. The function providing for VoIP

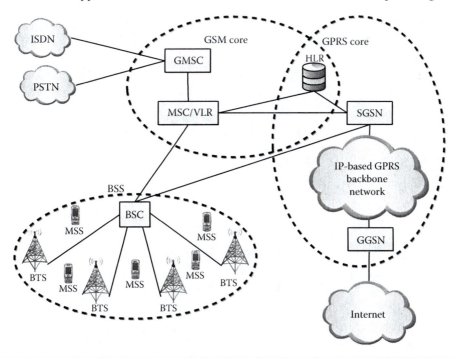

Figure 13.9 GPRS network.

capability is named the call state control function (CSCF), which is a function analogous to that used for call control in a circuit-switched environment. Besides the usual call-control functions, the CSCF also performs service-switching functions, address translation functions, and vocoder negotiation functions. The communications between this CN and the PSTN, ISDN, and other legacy networks are provided by a GMSC [15].

13.2.3.4 Universal Mobile Telecommunications System

In the UMTS network, the whole mobile telephone network evolves. Three basic blocks compose UMTS, as shown in Figure 13.10 [15]:

■ *User equipment (UE):* UE provides a means for the user to access UMTS services. It consists of the mobile equipment (ME) and of the UMTS SIM (USIM). The ME performs radio

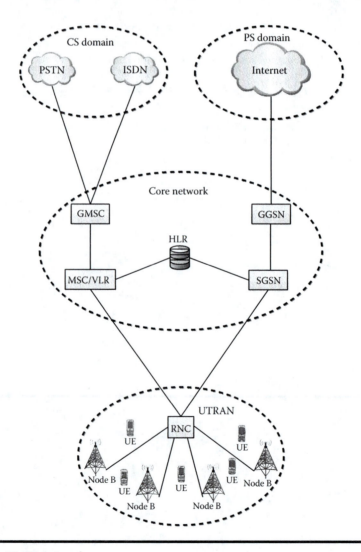

Figure 13.10 UMTS network.

communication with the network and contains applications for the services. The USIM is a smartcard performing functions to support user security and user services. It holds the subscriber's identity and some subscription information, performs authentication procedures, and stores authentication and encryption keys.

- *UMTS terrestrial radio access network (UTRAN):* A UTRAN performs functions to support communication with the MT and with the CN, acting as a bridge, router, and GW as required. It consists of a set of radio network subsystems (RNS), each of which contains a radio network controller (RNC) and one or more entities known as node B. As before, node B includes radio resources but with higher bit rates (BTS becomes node B). It performs channel coding and interleaving, rate adaptation, spreading, radio resource management operation as in inner loop power control, and others. The RNC supports radio access control, connection control, geographic positioning, and access link relay.
- *Core network (CN):* A CN consists of the circuit-switched domain (CS) and the packet-switched domain (PS). These two domains in CN are overlapping in some common elements. CS mode is the GSM mode of operation, and PS is the mode supported by the GPRS. The MSC/VLR and GMSC are used by the CS domain, and in the PS domain, their equivalents are the SGSN and the GGSN. Finally, the entities common to both domains in the CN are the HLR and the authentication center. A UMTS mobile is capable of communicating simultaneously via both domains [1].

Because UMTS offer large data rates, it is possible to make VoIP calls under different compression and decompression (CODEC) schemes in a mobile environment. This allows a flexible way to select the most suitable CODEC scheme according to the network state and guarantee some QoS level.

13.3 NGNs: Convergence of Networks

Telecommunications technology has a long history. To visualize the future, we need to understand the historical development leading to present-day and emerging technologies. Along this historical development, one may distinguish among three key events [3]:

- The introduction of automatic telephone exchange
- The digitalization of telecommunications systems
- The integration of circuit-switched technology and packet-switched technology (VoIP emergence)

The last two events were also followed by mobile systems [3]:

- First-generation (1G) mobile cellular systems appeared in the 1980s and provided only classical analog voice service.
- Second-generation (2G) in the 1990s introduced digitalization of the communication link end-to-end as well as additional ISDN-based services and modem-based data services.
- Third-generation (3G) mobile systems appeared in the 2000s and were created to support Internet connectivity and packet-switched services besides the traditional circuit-switched.
- Future mobile networks (4G and beyond) are expected to include heterogeneous access technologies, such as WLAN and 3G, as well as end-to-end IP connectivity (i.e., a wireless IP

network). IP is the dominant technology in the world of telecommunication networks and, arguably, the secret ingredient in all the new and traditional technologies.

In the telecommunications world, the development of technology and the enhanced services that can be offered to the end users have been captured in the concept of NGNs. This concept does not mean a particular network, but rather the process of development from present technology and services to new technologies enabling new services and applications.

On the other hand, within telecommunications, the historical separations between the fixed telephone network, IP network, and wireless and mobile networks have been diminishing. As a result of this fact, it presents the networks' convergence, where the convergent network is called the multiservice transport network. In this network, numerous types of access networks with different types of terminals must be integrated, for example, must interwork with legacy networks through GWs as shown in Figure 13.11. Access networks can use a variety of layer 1 and 2 protocols. In the multiservice transport network, packets should be transported transparently between endpoints without excessive protocol conversion and adaption. Also, all terminals should use the same network layer protocol to give a uniform end-to-end routing method. Currently, the ubiquitous internet protocol is seen as the unifying factor leading to a multiservice transport network [4].

The NGN concept encapsulates the product of the convergence process as well as ways of abstracting and modeling the complex networks and software systems. Like convergence, the NGN is an evolving concept, that is, there is no single NGN. Rather, the term captures the movement or development process to one or more future networks that exhibit convergence. Several developments have been or are candidates for consideration as NGNs; one of the most important is the deployment of VoIP (including H.323 and SIP) [4].

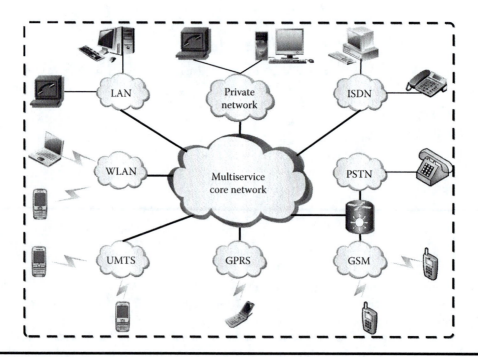

Figure 13.11 Multiservice transport network.

13.4 Voice-Over IP

VoIP is the real-time transmission of voice between two or more parties by using IP technologies. Current implementations of VoIP have two main types of architectures, which are based on H.323 [16,17] and session initiation protocol (SIP) frameworks [18,19], respectively. Regardless of their differences, the fundamental architectures of these two implementations are the same. They consist of three main logical components: terminal, signaling server, and GW. They differ in specific definitions of voice coding, transport protocols, control signaling, GW control, and call management.

13.4.1 H.323 Framework

ITU-T H.323 is a set of protocols for voice, video, and data conferencing over packet-switched networks, such as Ethernet LANs and the Internet, that do not provide a guaranteed QoS. The H.323 protocol stack is designed to operate above the transport layer of the underlying network. H.323 was originally developed as one of several videoconferencing recommendations issued by the ITU-T. The H.323 standard is designed to allow clients on H.323 networks to communicate with clients on other videoconferencing networks. The first version of H.323 was issued in 1996, designed for use with Ethernet LANs, and borrowed much of its multimedia conferencing aspects from other H.32.x series recommendations. H.323 is part of a large series of communications standards that enable videoconferencing across a range of networks. This series also includes H.320 and H.324, which address the ISDN and PSTN communications, respectively. H.323 is known as a broad and flexible recommendation. Although H.323 specifies protocols for real-time, point-to-point communication between two terminals on a packet-switched network, it also includes support of multipoint conferencing among terminals that support not only voice, but also video and data communications. This recommendation describes the components of H.323 architecture (see Figure 13.12). This includes terminals (Ts), GWs, gatekeepers (GKs), multipoint control units (MCUs), multipoint controllers (MCs), and multipoint processors (MPs).

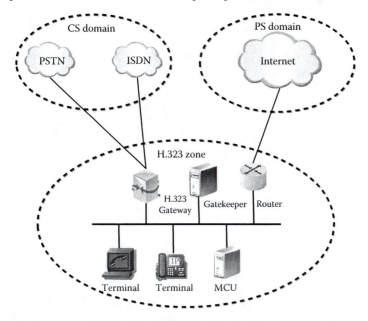

Figure 13.12 H.323 framework.

- *Terminal:* An H.323 terminal is an endpoint on the network, which provides real-time, two-way communications with another H.323 terminal, GW, or MCU. This communication consists of control, indications, audio, moving color video pictures, and/or data between the two terminals. A terminal may provide speech only; speech and data; speech and video; or speech, data, and video.
- *GW:* The GW is an H.323 entity on the network, which allows intercommunication between IP networks and legacy circuit-switched networks, such as ISDN and PSTN. They provide signal mapping as well as transcoding facilities. For example, GWs receive an H.320 stream from an ISDN line, convert it to an H.323 stream, and then send it to the IP network.
- *GK:* The GK is an H.323 entity on the network, which performs the role of the central manager of VoIP services to the endpoints. This entity provides address translation and controls access to the network for H.323 terminals, GWs, and MCUs. The GK may also provide other services to the terminals, GWs, and MCUs, such as bandwidth management and locating GWs.
- *MCU:* The MCU is an H.323 entity on the network, which provides the capability for three or more terminals and a GW to participate in a multipoint conference. It may also connect two terminals in a point-to-point conference that may later develop into a multipoint conference. The MCU consists of two parts: a mandatory MC and an optional MP. In the simplest case, an MCU may consist only of an MC with no MPs. An MCU may also be brought into a conference by the GK without being explicitly called by one of the endpoints.
- *MC:* The MC is an H.323 entity on the network, which controls three or more terminals participating in a multipoint conference. It may also connect two terminals in a point-to-point conference that may later develop into a multipoint conference. The MC provides the capability of negotiation with all terminals to achieve common levels of communications. It may also control conference resources, such as who is multicasting video. The MC does not perform mixing or switching of audio, video, and data.
- *MP:* The MP is an H.323 entity on the network, which provides for the centralized processing of audio, video, and/or data streams in a multipoint conference. The MP provides for the mixing, switching, or other processing of media streams under the control of the MC. The MP may process a single media stream or multiple media streams depending on the type of conference supported.

The H.323 architecture is partitioned into zones. Each zone is composed of the collection of all terminals, a GW, and MCU managed by a single GK. H.323 is an umbrella recommendation that depends on several other standards and recommendations to enable real-time multimedia communications. The main ones are the following:

- *Call Signaling and Control:* Call control protocol (H.225), media control protocol (H.245), security (H.235), digital subscriber signaling (Q.931), generic functional protocol for the support of supplementary services in H.323 (H.450.1), and supplemental features (H.450.2-H.450.11)
- *H.323 Annexes:* Real-time facsimile over H.323 (annex D); framework and wire-protocol for multiplexed call signaling transport (annex E); simple endpoint types—SET (annex F); text conversation and text SET (annex G); security for annex F (annex J); hypertext transfer protocol (HTTP)-based service control transport channel (annex K); stimulus control protocol (annex L); and tunneling of signaling protocols (annex M)

- *Audio CODECs:* Pulse code modulation (PCM) audio CODEC 56/64 kbps (G.711), audio codec for 7 KHz at 48/56/64 kbps (G.722), speech CODEC for 5.3 and 6.4kbps (G.723), speech CODEC for 16 kbps (G.728), and speech CODEC for 8/13 kbps (G.729)
- *Video CODECs:* Video CODEC for ≥ 64 kbps (H.261) and video CODEC for ≤ 64 kbps (H.263)

13.4.2 SIP Framework

SIP was developed by IETF in reaction to the ITU-T H.323 recommendation. The IETF believed that H.323 was inadequate for evolving IP telephony because its command structure is complex, and its architecture is centralized and monolithic. SIP is an application layer control protocol that can establish, modify, and terminate multimedia sessions or calls. SIP transparently supports name mapping and redirection services, allowing the implementation of ISDN and intelligent network telephony subscriber services. The early implementations of SIP have been in network carrier IP-Centrex trials. SIP was designed as part of the overall IETF multimedia data and control architecture that supports protocols, such as resource reservation protocol (RSVP), real-time transport protocol (RTP), real-time streaming protocol (RTSP), session announcement protocol (SAP), and session description protocol (SDP). SIP establishes, modifies, and terminates multimedia sessions. It can be used to invite new members to an existing session or to create new sessions. The two major components in a SIP network are the user agent (UA) and network servers (registrar server, location server, proxy server, and redirect server), as shown in Figure 13.13.

- *UAs:* This is an application that interacts with the user and contains both a UA client (UAC) and a UA server (UAS). A UAC initiates SIP requests, and a UAS receives SIP requests and returns responses on the user's behalf.

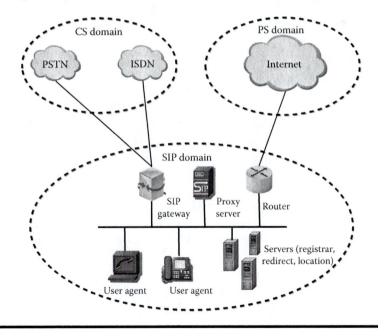

Figure 13.13 SIP framework.

- *Registrar server:* This is a SIP server that accepts only registration requests issued by UAs for the purpose of updating a location database with the contact information of the user specified in the request.
- *Proxy server:* This is an intermediary entity that acts both as a server to UAs by forwarding SIP requests and as a client to other SIP servers by submitting the forwarded requests to them on behalf of UAs or proxy servers.
- *Redirect server:* This is a SIP server that helps to locate UAs by providing alternative locations where the user can be reachable, that is, it provides address mapping services. It responds to a SIP request destined to an address with a list of new addresses. A redirect server does not accept calls, does not forward requests, and does not initiate any of its own.

The SIP protocol follows a web-based approach to call signaling, contrary to traditional communication protocols. It resembles a client–server model, where SIP clients issue requests and SIP servers return one or more responses. The signaling protocol is built on this exchange of requests and responses, which are grouped into transactions. All the messages of a transaction share a common unique identifier and traverse the same set of hosts. There are two types of messages in SIP: requests and responses. Both of them use the textual representation of the ISO 10646 character set with UTF-8 encoding. The message syntax follows HTTP/1.1, but it should be noted that SIP is not an extension to HTTP.

- *SIP responses:* Upon reception of a request, a server issues one or several responses. Every response has a code that indicates the status of the transaction. Status codes are integers ranging from 100 to 699 and are grouped into six classes. A response can be either final or provisional. A response with a status code from 100 to 199 is considered provisional. Responses from 200 to 699 are final responses.
- 1xx informational: Request received, continuing to process request. The client should wait for further responses from the server.
- 2xx success: The action was successfully received, understood, and accepted. The client must terminate any search.
- 3xx redirection: Further action must be taken in order to complete the request. The client must terminate any existing search but may initiate a new one.
- 4xx client error: The request contains bad syntax or cannot be fulfilled at this server. The client should try another server or alter the request and retry with the same server.
- 5xx server error: The request cannot be fulfilled at this server because of server error. The client should try with another server.
- 6xx global failure: The request is invalid at any server. The client must abandon search.

The first digit of the status code defines the class of response. The last two digits do not have any categorization role. For this reason, any response with a status code between 100 and 199 is referred to as a "1xx response," any response with a status code between 200 and 299 as a "2xx response," and so on.

- *SIP requests:* The core SIP specification defines six types of SIP requests, each of them with a different purpose. Every SIP request contains a field, called a method, which denotes its purpose.
- INVITE: INVITE requests invite users to participate in a session. The body of INVITE requests contains the description of the session. Significantly, SIP only handles the invitation

to the user and the user's acceptance of the invitation. All of the session particulars are handled by the SDP used. Thus, with a different session description, SIP can invite users to any type of session.

■ ACK: ACK requests are used to acknowledge the reception of a final response to an INVITE. Thus, a client originating an INVITE request issues an ACK request when it receives a final response for the INVITE.

■ CANCEL: CANCEL requests cancel pending transactions. If a SIP server has received an INVITE but has not returned a final response yet, it will stop processing the INVITE upon receipt of a CANCEL. If, however, it has already returned a final response for the INVITE, the CANCEL request will have no effect on the transaction.

■ BYE: BYE requests are used to abandon sessions. In two-party sessions, abandonment by one of the parties implies that the session is terminated.

■ REGISTER: Users send REGISTER requests to inform a server (in this case, referred to as a registrar server) about their current location.

■ OPTIONS: OPTIONS requests query a server about its capabilities, including which methods and which SDPs it supports.

SIP is independent of the type of multimedia session handled and of the mechanism used to describe the session. Sessions consisting of RTP streams carrying audio and video are usually described using SDP, but some types of session can be described with other description protocols. In short, SIP is used to distribute session descriptions among potential participants. Once the session description is distributed, SIP can be used to negotiate and modify the parameters of the session and terminate the session.

13.4.3 VoIP System

A basic VoIP system consists of three parts: the sender, the IP network, and the receiver, as shown in Figure 13.14.

Sender: The first component is the coder, which periodically samples the original voice signal and assigns a fixed number of bits to each sample, creating a constant bit rate stream.

The voice stream from the voice source is first digitized and compressed by using a suitable coding algorithm, such as G.711, G.729, etc. Various speech CODECs differ from each other in terms of features such as coding bit-rate (kbps), algorithmic delay (ms), complexity, and speech quality (mean opinion score or MOS). In order to simplify the description of speech CODECs, they are often broadly divided into three classes: waveform coders, parametric coders or vocoders, and hybrid coders (as a combination thereof).

Typically waveforms CODECs are used at high bit rates and give very good quality speech. Parametric CODECs operate at very low bit rates but tend to produce speech that sounds

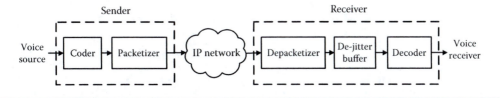

Figure 13.14 VoIP system.

synthetic. Hybrid CODECs use techniques from both parametric and waveform coding and give good quality speech an intermediate bit rates.

After compression and encoding into a suitable format, the speech frames are packetized. The packetized process is implemented for gathering a set of voice data to be transmitted and adding the information needed for routing and handling those voice data across the IP network. The added bits are referred as the header, and the voice data to be delivered are referred to as the payload. The voice data length can be changed according to the VoIP transmission efficiency. The media access control (MAC) header, IP header, user datagram protocol (UDP) header, RTP header, and frame check sequence (FCS) are necessary for transmitting voice data over the Ethernet while the preamble and inter packet gap (IPG) should be considered as occupied bandwidth on the transmission line. For instance, the total occupied bandwidth is 98 bytes, including IPG, preamble, MAC header, IP header, UDP header, RTP header, and FCS, when transmitting 20-byte voice data. The 78 bytes thus correspond to the overhead of IP transmission, so the ratio of voice data to the total is less than 25%.

The voice data length of an IP packet usually depends on the coding algorithm used. Eighty-byte voice data are often used for G.711, whereas 20-byte voice data are used for G.729 in conventional VoIP communication. Table 13.1 shows the relationship between the voice data length in milliseconds and the voice data length in bytes.

IP network: Because of the shared nature of IP networks, guaranteeing the QoS of Internet applications from end-to-end is difficult. Because current IP networks are based on best-effort services, the packet may suffer different network impairments (e.g., packet loss, delay, and jitter), which directly impact the quality of VoIP applications.

Receiver: The packet headers are stripped off, and voice samples are extracted from the payload by a depacketizer. The voice samples must be presented to the decoder in such a way that the next sample is present for processing when the decoder has finished with its immediate predecessor. Such a requirement severely constrains the amount of jitter that can be tolerated in a VoIP system

Table 13.1 Voice Data Length of VoIP Packets

Voice Data Length (ms)	Voice Data Length (Bytes)	
	G.711	G.729
10	80	10
20	160	20
30	240	30
40	320	40
50	400	50
60	480	60
70	560	70
80	640	80
90	720	90
100	800	100

without having to gap the samples. When jitter results in an interarrival time (IAT) that is greater than the time required to re-create the waveform from a sample, the decoder has no option but to continue to function without the next sample information. Therefore, the effects of jitter will be manifested as an increase in the packet loss rate.

The buffer that holds the queued segments is called a de-jitter buffer. The employment of such de-jitter buffers defines the relationship between jitter and packet loss rate on the receiver side. The delay variation that can be tolerated becomes, therefore, the essential descriptor of intrinsic quality that supplants jitter.

Therefore, an important design parameter at the receiver side is the de-jitter buffer size or play-out delay of a de-jitter buffer because the de-jitter buffer is used to compensate for network jitter at the cost of further delay (buffer delay) and loss (late arrival loss). Finally, the de-jittered speech frames are decoded to recover the original voice signal.

13.4.4 QoS in VoIP

Figure 13.15 shows a basic two-user VoIP network, where the end users represent the terminal devices, such as a VoIP application based on software (softphone) or an IP telephone. The network is an IP network that connects the end users.

Referring to Figure 13.15, QoS can be defined from three different points of view: QoS experienced by the end user, the QoS from the point of view of the application, and from the network.

From the end user's perspective, QoS is the end user's perception of the quality that he receives from the VoIP services provider. The end user's perception of the voice quality is determined by subjective and objective testing as a function of some impairments (OWD, PLR, CODEC type).

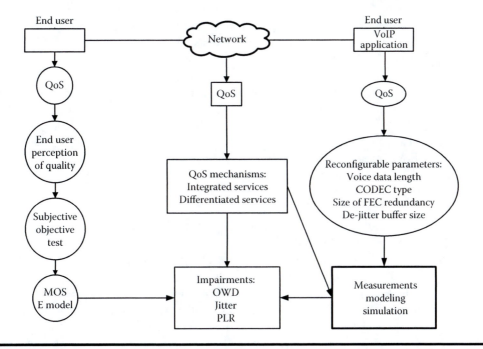

Figure 13.15   QoS in a basic two-user VoIP network.

From the point of view of the application, the QoS refers to the application's capabilities to reconfigure some parameters (voice data length, CODEC type, size of FEC redundancy, de-jitter buffer size, etc.) to values based on network conditions in order to meet good levels of voice quality.

From the network's perspective, the term QoS refers to the network's capabilities to provide the QoS perceived by the end user as defined above. A QoS mechanism has the capability to provide resource assurance and service differentiation in the IP network. Without a QoS mechanism, an IP network provides best-effort service. Two main QoS mechanisms are available for the IP networks [20]: integrated services (IntServ) [21] and differentiated services (DiffServ) [22].

In best-effort service, all packets are lumped into a single mass regardless of the source of the traffic. In IntServ, individual flows are distinguished end-to-end, and the applications use the RSVP [23] to request and to reserve resources through a network. In DiffServ, individual flows are not identified end-to-end. Rather, they are aggregated into a smaller number of classes. Furthermore, these classes of traffic are given differential treatment per hop, and there is no end-to-end treatment of these traffic classes.

Because the NGN carries diverse traffic types with different performance requirements, one type of impairment can be important to a particular service or application, and for other applications, it may not be as important and vice versa. Therefore, a QoS mechanism implemented in an IP network must consider various performance requirements and optimize the trade-off between multiple impairments.

The designing of VoIP applications with reconfigurable parameters and the implementation of QoS mechanisms in convergent networks involves traffic measurements, traffic simulations, and traffic modeling.

13.5 Future Research Directions

VoIP in next-generation converged networks has emerged as an important application, which demands strict QoS levels. However, the convergent networks have an IP network core. The IP network provides best-effort services in most of the cases and cannot guarantee the full QoS of real-time applications, such as VoIP.

An interesting fact is that sending wireless phone calls over IP networks is considerably less expensive than that of sending them over cellular voice networks. However, such types of communications must ensure good performance and quality of voice transmissions.

On the other hand, the converged networks carry diverse traffic types with different performance requirements; one type of impairment can be important to a particular service or application, and for other applications, it may not be as important and vice versa.

Consequently, the above points direct us to consider the follow actions to pursue as future research areas:

■ Accomplish a detailed characterization and accurate modeling of the main QoS parameters in the next-generation converged networks.
■ Current converged networks must be enhanced by means of robust QoS mechanisms in order to ensure acceptable QoS levels of VoIP applications.
■ The VoIP applications must be designed with reconfiguration capability of some parameters (voice data length, CODEC type, size of FEC redundancy, de-jitter buffer size, etc.) in order to guarantee an acceptable voice quality.
■ The design of powerful mobile telephones and faster wireless/mobile IP networks is needed in order to ensure good performance and quality of mobile voice over IP.

13.6 Conclusions

Within telecommunications, there are many heterogeneous networks (PSTN, SS7, ISDN, IP Network, WLAN, GSM, GPRS, UMTS, etc.) corresponding to services that are initially distinct and then converge into a single network. VoIP is a clear example of such convergent services, pushing toward adoption of NGNs. The concept of NGNs does not mean a particular network, but rather the process of development from present technology and services to new technologies enabling new services and applications.

On the other hand, within telecommunications, the historical separations between the above heterogeneous networks have been diminishing. As a result of this fact, it presents the networks' convergence, where the convergent network is called a multiservice transport network. In this multiservice transport network, numerous access networks with different terminals must be integrated. Access networks can use a variety of layer 1 and 2 protocols; however, all terminals should use the same network layer protocol. Currently, the ubiquitous Internet protocol is seen as the unifying factor leading to a multiservice transport network.

However, with this convergence, a new technical challenge has emerged. Essentially, the converged network has an IP network core. The IP network is based on best-effort service, and it does not provide a QoS level to meet the requirements of real-time applications, such as VoIP. Therefore, to guarantee an acceptable voice quality, the VoIP applications must be designed with reconfiguration capability of some parameters, and the convergent network must consider the implementation of robust QoS mechanisms.

References

1. G. Fiche and G. Hébuterne, *Communicating Systems and Networks: Traffic and Performance*. London and Sterling, VA: Kogan Page Science, 2004.
2. A. F. Ibikunle, J. A. Sarumi, and E. O. Shonibare, "Comparative analysis of routing technologies in next generation converged IP network." *International Journal of Engineering and Technology* 11, no. 2 (2011): 158–169.
3. T. Janevski, *Traffic Analysis and Design of Wireless IP Networks*. Norwood, MA: Artech House, Inc., 2003.
4. H. Hanrahan, *Network Convergence: Services, Applications, Transport, and Operations Support*. England: John Wiley & Sons, Ltd., 2007.
5. J. Jo, G. Hwang, and H. Yang, "Characteristics of QoS Parameters for VoIP in the Short-Haul Internet." *Proc. International Conferences on Info-tech and Info-net (ICII), IEEE*, Beijing, China, 29 October—01 November, 2001, 498–502.
6. H. Toral, "QoS Parameters Modeling of Self-Similar VoIP Traffic and an Improvement to the E Model," Ph.D. thesis, electrical engineering, telecommunication section, CINVESTAV, Guadalajara, Jalisco, Mexico, 2010.
7. S. Kashihara, *VoIP Technologies*. Croatia: INTECH, 79–94.
8. J. Davidson, J. Peters, and B. Gracely, *Voice over IP Fundamentals*. Indianapolis, IN: Cisco Press, 2000.
9. K. I. Park, *QoS in Packet Networks*. Boston: Springer Science + Business Media, Inc., 2005.
10. J. F. Kurose and K. W. Ross, *Computer Networking: A Top-Down Approach*. Boston: Addison-Wesley, 2010.
11. A. S. Tanenbaum, *Computer Networks*. Upper Saddle River, NJ: Prentice Hall, 2003.
12. A. Ahmad, *Wireless and Mobile Data Networks*. Hoboken, NJ: John Wiley & Sons, Inc., 2005.
13. L. Cai, Y. Xiao, X. (S.) Shen, L. Cai, and J. W. Mark, "VoIP over WLAN: Voice Capacity, Admission Control, QoS, and MAC," *International Journal of Communication Systems* 19, (2006): 491–508.
14. R. Noldus, *CAMEL: Intelligent Networks for the GSM, GPRS and UMTS Network*. England: John Wiley & Sons, Ltd., 2006.

15. M. D. Yacoub, *Wireless Technology: Protocols, Standards, and Techniques*. Boca Raton, FL: CRC Press, 2002.
16. ITU-T Recommendation H.323, Packet-Based Multimedia Communications Systems, International Telecommunications Union, Geneva, Switzerland, 2007.
17. A. Sulkin, *PBX Systems for IP Telephony: Migrating Enterprise Communications*. New York: McGraw-Hill Professional, 2002.
18. J. Rosenberg, H. Schulzrinne, G. Camarillo, A. Johnston, J. Peterson, R. Sparks, M. Handley, and E. Schooler, SIP: Session Initiation Protocol (RFC 3261), Internet Engineering Task Force, 2002.
19. G. Camarillo, *SIP Demystified*. USA: McGraw-Hill Companies, Inc, 2002.
20. Z. Wang, *Internet QoS: Architectures and Mechanisms for Quality of Service*. San Francisco: Morgan Kaufmann Publishers, 2001.
21. R. Braden, D. Clark, and S. Shenker, Integrated Services in the Internet Architecture: An Overview. (RFC 1633), Internet Engineering Task Force, 1994.
22. D. Black, S. Blake, M. Carlson, E. Davies, Z. Wang, and W. Weiss, An Architecture for Differentiated Services (RFC 2475), Internet Engineering Task Force, 1998.
23. R. Braden, L. Zhang, S. Berson, S. Herzog, and S. Jamin, Resource Reservation Protocol-RSVP (RFC 2205), Internet Engineering Task Force, 1997.

INFORMATION INFRASTRUCTURE AND CLOUD COMPUTING

Chapter 14

Quality-of-Service Provisioning for Cloud Computing

Ahmed Shawish and Maria Salama

Contents

14.1 Introduction

Cloud computing has gained considerable attention in the past few years as one of the important paradigms in distributed computing and Internet technologies, where hardware and software are delivered as a service on demand through the Internet, following a simple pay-as-you-go financial model. The source of the cloud power is its ability to stand at different levels of service: (1) infrastucture-as-a-service (IaaS), where the cloud enables access to hardware resources, such as servers and storage devices; (2) platform-as-a-service (PaaS), where the cloud allows the access to software resources, such as operating systems and software development environments; and (3) software-as-a-service (SaaS), where classical software applications running on local computers are instead provided remotely by the cloud. Such service-oriented diversity makes the cloud a very promising technology, and it becomes crucial to understand and explore its methodologies as well as its quality of service (QoS) management criteria.

With the rapid growth of cloud computing, its QoS still poses significant challenges. Various QoS models have been adopted by different service providers and consumers for describing QoS attributes. This diversity led to the usage of different QoS descriptions as well as different concepts, scales, and measurements of, sometimes, the same QoS factor. Unfortunately, most of these QoS models are scattered among scientific papers, literatures, and research groups without being combined in a single source to help cloud researchers to build a new edge in this perspective based on a concrete background. Therefore, the necessity to combine all of the QoS-related factors with a unified description, scale, and metric becomes very crucial.

This chapter combines and reviews most of the quality parameters and metrics that have been innovated or improved for the cloud as well as the basic attributes for defining QoS in a well-defined ontology. This chapter also presents QoS models that were particularly developed for different service models (SaaS, PaaS, IaaS). Furthermore, service-selection approaches, quality-assurance techniques, and service-level agreements (SLAs) in cloud services are also addressed in this chapter.

The rest of this chapter is organized as follows. First, some background about the common quality attributes of both providers and consumers is discussed in Section 14.2. Section 14.3 presents the QoS ontology for unified descriptions of quality parameters. Section 14.4 presents the QoS aspects of cloud computing. In Section 14.5, QoS models for different service platforms are presented. Service-selection approaches, quality assurance, and SLAs for cloud computing are presented in Sections 14.6 and 14.7, respectively. Finally Section 14.8 elaborates on possible trends in QoS research and proposes some open research questions formulated from the material discussed in the chapter congruous to cloud computing. The whole chapter is finally concluded in the last section.

14.2 Background

Research work on the QoS models that have been developed in the context of service computing, as well as those developed for cloud computing, were primarily trending a certain perspective that emphasizes common quality attributes of both the provider and consumer.

A cloud service QoS model taking into account the resource process and service provision capabilities has been proposed, considering common QoS attributes of service to the consumer and provider [1].

As for cloud service providers, cloud service QoS provided by the physical device and virtual resources layer mainly focus on the data center's performance, reliability, and stability; cloud service QoS provided by IaaS likely emphasizes response time, resource utilization, prices, and so on. As for cloud service consumers, they are very important, such as response time, price, availability, reliability, and reputation, and they can also be provided by the service provider. Thus, considering QoS of cloud services to providers and consumers, the most common attributes of QoS are the following:

- *Response time* represents the interval from the requirement sending of cloud service consumers to cloud service implements competition.
- *Cost* represents fees paid when customers use the service provided by the cloud service provider, that is, pay-per-use.
- *Availability* represents the probability that cloud services can be accessed.
- *Reliability* shows the capacity of cloud service to accurately implement its function, and the times of validation and invalidation can be acquired by a cloud service monitor.
- *Reputation* expresses the creditability of cloud services. Reputation can be seen as the sum of subjective customer ratings and objective QoS advertising messages' credibility.

This model presents a good perspective for only starting negotiations between a provider and a consumer for SLAs. Yet more quality parameters should be considered in detail along with their measuring parameters and how to assess them. Also, going profoundly with the technical details of different service platforms of the clouds, specialized QoS models should also be discussed. Such QoS aspects and models reflect directly on the selection approaches and SLA processing. Since different service providers and consumers use different QoS descriptions as well as different concepts, scales, and measurements for QoS advertisements, requirements, and SLAs, it is necessary to find a unified QoS description.

14.3 QoS Ontology

An ontology that can provide a unified description, scale, and metric for all QoS factors would solve the diversity of concepts held between providers and consumers. Therefore, the development of such an ontology that can deal with QoS factors' interoperability becomes significantly crucial.

In this section, we present a general and flexible QoS ontology to express QoS information in a unified way, to cover all quality aspects of both service providers and consumers of different cloud platforms: IaaS, PaaS, and SaaS.

The development of the QoS ontology reports the necessity of putting clear and unified definitions for different QoS concepts, such as QoS properties, metrics, and units that are employed in QoS advertisements, QoS requirements, or SLAs. This is why ontology is needed in order to solve the interoperability of QoS. The QoS ontology offers a unified QoS description for both providers and consumers. Moreover, it solves semantic descriptions and ambiguity issues. The QoS ontology specifies how the information of service quality will be described, that is, it specifies the basic attributes that will be used to describe QoS factors in a unified and consistent manner.

In the following subsections, the QoS ontology is presented in two layers. First, the basic layer encloses the definitions of the basic attributes (e.g., QoS properties, metrics, and units) that will be utilized in defining the QoS factors. Second, the support layer presents some auxiliary means (e.g., transformation, values comparison, and others) to support the modeling of the QoS properties.

14.3.1 QoS Ontology Basic Layer

The QoS ontology basic layer encloses definitions of the basic attributes that will be employed to define all QoS factors. Such attributes depict how to define, describe, and measure any QoS factor. They are listed as follows:

- *QoS property* represents a measurable, non-functional requirement of a service within a given domain. A QoS property is a set of attributes of cloud service quality characteristics and may be refined into multiple levels of subcharacteristics, also called QoS parameter.
- *Metric* defines the way each QoS property is assigned with a value [2]. Characteristics of a metric can be simple, complex, or static.
- *Unit* allows for various types of units, their equivalence, and synonyms. Custom units should be easily added to QoS ontology [3].
- *Value type:* QoS ontology should include various data type definitions for specifying values of QoS metrics, for example, string, numeric, Boolean, list, etc. [2].
- *Constraints* permit the consumer to specify the boundary levels of the required service quality, that is, minimums or maximums. This requirement relates to the question of how to specify a constraint on a QoS property. Almost existing approaches assume simple operators, such as >, <, =, >=, =< for expressing QoS constraints. However, other operators related to value types of string, list, etc., are supported [3].
- *Mandatory:* This requirement allows specifying which QoS properties are strongly required while others may be optional [4].
- *Aggregated:* A quality property is said to be an aggregated one if it is composed from other qualities. For instance, the price performance ratio aggregates price and performance [4].
- *Levels* specify different quality levels of a service, so the most appropriate quality levels for the user demands can be chosen. This way of organization helps in creating different usage modes [2].
- *QoS dynamism:* A QoS property can be specified once (static property) or require periodically updating its measurable value (dynamic property).
- *Valid period:* In fact, the value of a QoS property is not fixed all the time. Thus, we need specify its valid period, so other parties can correctly evaluate it.
- *Weight:* QoS properties often have different importance levels according to different service consumers or providers. A float range ([0,1]) specifies the preferences toward which properties carry higher importance while others may be less important [2].
- *Impact direction* enables the system to estimate the degree of user satisfaction with regards to a given QoS parameter value by representing the way the QoS property value contributes to the service quality provided by the consumer or perceived by the user. A QoS property can have one of five impact directions: negative, positive, close, exact, and none [3].

14.3.2 QoS Ontology Support Layer

Modeling QoS properties using the previously cited attributes entailed a concrete support because of the current diversity in metrics and units for the same QoS property, the interdependence between the provider and consumer, as well as the fact of correlations between QoS properties. Such kinds of support emerge in grouping factors of similar characteristics, transformation of values and metric units, and others. The support layer defines auxiliary means to sustain modeling QoS using the basic attributes previously defined in the basic layer as follows:

- *Roles:* Besides providers' and consumers' service quality evaluation, other third-party participants, such as certificate authorities and security providers, should also be supported in a process of measurement and evaluation of QoS information.
- *Transformation:* Providers and consumers may be familiar with, use, or understand different metrics for the same QoS properties. Therefore, QoS ontology supports converting and transforming QoS metrics as well as values and units of related metrics.
- *QoS interdependence:* Not only does a service consumer require QoS from a service provider, but a service provider can also specify its QoS demands that a requester must guarantee in order to get expected QoS from executing provider service [5].
- *QoS value comparison:* Not all QoS properties have the same mechanism for comparing their values. For example, numeric-based QoS properties are compared differently from string-based ones.
- *QoS grouping:* Allow for grouping of QoS properties that share similar characteristics or impact in order to facilitate the evaluation and computing of the whole QoS value.
- *Concrete QoS:* A minimum set of common and domain-independent QoS properties should be defined as the model presented in the following section.

14.4 QoS Aspects for Cloud Computing

In this section, we present QoS aspects for cloud computing along with their measuring parameters and the assessment methods. These include performance, transparency, information assurance, security risks, and trustworthiness. Readiness, as a special aspect related to SaaS only, is also presented.

14.4.1 Performance

With respect to performance evaluation, response time, latency, and throughput factors have been considered.

14.4.1.1 Response Time and Latency

Response time and latency have been widely adopted in the distributed computing field. Defining the response time as the time a system or functional unit takes to react to a given input as well as the latency as a measure of time delay experienced in a system provide a straightforward quality measure for the cloud service. Other instances of response time are now considered for cloud computing, such as percentile of the response time.

The response time from the perspective of the customer is more inclined to require a statistical bound on its response time than an average response time [6]. For instance, a customer can request that 95% of the time its response time should be less than a given value. Therefore, that research is concerned with a percentile of the response time. That is, the time to execute a service request is less than a predefined value with a certain percentage of time [6]. This metric has been used by IBM's researchers. The metric was named "percentile delay" by scientists at Cisco and MIT Communications Future Program [7].

Assume that $f_T(t)$ is the probability distribution function of a response time T. T_D is a desired target response time that a customer requests and agrees to with the service provider based on a fee paid by the customer. The SLA performance metric that a $\gamma\%$ SLA service is guaranteed is as follows:

$$\int_0^{T^D} f_T(t)\,dt \geq \gamma\% \tag{14.1}$$

That is, $\gamma\%$ of the time a customer will receive his or her service in less than T_D.

The calculation of the percentile of response time plays a key role in answering the following performance questions:

■ For a given arrival rate of service requests, service rates at the web server and the service center, what level of QoS services can be guaranteed?
■ What are the minimal service rates required at the web server and the service center, respectively, so that a given percentile of the response time can be guaranteed for a given service arrival rate from customers?
■ How many customers can be supported so that a given percentile of the response time can be still guaranteed when service rates are given at the web server and the service center, respectively?

14.4.1.2 Throughput

The throughput is also considered as an important performance factor in a wide range of applications, such as content delivery networks [8,9], where the service deployment is optimized based on average throughput as well as its utility being measured as the fraction of processed requests (throughput) or the total evaluation (weighted throughput).

MetaCDN, which is an integrated overlay that utilizes cloud computing to provide content delivery services to Internet end users, allows users to deploy files either directly (uploading a file from their local file system) or from an already publicly accessible origin website (side-loading the file, where the backend storage provider pulls the file). Given that not all providers support side-loading, the MetaCDN system can perform this feature on behalf of the user and subsequently upload the file manually. When accessing the service via the web portal or web services, MetaCDN users are given a number of different deployment options depending on their needs, including the following:

■ Maximize coverage and performance, where MetaCDN deploys as many replicas as possible to all available providers and locations.
■ Deploy content in specific locations, where a user nominates regions and MetaCDN matches the requested regions with providers that service those areas.
■ Cost-optimized deployment, where MetaCDN deploys as many replicas in the locations requested by the user as their transfer and storage budget will allow.
■ QoS-optimized deployment, where MetaCDN deploys to providers that match specific QoS targets that a user specifies, such as average throughput or response time from a particular location, which is tracked by persistent probing from the MetaCDN QoS monitor.

Once a user deploys using the options above, all information regarding the deployment is stored in the MetaCDN database, and the user is returned a set of publicly accessible URLs pointing to the replicas deployed by MetaCDN and a single MetaCDN URL that can transparently redirect end users to the best replica for their access location.

14.4.2 Transparency

An empirical evaluation of cloud provider transparency has been conducted [10]. The purpose of this study was to develop an instrument for evaluating a cloud provider's transparency of security, privacy, and service level competencies via its self-service web portals and web publications and then to empirically evaluate cloud service providers to measure how transparent they are by using the instrument. The instrument is referred to as the cloud provider transparency scorecard (CPTS). CPTS includes a section on the offer for preassessment, security policies and procedures, privacy policies and procedures, and service levels. The methodology developed segmented the evaluation into four key domains: security, privacy, audit and service level agreement. Each domain included a series of questions based on key areas outlined by the Cloud Security Alliance (CSA), NIST, and the European Network and Information Security Agency (ENISA). Each question equated to a 0 = no, 1 = yes value with each domain totaled and an overall score based on the total of all the scores. The domain-based scores were then divided by the total possible to provide a simple percentile equivalent. The total score was also divided by the total score possible to derive a percentile equivalent. The methodology provides the evaluator a simple method to compare the differences between cloud providers based on each domain and an overall score.

Table 14.1 illustrates the preassessment and Table 14.2 shows the full assessment.

Table 14.1 Transparency Instrument: Preassesment

Cloud Provider Transparency Scorecard Preassessment			
Business Factors	1	Length in business in years > 5?	0 = <5, 1 = >5
	2	Published security or privacy breaches?	0 = Y, 1 = N
	3	Published outages?	0 = Y, 1 = N
	4	Published data loss?	0 = Y, 1 = N
	5	Similar customers?	0 = Y, 1 = N
	6	Member of ENISA, CSA, CloudAudit, OCCI, or other cloud standards groups?	0 = Y, 1 = N
	7	Profitable or public?	0 = Y, 1 = N
		Preassessment total score	Total
		Percentile score	Score/7

Table 14.2 Transparency Instrument: Full Assesment

Cloud Provider Transparency Scorecard Full Assessment		
Security	1	Portal area for security information?
	2	Published security policy?
	3	White paper on security standards?
	4	Does the policy specifically address multitenancy issues?
	5	Email or online chat for questions?
	6	ISO/IEC 27000 certified?
	7	COBiT certified?
	8	NIST SP800-53 security certified?
	9	Offer security professional services (assessment)?
	10	Employees CISSP, CISM, or other security certified?
		Security subtotal score
Privacy	11	Portal area for privacy information?
	12	Published privacy policy?
	13	White paper on privacy standards?
	14	Email or online chat for questions?
	15	Offer privacy professional services (assessment)?
	16	Employees CIPP or other privacy certified?
		Privacy subtotal score
External Audits or Certifications	17	SAS 70 type II
	18	PCI-DSS
	19	SOX
	20	HIPAA
		Audit subtotal score
Service-Level Agreements	21	Do they offer an SLA?
	22	Does the SLA apply to all services?
	23	99.9 = 1, 99.95 = 2, 99.99 = 3, 99.999 = 4, 100 = 5
	24	ITIL certified employees?
	25	Published outage and remediation?
		SLA subtotal score
		Total score

14.4.3 Information Assurance

An organization adopting cloud-computing services should assure itself that it is sufficiently protecting the information entrusted to the cloud provider. An information assurance framework has been developed as a set of assurance criteria designed to assess the risk of adopting cloud services and to compare different cloud provider offers [11]. The framework provides a set of questions that an organization can ask a cloud provider. These questions are intended to provide a minimum baseline; the customer may, therefore, have additional specific requirements not covered within the baseline.

The information assurance framework covers the following security aspects:

- Personnel security
- Supply-chain assurance
- Operational security
- Identity and access management
- Asset management
- Data and service portability
- Business continuity management
- Physical security
- Environmental controls
- Legal requirements

14.4.4 Security Risks

Both security benefits and risks resulting from using cloud computing should be considered as an essential aspect when considering the QoS, covering the technical, policy, and legal implications [12].

Risk should always be understood in relation to the overall business opportunity and appetite for risk—sometimes risk is compensated by opportunity. The level of risk will, in many cases, vary significantly with the type of cloud architecture being considered.

The risks identified in the assessment are classified into three categories:

- Policy and organizational
- Technical
- Legal

Each risk is presented in tables, which include the following:

- Probability level
- Impact level
- Reference to vulnerabilities
- Reference to the affected assets
- Level of risk

Policy and organizational risks include the following:

- *Lock-in:* The extent and nature of lock-in varies according to the cloud type. SaaS application lock-in is the most obvious form of lock-in. SaaS providers typically develop a custom application tailored to the needs of their target market. SaaS customers with a large user base can incur very high switching costs when migrating to another SaaS provider as the end-user experience is impacted (e.g., retraining is necessary). Where the customer has developed programs to interact with the providers' API directly (e.g., for integration with other applications), these will also need to be rewritten to take into account the new provider's API. PaaS lock-in occurs at both the API layer (i.e., platform-specific API calls) and at the component level. For example, the PaaS provider may offer a highly efficient back-end data store. Not only must the customer develop code using the custom APIs offered by the provider, but they must also code data access routines in a way that is compatible with the back-end data store. This code will not necessarily be portable across PaaS providers, even if a seemingly compatible API is offered as the data-access model may be different. IaaS lock-in varies depending on the specific infrastructure services consumed. For example, a customer using cloud storage will not be impacted by non-compatible virtual machine formats. IaaS storage provider offerings vary from simplistic key- or value-based data stores to policy-enhanced, file-based stores.
- *Loss of governance:* The loss of governance and control could have a potentially severe impact on the organization's strategy and, therefore, on the capacity to meet its mission and goals. The loss of control and governance could lead to the impossibility of complying with the security requirements; a lack of confidentiality, integrity, and availability of data; and a deterioration of performance and QoS.
- *Compliance challenges:* Certain organizations migrating to the cloud have made considerable investments in achieving certification either for competitive advantage or to meet industry standards or regulatory requirements.
- *Loss of business reputation as a result of cotenant activities:* Resource sharing means that malicious activities carried out by one tenant may affect the reputation of another tenant. The impact can be deterioration in service delivery and data loss as well as problems for the organization's reputation.
- *Cloud service termination or failure:* Competitive pressure, an inadequate business strategy, or lack of financial support could lead some providers to go out of business or, at least, to force them to restructure their service portfolio offering. The impact of this threat for the cloud customer is easily understandable because it could lead to a loss or deterioration of service delivery performance and QoS as well as a loss of investment. Failures in the services outsourced to the provider may have a significant impact on the cloud customer's ability to meet his/her duties and obligations to his/her own customers. Failures by the cloud provider may also result in liability by the customer to its employees.
- *Cloud provider acquisition:* Acquisition of the cloud provider could increase the likelihood of a strategic shift and may impose non-binding agreements at risk. This could make it impossible to comply with the security requirements.
- *Supply chain failure:* When outsourcing certain specialized tasks to third parties, the level of security of the cloud provider may depend on the level of security of each one of the links and the level of dependency of the cloud provider on the third party. Any interruption or corruption in the chain or a lack of coordination of responsibilities between all the involved

parties can lead to services unavailability loss of data confidentiality, and integrity; economic and reputational losses resulting from failure to meet customer demand; violation of SLA; and cascading service failure.

Technical risks include the following:

- *Resource exhaustion (under or over provisioning):* From the cloud customer's perspective, a poor provider selection and lack of supplier redundancy could lead to service unavailability, a compromised access control system, as well as economic and reputational losses.
- *Isolation failure:* This class of risks includes the failure of mechanisms separating storage, memory, routing, and even reputation between different tenants of the shared infrastructure (e.g., so-called guest-hopping attacks, SQL injection attacks exposing multiple customers' data stored in the same table, and side channel attacks). The impact can be a loss of valuable or sensitive data, reputation damage, and service interruption for cloud providers and their clients.
- *Cloud provider malicious insider:* The malicious activities of an insider could potentially have an impact on the confidentiality, integrity, and availability of all kinds of data, IP, and all kinds of services and, therefore, indirectly on the organization's reputation, customer trust, and the experiences of employees. This can be considered especially important in the case of cloud computing because of the fact that cloud architectures necessitate certain roles, which are extremely high risk.
- *Management interface compromise (manipulation, availability of infrastructure):* The customer management interfaces of public cloud providers are Internet accessible and mediate access to larger sets of resources and, therefore, pose an increased risk, especially when combined with remote access and web-browser vulnerabilities. This includes customer interfaces controlling a number of virtual machines and, most importantly, CP interfaces controlling the operation of the overall cloud system.
- *Intercepting data in transit:* Cloud computing, being a distributed architecture, implies more data in transit than traditional infrastructures. For example, data must be transferred in order to synchronize multiple distributed machine images, images distributed across multiple physical machines, between cloud infrastructure and remote web clients.
- *Data leakage on upload or download intracloud:* This risk applies to the transfer of data between the cloud provider and the cloud customer.
- *Insecure or ineffective deletion of data:* When a request to delete a cloud resource is made, this may not result in true wiping of the data. Where true data wiping is required, special procedures must be followed, and this may not be supported by the standard API. If effective encryption is used, then the level of risk may be considered to be lower.
- *Distributed Denial of Service (DDoS):* This risk directly affects a customer's reputation, trust, and service delivery, especially in real time.
- *Economic Denial of Service (EDoS):* This happens when a cloud customer's resources may be used by other parties in a malicious way that has an economic impact. EDoS destroys economic resources; the worst-case scenario would be the bankruptcy of the customer or a serious economic impact.
- *Loss of encryption keys:* This includes disclosure of secret keys (SSL, file encryption, customer private keys, etc.) or passwords to malicious parties, the loss or corruption of those keys, or their unauthorized use for authentication and nonrepudiation (digital signature).

- *Undertaking malicious probes or scanning:* Malicious probes or scanning are indirect threats and can be used to collect information in the context of a hacking attempt. A possible impact could be a loss of confidentiality, integrity, and availability of service and data.
- *Compromise service engine:* Malicious probes or scanning, as well as network mapping, are indirect threats to the assets being considered. They can be used to collect information in the context of a hacking attempt. A possible impact could be a loss of confidentiality, integrity, and availability of service and data. An attacker can compromise the service engine by hacking it from inside a virtual machine (IaaS clouds), the runtime environment (PaaS clouds), the application pool (SaaS clouds), or through its APIs. Hacking the service engine may be useful to escape the isolation between different customer environments (jailbreak) and gain access to the data contained inside them to monitor and modify the information inside them in a transparent way (without direct interaction with the application inside the customer environment) or to reduce the resources assigned to them, causing a denial of service.
- *Conflicts between customer hardening procedures and the cloud environment:* The failure of customers to properly secure their environments may pose a vulnerability to the cloud platform if the cloud provider has not taken the necessary steps to provide isolation. Cloud providers should further articulate their isolation mechanisms and provide best-practice guidelines to assist customers in securing their resources. Customers must realize and assume their responsibility as failure to do so would place their data and resources at further risk.

Legal risks include the following:

- *Risks from changes of jurisdiction:* Customer data may be held in multiple jurisdictions, some of which may be high risk. If data centers are located in high-risk countries, sites could be raided by local authorities and data or systems subject to enforced disclosure or seizure. High-risk countries can include those lacking the rule of law and having an unpredictable legal framework and enforcement, autocratic police states, or states that do not respect international agreements.
- *Data protection risks:* Several data protection risks are posed on the customer. It can be difficult for the cloud customer to effectively check the data processing that the cloud provider carries out and thus be sure that the data is handled in a lawful way. There may be data security breaches that are not notified to the controller by the cloud provider. The cloud customer may lose control of the data processed by the cloud provider. This issue is increased in the case of multiple transfers of data.
- *Licensing risks:* Licensing conditions, such as per-seat agreements, and online licensing checks may become unworkable in a cloud environment.

14.4.5 Trustworthiness

The problem of assessing the trustworthiness of software service has been addressed recently in the cloud context as one of the important quality attributes [13].

The trustworthiness of software service providers is commonly measured by their reputations. Reputation systems not only record and track providers' behavior but create an incentive for good behavior by providing consumers with some control over market quality. In order for a reputation mechanism to be fair and objective, it is essential to compute reputation on the basis of fair and objective feedback.

An automated rating model, based on the expectancy-disconfirmation theory from market science, has been defined to overcome feedback subjectivity issues. The goal of the rating function is to provide objective feedback on a delivered service without human intervention. Feedback is a measure of a user's satisfaction with the service. Thus, it is necessary that the automated rating process provide feedback that corresponds to the level of satisfaction or dissatisfaction with service delivery. Consumer satisfaction is the outcome of the comparison between consumers' *preconsumption expectation* and *postconsumption disconfirmation*, where confirmed expectations lead to moderate satisfaction, positively disconfirmed (i.e., exceeded) expectations lead to high satisfaction, and negatively disconfirmed (i.e., under-achieved) expectations affect satisfaction more strongly than positive disconfirmation and lead to dissatisfaction. Rating is computed using a single scalar metric to quantify quality perception. This metric is basically the utility function of the delivered service. Utility expresses the conformance of service execution quality to the agreement. The utility function can be considered as the distribution function of the probability that the observed quality meets the agreed quality level during service execution. Thus, the utility function can be estimated from quality monitoring results.

Reputation is computed on the basis of past feedbacks. It helps consumers predict the credibility of the service offer and the trustworthiness of the service provider prior to reaching an agreement. Past feedback reflects the past behavior of a service and may give an indication of its future behavior; feedback may be randomly distributed when a service's behavior is not deterministic; they may follow a trend, for example, increasing feedback may reflect an improvement in service quality, or they may be cyclic when there is a periodicity in a service's behavior, for example, quality may decrease in rush hours leading to low feedback at those times. When the feedback does not show any trend, it is difficult to predict a service's future behavior. However, when feedback exhibits a trend, this should be taken into account by the reputation function as it could help to predict future behavior.

A *feedback forecasting* model translates service execution quality into feedback, so that any quality monitoring system can be enhanced with the rating function. The model is illustrated in Figure 14.1.

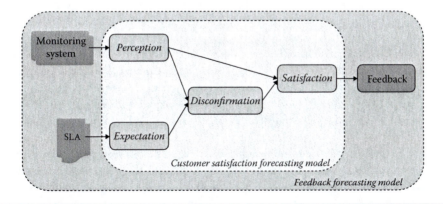

Figure 14.1 Feedback forecasting model.

14.4.6 SaaS Readiness

Another model for assessing the readiness of SaaS vendors specifically has also been introduced [14].

The following QoS model focused on enterprise software sourcing (SaaS) aimed at assessing the readiness of the SaaS vendors to deliver the service.

Vendor readiness was assessed based on three criteria:

1. Vendors' interpretation of SaaS as either disruptive or evolving innovation and implications of this assessment on their business processes.
2. Lessons learned from the ASP history.
3. Vendors' capabilities in terms of strength based on a new framework consisting of seven fundamental capabilities: the seven Fundamental Organizational Capabilities (FOCs) model.

The FOCs model has been developed, aiming at presenting a holistic and generic approach. It includes a parsimonious yet comprehensive set of organizational capabilities that are essential to all organizations regardless of size, industry, type of organization (e.g., for-profit or not), and phase in its life cycle (e.g., a start-up or an established organization). Furthermore, all seven capabilities are required for each organizational activity, internal or market-oriented, although their relative strength may change according to the circumstances.

This framework was based on the following concept of organizational fundamental capabilities. An organizational capability is defined as "a high-level routine (or collection of routines) that, together with its implementing input flows, confers upon an organization's management a set of decision options for producing significant outputs of a particular type." This definition pertains to organizational routines required to convert inputs to significant outputs, yet it is silent about the ability to consistently achieve the process goals, namely to maintain a reliable process.

The FOC framework defines seven organizational capabilities as fundamental to any organization in all industries for achieving solid performance, as illustrated in Figure 14.2. These capabilities are

1. Sensing the stakeholders
2. Sensing the business environment
3. Sensing the knowledge environment
4. Process control
5. Process improvement
6. New process development
7. Appropriate resolution capabilities

A process control requires examination of the process outputs (feedback), making appropriate decisions about the implications of the feedback, and implementation of a required intervention in the process. Accordingly, the seven FOCs are grouped into three categories: sensing capabilities required for the feedback phase, intervention capabilities, and a resolution capability.

Three sensing capabilities are proposed: sensing the stakeholders, sensing the business environment, and sensing the knowledge environment. Similarly, three intervention capabilities are outlined: process control, process improvement, and new process development. Finally, an appropriate resolution capability is linking and bridging the sensing and the intervention capabilities. All capabilities are embedded in each organizational process, which starts with inputting resources into the process and culminates in distributing the outputs back to the stakeholders while applying knowledge, interventions, and appropriate resolutions during the various subprocesses required during the conversion of inputs into more valuable significant outputs.

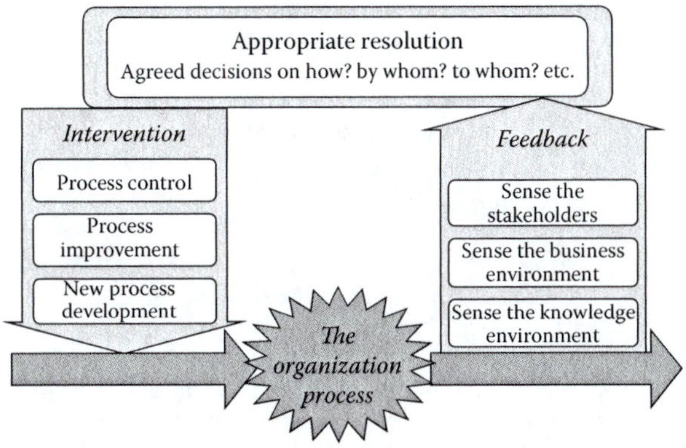

Figure 14.2   FOCs model.

Table 14.3 presents definitions of the terminology used, and the FOCs are defined in Table 14.4. Consequently, answers to the following research questions were pursued in this study:

1. What are the vendors' perceptions of SaaS as an innovation?
2. What have SaaS vendors learned from the past ASP experience?
3. What are the vendors' perceptions of the capabilities required to thrive in the SaaS market?

Table 14.3   Definitions of Terminology Used in the Model

Term	Definition
Organization	A temporary agreement among stakeholders to cooperate in conducting a certain process according to some rules under certain conditions
Stakeholders	"Any group or individual who can affect, or is affected by, the achievement of the organization's objectives" (e.g., stockholders, employees, customers, suppliers, community)
Organizational process	A mechanism that converts assets and capabilities, inputted by a collaborating group of stakeholders, into outputs that are redistributed to the same stakeholders
Business environment	The group of all organizations who can exchange stakeholders with the organization
Knowledge environment	The total knowledge that exists in the organization's environment and can be accessed by the organization
Fundamental capability	An organizational capability that meets all three criteria: • It exists (to a certain level) in each organization and every organizational process. • It exists throughout the organization's lifecycle. • All other organizational capabilities are based on it.

Table 14.4 Definitions of FOCs

Capability	Definition
Sensing the stakeholders	The ability to identify the stakeholders and to understand and map their requirements and expectations from the organization
Sensing the business environment	The ability to identify, understand, and map risks and opportunities that are relevant to the organization
Sensing the knowledge environment	The ability to identify, understand, and map opportunities to acquire relevant organizational knowledge
Appropriate resolution	The ability of the stakeholders to reach an agreed upon resolution concerning the sources of the inputs, its conversion processes, and the redistribution of the outputs back to the stakeholders
Process control	The ability to achieve and sustain a desirable standard in the organizational process and in each of its subprocesses
Process improvement	The ability to increase the output to each unit of input in the organizational process and in each of its subprocesses
New process development	The ability to develop new processes as part of the organizational process or its subprocesses, whose output per unit of input is significantly higher than in the existing processes in the organization or its accessible environment

Examination of these questions can shed light on vendors' readiness to successfully deliver SaaS.

14.5 QoS Models for Cloud Platforms

QoS models have been developed for SaaS only focusing on quality properties in the software development while others focused on IaaS and PaaS by considering computing resource performance.

14.5.1 QoS Models for SaaS

For a SaaS cloud service model, a model for high-quality SaaS cloud service has been developed focusing on several quality properties in the development process [15]. This QoS model has been developed for high-quality SaaS cloud services. The model focused on reusability, availability, and scalability quality properties only in the development process. The model defined two main design criteria for SaaS cloud services. One criterion is to reflect the intrinsic *characteristics* of SaaS. Because every well-defined development process should reflect key characteristics of its computing paradigm, this is considered as the main criterion. The other criterion is to promote SaaS with the *desired properties*, which are defined as the requirements that any SaaS should embed in order to reach a high level of QoS.

Key characteristics and desired properties of SaaS are defined in Table 14.5.

Table 14.5 Characteristics and Desired Attributes of SaaS

Characteristics	Desired Properties
• Supporting commonality • Accessible via the Internet • Providing complete functionality • Supporting multitenants' access • Thin client model	• High reusability • High availability • High scalability

The key characteristics are described as follows:

■ *Supporting commonality:* As an extreme form of reuse approaches, a SaaS provides software functionality and features that are common among and so reused by potentially a number of service consumers. Services with high commonality would yield high profits and return on the investment (ROI).

■ *Accessible via the Internet:* All the current reference models of cloud computing assume that cloud services deployed are accessed by consumers through the internet.

■ *Providing complete functionality:* SaaS provides the whole functionality of certain software in the form of service. This is in contrast to a mash-up service, which provides only some portion of the whole software functionality.

■ *Supporting multitenants' access:* SaaS deployed on the providers' side is available to the public. And a number of service consumers may access the services at the given time without advanced notices. Hence, SaaS should be designed in a way to support concurrent accesses by multiple tenants and handle their sessions in isolation.

■ *Thin client model:* SaaS services run on the providers' side, and service consumers use browsers to access the computed results. Moreover, consumer-specific data sets that are produced by running SaaS are stored and maintained on the providers' side. Hence, there will be nothing on the browser-like user interaction tool installed and run on the client or consumer side.

The desired properties are described as follows:

■ *High reusability:* Service providers develop and deploy cloud services and expect that the services would be reused by a large number of consumers. Services that are not very reusable by consumers would lose the justification for investment, and services that can be reused by many consumers would return high enough on the investment. Therefore, it is highly desirable for cloud services to embed a high level of reusability.

Commonality is the main contributor to reusability. That is, cloud services with high commonality will yield higher reusability. Variability is a minor difference among applications or consumers within a common feature.

Commonality in SaaS: Commonality, in general, denotes the amount of potential applications that need a specified feature, such as a component or a service. To derive the common features that will be realized in SaaS, the relationships among requirement-related elements is defined as shown in Figure 14.3.

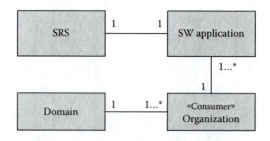

Figure 14.3   Criteria on commonality determination.

Domains such as finance and telecommunications consist of several organizations, and an organization needs one or more software applications. Hence, a commonality of a feature can be computed as follows:

$$\text{Commonality}(i) = N/T, \tag{14.2}$$

where N expresses the number of applications needing the feature i, and T denotes the total number of target applications.

If every application in the domain needs the given feature, the value of commonality will be 1. It would be desirable to include features with high commonality into the target SaaS. The range of the metric is between 0 and 1.

Variability in SaaS: The notion of variability is represented in SaaS using three aspects: persistency, variability type, and variability scope. In SaaS, there can be places where variability occurs, called the variation point. Each variation point is associated with a set of values that can fill in, called variants.

- Persistency of variant settings: Variability-related elements can have three different persistencies: no persistency, permanent persistency, and near-permanent persistency.

 A variable in a program can hold a value at a given time, and its value can be changed as a new value is assigned. Hence, the value stored in a variable does not have an intended persistency, which is indicated as *no persistency.*

 A variation point is a means for users to set a valid variant, and a variant set in the variation point is persistent, that is, "once set not changed." This level of persistency is called *permanent persistency.*

 A variation point in SaaS adds *near-permanent persistency*, where the variant set in a variation point may be changed over time but in a limited way. SaaS is for potentially many consumers, and it has to consider the consumer-specific context and environment in running SaaS application. The consumer may change its context, such as current location/time zone, and various units used, such as currency. Once a variant is set, its value must be stored until a new variant is set within a session or across multiple sessions. Hence, the persistency of this variant is *near-permanent.* For example, SaaS consumers using mobile-internet devices often travel, and their locations get changed and noticed by the SaaS application. Then, SaaS may set new variants for the new location and provide services with the right mobile network protocol and its associated services and contents.

 In SaaS, only the second and third types of variant persistency are considered.

- *Variability types:* The variability embedded in a variation point can be classified into several types: attribute, logic, workflow, interface, and persistency. In addition to these, two additional types are defined for SaaS: context and QoS. Variability can occur on the consumer's current context, such as location and time zone, which is called context variability type. For the same SaaS, different consumers may require different levels of QoS attributes. Hence, variability can also occur on the QoS required by each consumer, which is called QoS type.
- *Variability scope:* Variability scope is the range of variants that can be set into a variation point. Typically, there are three scopes: binary, selection, and open. When service consumers want to customize SaaS for their requirements, they can choose a variant in binary or selection scopes. However, there is a different implication of open scope in SaaS. Because of the thin-client characteristic, service consumers have a limitation on implementing and containing plug-in objects as variants on their client device.
- ■ *High availability:* Cloud services are not just for specific users; rather they are for any potential unknown consumers who may wish to use the services anytime and anywhere. Therefore, it is highly desirable for the service to be highly available if not always available. Services with low availability would cause inconvenience and negative business impacts to consumers, and, as a result, they will suffer on reliability and reputations.
- ■ *High scalability:* In cloud computing, the amount of service requests from consumers, that is, service load, is dynamic and hard to predict. Therefore, cloud services should be highly scalable even in the situation when an extremely high number of service invocations and their associated resource requests are requested. Services with low scalability would suffer at the time of peak requests and so lose their reputations by consumers.

14.5.2 QoS Models for PaaS and IaaS

Regarding IaaS and PaaS, QoS models were mainly concerned with the management and performance of resources [16].

Resource management encompasses the dynamic allocation of tasks to computational resources and requires the use of a scheduler (or broker) to guarantee performance. QoS is enabled by the efficient scheduling of tasks; this guarantees the resource requirements of an application are strictly supported, but resources are not over provisioned and used in the most efficient manner possible. Sequences of tasks are represented as workflows, directed graphs comprised of precedent constrained nodes, which each represent the specific ordered invocation of a service on computation resources to process a given task.

Monitoring tools are essential in ascertaining the availability of resources and providing feedback to schedulers. Monitoring tools enable guarantees to be made on the performance of any given resource by making sure that the computational resource in question is not over utilized and is online. Performance is characterized by the amount of useful work accomplished by a computer system in comparison to the time and resources used. Monitoring tools are also essential in providing fault tolerance and the migration of tasks in the event of a resource failure. Fault tolerance involves the identification of a resource failure via monitoring tools, the rescheduling of the task to an alternative available resource, and migration of the state of the task to the newly allotted resource, at which point the task continues execution. The state of a task in execution must be regularly saved for fault tolerance to function; this process is known as check pointing.

The current state-of-the-art technology in cloud computing centers on the virtualization of resources at the lowest level. The main technology enabling virtualization is the hypervisor, a Virtual Machine Manager (VMM) that partitions a physical host server transparently via emulation or hardware-assisted virtualization. This provides a complete simulated hardware environment, known as a virtual machine, in which a guest operating system can execute in complete isolation. There are several benefits of utilizing virtual machines. Hardware can be consolidated when several servers are underutilized and provisioned as needed, endowing a organization with reductions in the up-front cost of hardware purchases, and virtual machines can be migrated from one physical location to another with ease as the need arises. There are no such limitations on the availability of software that can be installed into virtual machine images.

There are five types of virtualization:

■ *Full virtualization* involves simulating enough hardware to allow an unmodified guest operating system to run in isolation at a considerable performance penalty because of the overhead associated with emulating hardware.

■ *Hardware-assisted virtualization* utilizes the additional hardware capabilities, in the form of additional Virtual Machine Extensions (VMXs) within the host processor instruction set, to accelerate and isolate context switching between processes running in different virtual machines. This increases the computational performance of a virtual machine as instructions can be directly passed to the host processor without having to be interpreted and isolated at the expense of limiting guest operating systems to using the same instruction set as the host machine.

■ *Partial virtualization* involves the simulation of most but not all the underlying hardware of host and supports resource sharing but does not isolate guest operating system instances. This basic approach is utilized in paravirtualization, hybrid virtualization, and operating system–level virtualization.

■ *Paravirtualization* simulates all or most hardware by providing software interface or APIs that are similar to that provided for the underlying hardware of the host. These can be utilized to create hardware device drivers for guest operating systems that achieve near native performance to that of the host. The downside of this approach is that the operating system must be modified to run on paravirtualized VMMs.

■ *Hybrid virtualization* combines the principles of both hardware-assisted virtualization and paravirtualization to obtain near native performance from guest operating systems but with the disadvantages of both. Although these disadvantages prevent the consolidation of an organization's current hardware, they do provide an excellent foundation for the creation of new cloud-based systems, reducing the number of physical machines needed at peak demand and thus hardware running and setup costs. Most VMMs support multiple types of virtualization, so the disadvantages can be somewhat mitigated.

■ *Operating system–level virtualization* is achieved through multiple isolated user space instances. A disadvantage of this virtualization technique is that the guest operating system of the virtual machine must be the same as the host, but the guests run at native performance.

14.6 Selection Approaches in Clouds

Selection approaches have been addressed in the context of cost–benefit analysis or focusing only on the calculation of the cost metric and estimating the value of the cloud in terms of opportunity costs. Meanwhile, these approaches are not only used for service selection but might also be used for service composition and resource allocation based on QoS. Still, such different directions and purposes in the selection approaches never deny the fact of the need of QoS levels' formalization.

14.6.1 QoS Levels' Formalization

The formalization of QoS levels is a main precondition for the negotiation of SLAs and QoS-based selection approaches, dealing with different QoS parameters as Non-Functional Properties (NFPs) and their enforcement in cloud computing environments [17].

Figure 14.4 shows the structure for formalization of business process Service-Level Objectives (SLOs) and technical service capabilities. The figure also contains sample service-level statements about response time, throughput, and transaction rate. These statements are then stored with the service description (service capabilities), respectively, with the business process definition (business process SLO) and are the starting point for the negotiation of SLAs. An example for a statement about the service capability is, "The transaction rate of the service is higher than 90 transactions per second in 98% of the cases as long as throughput is higher than 500 kb/s." An example for a requirement business process level is, "The transaction rate of the process should be higher than 50 transactions per second in 97% of the cases while throughput is higher than 500 kb/s."

A successful service offering has two main objectives: to provide the needed functionality and to provide the needed QoS. QoS parameters are part of the run-time related NFPs of a service. Run-time related NFPs are performance oriented (e.g., response time, transaction rate, availability) and can change during run time—when times of extensive concurrent usage by many users are followed by times of rare usage or when failures occur.

Performance is to observe the system output $\omega(\delta)$ that represents the number of successfully served requests (or transactions) from a total of input $\iota(\delta)$ requests during a period of time.

$$\omega(\delta) = f(\iota(\delta)). \tag{14.3}$$

- This definition of performance corresponds to transaction rate as NFP—the system guarantees to process n requests during time period t.
- Dependability integrates several attributes: availability, reliability, safety, integrity, and maintainability. These are defined as follows:
- Availability denotes the readiness to provide a correct service.
- Reliability denotes the continuity of service provision.

	NFP	Predicate	Metric (value, unit)	Percentage		Qualifying conditions (QC)		
SLO pattern						NFP	Predicate	Metric
	Response time	less than	100 ms	in 95% of the cases	**if**	transaction rate	less than	10 tps
SLO examples	Throughput	higher than	1000 kb/s	in 95% of the cases	**if**	transaction rate	less than	10 tps
	Transaction rate	higher than	90 tps	in 98% of the cases	**if**	throughput	higher than	500 kb/s

Figure 14.4 Structure of service levels.

- Safety is an attribute that assures there are no catastrophic consequences on the user and the environment.
- Integrity denotes that there are no improper changes to the system.
- Maintainability denotes that a system can undergo changes and repairs.

14.6.2 Cost–Benefit Analysis

On the other hand, cost–benefit analysis has been also conducted on the cloud computing services. A comparison between clouds and grids has been conducted in terms of performance and resource requirement in the context of cost–benefits [18].

As for server hosting on a cloud, it is advantageous for variable workloads as the infrastructure can scale with rapid increases (or decreases). Moreover, costs are variable (and, in total, less than the fixed costs). Cloud computing is effective for small to medium-sized applications. For large projects, the costs are simply too high to host on a cloud. For example, with the Folding@home project, the storage requirements are about 500 TB. If stored on Amazon's S3, the 500 TB of Folding@home's data would cost more than $50,000 per month. Moreover, Folding@home data analysis requires data access and manipulation, so the potential costs of inbound and outbound data transfers from S3 would make the estimate even higher. In comparison to Folding@home, SETI@home and XtremLab have fewer demands in terms of the computing infrastructure.

In general, the SETI@home server uses about 3 TB of storage and 100 Mb/s of bandwidth and has modest IO rates. The server serves about 318,380 active clients. The scheduler and download outbound data transfer rate is much greater than the inbound. The download throughput is constrained by a 100-Mb limit. The XtremLab server uses about 65 GB of storage, 11 kb/s of bandwidth, and very light IO rates. The server serves about 3000 active clients. Monthly resource usage of those projects is shown in Table 14.6.

Figure 14.5a shows the costs for SETI@home on Amazon's cloud. In total, it would cost about $7000 per month to host the SETI@home server on cloud. The majority of costs (~60%) are a result of bandwidth alone. About 25% of costs are a result of CPU time of the six instances.

Table 14.6 Projects Resource Usage

Component	SETI@home Project	XtremLab Project
Upload (result) storage	200 GB	Negligible
Download (work unit) storage	2500 GB	0.14 GB
Database storage	200 GB	1 GB
Science results/database storage	1000 GB	64 GB
Scheduler throughput	6 Mb/s outbound	Negligible
Upload throughput (peak)	10 Mb/s inbound	9.3 kb/s
Download throughput (peak)	92 Mb/s outbound	1.7 kb/s
IO transactions	141.9 million	Negligible

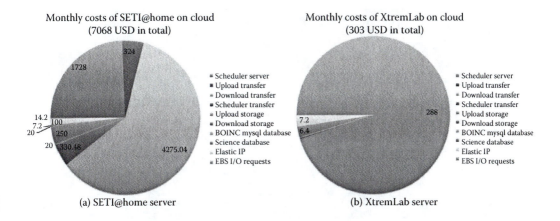

Figure 14.5   Server costs on cloud.

Nevertheless, the cloud costs are less than 60% of SETI@home's current costs. So surprisingly, clouds may be cost-effective for even a large project, such as SETI@home. However, one has to consider that staff costs for maintenance, etc., are variable across projects, and that we assume these costs would be subsumed by the cloud. Figure 14.5b shows the costs for XtremLab. Monthly costs amount to about $300 per month. Ninety-five percent of costs are a result of the CPU time of the instance. The cloud costs are about 6% of XtremLab's standalone costs. Clearly, clouds are advantageous for smaller, less bandwidth-intensive projects.

It has been concluded that cost efficiency varies depending on the platform size, where a minimum number of nodes are sufficient for a grid to become cost-effective for a short period of time while a cloud of the same size is more efficient to support a long-term scientific project.

14.6.3 Cost Calculation

Calculating a cost metric and estimating the value of the cloud in terms of opportunity costs have been proposed by developing a basic framework for estimating value and determining benefits from cloud computing [19].

Valuation is an economic discipline of estimating the value of projects and enterprises. The approach to estimating the value of cloud computing services uses ideas from relative valuation with market comparables by calculating a *cost metric* instead of a *valuation metric* and then estimating the value in terms of opportunity costs. Thereby, one can valuate the benefits that cloud computing services provide over the best alternative technology.

Figure 14.6 illustrates our framework for estimating the value of IaaS offerings.

■ Step 1: Business Scenario
In a first step, a business scenario is modeled in order to evaluate the use of cloud computing services within the scope of a specific project. Decision makers create a matrix with available cloud computing service solutions in the rows and a set of scenario criteria in the columns. The matrix can be evaluated like a checklist that outputs the most suitable cloud computing services for the scenario.

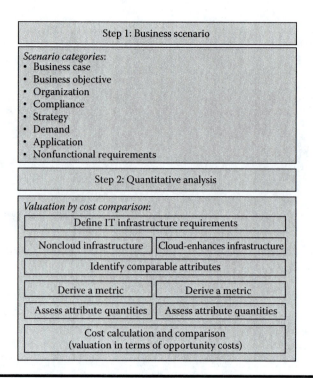

Figure 14.6   Framework for estimating the value of cloud computing.

Business case: It is important to understand which types of applications and services can be built on a cloud computing infrastructure.

Business objective: Typical business benefits mentioned in the context of cloud computing are high responsiveness to varying, unpredictable demand behavior, and shorter time to market.

Organization: An important question is to what extent a cloud computing service must and can be integrated into the organizational structure of an enterprise and how a cloud might be used to leverage communication and collaboration with business partners and customers.

Compliance: The shift from personal and business data into the cloud where it is processed and stored must comply with privacy laws and regulations. Legal aspects could also make a cloud application more expensive because of different security policies that must be followed.

Strategy: Developing an application for the cloud or migrating an existing application into the cloud comes with the risk of vendor lock-in and dependency on the vendor's pricing policy.

Demand: Services and applications on the web and in corporate networks can roughly be divided into two categories: services that deal with somewhat predictable demand behavior and those that must handle unexpected demand volumes, respectively. Services from the first category must be built on top of scalable infrastructure in order to be adopted to changing demand volumes. The second category is even more challenging because increase and decrease in demand cannot be forecast and sometimes occur within minutes or even seconds.

Application: Application-specific requirements comprise, for example, the run-time environment, database technology, web-server software and additional software libraries, as well as load-balancing and redundancy mechanisms.

Non-functional requirements: In addition to application-specific requirements, nonfunctional requirements must be addressed as well: security, high availability, reliability, and scalability.

■ Step 2: Quantitative Analysis

In the second step of the framework, a quantitative analysis of the business scenario is conducted by means of valuation methods. The valuation of cloud computing services must take into account costs as well as cash flows resulting from the underlying business model. Based on the cost comparison, the benefits of cloud computing technology can be measured as opportunity costs.

Define IT infrastructure requirements: First of all, the decision maker must identify mission-critical requirements of the IT infrastructure according to the criteria of the business scenario.

Cloud-enhanced IT infrastructure: The decision maker must model the infrastructure environment that is suitable to fulfill the criteria of the business scenario.

Identify comparable attributes: It is necessary to identify key attributes that indicate fulfillment of the business scenario requirements and that can be measured or estimated. These comparable attributes can be derived from a subset of the business scenario criteria.

Derive a metric based on comparable attributes: A *cost metric* can then be calculated by dividing the cumulative infrastructure costs by the amount of *attribute units* used over time. A scorecard method could be used in order to map the qualitative criteria of the business scenario into quantitative "costs." Thereby, risk factors can be captured by the calculation.

Assess attribute quantities: In order to calculate the costs, one needs to measure or estimate each attribute. Then multiply the measured units of an attribute with the corresponding cost metric and compute a cost estimate. Since there can be more than one relevant attribute, it is necessary to repeat the calculation for each attribute and then aggregate the value of all cost estimates.

Cost calculation and comparison: By means of a comparison between the cost aggregates, it is finally possible to calculate the value of using cloud computing services in terms of opportunity costs.

Such framework helps decision makers to estimate and compare the cloud computing costs for conventional IT solutions. In general, despite the effort done to analyze the costs and benefits, consumer utility and QoS are not considered.

14.7 Service-Level Agreements

In this section, we present an overview about SLAs in the cloud computing context. SLA formalization is then discussed by mapping business processes and cloud infrastructures. Special attention is paid to the negotiation process as one of the important stages performed between the provider and consumer in SLA formalization.

The cloud constitutes a single point of access for all services that are available anywhere in the world on the basis of commercial contracts that guarantee satisfaction of the QoS requirements of customers according to specific SLAs. The SLA is a contract negotiated and agreed to between a customer and a service provider. That is, the service provider is required to execute service requests from a customer within negotiated QoS requirements for a given price [6]. The purpose of using SLAs is to define a formal basis for performance and availability the provider guarantees to deliver. SLA contracts record the level of service, specified by several attributes, such as availability, serviceability, performance, operation, billing, or even penalties in the case of violation of the SLA. Also, a number of performance-related metrics are frequently used by Internet Service Providers (ISPs), such as service response time, data transfer rate, round-trip time, packet loss ratio, and delay variance. Often, providers and customers negotiate utility-based SLAs that determine the cost and penalties based on the achieved performance level. A resource allocation management scheme is usually employed with a view to maximizing overall profit (utility) which includes the revenues and penalties incurred when QoS guarantees are satisfied or violated, respectively. Usually, step-wise utility functions are used where the revenue depends on the QoS levels in a discrete fashion [20].

14.7.1 SLA Mapping

When a company is in control of its internal IT infrastructure, business analysts and developers can define service-level requirements during design time and can actively select and influence components in order to meet these requirements. However, in cloud computing environments, SLAs are typically provided for basic platform services (e.g., system uptime, network throughput). Business processes typically expect service levels for the technical services they integrate (e.g., order submission in less than 1 s).

To bring these two worlds together, an approach for SLA mapping between business processes and IT infrastructures has been proposed. It is based on a method for the assurance of NFPs and includes three major tasks as depicted in Figure 14.7:

1. *Formalization of business process requirements on the business side and of service capabilities at the IT infrastructure*: Both are specified in a formal way, using a predefined SLO structure and predefined NFP terms.
2. *Negotiation of service capabilities at the IT infrastructure that correspond to the formalized business process requirements:* Here, we assess whether the aggregated technical services provide the expected service levels to meet business process requirements under different load hypotheses. Within this comparison, we also calculate the aggregated service level using the

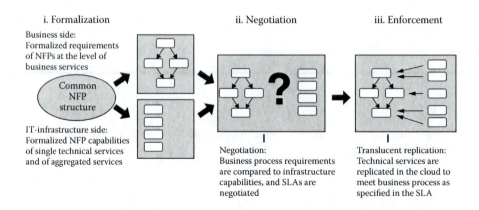

Figure 14.7 SLA mapping approach.

performance metrics of the individual technical services. Based on the result of this comparison, we can decide where to apply replication in the next step. A reasoner or comparing unit must understand the structure of the statements and used NFPs of both the business and the infrastructure side.

3. *Enforcement of business process SLAs at the IT infrastructure level:* Here, we apply translucent parallelization of service processing using multiple nodes in a data-center environment. Replication can be enacted to improve service levels regarding response time; transaction rate; throughput and availability, respectively, and reliability.

In this approach, the process owner specifies service-level requirements as expected from the business perspective. On the IT infrastructure side, different replication configurations of technical services in grid and cloud computing environments, such as Amazon EC2, are evaluated. Both requirements and service capabilities of the different configurations are then formalized and compared [17].

14.7.2 SLA Negotiation

The service providers and their customers negotiate utility-based SLAs to determine costs and penalties based on the achieved performance level. The service provider needs to manage its resources to maximize its profits. Recently, the problem of maximization of SLA revenues in shared service-center environments implementing autonomic self-managing techniques has attracted vast attention by the research community. The SLA maximization problem considers (1) the set of servers to be turned *on* depending on the system load, (2) the application tiers to servers' assignment, (3) the request volumes at various servers, and (4) the scheduling policy at each server as joint control variables [21].

Negotiation is the process whereby the service provider and consumer discuss, debate, and agree upon expectations of the service provision. Service consumers will require a specific QoS; for example, consumers with critical operations will desire faster response time, and their expectations may change over time as a result of continuing changes in the environment. Service providers have to ensure such flexibility in their service delivery in order to bring about greater consumer satisfaction by allowing for customized and dynamic SLA while taking into consideration factors such as feasibility and profitability. The providers should also adopt a flexible pricing strategy that differentiates the service requests based on their requirement in order to provide financial incentives for both the customer and themselves. Gridbus Broker and Aneka represent the tools used by the consumer and provider, respectively, to determine the negotiation strategy, which is the logic used by a partner to decide which provider or consumer satisfies his needs best. The process may begin with the service providers publishing their prerequisites, which will be translated into a common format and displayed in a registry. When a consumer wants to obtain a particular service, he submits his request for service into the registry, and the registry will return a list of documents published by the providers that matches the consumer's needs. The consumer then selects the appropriate service based on his negotiation strategy and a negotiation session can be established between the two parties. In fact, most providers of cloud computing applications, such as Amazon, Google, and IBM, do not even support the dynamic negotiation of SLAs, that is, all consumers adopt a SLA predetermined by the provider [8].

At the end of the negotiation process, provider and consumer commit to an agreement. This SLA serves as the foundation for the expected level of service between the consumer and the provider. The QoS attributes that are generally part of an SLA (such as response time and throughput), however, change constantly, and to enforce the agreement, these parameters need to be closely monitored. Because of the complex nature of consumer demands, a simple measure and trigger process may not work for SLA enforcement. Four different types of monitoring demands made by consumers are mentioned. One scenario is that a consumer demands the data exposed by a service provider without further refinement, such as transaction count, which is a raw metric. The second scenario is that the consumer requests that collected data should put into a meaningful context. This scenario creates the requirement for a process that collects data from different sources and applies suitable algorithms for calculating meaningful results. Such metrics include statistical measures, such as average or standard deviation, that need to be computed from a raw set of numbers. The third scenario is that the consumer requests certain customized data to be collected. In the fourth scenario, the consumer even specifies the way data should be collected. Both the latter-mentioned scenarios imply an advanced consumer who would have a knowledge of the inner workings of a provider and somewhat rare in practice. Other issues, such as trust, also need to be considered during SLA enforcement. For example, consumers may not completely trust the certain measurements provided solely by a service provider and regularly employ third party mediators. These mediators are responsible for measuring the critical service parameters and reporting violations of the agreement from either party [22].

14.8 Challenges and Future Research Directions

In this section, various QoS challenges of cloud computing are highlighted. Such challenges provide a good foundation to tackle interesting research directions aiming to guarantee further success of the cloud computing paradigm.

- *Fault management:* Fault management starts to take place at the time of fault detection and service restoration. Such issues directly affect the QoS guarantees. Hence, fault detection and notification mechanisms as well as the performance should be an essential part of the SLA between the provider and the consumer. Fault isolation, troubleshooting, and sharing troubleshooting information require special attention from the researchers to eliminate the negative impact upon the whole cloud paradigm [23].
- *Transparency:* Transparency is always related to the management of resources, used by virtual machines to perform jobs, service migration, fault tolerance, and check pointing of tasks. The transparency is an open research point related to QoS in cloud computing. Problems surrounding transparency are inherently complicated because of the large quantities of volatile data associated with virtual machines that need to be transferred, stored, or backed up for such tasks to be accomplished within the cloud [16].
- *Simulation and modeling:* Advancements in simulation and modeling techniques will provide a better understanding, usability, and streamlining of cloud environments. Currently, the need for such a tool to support the performance evaluation of cloud environments has become crucial. Some progression has been done in terms of developing a simulation tool named "CloudSim," but it is still in its early phase of development [16].
- *QoS monitoring:* QoS is the key to the provider's success. Assessing the quality of what service consumers are paying for has become a mission-critical business necessity. Thus, implementing a pilot application for QoS monitoring for services running in the cloud is highly necessary. A number of issues must be addressed because the increasing complexity of

individual components, middleware, and interconnection infrastructures has impaired the ability to measure the performance indicators of the monitored service [24].

■ *Interoperability between grids and clouds:* Interoperability and integration between grids and clouds have recently attracted a lot of interest. This trend rose from the similar nature of technical problems that have been solved in grid computing and still need to be solved in cloud computing. Nimbus, as a prime example, is an IaaS cloud that enables the use of virtualized resources within grids [16].

14.9 Conclusion

This chapter provides a comprehensive description of cloud computing QoS. First, the QoS ontology is presented, to be used for expressing various QoS information in a unified manner. It consists of both basic and support layers. The basic layer encloses the definitions of the basic attributes (e.g., QoS properties, metrics, and units). The support layer presents some auxiliary means (e.g., transformation, values comparison, and others) that are used to model the QoS properties. The ontology presented in this chapter is significantly eliminating the diversity of concepts held between providers and consumers because it dealt with QoS interoperability.

Then, the QoS aspects of cloud computing are presented along with their measuring parameters and assessments methods. Varying between technical, economic, and business ones, these aspects included performance, transparency, information assurance, security risks, trustworthiness, and SaaS readiness.

The chapter then discusses the QoS models that have been particularly tailored for cloud computing platforms: IaaS, PaaS, and SaaS. QoS models for IaaS and PaaS are mainly concerned with computational resources management and virtualization technology while SaaS ones focus on properties in the development process as well as reusability, availability, and scalability.

Finally, the chapter pointed out how to formalize the QoS levels to be used for cost–benefit analysis and cost calculation as two different service selection approaches. SLA formalization was then discussed by mapping business processes and cloud infrastructures. Special attention was paid to the negotiation process as one of the important stages performed between the provider and consumer in SLA formalization. A vision of some future research directions in cloud QoS is also constructed through the discussion of some open research questions.

References

1. B. Cao and B. Li, "A Service-Oriented Qos-Assured and Multi-Agent Cloud Computing Architecture." In *Cloud Computing*, vol. 5531, edited by M. Jaatun, G. Zhao, and C. Rong, 644–649. Heidelberg: Springer Berlin, 2009.
2. "PerfCloud: GRID Services for Performance-oriented Development of Cloud Computing Applications." Proceedings of the 2009 18th IEEE International Workshops on Enabling Technologies: Infrastructures for Collaborative Enterprises (WETICE 09), Groningen, The Netherlands, 2009, 201–206.
3. I. Foster, "What Is the Grid? A Three Point Checklist." *GridToday*, July 2002, http://www-fp.mcs.anl.gov/foster/Articles/WhatIsTheGrid.pdf.
4. I. Foster, C. Kesselman, and S. Tuecke, "The Anatomy of the Grid: Enabling Scalable Virtual Organizations." *International Journal of High Performance Computing Applications* 15, no. 3, (2001): 200–222.
5. R. Buyya, D. Abramson, and J. Giddy, "Nimrod/G: An Architecture for a Resource Management and Scheduling System in a Global Computational Grid." Proceedings of the 4th International Conference and Exhibition on High Performance Computing in Asia-Pacific Region, Beijing, China, May 2000.

6. K. Xiong and H. Perros, "Service Performance and Analysis in Cloud Computing." Proceedings of the 2009 Congress on Services – I (SERVICES 09), IEEE Computer Society, Washington, 2009, 693–700.

7. P. Jacob and B. Davie, "Technical Challenges in the Delivery of Interprovider QoS." *IEEE Communications Magazine* 43, no. 6 (2005): 112.

8. R. Buyya, C. Yeo, S. Venugopal, J. Broberg, and I. Brandic, "Cloud Computing and Emerging IT Platforms: Vision, Hype, and Reality for Delivering Computing as the 5th Utility." *Future Generation Computer Systems* 25, no. 6 (2009): 599–616.

9. M. Pathan, J. Broberg, and R. Buyya, "Maximizing Utility for Content Delivery Clouds." Proceedings of the 10th International Conference on Web Information Systems Engineering (WISE 09), Berlin: Springer-Verlag, 2009, 13–28.

10. W. Pauley, "Cloud Provider Transparency: An Empirical Evaluation." *IEEE Security and Privacy* 8, no. 6 (2010): 32–39.

11. D. Catteddu and G. Hogben, *Cloud Computing Risk Assessment.* Greece: European Network and Information Security Agency (ENISA), 2009.

12. D. Catteddu, "Cloud Computing: Benefits, Risks and Recommendations for Information Security." In *Web Application Security*, vol. 72, edited by C. Serrão, V. A. Díaz, and F. Cerullo, 17. Heidelberg: Springer, 2010.

13. N. Limam and R. Boutaba, "Assessing Software Service Quality and Trustworthiness at Selection Time." *IEEE Transactions on Software Engineering* 36, no. 4 (2010) 559–574.

14. T. Heart, N. S. Tsur, and N. Pliskin, "Software-as-a-Service Vendors: Are They Ready to Successfully Deliver?" In *Global Sourcing of Information Technology and Business Processes*, vol. 55, edited by I. Oshri, and J. Kotlarsky, 151–184. Heidelberg: Springer, 2010.

15. H. J. La and S. D. Kim, "A Systematic Process for Developing High Quality SaaS Cloud Services." Proceedings of the 1st International Conference on Cloud Computing (CloudCom 2009), Beijing, 2009, 278–289.

16. D. Armstrong and K. Djemame, "Towards Quality of Service in the Cloud." Proceedings of the 25th UK Performance Engineering Workshop, Leeds, 2009.

17. V. Stantchev and C. Schröpfer, "Negotiating and Enforcing QoS and SLAs in Grid and Cloud Computing." In *Advances in Grid and Pervasive Computing*, vol. 5529, edited by N. Abdennadher, and D. Petcu, 25–35. Heidelberg: Springer, 2009.

18. D. Kondo, B. Javadi, P. Malecot, F. Cappello, and D. Anderson, "Cost–Benefit Analysis of Cloud Computing versus Desktop Grids." Proceedings of the IEEE International Symposium on Parallel and Distributed Processing (IPDPS 09), IEEE Computer Society Washington, May 2009, 1–12.

19. M. Klems, J. Nimis, and S. Tai, "Do Clouds Compute? A Framework for Estimating the Value of Cloud Computing." In *Designing E-Business Systems: Markets, Services, and Networks*, vol. 22, edited by C. Weinhardt, S. Luckner, and J. Stober, 110–123. Heidelberg: Springer-Verlag, 2009.

20. C. Yfoulis and A. Gounaris, "Honoring SLAs on Cloud Computing Services: A Control Perspective." European Control Conference, Budapest, Hungary, August 2009, 184–189.

21. D. Ardagna, M. Trubianb, and L. Zhangc, "SLA Based Resource Allocation Policies in Autonomic Environments." *Journal of Parallel and Distributed Computing* 67, no. 3 (2007): 259–270.

22. P. Patel, A. Ranabahu, and A. Sheth, "Service Level Agreement in Cloud Computing." Cloud Workshops (OOPSLA 09), 2009, 1–10.

23. MIT Communications Futures Program (CFP) Quality of Service Working Group, "Inter-provider Quality of Service," Quality of Service Working Group, November 2006.

24. L. Romano, "QoS Monitoring in the Cloud: Goals and Open Issues." Internet of Services 2010: Collaboration Meeting for FP6 & FP7 projects, Europe's Information Society, 2010.

Chapter 15

Service-Oriented Network Virtualization for Convergence of Networking and Cloud Computing in Next-Generation Networks

Qiang Duan

Contents

15.1 Introduction

One of the major recent developments in the field of information technology is cloud computing, which may significantly change the way people do computing and manage information. Cloud computing is a large-scale distributed computing paradigm that is driven by economies of scale in which a pool of abstracted, virtualized, dynamically scalable computing functions and services are delivered on demand to external customers over the Internet [1].

Networking plays a crucial role in cloud computing. Cloud services normally represent remote delivery of computing resources, whether hardware or software, most often via the Internet. This is especially relevant in public cloud environments where customers obtain cloud services from a third-party cloud provider. Usually this means data crosses multiple networks before it is delivered to the end user. From a service-provisioning perspective, cloud services consist of not only computing functions provided by the cloud infrastructure, but also data communications functions offered by the Internet. In addition, networking is also a key element of the cloud infrastructure that provides data communications both inside a cloud data center and among data centers distributed at different locations. Results obtained from a recent study on cloud-computing performance [2,3] have indicated that networking performance has a significant impact on the quality of cloud services, and in many cases data communications become a bottleneck that limits clouds from supporting high-performance applications. Therefore networks with quality-of-service (QoS) capabilities become an indispensable ingredient for high-performance cloud computing.

The significant role that networking plays in cloud computing calls for a holistic vision of both computing and networking resources in a cloud environment. Such a vision requires the underlying networking infrastructure to be opened and exposed to upper-layer applications in clouds, thus enabling combined control, management, and optimization of computing and networking resources for cloud service provisioning. This leads to a convergence of networking and cloud computing systems toward a composite network–cloud service provisioning system. Because of the complexity of networking technologies and protocols, exposure of network functionalities in a cloud environment is only feasible with appropriate abstraction and virtualization of networking resources.

On the other hand, telecommunication and networking systems are facing the challenge of rapidly developing and deploying new functions and services for supporting the diverse requirements of various computing applications. In addition, fundamental changes are also required in the Internet architecture to allow heterogeneous networking systems to coexist and cooperate for supporting the wide spectrum of applications. A promising approach that the networking research community takes for addressing these challenges lies in virtualization of networking resources, namely decoupling service provisioning from network infrastructure and exposing underlying network functionalities through resource abstraction. Such an approach, in general, is described by the term *network virtualization*, which is expected to become a fundamental attribute of the future networking paradigm and play a crucial role in next-generation networks (NGNs).

As a potential enabler of profound changes in both the communication and computing domains, virtualization is expected to bridge the gap between these two fields that traditionally live quite apart and, particularly, enable a convergence of networking and cloud computing. Network virtualization in a cloud environment enables a holistic vision of both computing and networking resources as a single collection of virtualized, dynamically provisioned resources for composite network–cloud service delivery. The convergence of networking and cloud computing is likely to open up an immense field of opportunities to the IT industry and allow the next-generation Internet to provide not only communication functions, but also various computing services. Various telecommunication and Internet service providers (SPs) around the world have already shown a great deal of interest in providing cloud services based on their network infrastructure.

Convergence of networking and cloud computing can be viewed from both vertical and horizontal aspects. From the vertical view aspect, resources and functionalities in the network infrastructure are opened and exposed through an abstract virtualization interface to upper-layer functions in a cloud environment, including resource management and control modules and other functions for offering cloud services. From the horizontal view aspect, cloud data centers that offer computing functions and the network infrastructure that provides data communications converge into a composite network–cloud service provisioning system. From both aspects, such a convergence enables combined control, management, and optimization of networking as well as computing resources in a cloud environment.

Some technical issues must be addressed for realizing the notion of convergence between networking and cloud computing. Key requirements for network-cloud convergence include networking resource abstraction and exposure to upper-layer applications and collaboration among heterogeneous systems across the networking and computing domains. Therefore, an important research problem is to develop the mechanism for supporting effective, flexible, and scalable interaction among all key players in a converged networking and cloud computing system, including networking and computing infrastructure providers (InPs), networking and computing SPs, and various applications as the customers of composite network-cloud services. The service-oriented architecture (SOA), when applied in network virtualization, offers a promising approach toward enabling a network-cloud convergence.

SOA provides effective architectural principles for heterogeneous system integration. Essentially, service orientation facilitates virtualization of computing systems by encapsulating system resources and capabilities in the form of *services* and provides a loose-coupling interaction mechanism among these services. SOA has been widely applied in cloud computing via the paradigms of infrastructure as a service (IaaS), platform as a service (PaaS), and software as a service (SaaS). Applying the SOA in the field of networking supports encapsulation and virtualization of networking resources in the form of SOA-compliant *network services*. Service-oriented network virtualization enables a *network-as-a-service* (NaaS) paradigm that allows network infrastructure to be exposed and accessed as network services, which can be composed with computing services in a cloud computing environment. Therefore, the NaaS paradigm may greatly facilitate a convergence of networking and cloud computing.

Currently web services provide the main implementation approach for SOA. Key web service technologies, including service description, discovery, and composition were mainly developed in the distributed computing area; therefore, evolution to these technologies is needed to meet the requirements of NaaS toward network-cloud convergence. Application of SOA in networking has recently formed an active research area that has attracted attention from both industry and academia. A great amount of research efforts have been made on key technologies for NaaS, including network service description, discovery, and composition. These works are conducted in various fields scattered across telecommunications, computer networking, web services, and distributed computing. Although some relevant surveys have been published (e.g., [4–8]), they either missed some of the latest progress or focused on general web services instead of NaaS and also lack discussion on integration of network services in a cloud computing environment. This motivates a comprehensive overview and survey in the literature that reflects the current status of service-oriented network virtualization for network and cloud convergence, which is the main objective of this chapter.

In this chapter, the author first introduces the SOA concept and service-orientation principle, gives an overview of the latest developments of applying SOA in realizing virtualization in both telecommunications and Internet infrastructure, and discusses the convergence of networking and cloud computing based on the NaaS paradigm. Then a framework of service-oriented network virtualization for network-cloud convergence is presented in this chapter followed by a survey on key enabling technologies for realizing the NaaS paradigm, mainly focusing on network service description, discovery, and composition technologies. Challenges and opportunities that are brought in by network-cloud convergence to these technologies are also discussed. A particular challenge is to evaluate performance of composite network-cloud service provisioning. A new modeling and analysis method for composite service performance evaluation is reported in the last section of this chapter.

15.2 Service-Oriented Architecture

The term "service-oriented" means that logic required to solve a large problem can be better constructed, carried out, and managed if it is decomposed into a collection of smaller, related pieces. Each of these pieces addresses a concern or a specific part of the problem. SOA encourages individual units of logic to exist autonomously yet not isolated from each other. Units of logic are required to conform to a set of principles that allow them to evolve independently while still maintaining a sufficient amount of commonality and standardization. Within SOA these units are known as services [9].

SOA provides an effective solution to coordinating computational resources across heterogeneous systems to support various application requirements. As described in [10], SOA is an architecture within which all functions are defined as independent services with interfaces that can be invoked, called in defined sequences to form business processes. SOA can be considered as a philosophy or paradigm to organize and utilize services and capabilities that may be under the control of different ownership domains. Essentially, SOA enables virtualization of various computing resources in the form of services and provides a flexible interaction mechanism among services.

A service in SOA is a module that is self-contained (i.e., the service maintains its own states) and platform-independent (i.e., interface to the service is independent from its implementation platform). Services can be described, published, located, orchestrated, and programmed through standard interfaces and messaging protocols. All services in SOA are independent of each other, and service operation is perceived as opaque by external services, which guarantees that external components neither know nor care how services perform their functions. The technologies providing the desired functionality of the service are hidden behind the service interface.

A key feature of SOA is the *loose-coupling* interaction among heterogeneous systems in the architecture. The term "coupling" indicates the degree of dependency any two systems have on each other. In a loosely coupled interaction, systems need not know how their partner systems behave or are implemented, which allows systems to connect and interact more freely. Therefore, loose coupling of heterogeneous systems provides a level of flexibility and interoperability that cannot be matched using traditional approaches for building highly integrated, crossplatform, interdomain communication environments. Other features of SOA include reusable services, formal contract among services, service abstraction, service autonomy, service discoverability, and service composition. These features make SOA a very effective architecture for resource virtualization and heterogeneous system integration to support various application requirements.

Though SOA can be implemented with different technologies, web services are the preferred environment for realizing SOA. A web service is an interface that describes a collection of operations that are network accessible through standardized XML messaging. A web service is described using a standard, formal XML notion, called its service description. It covers all the details necessary to interact with the service, including message formats, transport protocols, and location. The interface hides the implementation details of the service, allowing it to be used independently of its implementation. This enables web services-based systems to be loosely coupled and component-oriented with crosstechnology implementations.

The key elements of a web service-based implementation of SOA include the SP, service broker/registry, and service customer. The basic operations involved in the interaction among these elements are service description publication, service discovery, and service binding/access. In addition, service composition is also a key operation for meeting customers' service requirements. The key web services elements and their interaction are shown in Figure 15.1. An SP makes its service available in the system by publishing a service description at a service registry. Service discovery, typically performed by a broker, is the process that responds to a customer request for discovering a service that meets specified criteria. Multiple services may be composed into a composite service to meet the customer's requirements.

The SOA principle and its web services implementation technologies have become the state-of-the-art of information service delivery and have been widely applied in various distributed computing areas. SOA enables more flexible and reusable services that may be reconfigured and augmented more swiftly than traditional construction of applications systems and thus can accelerate the time-to-business objective and result in better business agility. SOA also provides a standard way to represent and interact with application functionalities, thus improving interoperability and integration across an enterprise and value chain. SOA has also been widely adopted by cloud computing as the main model for cloud service provisioning. Following this model, various virtualized computing resources, including both hardware (e.g., CPU capacity and storage space) and software applications are delivered to customers as services through the infrastructure-as-a-service, platform-as-a-service, and software-as-a-service paradigms.

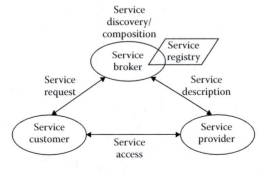

Figure 15.1   Key elements of web services and their interaction.

15.3 Service-Oriented Network Virtualization in Telecommunications

A key aspect of telecommunications that has been pushing its evolution in the past decades is to create new market-driven applications by reusing an extensible set of existing service components. The methodology taken by the telecom research and development community for achieving this objective is based on the idea of separating service-related functions and data transport infrastructure. Such a separation allows underlying transport functionalities and capacity to be virtualized and shared by service-related functions to create various applications. This is essentially the notion of virtualization in the telecommunications domain. In recent years, the SOA principle and web service technologies have been applied to facilitate virtualization in telecommunication systems.

The intelligent network (IN) that was initially developed in the 1980s was an example of the early research efforts toward making the telecommunication network a programmable environment for delivering value-added services [11]. The idea was to define overlay service architecture on top of physical network infrastructure and extract service intelligence into dedicated service control points. In the middle 1990s, some telecom API standards, including Parlay, Open Service Architecture (OSA), and Java API for Integrated Networks (JAIN), were introduced for achieving basically the same objective as IN but with easier service development. These APIs simplified telecom service development by abstracting signaling protocol details of the underlying networks.

Although these technologies were promising, their market acceptance was slow, partially because most network operators at that time were not yet ready to open up their infrastructure. In addition, both IN and the telecom APIs such as Parlay/OSA and JAIN lack an effective mechanism for realizing the separation of service provisioning and network infrastructure. Remote procedure call and functional programming conceptually drive IN implementation. Parlay/OSA and JAIN are based on distributed object-oriented computing technologies such as Common Object Request Broker Architecture (CORBA) and Java Remote-Method Invocation (RMI). Such distributed computing technologies are essentially tightly coupled between system modules and lack full support for networking resource abstraction and virtualization.

In the early 2000s, a simplified version of the Parlay/OSA API called Parlay X [12] was developed jointly by the Parlay Group, ETST, and 3GPP. Parlay X is based on the emergence of web services technologies. The objective of Parlay X is to offer a higher level of abstraction than Parlay/OSA in order to allow a large community of developers to design and build value-added applications in a complex telecommunication network without knowing the details of networking protocols and technologies. Web service technologies in Parlay X expose networking capabilities to upper-layer applications, which opens a door for applying the SOA principle to realize separation of service provisioning from network infrastructure. The resource abstraction and loose-couple interaction features of SOA make it a promising approach to realizing the benefits of network virtualization.

The traditional telecommunication system, under the pressure from Internet technology advanced and new network application providers, is undergoing a fundamental transition toward a multiservice, packet-switching, IP-based architecture. Two representative developments in such a transition are NGN and IP-based multimedia subsystems (IMS). An NGN is defined by ITU-T as a packet-based network able to provide services including telecommunication services and able to make use of multiple broadband, QoS-enabled transport technologies and in which service-related functions are independent from underlying transport-related technologies [13]. IMS is an effort of telecom-oriented standard bodies, such as 3GPP and ETSI, to realize the NGN concept that presents an evolution from the traditional closed signaling system to the NGN service control system [14].

A key feature of NGN is the decoupling of network transport and service-related functions, thus enabling virtualization of network infrastructure for flexible network service provisioning.

Recently, with the expansion of various multimedia applications supported by telecom systems, the capability of rapid development and deployment of new services has become a crucial requirement to telecomm operators. However, telecom systems have been designed specifically to support a narrow range of precisely defined communication services, which are implemented on a fairly rigid infrastructure with minimal capability for ad hoc reconfiguration. Operations, management, and security functions in traditional networks are also specifically designed and customized to facilitate particular types of services. Tightly coupling between service provisioning and network infrastructure becomes a barrier to rapid and flexible service development and deployment.

In order to resolve the problem of the "silo" mode of service offering in the current telecommunication systems, research and development efforts have been made for building a service delivery platform (SDP). At a high level, SDP is a framework that facilitates and optimizes all aspects of service delivery, including service design, development, provisioning, and management. The core idea is to have a framework for service management and operation by aggregating the network capabilities and service management functions in a common platform. Main SDP specifications include OMA Open Service Environment (OSE) [15] and TM Forum Service Delivery Framework (SDF) [16]. The objective of SDP is to provide an environment in which upper-layer applications can be easily developed by combining underlying networking capabilities and also enable collaboration across network SPs, content providers, and third-party SPs. The notion of virtualization and SOA principle play a key role in both OSE and SDF specifications to achieve this objective. The method taken by both specifications is to define a set of standard service components called *service enablers* and develop a framework that allows new services to be built by composing service enablers. The service enablers support virtualization of networking resources by encapsulating underlying networking functionalities and capabilities through a standard abstract interface. The web services approach has become a de facto standard for communications among system components in SDP. Web service orchestration technologies, such as BPEL, are also being adopted in SDP for enabling services to be composed with both telecom functional blocks and business logic/applications in the computing domain.

As users consume services offered through various networks, they have pushed for blending the service offerings of various providers for a richer experience. In order to allow network and computing SPs, content providers, and end users to offer and consume collaborative services, there is a need for an efficient way of service and application delivery that at the same time is customer-centric. This is a challenge that has not been sufficiently addressed by the aforementioned developments, such as IMS, NGN, and SDP. Therefore, there has been a motivation to organize the services/applications offered by various networks on an overlay that allows SPs to offer rich services. Toward this objective, IEEE has recently developed the Next Generation Service Overlay Network (NGSON) [17]. NGSON specifies context-aware, dynamically adaptive, and self-organizing networking capabilities, including both service-level and transport-level functions that are independent of the underlying network infrastructure. NGSON aims to bridge the service layer and transport network over IP infrastructure to address the accommodation of highly integrated services. NGSON particularly focuses on composing new collaborative services by using existing components (from IMS, NGN, SDP, etc.) or defining new NGSON components and delivering them to end users. Some key functional entities of NGSON, including service discovery and negotiation, service registry, and service composition, have been implemented based on web service technologies [18].

Recent evolution of telecommunications has been following a path toward network virtualization—decoupling service provisioning from transport infrastructure and exposing a networking platform through resource abstraction. The latest development of NGSON particularly focuses on collaborative service across different types of SPs. The SOA principle has been employed in these technologies for facilitating realization of the notion of virtualization.

15.4 Service-Oriented Network Virtualization in Future Internet

The stunning success of the Internet becomes an obstacle of its own development. A wide variety of applications have been deployed upon the Internet. On the other hand, various heterogeneous networking technologies have been developed for supporting diverse application requirements. The future Internet must be able to coordinate heterogeneous networks to support diverse applications, which requires fundamental changes in the Internet architecture and service delivery model. However, the current IP-based Internet protocol with the end-to-end design principle, along with the huge amount of investment in current Internet infrastructure, make any disruptive innovation in the Internet architecture very difficult (if not impossible). In order to fend off the ossification of the current Internet, network virtualization has been proposed as a key attribute of the future internetworking paradigm and is expected to play a crucial role in NGNs.

Network virtualization on the Internet can be described as a networking environment that allows one or multiple SPs to compose heterogeneous virtual networks that coexist together but in isolation from each other and to deploy customized end-to-end services as well as manage them on those virtual networks by effectively sharing and utilizing underlying network resources leased from InPs [6].

Essentially, network virtualization follows a well-tested principle—separation of policy from mechanism—on the Internet. In this case, network service provisioning is separated from data transportation mechanisms, thus dividing the traditional role of Internet SPs into two entities: InPs who manage the physical infrastructure and network SPs who create virtual networks for offering end-to-end network services by utilizing resources obtained from InPs. Key attributes of network virtualization include abstraction (details of the network resources are hidden), indirection (indirect access to network resources may be combined to form different virtual networks), resource sharing (network elements can be partitioned and utilized by multiple virtual networks), and isolation (loose or strict isolation between virtual networks). Physical network infrastructure, consisting of links and nodes, are virtualized and made available to virtual networks, which can be set up and torn down dynamically according to customer needs. Figure 15.2 illustrates a network virtualization environment in which the SPs SP1 and SP2 construct two virtual networks by using resources obtained from the InPs InP1 and InP2.

Network virtualization has a significant impact on NGNs. By allowing multiple virtual networks to cohabitate on shared physical infrastructure, network virtualization provides flexibility, promotes diversity, and promises increased manageability. The best-effort Internet today is basically a commodity service that gives network SPs limited opportunities to distinguish themselves from competitors. A diversified Internet enabled by network virtualization offers a rich environment for innovations, thus stimulating the development and deployment of new Internet services. In such an environment, SPs are released from the requirement of purchasing, deploying, and maintaining physical network equipment, which will significantly lower the barrier to entry of the Internet service market. Network virtualization enables a single SP to obtain control over the entire end-to-end service delivery path across network infrastructure that belongs to different domains, which will greatly facilitate the end-to-end QoS provisioning.

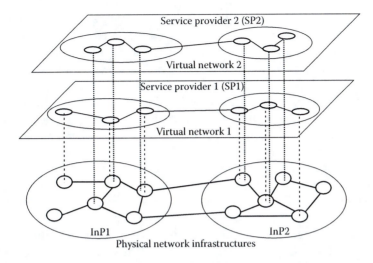

Figure 15.2   Illustration of a network virtualization environment.

Network virtualization has attracted extensive research interest from both academia and industry. Network virtualization was first employed as an approach to developing virtual test beds for investigating new network architecture and protocols. For example, in PlanetLab [19] and Global Environment for Network Innovation (GENI) [20] projects, virtualization was employed to build open experimental facilities for researchers to create customized virtual networks to evaluate new networking technologies. Then the role of virtualization on the Internet has evolved from a research method to a fundamental attribute of the internetworking paradigm [21]. Concurrent Architectures are Better than One (CABO) is an Internet architecture proposed in [22] that decouples network SPs and InPs to support virtual networks over a shared physical substrate. In 4WARD (Architecture and Design for the Future Internet), a large EU FP7 project, network virtualization is employed as a key technology to allow virtual networks to operate in parallel in the future Internet [23]. FEDERICA is another FP7 project with a core objective to support research on the future Internet by creating a Europe-wide infrastructure of network resources that can be sliced to provide virtual Internet environments for research [24]. AGAVE (A lightweight Approach for Viable End-to-end IP-based QoS Services) was an EU sixth framework project that developed solutions for open end-to-end service provisioning based on the concept of *network planes*, which can be described as slices of networking resources that may be interconnected to create parallel Internets tailored to service requirements [25]. User Controlled Light Paths (UCLP) was a Canadian research project with the main goal of providing a network virtualization framework upon which communities of users could build their own middleware or applications [26].

Recently, some standard organizations have also started their work on network virtualization specifications. In July 2009, ITU-T established the Focus Group on Future Network (FG FN), in which network virtualization is one of the fundamental study topics. In addition, the Internet Research Task Force (IRTF) created the Virtual Networks Research Group (VNRG) in early 2010, which specifically focuses on network virtualization.

Some of the recent developments in telecommunications, such as the Parlay X, NGN, SDP, and NGSON discussed in the previous section, are all based on the principle of separating service provisioning functions from network infrastructure, which is essentially network virtualization in the telecom domain. Comparison between virtualization in telecommunications and on the Internet shows some difference in the perspective and emphasis of the telecom and Internet communities regarding embracing the notion of virtualization in their domains. Application of virtualization in telecom systems, such as SDP and NGSON, focuses on exposing networking platforms to upper-layer applications for facilitating rapid development of value-added services. Therefore, virtualization is typically realized above the network layer through standard APIs with the abstraction of networking resources and functionalities. Virtualization on the Internet aims at adopting virtualization as a key attribute of the core network architecture rather than just using it as an approach for exposing the network platform. Therefore, virtualization tends to be realized on or below the network layer. In addition, Internet virtualization has been developed with an important objective to enable heterogeneous network architecture, including both IP-based and non-IP-based architecture, to coexist and cooperate in the future Internet. Though virtualization in telecom systems in principle supports an SDP that is independent of the underlying networking technologies, most of the current specifications, for example, NGN, IMS, and SDP, assume IP-based, packet-switching architecture for the physical network infrastructure.

Although significant progress has been made toward network virtualization, there are still many challenges that must be addressed before this notion can be realized in the future Internet. One of the primary challenges centers on the design of mechanisms and protocols needed to enable automated creation of virtual networks for service provisioning over a global, multidomain substrate comprising a heterogeneous network infrastructure. In order to create and provision virtual networks for meeting users' requirements, the first step is to discover a set of available network resources in network infrastructure that may belong to multiple administrative domains; then the appropriate (or optimal) resources need to be selected and composed to form virtual networks. Therefore, a key to realizing the network virtualization lies in flexible and effective collaboration among InPs, SPs, and virtual network end users (applications).

SOA, as a very effective architecture for coordinating heterogeneous systems to support various application requirements, offers a promising approach to facilitating network virtualization in the future Internet. A layered structure for service-oriented network virtualization is shown in Figure 15.3. Following the SOA principle, resources in network infrastructure can be encapsulated into *infrastructure services*. SPs access underlying networking resources through an infrastructure-as-a-service paradigm and compose infrastructure services into end-to-end *network services*. Applications, as the end users of virtual networks, utilize the underlying networking platform by accessing the network services offered by SPs, which is essentially a NaaS paradigm.

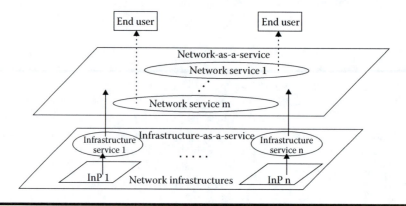

Figure 15.3   Layered structure for service-oriented network virtualization.

Applying SOA in network virtualization makes loose-coupling a key feature of interaction among InPs, SPs, and end users. Therefore, the NaaS paradigm inherits the merit of SOA that enables flexible and effective collaboration across heterogeneous networking systems for providing services that meet diverse application requirements. Service-oriented network virtualization also provides a means to present abstracted networking capabilities to upper-layer applications. Because of the heterogeneity of network protocols, equipment, and technologies, exposure of networking capabilities to applications without virtualization, would lead to unmanageable complexity. The abstraction of networking resources through service-oriented network virtualization can address the diversity and significantly simplify the interaction between applications and the underlying network platform.

15.5 Service-Oriented Network Virtualization for Convergence of Networking and Cloud Computing

Network virtualization will play a crucial role in both telecommunications and Internet service provisioning in NGNs. As a potential enabler of profound changes in both the communications and computing domains, virtualization is expected to bridge the gap between these two fields that traditionally live quite apart. In particular, the convergence of cloud computing and network virtualization is likely to open up an immense field of opportunities to the IT industry, including network SPs, cloud SPs, content providers, and third-party application developers.

With the advent of virtualized networking technology, cloud service delivery could be significantly improved via giving SPs options to implement virtual networks that offer customized networking solutions for cloud services. Convergence of networking and cloud computing presents a vital part of an overall cost-saving solution. With network–cloud convergence, business processes are enhanced and time for making decision shortened, saving time and money. Such a convergence also allows a single point of visibility of computing and networking operations, providing the opportunity to manage both more effectively. There is also a practical standpoint to network–cloud convergence, that is, dealing with one single supplier for a solution rather than two separate providers, which reduces costs and saves time through simplified interaction.

ITU-T started a focus group on cloud computing (FG Cloud) in May 2010, which aims at contributing with the networking aspects for flexible cloud infrastructure in order to better support cloud services/applications that make use of communication networks and Internet services [27].

Serving as a key enabler of virtualization, SOA forms a core element in the technical foundation for cloud computing. Recent research and development have been bridging the power of SOA and virtualization in the context of a cloud computing ecosystem [28]. The open grid forum (OGF) is working on the open cloud computing interface (OCCI) standard [29], which defines SOA-compliant open interfaces for interacting with cloud infrastructure. Taking a look at some of the most important cloud providers, we can see that the SOA principle has strongly influenced cloud service provisioning. For example, Amazon, a well-known provider that offers a complete ecosystem of cloud services including virtual machines (Elastic Compute Cloud EC2) and plain storage (Simple Storage Service S3), exposes its cloud services via web service interfaces.

The service-orientation principle, applied in both cloud computing and network virtualization, offers a promising approach to facilitate convergence of cloud computing and networking. Applying SOA in networking allows virtualization of networking resources in the form of SOA-compliant *network services*. This enables a NaaS paradigm that exposes networking resources and functionalities as services that can be composed with computing services in a cloud environment. From a service provisioning perspective, cloud services delivered to end users are composite services that comprise both computing services provided by cloud infrastructure and network services offered by network infrastructure.

Figure 15.4 shows a layered framework for SOA-based convergence of networking and cloud computing. In this framework, networking and computing resources are virtualized into services by following the same SOA principle, which offers a uniform mechanism for coordinating networking and computing systems for cloud service provisioning. Service-oriented virtualization enables a holistic vision of both networking and computing resources as a single collection of virtualized, dynamically provisioned resources, which allows coordinated control, management, and optimization of resources across the networking and computing domains. In this convergence framework, NaaS enables matching cloud service requirements with networking capabilities by discovering the appropriate network services. Composition of networking and computing services expands the spectrum of cloud services that can be offered to users. The loose-coupling feature of SOA provides a flexible and effective mechanism in this network–cloud convergence framework that supports interaction between networking/computing resource providers and service provisioning functions as well as collaboration across heterogeneous networking and computing systems.

As the main approach to implementing SOA, web services serve as the key enabling technologies for NaaS, that is, service-oriented network virtualization, for network and cloud convergence. Figure 15.5 shows the structure of a web service-based delivery system for composite network–cloud services. In this system, both network SPs and cloud SPs publish their service descriptions at a service registry. When a service consumer, typically an application, needs to utilize a cloud service, it sends a service request to the service broker. The service broker discovers available cloud and network services by searching the registry and then composes the appropriate network and cloud services into a composite service that meet the consumer's requirements.

For example, in a cloud-computing environment there are multiple providers for cloud services, such as Amazon EC2, Amazon S3, Google App, etc., and network services, such as AT&T Internet service and Verizon network service. When a user needs to send a data file to a cloud for processing, it sends a request to the service broker and specifies the requirements for both data

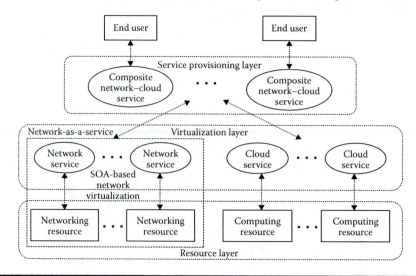

Figure 15.4 Framework for SOA-based convergence of networking and cloud computing.

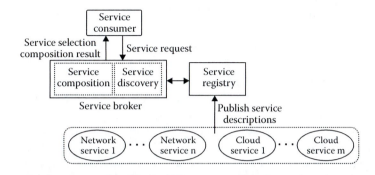

Figure 15.5   Web service-based delivery system for composite network–cloud services.

processing and transmission. The broker discovers available network and cloud services and selects the Amazon EC2 for data processing and the Verizon network service for data transmission and then composes these two services into one composite service that meets the requirements for both data-processing capacity (offered by Amazon EC2) and data-transmission bandwidth (provided by Verizon). In this example, the network service provides communications between a cloud SP and a service consumer. Data communications inside cloud data centers can also be virtualized as network services and composed with cloud services. A cloud infrastructure could be distributed across different geographic locations interconnected through networks. In such cases, the services offered by cloud infrastructure are also composite services comprising both networking and computing functions.

Service-oriented composition of network and cloud services allows provisioning of network services and cloud services, which used to be offered separately by different providers, to merge into the provisioning of composite network–cloud services. This convergence enables a new service delivery model in which the roles of traditional network SPs, such as AT&T and Verizon, and computing SPs, such as Amazon and Google, merge together into one role of composite network–cloud SP. This new service delivery model may stimulate innovations in service development and create a wide variety of new business opportunities.

15.6 Key Technologies for SOA-Based Convergence of Networking and Cloud Computing

As service description, discovery, and composition are key elements of web services for implementing SOA, network service description, discovery, and composition are key enabling technologies for the NaaS paradigm, which forms the foundation of service-oriented network–cloud convergence. In this section, a survey on the state-of-the-art of network service description, discovery, and composition technologies is presented and the challenges and research opportunities brought in by the convergence of networking and cloud computing to these technologies are also discussed.

15.6.1 Network Service Description

Network service description forms the basis of NaaS for network–cloud convergence as it determines the information that a network service needs to expose for enabling its unambiguous identification and usage in a cloud environment.

Web service description language (WSDL) [30] is the basic standard for web service descriptions. WSDL focuses on syntactic description and lacks the ability to provide semantic information. Numerous technologies have been developed for enhancing WSDL with semantic and nonfunctional (such as QoS) service information. These extensions include SAWSDL [31] and WS-QDL specifications recommended by W3C and the web service quality model (WSQM) [32] developed by OASIS. However, these technologies are mainly for general web services without particularly considering NaaS.

Efforts for applying web service descriptions to networking systems have been made by the telecom community. Parlay X and OMA Service Environment (OSE) both employ WSDL for describing network services for exposing telecommunications functionalities to upper-layer applications through a service abstraction layer. The European Computer Manufacturer Association (ECMA) also published a WSDL description of computer supported telecommunication applications [33]. By describing networking functions as web services with WSDL, these standards facilitate the creation and deployment of web service-based telecom applications. However, adoption of basic WSDL in these standards limits their capabilities of describing rich semantic and QoS information of network services. In addition, these specifications mainly aim at telephony-related functions, and further development is needed for supporting the wide variety of IP-based multimedia network services in a cloud environment.

Recently, researchers have tended to apply established semantic web tools in the networking realm and develop ontology specifically for describing network services. Network description language (NDL) developed in [34] is an ontology designed based on the resource description framework (RDF) to describe network services. NDL is basically an RDF vocabulary defined to describe network elements and topologies. In order to facilitate abstraction of networking resources, Campi and Callegai [35] proposed network resource description language (NRDL), which is also based on RDF but focuses on expressing the communication interaction relationship between network elements.

In addition to topology and connectivity information, networking capabilities and QoS properties are important aspects of network service descriptions. Research efforts in this direction include an abstraction algorithm [36] that presents network connectivity in a full-mesh topology associated with performance metrics and a capability matrix developed in [37] for modeling network service capabilities in a general form.

Despite the aforementioned progress, a network service description for NaaS is still a challenging open problem that offers research opportunities. The large-scale dynamic networking systems in cloud environments require a balance between the richness of information for accurate service descriptions and the abstraction and aggregation of service information for scalable networking. Describing QoS information is particularly important to network services but is very challenging, partly because of a lack of a standard specification of QoS attributes across heterogeneous network domains and the difficulty in measuring QoS performance. Network and cloud convergence also calls for research on general service description approaches that are applicable to both networking and computing services.

15.6.2 Network Service Discovery

As a core part of SOA, service discovery plays a key role in NaaS by discovering and selecting the most suitable network services that match user requirements for cloud computing and services.

Early efforts for service discovery in networking environments include IETF service location protocol (SLP) and industry standards, such as Jini, UPnP, Salutation, and Bluetooth. These protocols are mainly designed for personal/local or enterprise-computing environments and thus may not scale well to the Internet-based cloud environment. Service discovery is also an integral part in peer-to-peer (P2P) networks. Various technologies have been developed for achieving scalable and reliable service discovery in large-scale P2P overlay networks. A summary and comparison of these technologies are given in [38]. Mobile ad hoc networks bring special challenges to service discovery and trigger an extensive study. A survey of research progress in this area can be found in [39].

Most of the aforementioned technologies focus on locating devices that host functions or contents (e.g., data or files) in a networking environment, which is essentially the discovery of computing services in networks rather than the discovery of network services. Therefore, these technologies may not be directly applicable to a NaaS that offers virtualization of networking resources. Nevertheless, the obtained results in these areas provide insights on various aspects of service discovery that are valuable to network service discovery for supporting the NaaS paradigm.

The baseline approach to web service discovery is the OASIS standard UDDI [40], which specifies a data model for organizing service information and APIs for publishing and querying service descriptions. Both Parlay X and OSE adopted UDDI as the technology for network service discovery. Though serving as the de facto standard for service discovery during a certain period of time, UDDI lacks sufficient semantic support for service discovery. Research efforts have been made to enhance the UDDI data model with semantically rich metadata. Various distributed registry organizations and searching protocols have also been proposed for improving the scalability of web service discovery. A summary and comparison of these technologies can be found in [7]. However, these methods are developed for general web service and their applications to NaaS for cloud computing need further investigation.

Heterogeneity and scalability are two key requirements for network service discovery. In order to enable service discovery across heterogeneous network domains that adopt different service registry architecture and searching protocols, open service discovery architecture (OSDA) is proposed in [41], and it provides interdomain distributed storage and query of service information. PYRAMID-S architecture developed in [42] uses a hybrid P2P topology to organize service registries and provides a scalable framework for unified service publication and discovery over heterogeneous network domains. In [43], Cheng et al. propose a holistic service management framework with a P2P overlay broker network and a distributed service registry for enabling scalable service discovery across heterogeneous networks.

Adaptive service discovery is also important to NaaS in a cloud environment. Papakos et al. [44] propose an approach to dynamically adapt cloud service requests and provisioning according to user context and resource availability and develop a declarative language for specifying adaptation policy. Effectiveness and performance of the proposed method still need to be fully evaluated, and its application to network services needs further study.

Though encouraging progress has been made toward network service discovery for NaaS-based network–cloud convergence, this is still an open area that is rich in challenges and research opportunities. The wide variety of networking systems utilized by cloud computing, which scale from LANs to the Internet, make development of service discovery mechanisms meeting heterogeneity and scalability requirements a challenging but important research problem. Discovery of networking and computing services should be combined in a cloud environment in order to meet the user's requirements on cloud services. Therefore, discovery and selection of composite network–cloud services would also be an important research topic. Special features of cloud computing, such as elastic on-demand self-services, bring new challenges to dynamic and adaptive network service discovery. Therefore, research on effective and efficient methods for updating service information and adapting service requests is also crucial to achieving high-performance network service discovery in clouds.

15.6.3 Network Service Composition for Network–Cloud Convergence

Service composition plays an essential role in the NaaS-based convergence of networking and cloud computing. Cloud service provisioning consists of not only computing functions provide by cloud infrastructure, but also data communications offered by network infrastructure; therefore, in this sense, all the cloud services delivered to end users are essentially composite computing and networking services.

Web service composition has been an active research area for years. Numerous technologies have been developed to achieve functional and/or performance requirements of web service composition. Most of the technologies are based on either workflow management or AI-planning and employ heuristic search, linear programming, or automatic reasoning algorithms. Surveys on web service composition technologies are given in [44–46]. A recent survey, particularly on QoS-aware service composition, can be found in [8]. The aforementioned research works mainly aim at computing services instead of network services. Network service composition, a relatively new concept in the networking domain, recently started attracting the attention of the research community in this field.

Network composition is a core feature of the ambient network (AN) developed in the EU sixth framework project. Dynamic network composition in ANs enables transparent, on-demand cooperation across heterogeneous networking systems [47]. However, realization of network composition developed in the AN project was based on the generic ambient signaling system that lacks compatibility with the SOA principle, which limits its application to the NaaS paradigm in a cloud environment.

The OMA OSE supports service composition through a mechanism defined as a policy enforcer. Service enablers in OSE can be implemented in the form of web services as specified in the OSE Web Service Enabler Release (OWSER) [48]. Though service composition is not explicitly standardized in the OWSER specification, BPEL is adopted in OSE as a technology choice to express policy for service orchestration. This opens a door to apply web service composition technologies to composing network services in OSE, thus allowing network services to be composed through the same mechanism as web–cloud services in a cloud environment.

Network functional composition (NFC) has been proposed as a clean-slate approach to a flexible future Internet architecture. Technologies have been developed in various projects toward NFC, but they all share the idea of decomposing the layered network stack to functional building blocks and organizing the functionalities in a composition framework [49]. Through NFC, network services can be composed based on specific application requirements, therefore enabling customized functional networks to be assembled by different network services to provide the best service for supporting cloud computing. However, because of its disruptive approach, intensive investigation on NFC is still needed before it can be widely adopted into the Internet architecture and cloud computing environments.

Progress has been made in both web–cloud service and network service compositions; therefore, composition across these two types of services would be naturally the next step toward a convergence of networking and cloud computing. In NGSON recently standardized by IEEE, function entities for service composition and service routing are defined to establish service routes that combine composition of computing services and routing for data communications. The NGSON standard provides a protocol framework for these functional entities without specifying any technology for realizing the protocols. The problem of combined service composition and network routing was studied in [50]. The goal is to find the optimal service composition in a network that will lead to the minimum routing cost. A decision-making system was developed to solve this problem with the AI-planning technique. Though the authors studied the problem in the context of a general networking environment without considering cloud computing in particular, essentially the same problem exists in the foundation of networking and computing service composition for cloud computing. Therefore the solution proposed in [50] may be applied for network–cloud convergence.

Recent research on service composition has been conducted separately in the areas of web–cloud services and networking, each of which has its own features and requirements that lead to different technologies. Convergence between networking and cloud computing through the NaaS paradigm calls for bridging these two areas toward a generic composition framework in which different types of services, including both networking and computing services, are orchestrated to meet user requirements. This forms an interesting research topic with a lot of opportunities. QoS-based network service composition is very important to meet the performance requirements of cloud service provisioning. This is particularly challenging for composition across heterogeneous computing and networking services in a cloud environment, which is basically an unexplored area that deserves thorough investigation.

15.7 Modeling and Analysis for Converged Networking and Cloud Computing Systems

The end user's perception of cloud service quality, which has a direct impact on the operations of cloud-based applications, is determined by the performance of composite networking-computing service provisioning. Modeling and analysis provide a method to obtain a thorough understanding of, and deep insights about, composite network–cloud service performance. However convergence of networking and cloud computing brings new challenges to system modeling and analysis, which mainly come from the heterogeneity of SPs and virtualization of system resources. Composite network–cloud service provisioning systems comprise both networking and computing systems with diverse implementations. Therefore, modeling and analysis techniques for such systems must be general and applicable to the heterogeneous networking and computing systems. Both networking and computing systems are encapsulated into services through virtualization, which requires modeling and analysis techniques to be agnostic to implementation technologies. In this section, a recently developed modeling and analysis approach for converged network–cloud service provisioning systems is reported.

A typical provisioning system for composite network–cloud services is shown in Figure 15.6, which consists of both a cloud infrastructure that offers a cloud service and the communication system that provides a network service. In order to analyze the composite service performance, one must understand the communication capability offered by the network service as well as the computing capability provided by the cloud service. The methodology taken in this approach is to first develop a general capability profile that can model the service capabilities of both network and cloud services and then compose the capability profiles of the two service components into one profile that models the service capability of the composite system.

A capability profile for a service component can be defined as follows, which is based on the *service curve* concept from *network calculus* theory [51]. Let $R(t)$ and $E(t)$, respectively, be the accumulated amount of traffic that arrives at and departs from a service component by time t. Given a

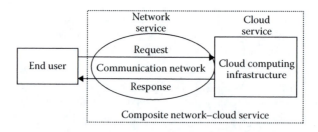

Figure 15.6   Composite networking and computing system for cloud service provisioning.

nonnegative, nondecreasing function, $P(\cdot)$, where $P(0) = 0$, we say that the service component has a capability profile $P(t)$ if for any $t \geq 0$ in the busy period of the service component, $E(t) \geq R(t) * P(t)$, where $*$ denotes the convolution operation defined in min-plus algebra.

The capacity of a composite network-cloud service provisioning system can be characterized by the model shown in Figure 15.7. Consider the general case that the two directions of data communications between the end user and the cloud infrastructure take two network services, denoted as S_{net1} and S_{net2}, respectively. The cloud service component is denoted as S_{cloud}. It is assumed that the forward communication (from user to cloud), the cloud service component, and the backward communication (from cloud to user), respectively, have the capability profiles $P_{net1}(t)$, $P_{Cloud}(t)$, and $P_{net2}(t)$. It is known from network calculus theory that the service curve of a system consisting of a series of tandem servers can be obtained from the convolution of the service curves of all these servers. Therefore, the capability profile for the composite service, denoted by $P_{Comp}(t)$, can be determined as

$$P_{Comp}(t) = P_{net1}(t) * P_{Cloud}(t) * P_{net2}(t).$$

The *latency-rate* (*LR*) profile, defined as follows, provides a more tractable profile for characterizing capabilities of typical network and cloud services. If a service component S has a capability profile $P[r, \theta] = \max\{0, r(t - \theta)\}$, then we say that the service component S has an *LR* profile, where the θ and r are, respectively, called the *latency* and *rate* parameters of the profile.

The LR profile can serve as the capability model for typical network services. The QoS expectation of a typical network service includes a certain amount of data transport capacity (the minimum bandwidth) guaranteed to a service user. Such a minimum bandwidth guarantee is described by the rate parameter r in the LR profile. Data communication in a network infrastructure also experiences a fixed delay that is independent with traffic queuing behavior; for example, signal

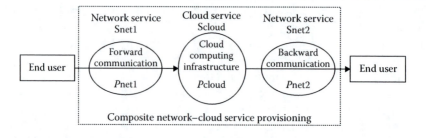

Figure 15.7   Capability model for a converged communication and computing system.

propagation delay, link transmission delay, router/switch processing delay, etc. The latency parameter θ of LR profile is to characterize this part of a fixed delay of a network service.

The LR profile can also characterize service capabilities of typical cloud computing systems.
Cloud SPs typically offer a certain amount of service capacity to each user. For example, according
to Amazon, each type of virtual machine (called instance) in Amazon EC2 provides a predictable
amount of computing capacity and I/O bandwidth. Each EC2 computer unit provides the equivalent CPU capacity of 1.0–1.2 GHz 2007 Opteron or 2007 Xeon processor. Amazon also claims an
internal I/O bandwidth of 250 Mb/s regardless of instance type. The latency and rate parameters
of the LR profile for a cloud service can be derived from the information of I/O bandwidth and
computing capacity provided by the cloud SP.

Suppose each service component in the composite network–cloud system has an LR profile,
that is

$$P_{net1} = P[r_1, \theta_1], P_{Cloud} = P[r_C, \theta_C], P_{net2} = P[r_2, \theta_2].$$

It can be proved that the capability profile of the composite service provisioning system is

$$P_{Comp} = P[r_1, \theta_1] * P[r_C, \theta_C] * P[r_2, \theta_2] = P[r_e, \theta_e],$$

where $r_e = \min\{r_1, r_C, r_2\}$ and $\theta_e = \theta_1 + \theta_C + \theta_e$. This implies that if each network and cloud service
component in a composite service provisioning system can be modeled by an LR profile, then the
service capability of the entire provisioning system can also be modeled by an LR profile. The
latency parameter of the composite LR profile is the summation of latency parameters of all service
components in the system, and the service rate parameter of the composite profile is the minimum
service rate of all the service components.

The performance analysis presented in this section focuses on the maximum response delay
(the delay between when a user sends out a request to the cloud and when the user receives the
corresponding response from the cloud). Delay analysis requires a method to characterize the
traffic that end users load onto a composite network–cloud service system. The concept of *arrival
curve* in network calculus is adopted here to define a general load profile as follows. Let $R(t)$ denote
the accumulated amount of traffic that arrives at the entry of a composite service system by time
instant t. Then given a nonnegative, nondecreasing function, $L(\cdot)$, if for all time instants s and t
such that $0 < s < t$, $R(t) - R(s) \le L(t-s)$, the service system is said to have a *load profile* $L(t)$.

Most QoS-capable networking systems apply traffic regulation mechanisms at network boundaries to shape arrival traffic from end users. The traffic regulators that are most commonly used
in practice are leaky buckets. A networking session constrained by a leaky bucket controller has a
traffic load profile $L[p, \rho, \sigma] = \min\{pt, \rho t + \sigma\}$, where p, ρ, and σ are, respectively, called the peak
rate, sustained rate, and maximal burst size for the traffic load.

The maximum response delay performance is associated with both the guaranteed service
capacity of the system, which is modeled by a capability profile, and the characteristic of the traffic load of the system, which is described by a load profile. It can be shown, by following network
calculus, that for a service system with a capability profile $P(t)$ under traffic described by a load
profile $L(t)$, the maximum delay d_{e2e} guaranteed by the system to its end user can be determined as

$$d_{e2e} = \max_{t:\, t \ge 0}\{\min\{\delta: \delta \ge 0, L(t) \le P(t + \delta)\}\}.$$

Because the LR profile can model typical network and cloud service capabilities, and a leaky
bucket traffic regulator is widely deployed at the entries of QoS-enabled networks, the rest of this

section focuses on analyzing the end-to-end response delay of composite network–cloud service systems with an LR profile and a leaky bucket load profile. Suppose the capability profile of the composite service system is $P_{Comp} = P[r_e, \theta_e] = \max\{0, r_e(t - \theta_e)\}$; then the maximum response delay guaranteed by this composite network–cloud service under a load profile $L[p, \rho, \sigma]$ can be determined as

$$d_{e2e} = \theta_e + (p/r_e - 1)(\sigma/(p - \rho)) = \theta_\Sigma + d_C + (p/r_e - 1)(\sigma/(p - \rho))$$

where $\theta_\Sigma = \theta_1 + \theta_2$, which is the round-trip communication latency of network services, and d_C is the computing latency of the cloud service.

Numerical examples are given in this section for illustrating application of the analysis technique. Considering a composite service provisioning system as shown in Figure 15.6 and suppose the round-trip data transmission is supported by the same network, then the forward and backward network services have an identical capability profile. We assume that each of the network services and the cloud services has an LR profile. The latency parameter of the cloud service profile is assumed to be 150 ms, and traffic parameters of the load profile are 320 Mb/s, 120 Mb/s, and 200 kb for the peak rate, sustained rate, and burst size, respectively.

The first example is a scenario in which a user accesses the cloud infrastructure through a high-speed network with a link transmission rate up to 10 Gb/s. The maximum delay performance of the composite network–cloud service is determined with different amounts of network service capacity (bandwidth). The obtained results are given in Table 15.1, which also gives the percentage of networking delay in the entire service response delay (including both networking and computing delay).

Table 15.1 shows that when the available network bandwidth is less than the peak load rate (325 Mb/s in this example), delay of the composite service decreases significantly with the increment of network service capacity. This implies that, in this case, networking forms the performance bottle for the composite service; therefore the network SP can contribute to delay performance improvement by leasing more bandwidth from underlying InPs. The table also shows that when the network service offers a transport capacity that is greater than the peak load rate, networking delay only contributes less than 30% of the total service delay, and computing delay in the cloud infrastructure becomes the major part of the total service delay. In this case, leasing more network bandwidth has a minor contribution for improving delay performance of the composite service.

In the other service scenario analyzed in this section, a user accesses the cloud infrastructure through a network with a moderate link rate up to 300 Mb/s. The results for the maximum service delay with different amounts of capacity (bandwidth) in the network service are given in Table 15.2, which also includes the percentage of networking delay in the entire service delay. This table shows that networking delay is a major part (more than 50%) of the total service delay, which decreases with the increment of available network capacity. This implies that when a user accesses

Table 15.1 Maximum Delay of Composite Network–Cloud Service with High-Speed Network

Available Bandwidth (Mb/s)	125	175	225	275	325	375	425	475
Service Delay (ms)	500	372	301	256	212	203	197	192
Networking Delay Percentage (%)	69.9	59.7	50.2	41.2	29.1	25.9	23.7	21.9

Table 15.2   Maximum Delay of Composite Network–Cloud Service with Moderate Speed Network

Available Bandwidth (Mb/s)	125	150	175	200	225	250	275	300
Service Delay (ms)	578	503	450	410	379	354	334	317
Networking Delay Percentage (%)	74.0	70.2	66.7	63.4	60.3	57.6	54.9	52.6

a cloud infrastructure through a bandwidth-constrained networking environment, such as a wireless network or a cellular communication system, the networking system may have a more significant impact on the user's perception of service performance than the cloud infrastructure does; that is, the network forms a performance bottleneck for cloud service provisioning.

15.8 Conclusions

The significant role that networking plays in cloud computing calls for a holistic vision of both networking and computing resources that allows combined control, management, and optimization in a cloud environment. This leads to a convergence of networking and cloud computing in NGNs. Network virtualization, which essentially decouples network service provisioning from data transport infrastructure, is being adopted in both telecommunication and Internet architecture and is expected to be a key attribute for NGNs. Virtualization, as a potential enabler of profound changes in both the communications and computing domains, is expected to bridge the gap between these two fields and enables the convergence of networking and cloud computing. The SOA offers an effective architectural principle for heterogeneous system integration. An overview presented in this chapter about recent research progress in the fields of both telecommunications and the Internet indicates that SOA has been adopted as a key mechanism for realizing network virtualization. Therefore SOA, applied in both cloud computing and network virtualization, may greatly facilitate the convergence of networking and cloud computing through the NaaS paradigm. Web services technology, as the main implementation method for SOA, will form the technical foundation of NaaS for network–cloud convergence. The survey given in this chapter on the key technologies for NaaS, mainly focusing on network service description, discovery, and composition, shows that although significant progress has been made toward NaaS for cloud computing, this field is still in an early stage, facing a lot of challenges and thus offering rich opportunities for future research. A new modeling and analysis method for performance evaluation of composite network–cloud services is also reported in this chapter, which employs network calculus techniques to address some challenging issues. Crossfertilization among multiple areas, including telecommunications, computer networking, web services, and cloud computing, may provide innovative solutions to network–cloud convergence, that will significantly enhance the performance of not only next generation networks, but also the entire future information infrastructure.

References

1. I. Foster, Y. Zhao, I. Raicu, and S. Lu, "Cloud Computing and Grid Computing 360-Degrees Compared." Proceedings of the 2008 Grid Computing Environment Workshop, Nov. 2008.

2. K. R. Jackson, K. Muriki, S. Canon, S. Cholia, and J. Shalf, "Performance Analysis of High Performance Computing Applications on the Amazon Web Services Cloud." Proceedings of the 2nd IEEE International Conference on Cloud Computing Technology and Science (CLOUDCOM 2010), Nov. 2010.

3. G. Wang and T. S. Eugene Ng, "The Impact of Virtualization on Network Performance of Amazon EC2 Data Center." Proceedings of IEEE INFOCOM 2010, March 2010.

4. T. Magedanz, N. Blum, and S. Dutkowski, "Evolution of SOA Concepts in Telecommunications." *IEEE Computer Magazine* 40, no. 11 (2007): 46–50.

5. D. Griffin and D. Pesch, "A Survey on Web Services in Telecommunications." *IEEE Communications Magazine* 45, no. 7 (2007): 28–35.

6. N. M. M. K. Chowdhury and R. Boutaba, "Network Virtualization: State of the Art and Research Challenges." *IEEE Communications Magazine* 47, no. 7 (2009): 20–26.

7. M. Rambold, H. Kasinger, F. Lautenbacher, and B. Bauer, "Toward Automatic Service Discovery: A Survey and Comparison." Proceedings of the 2009 IEEE International Conference on Services Computing (SCC 2009), Sept. 2009.

8. A. Strunk, "QoS-Aware Service Composition: A Survey." Proceedings of the 8th IEEE European Conference on Web Services, Dec. 2010.

9. T. Erl, *Service-Oriented Architecture: Concepts, Technology, and Design.* Prentice Hall, New Jersey, 2005.

10. K. Channabasavaiah, K. Holley, and E. Tuggle, "Migrating to a service-oriented architecture." *IBM Developer Works,* 2003.

11. T. Magedanz, "IN and TMN: Providing the Basis for Future Information Networking Architectures." *Computer Communications* 16, no. 5 (1993): 267–276.

12. ETSI, Parlay X 2.1 Web Service Specification. Available at http://docbox.etsi.org/TISPAN/Open/OSA/ParlayX30.html.

13. ITU-T, "Functional Requirements and Architecture of the NGN Release 1." Recommendation Y.2012, Sept. 2006.

14. K. Knightson, N. Morita, and T. Towl, "NGN Architecture: Generic Principles, Functional Architecture, and Implementation." *IEEE Communications Magazine* 43, no. 10 (2005): 49–56.

15. The Open Mobile Alliance, "OMA Enabler Releases and Specifications: OMA Service Environment Architecture Document." Nov. 2007.

16. TM Forum, TMF061 Service Delivery Framework (SDF) Reference Architecture, July 2009.

17. IEEE Standard 1903, "Functional Architecture of Next Generation Service Overlay Networks." October 2011, available at http://grouper.ieee.org/groups/ngson/.

18. C. Makaya, A. Dutta, B. Falchuk, D. Chee, S. Das, F. Lin, M. Ito, S. Komorita, T. Chiba, and H. Yokota, "Enhanced Next-Generation Service Overlay Networks Architecture." Proceedings of the 2010 IEEE International Conference on Internet Multimedia Systems Architecture and Application, December 2010.

19. T. Anderson, L. Peterson, S. Shenker, and J. Turner, "Overcoming the Internet impasses through virtualization." *IEEE Computer Magazine* 38, no. 4 (2005): 34–41.

20. GENI Planning Group, "GENI design principles." *IEEE Computer Magazine* 39, no. 9 (2006): 102–105.

21. J. Turner and D. E. Taylor, "Diversifying the Internet." Proceedings of IEEE Globecom 2005, Nov. 2005.

22. N. Feamster, L. Gao, and J. Rexford, "How to lease the Internet in your spare time." *ACM SIGCOMM Computer Communications Review* 37, no. 1 (2007): 61–64.

23. L. M. Correia, H. Abramowicz, M. Johnsson, and K. Wunstel, *Architecture and Design for the Future Internet.* Springer, London, 2010.

24. FEDERICA Project, Deliverable DSA1.1, "FEDERICA Infrastructure version 7.0." Available at http://www.fp7-federica.eu/documents/FEDERICA-DSA1.1.pdf.

25. M. Boucadair, P. Georgatsos, N. Wang, D. Driffin, G. Pavlou, and A. Elizondo, "The AGAVE approach for network virtualization: Differentiated services delivery." *Annals of Telecommunications* 64, no. 5–6 (2009): 277–288.

26. E. Grasa, G. Junyent, S. Figuerola, A. Lopez, and M. Savoie, "UCLPv2: A network virtualization framework built on web services." *IEEE Communications Magazine* 46, no. 3 (2008): 126–134.

27. ITU-T Focus Group on Cloud Computing (FG Cloud). http://www.itu.int/en/ITU-T/focusgroups/cloud/Pages/default.aspx.

28. L.-J. Zhang and Q. Zhou, "CCOA: Cloud Computing Open Architecture." Proceedings of the 1st Symposium on Network System Design and Implementation (NSDI '09), April 2009.
29. Open Grid Forum (OGF), "Open Cloud Computing Interface." http://occi-wg.org/, May 2010.
30. World Wide Web Consortium (W3C), "Web Service Description Language (WSDL) version 2.0." June 2007.
31. World Wide Web Consortium (W3C), "Semantic Annotation for WSDL and XML-Schema." August 2007, available at http://www.w3.org/TR/sawsdl/.
32. OASIS. "Web Services Quality Model (WSQM)," August 2004, available at http://www.oasis-open.org/committees/.
33. ECMA, "Services for Computer Supported Telecommunications Applications (CSTA), the 9th edition." December 2011, available at http://www.ecma-international.org/publications/standards/Ecma-269.htm.
34. J. Ham, P. Grosso, R. Pol, A. Toonk, and C. Laat, "Using the Network Description Language in Optical Networks." Proceedings of the 10th IFIP/IEEE International Symposium on Integrated Network Management, May 2007.
35. A. Campi and F. Callegai, "Network Resource Description Language." Proceedings of the 2009 IEEE Global Communication Conference, Dec. 2009.
36. C. E. Abosi, R. Nejabati, and D. Simeonidou, "A novel service composition mechanism for the future optical Internet." *Journal of Optical Communications and Networking* 1, no. 2 (2009): A106–A120.
37. Q. Duan, "Network service description and discovery for high performance ubiquitous and pervasive grids." *ACM Transactions on Autonomous and Adaptive Systems* 6, no. 1 (2011): 3:1–3:17.
38. E. Meshkova, J. Riihijarvi, M. Petrova, and P. Mahonen, "A survey on resource discovery mechanisms, peer-to-peer and service discovery frameworks." *Computer Networks Journal* 52, no. 11 (2008): 2097–2128.
39. A. Mian, R. Baldoni, and R. Beraldi, "A survey of service discovery protocols in multihop mobile ad hoc networks." *IEEE Pervasive Computing Magazine* 8, no. 1 (2009): 66–74.
40. OASIS, "Universal Description, Discovery and Integration (UDDI) version 3.0.2." Feb. 2005.
41. N. Limam, J. Ziembicki, R. Ahmed, Y. Iraqi, D.-T. Li, R. Boutaba, and F. Cuervo, "OSDA: Open service discovery architecture for efficient cross-domain service provisioning." *Journal of Computer Communications* 30, no. 3 (2007): 546–563.
42. T. Pilioura and A. Tsalgatidou, "Unified publication and discovery of semantic web services." *ACM Transactions on the Web* 3, no. 3 (2009): 11:1–11:44.
43. Y. Cheng, A. Leon-Garcia, and I. Foster. "Toward an Automatic Service Management Framework: A Holistic Vision of SOA, AON, and Autonomic Computing." *IEEE Communications Magazine* 46, no. 5 (2008): 138–146.
44. P. Papakos, L. Capra, and D. S. Rosenblum. "VOLARE: Context-Aware Adaptive Cloud Service Discovery for Mobile Systems." Proceedings of the 9th International Workshop on Adaptive and Reflective Middleware, Nov. 2010.
45. S. Dustdar, and W. Schreiner. "A Survey on Web Services Composition." *International Journal of Web and Grid Services* 1, no. 1 (2005): 1–30.
46. J. Rao, and X. Su. "A Survey of Automated Web Service Composition Methods." Proceedings of 1st Int. Workshop on Semantic Web Services and Web Process Composition, 2004.
47. F. Belqasmi, R. Glitho, and R. Dssouli, "Ambient network composition." *IEEE Network Magazine* 22, (2008): 6–12.
48. Open Mobile Alliance, "OMA Web Services Enabler version 1.1," March 2006.
49. C. Henke, A. Siddiqui, and R. Khondoker, "Network Functional Composition: State of the Art." Proceedings of the 2010 IEEE Australasian Telecommunication Networks and Applications Conference, Nov. 2010.
50. X. Huang, S. Shanbhag, and T. Wolf, "Automated Service Composition and Routing in Networks with Data-Path Services." Proceedings of the 19th IEEE International Conference on Computer Communication Network (ICCCN 2010), Aug. 2010.
51. J. L. Boudec and P. Thiran, *Network Calculus: A Theory of Deterministic Queuing Systems for the Internet,* Springer Verlag, London, 2003.
52. Q. Duan, "Service-oriented network virtualization for composition of cloud computing and networking." *International Journal of Next Generation Computing* 2, no. 2 (2011): 123–138.

Chapter 16

Rule-Driven Architecture for Managing Information Systems

Juan M. Marín Pérez, Jorge Bernal Bernabé,
José M. Alcaraz Calero, Jesús D. Jiménez Re,
Félix J. García Clemente, Gregorio Martínez Pérez,
and Antonio F. Gómez Skarmeta

Contents

16.1 Introduction

The definition of an advanced and distributed management framework is one of the key research issues that still need to be improved in the network and service management research field. To deal with all aspects of the system management, the researchers mainly follow a policy-based management approach. This approach enables the usage of policies defined in a well-known language to

417

describe the system behavior. Thus, it permits the specification of high-level policies (rules) with the aim of becoming low-level configurations suitable to be directly applied into final devices. Therefore, the architecture described in this chapter makes use of a rule-driven approach to manage the behavior of a distributed system.

So far, important efforts have been made in order to standardize a framework enabling the management of information systems in different application domains. Inside the Distributed Management Task Force (DMTF) [1] standardization organism, several works have been done in the definition of the basic components of generic policy-based management architecture. In this sense, the DMTF defines the Common Information Model (CIM) [2] to be used by the different components building the management architecture. On the other hand, the Organization for the Advancement of Structured Information Standards (OASIS) has been also working in the same direction and provides another framework proposal for resource management based on service-oriented architecture (SOA). Whereas both DMTF and OASIS are endowed with a set of components that set up a management framework based on web standards, they are lacking, at the same time, in some advanced features, such as support to multidomain environments, extension capability, policy checking, or reconfiguration capabilities.

In this context, this chapter proposes a novel architecture to overcome these lacks and which evolves components of the management architecture using XML technologies and web standards. This proposal allows the integration of heterogeneous management applications running on different platforms as well as allowing new components to be plugged into the architecture using standard interfaces. Moreover, it provides a monitoring mechanism to assure the system is fulfilling the management policy defined by the administrator and detecting possible attacks or misbehavior. A status control mechanism provides reconfiguration capabilities, being able to automatically change the system configuration according to monitoring events.

Different basic areas have been identified in order to provide the architecture with the desired management functionality. Thus, a *requirement management area* manages the context information and the requirements defined by the administrator. A *common model management area* is in charge of managing the common information model used by the different architecture components. An *analysis area* performs a validation process to assure that the defined requirements do not contain any inconsistency or possible conflicting semantics. A *configuration area* uses the information defined in the common model to produce and enforce the specific configurations for the actual system devices and services. A *monitoring area* monitors the managed system behavior to detect possible anomalies. Finally, a *status control area* controls the system status processing the different alerts that may arise from the monitoring system.

The proposed architecture has been tested and implemented in a prototypical framework, which offers a set of automatic tools to manage security in protecting networked infrastructures and applications. Making use of these tools, the framework solves some of the limitations of the previous models, providing the administrators the ability to manage a wide range of different and heterogeneous resources. Moreover, the web service-based communication between all components makes possible the distributed execution of the architecture, allowing each component to be executed in different locations. In order to provide a consistent cooperation and a seamless integration among the different components, the management architecture follows a well-known defined workflow. It describes in detail how the different components should cooperate in order to achieve a whole and correct system operation.

The chapter is organized as follows: The first section presents a review of both the DMTF and OASIS proposals for systems management. The second section presents our proposal describing a novel architecture and is composed of a set of management areas and modules. The third section is focused on the implementation and details about the extensibility and the workflow of the proposal. The fourth section presents new challenges of cloud computing management. Finally, the concluding section presents some concluding remarks.

16.2 Background

In 2004, the DMTF devised a framework based on web technologies, which was called the Web-Based Enterprise Management (WBEM) [3] to provide a unifying mechanism for describing and sharing management information. WBEM combines the usage of the CIM as a data model with several web technologies as transport and representation mechanisms for the information defined by the model CIM.

With the arrival of the web services technology, the Web Services Distributed Management (WSDM) [4] from OASIS appears with the goal of evolving the current management infrastructure to a management approach based on web services in which the infrastructure is independent of vendors and platforms, allowing standard messaging protocols between a resource and the client that manages it.

In this direction, the DMTF evolved WBEM to a new approach also based on web services, including the Web Services Management (WS-Management) specification [5] and resulting in the first specification that exposes CIM resources using a set of web-service protocols. Moreover, there is even a proposal by the DMTF to include WSDM as part of the WBEM set of standards, unifying the DMTF and OASIS management architectures with the CIM.

The following two subsections provide a brief overview of these standards.

16.2.1 Web-Based Enterprise Management

WBEM is a set of standard networking and management technologies developed by the DMTF to unify the management of distributed systems. It uses CIM as an information model and provides a mechanism of exchanging CIM information in an efficient and interoperable way. The WBEM standard includes protocols, query languages, discovery mechanisms, mappings, and all the needed resources to exchange CIM information. DMTF technologies are defined independently in order to provide the maximum flexibility. CIM specifies the syntax that is used to define the structure of the management information, and WBEM provides an interoperable and extensible way to manage CIM information. The set of standards defined by WBEM together with the model CIM provides the information management infrastructure. Separately, each of these DMTF technologies is interesting by itself, but when they are used together, they provide a powerful enterprise management solution.

The WS-Management specification of WBEM deals with the cost and complexity of IT systems by providing a common mechanism to access and exchange management information. Using web services to manage the systems, deployments with WS-Management support will allow the administrators to access remotely all kind of devices making use of web-service protocols.

This specification provides mechanisms to do the following:

■ Obtain, update, create, and delete resource instances as well as their properties and values
■ Enumerate the contents of containers and collections as long tables and logs
■ Subscribe to events sent by the managed resources
■ Execute specific management commands

For each of these areas, the specification defines a set of minimal requirements that should be implemented to fulfill the web-service standards. A particular implementation is allowed to extend its functionality beyond the defined set of operations, and it even could choose to provide no support for one or more functionality areas from the above list if such functionality is not appropriate for the managed resource.

WS-Management makes use of endpoint references (EPRs) as defined by the WS-Addressing standard as an addressing model for single instances of the resources. It also defines an EPR format to be used in the addressing of resources. The access to resources implies synchronous operations in order to get, set, and enumerate values. The WS-Transfer specification is used for unitary resources. For operations that imply multiple instances, it makes use of WS-Enumeration messages. If the service is able to send events, it should publish these events using the WS-Eventing standard. WS-Management also imposes a set of additional restrictions to the general specification of WS-Eventing.

16.2.2 Web Services Distributed Management

WSDM is a standard whose main goal is to unify the management infrastructures by providing a framework independent from the platform, network, and protocols to allow the management technologies to access and get notifications from resources with management capabilities. It is based on an XML standardized suite and can be used to standardize the management of a wide range of devices, from network devices to electronic devices, such as televisions, video players, and PDAs.

WSDM has been developed based on the set of web service standards and the SOA architecture. It is a specification and a set of standards. It defines two main standards: *MUWS* and *MOWS*. The former deals with the management of any resource using web services, and the latter considers a web service itself as a manageable resource, defining a way to manage it also using web services.

The WSDM standard specifies how to make a resource management available to the clients by means of web services. The WSDM architecture is based on what is called a *manageable resource*. A *manageable resource* is represented by a web service. In other words, the management information regarding the resource should be accessible through a web-service endpoint. To provide access to a resource, this endpoint should be able to be referenced by means of an EPR defined by the WS-Addressing standard. The EPRs that give access to a manageable resource are called *manageability endpoints*, and their implementation should be able to recover and manage the information from the corresponding resource.

An EPR provides the point to which a management client should send its messages. The manageable resource can also launch event notifications, which could result from interest for the client, provided that the client has been previously subscribed to receive such notifications. This way, WSDM provides three interaction modes between a manageable resource and the management client. These three interaction modes are the following:

■ A client can recover management information about a resource. For instance, the client can recover the current status of the resource or the current status of a process running in the resource.
■ A client can affect the status of a resource by changing its management information.
■ A resource can inform or notify a client with a relevant event. This interaction model requires the previous subscription of the client to receive events from a particular topic.

WSDM tries to define a common management structure and a message exchange format through which a manageable resource and a client could communicate regardless of their implementation or platform. The compliance to the standard requires that both the client and the resource should be able to generate messages with the specified format and also to fulfill several requirements. Thus, WSDM defines a messaging protocol for information management that should be shared by the client and the resource regardless of their platform and implementation.

A WSDM service is a management interface for web resources. However, except for some metrics, WSDM does not specify the contents of any accessible management information. The standard only specifies the format to recover and manipulate management information.

Both WBEM and WSDM lack of some mechanisms to cover advanced framework features, such as policy checking for conflict detection, monitoring, or reconfiguration capabilities as well as some features, such as the ability to deal with a secured multidomain environment. Nevertheless, the concepts and ideas described in these two frameworks have served as a reference for the construction and design of the management architecture described in this chapter.

16.3 Novel Management Architecture

This section describes the proposed architecture for information systems management. The proposal evolves the work carried out previously in the European project Policy-based Security Tools and Framework (POSITIF) [6] and lately in the author's research [7,8].

Before defining the architecture itself, the following set of main requirements that it should fulfill have been identified. Most of these requirements have been taken from [9].

- *Interdomain.* The system should be able to interoperate with other architectures that may exist in other administrative domains.
- *Formal underlined model.* Policy architectures need to have a well-defined model independent of the particular implementation in use. In it, the interfaces between the components need to be clear and well defined.
- *Flexibility to be able to deal with a wide variety of device types.* The system architecture should be flexible enough to allow the addition of new types of devices with minimal updates and recoding of existing management components.
- *Conflict detection and resolution.* It has to be able to check that a given policy does not conflict with any other existing policy. If possible, the system should be able to suggest some solution in case a conflict has been detected.
- *Integrating the business and networking worlds.* The system should be able to map business policies (i.e., high-level policies) into middleware or network policies (i.e., low-level policies).
- *Scalability.* It should maintain quality performance under an increased system load.
- *Monitoring.* It should perform intensive and extensive monitoring over all the managed elements using various detection mechanisms to ensure fast detection of severe incidents and avoiding any impact propagation.
- *Reaction upon incident.* The system should be able to respond in a quick and appropriate way to a large range of incidents to mitigate the threats to the dependability and thwart the problem.
- *Reconfigurability.* The system should be able to change its behavior, adapting itself to changes in the environment. Critical activities should be prioritized by means of a fast reconfiguration.

The proposed management architecture tries to cover the aforementioned requirements. It provides rule-based management and uses a standard data model to represent information, resulting in a distributed and multiplatform, resource-independent management system. The architecture is capable of providing monitoring mechanisms to assure proper device functionality as well as possible attack detection. It also provides a status control mechanism, with reaction capabilities against possible attacks or failures detected by the monitoring system, that is able to make an automatic system reconfiguration.

In this framework, the administrator defines the system architecture and behavior in a high-level way, expressing management requirements in a definition language less detailed and with a higher level of abstraction than that used by the standard models introduced in the previous section. It uses some context information to detach specific and detailed data from high-level requirements.

The proposed management architecture is composed of the following basic areas:

- *Requirement management area.* It is in charge of managing the requirements defined by the administrator in a high-level language able to represent abstract concepts. It also manages the context information needed by those abstract concepts to be translated into a lower level and specific model representing all the details that could be later used to generate the configuration for final devices and services. This area is also in charge of the translation process from the high-level requirements with its context information to the common model with all the details that the rest of the framework components will use.
- *Common model management area.* It manages the standard common model information that will be used by the rest of the framework to enforce the configurations of the final devices, to monitor that the system fulfills the requirements defined by the administrator, and for any other additional functionality that could require the information that models the managed system.
- *Monitoring area.* It monitors the managed system to detect possible anomalies, failures, or possible intrusions that could compromise the security or availability of the system and reports it for the rest of the components to act accordingly.
- *Configuration area.* Based on the information defined by the common model, this area is in charge of the generation and enforcement of the appropriate specific configurations for every final device or service in order to make the global operation of the system to fit the requirements imposed by the administrator.
- *Status control area.* It controls the system status, listening for the possible alerts from the monitoring system and being able to react against them by applying new configurations to recover the system from possible failures or by applying higher security levels in order to protect the system from possible intrusions.

The usage of a common and standard data model, such as the CIM, allows all the modules that make up the architecture to use a single model to represent the needed management information. Moreover, as this model is independent from any implementation, and the usage of XML languages as well as communication protocols, such as web services that are also standards and platform independent, it makes the architecture easily extensible by adding new areas and modules to it.

Figure 16.1 shows the management architecture, depicting the above-mentioned areas with the different modules they contain. Circles depict the web services for the modules, which represent the architecture. Modules with a database icon are in charge of storing and retrieving the data whose management is their responsibility.

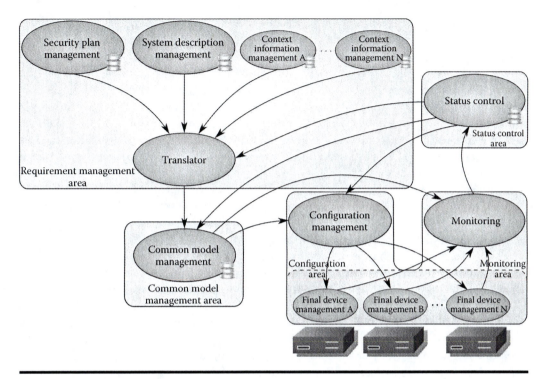

Figure 16.1   Proposed management architecture.

The *requirement management area* is composed of the following modules:

■ The *security plan management* module is in charge of managing the security requirements the administrator defines for the system. It provides methods to obtain, insert, delete, and modify the different requirements and security plans as well as methods to get meta-information, such as the different requirement types supported by the system and which it is able to translate to the common model and manage.

■ The *system description management* module manages the information describing the managed system. As in the *security plan management* module, it provides methods to obtain, insert, delete, and modify the descriptions for the different system components that are being managed.

■ A set of modules are also defined to manage the context information (*context information management A,..., context information management N*) that could be needed in order to generate the detailed model from the abstract concepts expressed in the higher level languages that define both the security requirements and the managed system description. Each of them is in charge of the administration of the concepts related to a context information area. The number of these modules will change for different architectures, depending on the different context information areas that are taken into account. Concepts of these areas do not necessarily need to be defined by administrators. Security experts in specific fields could provide detailed definitions of abstract concepts, which could then be used by nonexpert administrators to produce security plans that will be translated to good and secure models, taking into account those experts' knowledge. This enables nonexpert administrators to automatically perform the same decisions as security experts without having to become experts in specific security fields.

- A translation process is performed by the *translator* module. This module will send a request to the *security plan management*, the *system description management*, and the web services that manage the context information areas that need data to make the translation of the defined security plans to the common model, and it will send the resulting model to the *common model management* module [8].

The *common model management area* includes the *common model management* module, which is in charge of the CIM management and storage, providing methods to insert, modify, delete, and obtain this model by the rest of the elements that represent the system and that may require some information stored in this model in order to do their task.

To generate and enforce the configurations described by the common model to the final devices, the *configuration area* includes the *configuration management* module as well as several specific modules for managed devices (*final device management A,..., final device management N*), which will implement a common web service interface to be used by the *configuration management* service to perform the configuration enforcement. In this sense, these little modules can be considered as plug-ins, which should be provided for each kind of managed device in order to perform the transformation from the common model to the specific device-dependent configuration and to enforce it.

Notice that, although it is possible if they are wanted, these modules do not have to be installed into the final devices. In fact, there should be a module for each kind of device, not for every single device, because one of these modules can be in charge of one or more final devices, enforcing their configurations using any standard (or proprietary) protocol, such as uploading the configuration files via the File Transfer Protocol (FTP), the Secure Shell (SSH), or whatever. At this point, some standard management architectures, such as WBEM or WSDM, may be used by simply defining an adaptor web service with the proper interface used by the *configuration management* service, or even this module may also be able to interact with the web-service interfaces defined by those standards because, in the end, those standards are also based on web-services technology and are able to manage information models such as CIM.

In the *monitoring area*, the *monitoring* module permanently checks the system in order to detect any behavior that violates the requirements defined by the administrator. This module has a vulnerabilities database and is in charge of collecting the events received from a set of sensors situated in the final devices, comparing the monitored data with the defined requirements.

It should be noted that this monitoring service can go beyond the basic standard IDS functionality because it can understand the requirements defined by the administrator in the security plan and reflected in the CIM, being able to generate alerts when an event that violates the security plan is detected. This allows even the detection of unknown attacks, which cannot be found on standard vulnerabilities databases. Using the *security plan management* module, the administrator is able to define different security plans, which can be used to be applied in different situations. This can result in different models according to the situation of the system.

In the *status control area*, the *status control* module is in charge of being in control of the different situations or status, where the system can be and consequently is applying the appropriate security plan. For instance, different security plans could be defined for different security levels: *low, medium, high*. At the beginning, the system would be in one of these levels, but upon detection of any attack or intrusion by the monitoring system, the *status control* module would be able to react by applying a higher security level to try to counteract the attack with much more restrictive system security. In the same way, if it is considered that the threat has disappeared, the system could go back to its original status to avoid affecting possible operations, which could be restricted by the higher security level.

Finally, to keep all modules of the architecture synchronized, an event management system is needed for the different modules to communicate possible changes in their status or for them to be able to launch alerts asynchronously. For instance, the *monitoring* module should make use of such an event management system to launch alerts about possible anomalies it detects, and these alerts could be listened to by the *status control* module to react by applying another configuration if necessary. At the same time, the *configuration management* module will launch a new event every time it enforces a new configuration. This event will be listened to by the *monitoring* module to be noticed about the new requirements, which the system should now accomplish.

16.4 Implementation of Proposed Architecture

The components of the architecture have been designed using web-service standards to achieve the integration of heterogeneous management applications running on different platforms as well as allowing new components that could be plugged in to the architecture using standard interfaces. Web-service technology was designed to deal with application integration issues, more specifically heterogeneous applications from different platforms and implementation technologies. The widespread use of web service-based open standards gives the opportunity to use these technologies to integrate management applications for a set of heterogeneous resources.

In order to test the management architecture depicted in the previous section, a prototypical implementation has been developed as an extensible and interoperable framework, which may serve as a basis for managing systems based on the proposed architecture. This section provides an overview of the different components and technologies that are part of such a prototype.

16.4.1 Security Policy Language and System Description Language

For the *security plan* and *context information* definitions, two high-level languages have been developed. These languages allow the administrator to define policies and a description of the managed system using simple languages with abstract semantics. The language that allows the definition of the policies is called *security policy language* (SPL). This language is used to define the security plan and some context information. The language used to define the managed system description is called *system description language* (SDL), and it allows describing the underlying system elements. This prototype provides support for five different kinds of policies that can be defined by means of SPL: (1) *authentication* policies to define how identities are validated in the system; (2) *authorization* policies to specify access control; (3) *filtering* policies to define filtering criteria using a network element; (4) *channel protection* policies to define some requirements based on security associations, such as IPsec or SSL; and (5) *operational* policies that enable it to describe the behavior of the network upon the occurrence of an event.

The model CIM is used as common model that flows through the different components of the framework, which implements the architecture. Concretely, an XML representation of this model called xCIM has been used [6]. But CIM tries to cover a wide range of aspects related to IT systems. Therefore, because xCIM is a full implementation of CIM (including extended classes), it provides a large number of classes, which represent the model. A subset of xCIM is used with the needed classes to represent the concepts defined by languages SPL and SDL. Concretely, two submodels are used: xCIM-SDL and xCIM-SPL. The former allows the representation in the model of the concepts defined by the high-level language SDL with the system description information. The latter allows the representation in the model of the policies, groups, and some context information defined by the SPL.

16.4.2 Management Modules

Figure 16.2 shows the prototypical instantiation of the modules of the architecture presented in the previous section. Web services are used to provide the needed operations for the different modules of the architecture areas. They are represented by circles in the figure. Communication between the different web services uses the simple object access protocol (SOAP). These communications are depicted with arrows. Connections with the event manager module are not shown in the figure for clarity reasons. Any module uses the event mechanism provided by this module to stay tuned in about the framework status.

The *requirement management area* is mainly composed of the *security plan management* and *system description management* modules. They provide the needed functionality for the administrator to define the managed system description and the policies with the security requirements that should be satisfied. This module contains the functionality for managing the policies in SPL and the system description in SDL as well as two kinds of context information that are used for policy definition. One of them is the set of *common elements* that are defined by the SPL itself and that will be referenced by the security policies defined by this language. The second one contains the parameters to establish security associations for channel protection policies. This last kind of context information is not defined by the SPL, but it has its own management component, not depicted in the figure for clarity reasons. A console providing a graphical user interface to aid administrators in the system management has been developed. This console generates the corresponding descriptions and policies expressed in SDL and SPL. It also allows the administrator to control some other modules of the framework, such as seeing monitoring logs and alerts or applying a different configuration by means of the *status control* module.

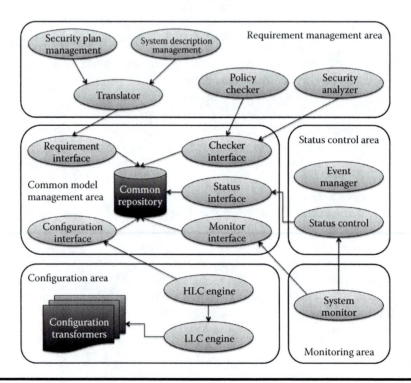

Figure 16.2 Management framework prototype.

The *translator* module is also inside the *requirement management area*. This module performs the translation of the security requirements defined by the administrator to the CIM formed by xCIM-SPL and xCIM-SDL. This translation process is done dynamically. Each time a requirement (i.e., a policy) is added, modified, or deleted from the security plan, the translation is done, and the common model is updated. The same applies to the context information defined by the *common elements* of the SPL, which are also dynamically translated to the CIM.

As noticed in the previous section, context information can be defined by security experts beforehand, allowing the administrator to use this information, which will lead to good models. For instance, when defining a *channel protection* policy, the administrator does not have to select which encryption or signature algorithms should be used in the security association that will secure the communication channel. That information will be defined in context information. Administrators can make use of concepts such as "high security," and the chosen algorithms are guaranteed to be secure enough because security experts have specified them for the high-security concept of the context information.

The different security plans defined in the architecture correspond to different security levels. Each security policy and policy group defined in SPL belongs to a specific security level, identifying in this way the security plan to which it belongs. These security levels are prioritized and used to change to a more restrictive security level if an intrusion or attack is detected by the monitoring system.

Regarding data storage, although the proposed architecture allows a totally distributed storage, this prototype framework implementation uses a *common repository* based on a native XML database for the storage of all the data used by the framework. This repository is represented by the central block in Figure 16.2, and many web services make use of it to store its managed information.

The *requirement management area* also contains the *policy checker* module. This module checks whether the defined security policies are semantically coherent and if the requirements defined by them could be achieved by the managed system specified in the system description. If this checking fails, it will be reported to the administrator, allowing him or her to solve the problem. Another module called the *security analyzer* is able to provide a theoretical security measurement of the security level that could be achieved by the system once the deployment of the configuration is done. Given that this analysis can be done as many times as needed before the actual enforcing to the real system, several alternatives can be evaluated for the system architecture and security requirements.

Once the security plan and the system description are translated to the final common standard model (CIM) and the security of the system has been successfully checked and evaluated, the required and detailed information is available. So, it is possible to transform this data model into the final configuration that will be applied in real system devices. This provides the administrator with a powerful way of having the managed system and all its network nodes configured according to the desired security requirements without dealing with the problems of having a multivendor and multiplatform architecture.

The first steps of this transformation are performed in the *configuration area*. The *HLC engine* in this area is in charge of creating the desired security configurations according to user requirements specified in the common data model definition in the xCIM format. These configurations are produced in a generic way into what is called *high-level configuration* (HLC) documents. This level of abstraction is considered because it is possible that diverse elements with different features have similar capabilities. So they may have to apply the same configuration. This can be the case of different host software elements and networking devices doing filtering capabilities of the same kind of malicious traffic.

These HLCs have to be then translated into specific configurations according to the actual systems where the enforcing takes place. This work is also done in the *configuration area* by another component called the *LLC engine*. This module produces *low-level configuration* (LLC) documents taking as input the previously generated HLCs.

LLC documents are generated by parsing the generic security elements defined in the HLCs and producing the particular configuration parameters. To do this, both XSLT (XSL transformations) and Java technologies have been used. XSLT provides the parsing of fixed parameters, and Java classes are used to support any dynamic feature that needs to be included in the configuration file. These XSLT transformations and Java classes used to translate from HLCs to LLCs are called *configuration transformers*. Different *configuration transformers* are used to generate device-specific configurations for different kinds of system elements and devices.

At the end of the configuration generation, the *LLC enforcer* module in the *configuration area* applies the configuration on the target system devices or services. Until this point, the described processes affect the whole set of policies for all the defined security levels. The security requirements and system definitions are checked and evaluated for each security level, and the configurations are also generated for all of them. This allows a quick deployment in order to provide a quick reaction if an intrusion or security hazard is detected. The *LLC enforcer* takes the configurations for the current security level and deploys them to the final devices and services. This deployment can be done using different protocols, such as Simple Network Management Protocol (SNMP), Common Open Policy Service (COPS/COPS-PR), or Hypertext Transfer Protocol (HTTP/HTTPS). Proprietary protocols are also supported thanks to the plug-in interface provided by the *LLC Enforcer* module.

The *monitoring area* is composed of the *system monitor* and a set of security modules called *security watchers*. These modules are deployed in the managed system through a set of lightweight and small-footprint modules installed on the system devices. They provide monitoring capabilities and cooperate with the *system monitor* module. This module has access to the defined security policies with the requirements that should be achieved and the desired configurations. Thanks to this knowledge, the module is able to detect and generate alerts when an event violating the requirements is detected. It also makes use of a database containing a set of well-known threats and vulnerabilities in order to detect possible attacks or vulnerabilities in the system.

If some strange behavior of the system that could violate the security policy is detected, the *system monitor* alerts the *status control* module. This module is able to change the security level according to the current system status by enforcing the configurations generated for that new security level. This provides a framework with the ability to react to possible hazards by increasing the security level. Applying a higher security level will enforce stronger security restrictions if some security hazard is detected. The module is also able to go back again to the previous (lower security level) status if the hazard disappears.

Finally, to deal with events, in the prototype, the choice has been to use the WS-Eventing standard, and the *event manager* module has been developed according to this standard to manage the event delivering inside the framework.

16.4.3 Extending Functionality

One of the main design goals borne in mind while specifying and developing the framework has been its extension capability, trying to use standards in order to provide support to future framework extensions.

XML is a widespread, platform- and technology-independent standard supported by a wide range of tools. Thus, the usage of such a technology in the design of the framework makes it quite extensible, allowing future modules or tools to parse the information managed by every other module of the framework.

At the same time, web-services technology also provides quite a good interoperability regarding component communication. Different modules communicate by making use of this technology inside the framework. Thus, any new component, module, or tool added to any of the areas could be able to communicate with the rest of the framework. Not only can a new component communicate with another one by using web services, but it can also listen to events thrown by any other component, making use of the WS eventing standard, which is the one used to manage events.

But using these technologies only solves the syntactical part of the problem of integrating new extensions to the framework. A common model is also needed in order to define common semantics for the information managed by the framework. Here is where the model CIM plays its role, defining in a common way the semantics for both the system description and the security policies.

Furthermore, inside the different areas, some extensions can be provided by third parties in order to support new functionality. This is the case of the *configuration area*. The *LLC enforcer* in this area is based on plug-ins providing the area with different ways of enforcing a configuration to a specific device. Developing a new plug-in for this module will give support to the framework to enforce a configuration for any new device that could not be configured using standard protocols, such as FTP, SSH, etc.

In the same way, different *configuration transformers* can be defined as extensions to the *configuration area*. This area uses both XSLT and Java technologies to translate HLCs with device-independent configurations into LLCS with device-dependent configurations. Providing a new extension to this area will make the *LLC engine* able to generate specific configurations for new kinds of devices not supported before.

New tools can be also integrated into the framework to provide improved functionality. For instance, new management consoles can be developed and included in the *requirement management area*. They can be included both at high (SDL, SPL) or low (xCIM-SDL, xCIM-SPL) levels to manage the system description and policies. They can be easily integrated because all languages are XML based. At low levels, tools can make use of the *requirement interface* of the repository, which is a web service providing methods to work with the model representation in xCIM. At a higher level, they can use the *security plan management* and *system description management* web-service interfaces. These are also web services, and they deal with SDL and SPL high-level languages, respectively.

The *requirement management area* can also be improved by adding new modules to detect more policy conflicts, applying new checking techniques, or analyzing the model to produce new metrics. These new modules can access the model that defines the system from the repository through the *checker interface* web service of the repository. Similarly, new monitoring modules can be added to extend the *monitoring area* functionality, being able to access the model through the *monitor interface* web service in order to monitor whether the system is properly running, according to the specified model and configurations.

All these features make the framework extensible, allowing the integration of improved tools and modules or the addition of new ones that can provide the framework with any domain-specific functionality required by the administrator to get the system managed according to his or her needs.

16.4.4 Management Workflow

Once the management architecture and the prototypical framework implementation have been described, this section provides an overview of the workflow followed by the different components of the management system.

As a first step, the administrator provides the system description and the security policies that will apply to it using the SDL and the SPL. This can be done by directly writing XML documents or making use of the graphical console tool. Different security levels can be defined containing different sets of policies specifying the security requirements that should be achieved for each level. The information model used by the core of the framework is based on xCIM. Therefore, a translation process generates such a model in xCIM-SDL and xCIM-SPL from the SDL and SPL high-level descriptions.

Once the system is described and the policies defined, the *policy checker* module takes this information and starts the security validation process. This process determines whether the system satisfies the desired security requirements. At this point, the security of the system is also evaluated by the *security analyzer*, which gives a theoretical measurement of the security level provided by the specified requirements.

If the defined system and policies fulfill the administrator's security needs, then the generic information described by the core information model in xCIM is ready to be used to generate the specific configuration that will be deployed to the actual system devices and services. The administrator can also redefine the system and the security policies again if they do not satisfy the security needs. This way, the framework provides a mechanism to validate and evaluate different alternatives before they are actually deployed to the actual system.

The configuration generation and further deployment process consists of a set of steps, beginning with the generation of *HLC* documents by the *HLC engine* and followed by the *LLC engine*, which produces the *LLC* documents. Finally, the *LLC enforcer* is the module in charge of deploying the configuration to the actual system, taking the configurations from the LLCs and deploying them to the final devices and services.

On the other hand, monitoring the network is the *system monitor*, looking for any behavior that violates the security requirements defined by the deployed security policies. Any hazard detected by this module is passed to the *status control* one, which is able to react by applying another configuration to lead the system to a different security level.

In order to make the framework components follow the specified workflow, a set of events has been defined. A state chart diagram has been defined, specifying the different statuses where the framework can be, and the status changes upon the different events. As stated in the previous section, the WS eventing standard is used in the framework. Thus, any interested module can listen to those events to keep track of status changes. Moreover, the *status control* module listens to those events and updates status information in the repository. Then, this information can be queried by other modules to get the current status of the framework. This component is always monitoring for status changes, allowing the rest of the components to know the current status at any time, even if they are not subscribed to the event system.

Figure 16.3 shows the stated chart diagram with the different statuses of the framework and the events that produce the transactions between them.

The workflow transactions are the following:

- The *SECURITY_VERIFICATION_OK* event signals that the security of the framework has been successfully verified by the *policy checker*.
- The *SECURITY_VERIFICATION_FAILURE* event is fired to signal that the *policy checker* module has encountered some problems while checking the security of the framework.
- The *SECURITY_EVALUATION_OK* event signals that the security of the framework has been successfully evaluated by the *security analyzer*.
- The *SECURITY_EVALUATION_FAILURE* event signals that the *security analyzer* has encountered some problems while evaluating the security of the framework.
- The *CONFIGURATION_GENERATED* event signals that the configuration has been generated, and it is ready to be deployed.

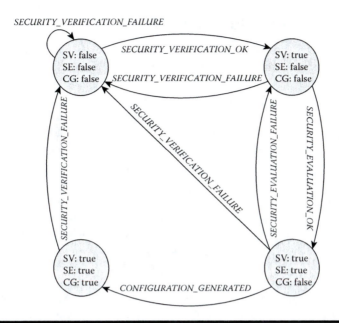

Figure 16.3   Workflow state chart diagram.

There are four possible statuses where the framework can be. In the figure, these statuses are tagged with three Boolean variable values: SV (*security verified*), SE (*security evaluated*) and CG (*configuration generated*).

Initially, no security has been verified nor evaluated. So there is no possible configuration generated, and the three variables take the *false* value. In this status, the only tool that could be launched is the *policy checker* because it is the first step that should occur in the workflow in order to have the defined system description and security policies validated. Once the verification process ends successfully, the *SECURITY_VERIFICATION_OK* event is fired, and the framework moves to another status where the security has been verified (SV variable set to *true*). In this status, the *security analyzer* is enabled, allowing the administrator to perform an evaluation of the system security. The end of this process fires a *SECURITY_EVALUATION_OK* event, moving the framework status to one with the SV and SE variables set to *true*.

Once the security has been verified and evaluated, the *HLC engine* tool is enabled. This tool generates the *HLC* documents with the generic configurations for every specific system device or service. Automatically, the *LLC engine* is started and takes these HLCs to generate the corresponding *LLC* documents with the specific system-dependent configuration. At this point, the configuration generation process has finished, and the configuration is ready to be deployed to the final system components. Then, the *CONFIGURATION_GENERATED* event is fired, and the framework switches to the status with the three variables set to *true*. In this status, the *LLC enforcer* is enabled because the configuration for a verified and evaluated security definition is generated and ready to be deployed to the system.

In the status with the three variables set to *true*, all the tools are available, and all the needed information is also available on the system. The execution of some of these tools can fire an event notifying a failure, moving the system back to the proper status and therefore disabling the corresponding applications.

16.5 Management in Cloud Computing

Current trends in system management lead to the externalizing of the IT management to third-party providers. This approach is called cloud computing [10] and is growing a lot thanks to the great number of advantages it provides. The interest in researching about system management on such application domains is increasing enormously because the current solutions are still in their early stages. Cloud computing is the new way of managing information systems enabling the efficient provisioning of virtual IT architectures to third parties. Such virtual resources are dynamically created and destroyed on demand according to the constantly changing requirements of the clients. This way of using elastic infrastructures enables new pay-as-you-go business models for the usage of infrastructures.

The cloud computing provides a well-known logical stack: infrastructure as a service (IaaS), platform as a service (PaaS), and software as a service (SaaS). IaaS is the delivery of computer hardware, that is, servers, networking technology, storage, and data center space as a service. This layer also includes the delivery of operating systems and on-demand virtualization infrastructures to manage the resources. The PaaS layers use the IaaS to provide middleware services that are used to implement the final cloud services. These middleware services include security services, distributed processing services, user management services, and so on. Therefore, a rule-base architecture for the management of the cloud distributed system similar to the one described in this chapter could be placed between the PaaS and IaaS layers.

The creation of infrastructures is only a part of the process for provisioning services in the cloud. Ongoing work is focusing on providing mechanisms to automatically deploy services on cloud infrastructures and deal with advanced features, such as installation, configuration, elasticity, monitoring, reconfiguration, and management of software components. These kinds of application domains will require an architecture, which, based on rules and a well-known model, describe a common repository to manage the deployment and configuration of these software components. Moreover, the architecture should be extensible because new software components can be easily added to the repository. The workflow of these components as well as the language and rules intended for describing the different services to be managed are key challenges that still need to be addressed.

16.6 Conclusion

The architecture exposed herein defines a whole policy-based management framework, providing administrators with a set of powerful tools to manage complex systems, defining their requirements with abstract semantics in a platform- and technology-independent way. As has been shown in the chapter, the management architecture complies with the set of requirements that any management framework should fulfill. These features are interdomain; formal underlined model; flexibility, conflict detection, and resolution; and scalability, monitoring, and reconfigurability.

The architecture follows a well-defined model, that is platform and technology independent, using XML and web-service technologies that are totally platform-independent standards by themselves. Interfaces provided by the different components and areas of the framework are well and clearly defined using web services.

Both the system description and the different policies are defined using abstract semantics, allowing the administrator to manage a wide range of devices without knowledge of the underlying specific platform or technology. The framework has been developed to provide flexibility and support new tools, plug-ins, or extensions as exposed in the section on framework extension in this chapter, making it able to deal with new device types and generating the specific configurations for the administrator.

Common semantics have been employed to define the model used by the framework and the information managed by the different components. These semantics are defined by the DMTF in its CIM standard. This makes the framework easily interoperable with other architectures and domains, especially if they are also based on CIM because the semantics would be the same, requiring only a syntactic translation of the model.

The checking and transforming area performs a validation and checking of the different policies defined by the administrator, detecting possible conflicts and providing a theoretical security measurement. This checking and measurement can be done as many times as the administrator wishes before the whole set of policies is applied to the final devices and services, allowing the administrator to define several models until he or she finds one that fulfills his or her needs and avoiding the deployment of possible conflicting configurations.

The framework is able to translate the policies defined by the administrator in abstract semantics to lower-level policies. Some of these policies can be specified using abstract concepts such as "high security," which will be translated to more detailed ones, according to some context information that could have been defined previously by security experts, avoiding the administrator having to define a lot of parameters and making the system more secure because experts may have more knowledge than the administrator in some specific fields. These lower-level configurations have been lately translated to specific device configurations by the *mapping area,* and they are even deployed to the final devices by the *enforcing area.*

Furthermore, this framework includes a powerful monitoring mechanism, which takes into account the requirements defined by the administrator and monitors whether the system is accomplishing them. The framework also manages the security level concept, allowing the administrator to define a different security plan for every single level and making the framework able to react in case the monitoring system detects an undesirable status.

References

1. Distributed Management Task Force. *DMTF Standards.* Available at http://www.dmtf.org/standards.
2. W. Bumpus, J. W. Sweitzer, P. Thompson, A. R. Westerinen, and R. C. Williams. *Common Information Model: Implementing the Object Model for Enterprise Management.* Wiley, New York, USA, 2000.
3. C. Hobbs. *A Practical Approach to WBEM/CIM Management.* CRC Press Company, Boca Raton, FL, 2004.
4. K. Wilson, and I. Sedukhin. Web Services Distributed, Management of Web Services (WSDM-MOWS) 1.1. OASIS Standard, 2006.
5. Distributed Management Task Force. *Web Services Management (WS-MAN).* Available at http://dmtf.org/standards/wsman.
6. C. Basile, A. Lioy, G. Martinez, F. J. Garcia, and A. F. Gomez. *POSITIF:* "A Policy-Based Security Management System." Proceedings of the Workshop on Policies for Distributed Systems and Networks (IEEE POLICY), Italy, 2007.
7. J. M. Marin, J. Bernal, D. J. Martinez, G. Martinez, and A. F. Gomez. "Towards the Definition of a Web Service Based Management Framework." Proceedings of the Second International Conference on Emerging Security Information, Systems and Technologies, France, 2008.
8. J. M. Marin, J. Bernal, J. D. Jimenez, G. Martinez, and A. F. Gomez. "A Proposal for Translating from High-Level Security Objectives to Low-Level Configurations." Proceedings of the First International DMTF Academic Alliance Workshop on Systems and Virtualization Management: Standards and New Technologies, France, 2007.

9. G. Martinez, F. J. Garcia, and A. F. Gomez. "Policy-Based Management of Web and Information Systems Security: An Emerging Technology." In E. Ferrari, and B. Thuraisingham (Eds.), *Web and Information Security,* 173–195. Idea Group, Inc., Hershey, PA, 2006.
10. B. Hayes. "Cloud computing." *Communications of the ACM* 51, no. 7 (2008): 9–11.

Additional Reading Section

Adi, K., Y. Bouzida, I. Hattak, L. Logrippo, and S. Mankovskii. "Typing for Conflict Detection in Access Control Policies." Proceedings of MCETECH, 212–226, 2009.

Aktug, I. and K. Naliuka. "ConSpec: A formal language for policy specification." *Electronic Notes in Theoretical Computer Science* 197, no. 1 (2008): 45–58.

Alcaraz, J. M., J. M. Marin, J. Bernal, F. J. Garcia, G. Martinez, and A. F. Gomez. "Detection of semantic conflicts in ontology and rule-based information systems." *Data and Knowledge Engineering* 11, (2010): 1117–1137.

Becker, M. Y., C. Fournet, and A. D. Gordon. "SecPAL: Design and Semantics of a Decentralized Authorization Language." Proceedings of the 20th IEEE Computer Security Foundations Symposium (CSF), 3–15, 2007.

Boutaba, R. and I. Aib. "Policy-based management: A historical perspective." *Journal of Network and Systems Management,* (2007): 447–480.

Cuppens, F., and N. Cuppens-Boulahia. "Modeling contextual security policies." *International Journal of Information Security* 7, no. 4 (2010): 285–305.

Cuppens, F., N. Cuppens-Boulahia, and M. B. Ghorbel. "High level conflict management strategies in advanced access control models." *Electronic Notes in Theoretical Computer Science* 186, (2007): 3–26.

Davy, S., B. Jennings, and J. Strassner. "The policy continuum–policy authoring and conflict analysis." *Computer Communications* 31, no. 13 (2008): 2981–2995.

Feeney, K., R. Brennan, J. Keeney, H. Thomas, D. Lewis, A. Boran, and D. O'Sullivan. "Enabling decentralized management through federation." *Computer Networks* 54, no. 16 (2010): 2825–2839.

Garcia, F. J., G. Martinez, A. Muñoz, J. A. Botia, and A. F. Gomez. "Towards semantic web-based management of security services." *Springer Annals of Telecommunications* 63, no. 3–4 (2008): 183–193.

Garcia, F. J., J. M. Alcaraz, J. Bernal, J. M. Marin, G. Martinez, and A. F. Gomez. "Semantic web-based management of routing configurations." *Journal of Network and System Management* 19, no. 2 (2011): 209–229.

Hilty, M., A. Pretschner, D. Basin, C. Schaefer, and T. Walter. "Policy Language for Distributed Usage Control." Proceedings of ESORICS, 531–546, 2007.

Kandogan, E., P. P. Maglio, E. Haber, and J. Bailey. "On the roles of policies in computer systems management." *International Journal of Human-Computer Studies* 69, no. 6 (2011): 351–361.

Karat, J., C. M. Karat, E. Bertino, N. Li, Q. Ni, C. Brodie, J. Lobo, S. B. Calo, L. F. Cranor, P. Kumaraguru, and R. W. Reeder. "Policy framework for security and privacy management." *IBM Journal of Research and Development* 53, no. 2 (2009): 4:1–4:14.

Kodeswaran, P. B., S. B. Kodeswaran, A. Joshi, and T. Finin. "Enforcing Security in Semantics Driven Policy Based Networks." Proceedings of ICDE Workshops, 490–497, 2008.

Li, N. and Q. Wang. Beyond Separation of Duty: An algebra for specifying high-level security policies." *Journal of the ACM (JACM)* 55, no. 3 (2008): 1–46.

Marin, J. M., J. Bernal, J. M. Alcaraz, F. J. Garcia, G. Martinez, and A. F. Gomez. "Semantic-based authorization architecture for grid." *Future Generation Computer Systems* 27, (2011): 40–55.

Martinelli, F. and I. Matteucci. "Idea: Action Refinement for Security Properties Enforcement." Proceedings of the 1st International Symposium on Engineering Secure Software and Systems, LNCS 5429, 37–42, 2009.

Martinez, G., A. F. Gomez, S. Zeber, J. Spagnolo, and T. Symchych. "Dynamic policy-based network management for a secure coalition environment." *IEEE Communications Magazine* 44, no. 11 (2006): 58–64, 2006.

Olmedilla, D. "Semantic Web Policies for Security, Trust Management and Privacy in Social Networks." Invited talk at the Workshop on Privacy and Protection in Web-Based Social Networks, Barcelona, 2009.

Qin, L. and V. Atluri. "Semantics-aware security policy specification for the semantic web data." *International Journal of Information and Computer Security* 4, no. 1 (2010): 52–75.

Satoh, F. and Y. Yamaguchi. "Generic Security Policy Transformation Framework for WS-Security." Proceedings of IEEE International Conference on Web Services, 513–520, 2007.

Swamy, N., B. J. Corcoran, and M. Hicks. "Fable: A Language for Enforcing User-defined Security Policies." Proceedings of IEEE Symposium on Security and Privacy, 369–383, 2008.

Yau, S. S. and Z. Chen. "Security Policy Integration and Conflict Reconciliation for Collaborations among Organizations in Ubiquitous Computing Environments." Proceedings of the 5th International Conference on Ubiquitous Intelligence and Computing, 3–19, 2008.

Zhang, X., J. P. Seifert, and R. Sandhu. "Security Enforcement Model for Distributed Usage Control." Proceedings of IEEE International Conference on Sensor Networks, Ubiquitous, and Trustworthy Computing, 10–18, 2008.

Zhou, J. and J. Alves-Foss. "Security policy refinement and enforcement for the design of multi-level secure systems." *Journal of Computer Security* 16, no. 2 (2008): 107–131.

Chapter 17

Pragmatic Approach to Performance Evaluation of MPI–OpenMP on a 12-Node Multicore Cluster

Abdelgadir Tageldin Abdelgadir and Al-Sakib Khan Pathan

Contents

17.1 Introduction

High-performance clusters can be considered as part of the underlying components for next-generation networks (NGNs). As cluster size and core numbers continue to expand, NGN infrastructure could heavily use such systems. It can be envisioned that the NGNs will generate huge amounts of data that may need to be processed by high-performance computing clusters. Therefore, an understanding of clusters and the different programming paradigms (that are available) is needed for optimal utilization. This chapter is written from the philosophy that the gained practical knowledge on an implemented cluster would help development of future technologies and mechanisms to deal with related issues.

Computer clusters play a vital role in the distributed computing- and distributed networking-related research fields because they are necessary to solve intensive computational tasks requiring parallelism. The earlier generations of clusters usually featured a huge number of nodes with a small number of cores on each node. Current processor designs have shifted from increasing the speed of each processor to increasing the number of cores, triggering the start of a new *multiprocessor* era. The recent trend in processor design targets the number of cores per chip with the advent of each new processor model. Existing cluster technologies provide solutions that are composed of compute nodes with multicores ready for deployment. The complex nature of these new clusters has always challenged programmers and researchers. This complexity arises from the distributed memory across different nodes on one hand while, on the other hand, from a nonuniform memory access within different nodes across the network. Networking communication systems and interconnections are equally important in these clusters; otherwise, underutilization of the cluster may result, and the full potential may never be realized. Furthermore, as the size of the high-performance clusters increases, N-core compute nodes with plenty of shared memory are now being connected via high-speed networks.

The word *core* means a processor in this new context and can be used interchangeably. Some of the well-known and common examples of these processors are the Intel quad-core and the AMD Opteron or Phenom quad-core processors. This aggregation of classical cores into a single *processor* has introduced the division of workload among multiple processing cores by utilizing the locality of computation, multi-threading, and parallelization techniques. This has also introduced the need for parallel and multi-threaded approaches in solving most types of problems in different fields of science [1].

When N-core processors are deployed in a cluster, three types of communications must be considered:

1) Between different processors on the same chip
2) Between the chips in the same node
3) Between different nodes

All these communication methods need to be considered on such a cluster in order to deal with the associated challenges when considering building a cluster with N-nodes and N-processors [2]. Yet from a programming point of view, the choice might be application-specific when it comes to choosing between a model that uses message passing between nodes for coarse-grained parallelism and a model that utilizes message passing between each node while using local threading as a different level of fine-grained parallelism [3].

The main objective of this work is to experimentally investigate and analyze various critical aspects of an *MPI–OpenMP* approach on a 12-node, multicore, high-performance cluster. To perform this task, we have set the cluster environment that we describe in later sections and discuss the findings along with future expectations that could be beneficial to other researchers working with a similar cluster. The main goal is to benefit such research works also in the future for the NGN models [4].

The outline of the chapter is as follows: After the introduction part, in Section 17.2, we present some basic terminologies and note down some background information. Section 17.3 describes the architecture of our cluster; Section 17.4 mentions the research methodology and our experimental settings in detail. In Section 17.5, we analyze various aspects of our findings, and finally, Section 17.6 concludes the chapter with some critical discussions.

17.2 Basic Terminologies and Background

17.2.1 MPI and OpenMP

The message-passing interface (MPI) and open multiprocessing (OpenMP) are currently the programming models used in parallel systems. MPI provides a method of communication among scattered processes in a parallel environment. These processes execute on different nodes in a cluster but interact by *passing messages*; that is why such a name is given. There can be more than a single process thread in each processor. The MPI [5] approach focuses on the process communication happening across the network, and OpenMP [6] targets interprocess communication between processors. With this in mind, it will make more sense to employ OpenMP parallelization for interprocess communications within the node and MPI for message passing and network communications between nodes [7]. It is also possible to use MPI for each core as a separate entity with its own address space; this will force us to deal with the cluster differently though. With these simple definitions of MPI and OpenMP, a question arises whether it will be advantageous to employ a hybrid mode where more than one OpenMP and MPI process with multiple threads are on a node so that there is at least some explicit intranode communications [8]. In an MPI–OpenMP hybrid system as simplified in Figure 17.1 (i.e., unlike MPI, where domain-based mapping of processes is used, nor like OpenMP, where it is dependent on different threads running in a parallel context within the same processor), the system uses two levels of process mapping, where each MPI process controls another spawned OpenMP process.

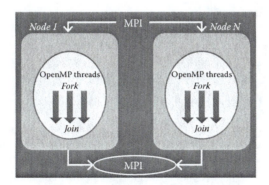

Figure 17.1 Process flow in a *hybrid MPI–OpenMP* that enables us to optimize instructions on a single address space used by any process sharing the same processor.

17.2.2 Performance Measurement with HPL

High-Performance Linpack (HPL) [9] is a well-known benchmark suitable for parallel workloads that are core limited and memory intensive. In a core-limited type of workload, data are needed to be *loose* to the processor in locations such as the cache, and the main factor that limits performance is the processor's clock frequency. HPL is a floating-point benchmark that solves a dense system of linear equations in parallel and attempts to measure the best performance of a cluster in solving a system of equations. The result of the test is a metric called *Gigaflops*, which translates to billions of floating point operations per second. HPL performs an operation called *LU factorization* [10]. This is a *highly* parallel process, utilizing the processor's cache up to the maximum limit possible, though the HPL benchmark itself may not be considered a memory-intensive benchmark. The processor operations that it performs are predominantly 64-bit floating-point vector operations and uses SSE [streaming SIMD (*single instruction, multiple data*) extensions] instructions. This benchmark is used to determine the world's top 500 fastest computers [11].

In the HPL benchmark, a number of metrics are used to rate a system. The metrics are generated using a pseudorandom number generator and are intended to force partial pivoting to be performed in Gaussian elimination [9]. One of these important measures is R_{max}, measured in *Gigaflops*; it represents the maximum performance achievable by a system. In addition to that, there is also R_{peak}, which is the theoretical peak performance for a specific system; this is obtained using the following formula:

$$[N_{proc} * \text{Clock frequency} * \text{FP/clock}], \tag{17.1}$$

where N_{proc} is the number of processors available, FP/clock is the floating-point operation per clock cycle, and clock frequency is the frequency of a processor in megahertz or gigahertz.

So far, different approaches have been taken to optimize the HPL benchmark using different methods. In [12], a method is discussed to use overlapping communication with a hybrid implementation. In [13], implementation of another hybrid version of HPL is discussed that utilizes existing GPU processors. In this work, HPL is used to measure the performance of a single node or a cluster of nodes through a simulated replication of scientific and mathematical applications by solving a dense system of linear equations.

17.3 Architecture of Cluster

In this section, we provide a description of the architecture of our cluster. Figure 17.2 shows the physical architecture consisting of 12 compute nodes and a head node.

17.3.1 Machine Specifications

Compute node specifications are shown in Table 17.1, which is same for all of the compute nodes. Each has an Intel Dual Xeon quad-core processor running at 3.00 GHz. Note that the system had eight of the mentioned Xeon processors. Having a high amount of cache reduces the latencies in accessing instructions and data; this generally improves performance for applications working on a large amount of data sets. The head node processor's specifications are shown in Table 17.2; it differs from other nodes as it has 16 cores from a different quad-core Xeon model running at 2.93 GHz and 128 GB of RAM.

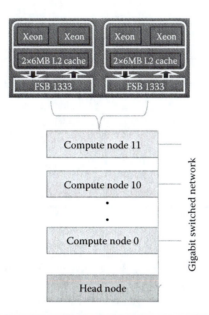

Figure 17.2 Cluster physical architecture.

Table 17.1 Processor Specifications of Compute Node

Element	Features
Processor	0 (up to 7)
CPU family	6
Model name	Intel(R) Xeon(R) CPU E5450 @ 3.00 GHz
Stepping	6
CPU MHz	2992.508
Cache size	6144 kB
CPU cores	4
FPU	Yes
Flags	fpu vme de pse tsc msr pae mce cx8 apic sep mtrr pge mca cmov pat pse36 clflush dts acpi mmx fxsr sse sse2 ss ht tm syscall nx lm constant_tsc pni monitor ds_cpl vmx est tm2 cx16 xtpr lahf_lm
Bogomips	6050.72
Clflush size	64
Cache_alignment	64
Address sizes	38 bits physical, 48 bits virtual
RAM	16 GB

Table 17.2 Processor Specification of Head Node

Element	Features
Processor	0 (up to 16)
CPU family	6
Model name	Intel(R) Xeon(R) CPU X7350 @ 2.93 GHz
Stepping	11
CPU MHz	2925.874
Cache size	4096 kB
CPU cores	4
FPU	yes
Flags	fpu vme de pse tsc msr pae mce cx8 apic sep mtrr pge mca cmov pat pse36 clflush dts acpi mmx fxsr sse sse2 ss ht tm syscall nx lm constant_ tsc pni monitor ds_cpl vmx est tm2 cx16 xtpr lahf_lm
Bogomips	5855.95
Clflush size	64
Cache_alignment	64
Address sizes	40 bits physical, 48 bits virtual

17.3.2 Configuration of Cluster

The cluster was built with Rocks 64-bit Cluster Suite. Rocks [14] is a Linux distribution based on CentOS [15], which is intended for high-performance computing systems. The Intel compiler suite was used for compilation; the Intel MPI implementation and the Intel Math Kernel Library were utilized as well to manage MPI jobs. The cluster was connected using two networks: one used for MPI-based operations and the other for the site's local data transfer.

17.4 Research Methodology and Details of Our Experiment

The tests were run on the compute nodes only as the head node was different both in capacity and in speed. In fact, adding it to the cluster would increase the complexity, which is beyond the scope of this experimental work.

Tests were executed in two main iterations; the first iteration was for each single-node performance measurement followed by an extended iteration that included all the 12 nodes. These tests took a long time to complete. Our main research point was focused on examining to what extent the cluster would scale as it was the first quad-core cluster deployed at the site. Note that in this chapter, we focus more on the successful test runs of the hybrid implementation of HPL by Intel for Xeon Processors. In each of the iterations, different configurations and setups were implemented; these included changing the grid topology used by HPL according to different settings.

Table 17.3   Performance of Separate Compute Nodes

Node Type	Average Gflops	N	NB	P × Q
Compute node	7.517e+01	40000	192	1 × 8

This was required because the cluster contained both an internal grid—between processors—and an external grid composed of the nodes themselves.

In each test trial, a configuration was set, and performance was measured using HPL. An analysis of the factors affecting performance was recorded for each different trial.

17.4.1 Single Node Test

The test for a single node was done for all nodes; this is a precautionary measure to check whether all nodes perform according to expectation because the cluster's performance in an HPL test run is limited by the slowest of the nodes. Table 17.3 shows the results of the single node test; the average is approximately 75.6 Gflops. The maximum *theoretical* value can be calculated using Equation 17.2. In each node, there are two quad-core processors, making the theoretical peak performance equal to

$$R_{peak} = 8*3*4 = 96 \text{ Gflops/node.} \tag{17.2}$$

However, the maximum performance obtained was at an approximate average of 75.6 Gflops/node; this is called the R_{max} value obtainable for a single node. The efficiency for a single node is therefore 78.8%.

Table 17.4 represents the parameters used for the single node test.

17.4.2 Overall Cluster Performance

The overall cluster performance test required several iterations to scale well and to reach an optimal performance within our experimentation period. The first thing that was put into consideration was the grid topology to be used in order to achieve optimal results. Several grid possibilities were anticipated depending on the knowledge gathered from previous experiences. When measuring a cluster, attainment of high performance is dependent on the number of cores and the frequency of the processor being used on each node. Granular distribution of processes across the grid is therefore crucial to obtain good performance results.

In general, HPL is controlled by two main parameters that describe how processes are distributed across the cluster's nodes; the values are P and Q. These two are benchmark-tuning parameters that are critical when producing a good performance is required. P and Q should be as close to equal as possible, but when they are not equal, P should be less than Q. That is because when Q *multiplies P*, it gives the number of MPI processes to be used and how they are distributed across the nodes. In this cluster, several choices can be used, such as 1 × 96, 2 × 48, 3 × 32, 4 × 24, 6 × 16, or 8 × 12, with each giving a totally different result. Moreover, the network as well as other low-level factors, such as the distribution of processes within a *domain* or a socket, can affect performance [16]. Therefore, different trials are usually needed to achieve best performance. Another parameter

Table 17.4   Complete HPL Parameters for Compute Node

Choice	Parameters
6	Device out (6 = stdout,7 = stderr,file)
1	# of problems sizes (*N*)
40,000	*N*s
1	# of NBs
192	NBs
0	PMAP process mapping (0 = Row-,1 = Column-major)
1	# of process grids (*P* × *Q*)
1	*P*s
8	*Q*s
16.0	Threshold
1	# of panel fact
0 1 2	PFACTs (0 = left, 1 = Crout, 2 = Right)
1	# of recursive stopping criterium
4 2	NBMINs (>= 1)
1	# of panels in recursion
2	NDIVs
1	# of recursive panel fact
1 0 2	RFACTs (0 = left, 1 = Crout, 2 = Right)
1	# of broadcast
0	BCASTs (0 = 1rg,1 = 1rM,2 = 2rg,3 = 2rM,4 = Lng,5 = LnM)
1	# of look ahead depth
0	DEPTHs (>= 0)
2	SWAP (0 = bin-exch,1 = long,2 = mix)
256	Swapping threshold
1	L1 in (0 = transposed,1 = no-transposed) form
1	*U* in (0 = transposed,1 = no-transposed) form
0	Equilibration (0 = no,1 = yes)
8	Memory alignment in double (>0)

that we need is N; it represents the size of the problem to be solved by HPL. We used the following formula to estimate the problem size:

$$\sqrt{\left[\left(\left(\sum M_{sizes}\right)MB*1{,}000{,}000{,}000\right)\Big/8\right]}.\qquad(17.3)$$

This gives a value that would approximately be N, for example,

$$N = \text{sqrt}(12*16*1{,}000{,}000{,}000) \sim = 154{,}919.$$

However, it is preferred not to take the whole result. In our case, we chose 140,000 as N, giving more than 25% to other local system processes to avoid the use of virtual memory that can severely degrade the results. An overloaded system would use the swap area; this would negatively influence the results of the benchmark. It is advisable to make full use of the main memory but, at the same time, to avoid the use of virtual memory. The optimal performance was achieved with the HPL input parameters shown in Table 17.5. In Figure 17.3, we show the results captured at different time intervals while the cluster performed the benchmark. We observe, at some points, that results reached a maximum of 760 Gflops for the benchmark. As this was not the final result, it was not considered as the overall measure. At these points, less communication happens between the nodes, and most of the calculations are done within nodes. It is an indication of how the network—which we opted to use—degraded the results obtainable from this cluster.

On the other hand, our first expectation was that one of the 3 × 32 or 4 × 24 parameters would produce the optimal performance. As we have noted, a 6 × 16 grid, such as shown in Figure 17.4, obtained the best overall performance at 662.2 Gflops. Performance increase was linear to some

Table 17.5 HPL Configuration for Overall Cluster Test

Parameter	Value
N	140,000
NB	192
PMAP	Row-major process mapping
P	6
Q	16
RFACT	Crout
BCAST	1ring
SWAP	Mix (threshold = 256)
L1	No-transposed form
U	No-transposed form
EQUIL	No
ALIGN	Eight double precision words

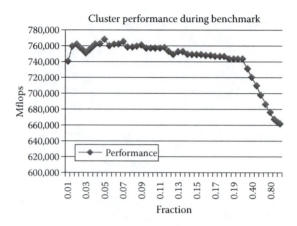

Cluster performance during benchmark

Figure 17.3 Performance of a cluster captured while working on different fractions of the problem.

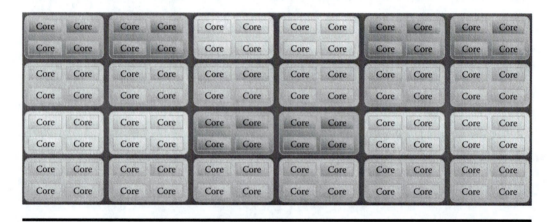

Figure 17.4 Physical view of 6 × 16 cores grid; each of the eight cores represent a single node.

extent but would not equal the overall absolute sum of the 12 nodes, which is about 907 Gflops (see Figure 17.6). This is acceptable though, as a cluster's performance does not scale linearly in reality; thus the overall efficiency of the cluster is calculated at approximately 60%, which is satisfactory for a gigabit-based cluster, although it might not be satisfactory for many applications. In Figure 17.6, we summarize the different results of several experiments. Comparison is done against optimal and expected results. Moreover, we include the results for an MPI-only execution. These results are a clear indicator from the experiments that an MPI–OpenMP approach can perform better in *N*-core clusters.

17.5 Observations and Analysis

By looking at the general topological structure of this cluster, it could be understood that different cores complete the same process in parallel. This leads to high network communications between the different nodes in such types of clusters. Moreover, processing speed tends to be faster than the gigabit network's communication link speed available for the cluster. This is translated into waiting time in which some cores may become idle [17]. In preliminary test runs, we opted to use an MPI-only approach based on our previous experiences with normal clusters, the results of which were disappointing, reaching a maximum of approximately 205 Gflops. The main thing we decided to change was the MPI-only implementation of HPL. Another factor that we noticed that affected the performance of this cluster was the network.

The setup of the cluster included two networks of which one was used solely for MPI traffic; it was the network that obtained the highest possible result. From our findings, it is recommended that multicore clusters deployed for MPI jobs should have a dedicated network to run parallel jobs. It was noticeable in the test-run phases that MPI processes in general can generate a huge amount of data to be transferred, which, in turn, requires substantial network bandwidth. This is mainly caused by the higher speed of multiprocessing in each node in relation to the current speed available in the test cluster.

The main advantage of using Intel's MPI implementation in this study is its ability to define network or device *fabrics* or, in other words, defining a cluster's physical connectivity. In this cluster, the fabric can be defined as a Transmission Control Protocol (TCP) network with shared-memory cores, which is synonymous to an Ethernet-based Symmetrical Multi Processing (SMP) cluster. When running a test using a hybrid implementation without explicitly defining the underlying fabric of the cluster, overall performance degradation was noticeable as the cluster's overall benchmark result was merely 240 Gflops, 40 Gflops more than the previously mentioned failed attempts with an MPI-only approach. This value is considered low when calculating the overall expected performance using formula 17.2 and multiplying that by 12 nodes. The main reason behind the degradation was caused by having MPI processes starting without previous *knowledge* of multiple-core architecture. In this scenario, each core would be treated as a single component with no communicative relationship with its neighboring cores within the same node, resulting in communication rather than processing, which leads to more idle time for that specific core. The fabric in such a system consists of different sockets and processors viewed as *domains*. The addition of the option leads to execution awareness of both communication types available for this cluster, which are the intercommunications between the nodes and the shared-memory communications within a node's cores. This essentially leads to the achieved better performance.

Another aspect of these tests was to investigate how the cluster was viewed or perceived physically and how that differs from our expectations. When dealing with multicore processors, an *abstract* view is needed as well, and the best method for this is to use diagrams, such as Figures 17.4 and 17.5. These figures depict how a 6 × 16 topology was chosen and how processes were distributed among nodes. It can be noticed from the figures that processes are passed in a round-robin way across different cores and not nodes. In this cluster, each node has eight processors, so it can be viewed as eight different single-core processor nodes. This distribution of processes affects the overall performance as well. Unexpectedly, and in contrast to previous hints that state having $P \times Q$ to be as close to equal as possible, the 6 × 16 grid performed well as a result of having more related processors on a single node. Also, less communication was needed between the processes across the cluster [18]. In this configuration, each of the running processes can heavily utilize the shared cache and local communication bridges to accomplish some of the tasks. On the other hand, network communication happens while processing cores are being utilized for processing.

Table 17.6 summarizes the best as well as the unexpected results obtained from several test runs.

From Table 17.6, we can observe the performance gain obtained by changing the way we deal with modern-day computer clusters. A high increase in performance was the result of a detailed understanding of how processes are distributed in the cluster.

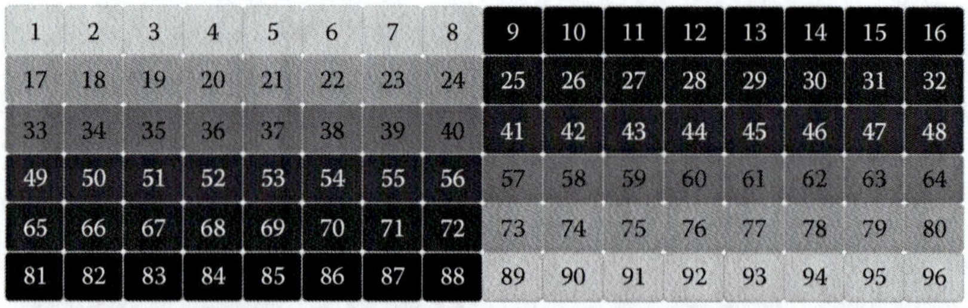

1 2 3 4 5 6 7 8 9 10 11 12 13 14 15 16
17 18 19 20 21 22 23 24 25 26 27 28 29 30 31 32
33 34 35 36 37 38 39 40 41 42 43 44 45 46 47 48
49 50 51 52 53 54 55 56 57 58 59 60 61 62 63 64
65 66 67 68 69 70 71 72 73 74 75 76 77 78 79 80
81 82 83 84 85 86 87 88 89 90 91 92 93 94 95 96

Figure 17.5 Abstract view of MPI processes distribution on the 6 × 16 grid; notice that each eight processes are within a single node. This reduces communications across the network.

Table 17.6 General Summary of Trials

Option Types	*Gflops Obtained*	*P × Q*	*Problem Size N*
OpenMPI, MPI	207	8 × 12	140,000
Intel MPI, default	204	8 × 12	140,000
Intel MPI, default	224.6	6 × 16	140,000
Intel MPI, TCP+Shared Mem.	662.6	6 × 16	140,000

In general, we can summarize the main experiences and observations gathered from a modern-day cluster in the following points:

1. The network significantly affects the cluster's performance. Thus, we believe separating the MPI network from the normal network will result in better overall performance of the cluster. Moreover, an Ethernet gigabit network should consider upgrading to 10GE, Infiniband, and Myrinet as it will be very hard to scale as observed in Figure 17.6. High-performance communication links are needed for such clusters [19].

Performance in Gflops
1400
1200
1000
800
600
400
200
0
0
5
10
15
Optimal
Linear
Actual results
MPI-only

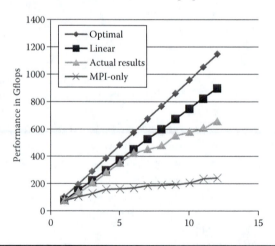

Figure 17.6 Different results expected and obtained when running the benchmark in a hybrid mode. Optimal, linear, and actual results depict those of a hybrid mode, and the MPI-only results show some significant difference.

2. The node's physical architecture and the MPI implementation in use must be considered. Not all provide the same features and perform similarly as shown in Table 17.6, though all can run MPI jobs. Some of these libraries allow much more control that is granular. An example of MPI libraries available are OpenMPI, MVAPICH2, and the Intel MPI implementation.
3. Both the physical view and logical view of a cluster are important. Details of how MPI applications process data must be known by programmers and the cluster administrator because these details will determine how a cluster performs.

17.6 Concluding Remarks and Future Research Direction

In this work, we have presented our experience in measuring the performance of a 12-node cluster consisting of 96 cores. While conducting the experiments, we have observed interesting behavior when applying some of the techniques usually practiced when conducting these experiments. From the results we presented in this chapter, we can observe the difference between a hybrid MPI–OpenMP approach when compared with an MPI-only approach and how it heavily affects the performance of the benchmark when executed on the cluster. Of the various issues, the fact we would like to highlight is how poor the cluster performed when an in-depth knowledge of the architecture was ignored. The main factors that we have detected are the processor type, socket layouts, internetwork communications, the programming module, and the different parameters used to run the benchmark.

We found that the scalability of the multicore cluster benchmark is doubtful with the gigabit Ethernet communication systems being utilized. Performance degradation caused by scaling up was relatively high; we assume a faster network could yield better performance in relation to scalability for these types of clusters. We hope that our work along with the critical observations and suggestions will benefit the researchers working with similar types of clusters. In future work, we will delve further into other types of implementations of hybrid applications and examine how these applications will behave in multicore cluster environments. Another direction of future study could be to examine different internetworking systems that were not available at the time of recording this work.

References

1. Huang, L., Jin, H., Yi, L., and Chapman, B. "Enabling Locality-Aware Computations in OpenMP. Sci. Program." 18, no. 3–4 (August 2010): 169–181. DOI: 10.3233/SPR-2010-0307. Available at http://dx.doi.org/10.3233/SPR-2010-0307.
2. Zhang, Y., Kandemir, M., and Yemliha, T. "Studying Inter-Core Data Reuse in Multicores." In Proceedings of the ACM SIGMETRICS Joint International Conference on Measurement and Modeling of Computer Systems (SIGMETRICS '11). ACM, New York, 25–36. DOI: 10.1145/1993744.1993748. Available at http://doi.acm.org/10.1145/1993744.1993748.
3. Wu, X., and Taylor, V. "Performance Characteristics of Hybrid MPI/OpenMP Implementations of NAS Parallel Benchmarks SP and BT on Large-Scale Multicore Supercomputers." SIGMETRICS Perform. Eval. Rev. 38, no. 4 (March 2011), 56–62. DOI: 10.1145/1964218.1964228, http://doi.acm.org/10.1145/1964218.1964228.
4. Paul, S., Pan, J., and Jain, R. "Architectures for the Future Networks and the Next Generation Internet: A Survey." *Computer Communications* 34, no. 1, 15 January 2011, 2–42.

5. Hochstein, L., Shull, F., and Reid, L. B. "The Role of MPI in Development Time: A Case Study." International Conference for High Performance Computing, Networking, Storage and Analysis, 2008. SC 2008, 1–10, 15–21 Nov. 2008. DOI: 10.1109/SC.2008.5213771.

6. Chen, C., Manzano, J. B., Gan, G., Gao, G. R., and Sarkar, V. "A Study of a Software Cache Implementation of the OpenMP Memory Model for Multicore and Manycore Architectures." In Proceedings of the 16th International Euro-Par Conference on Parallel Processing: Part II (Euro-Par '10), edited by Pasqua D'Ambra, Mario Guarracino, and Domenico Talia. Berlin, Heidelberg: Springer-Verlag, 341–352. DOI: 10.1007/978-3-642-15291-7_31.

7. Rabenseifner, R., Hager, G., and Jost, G. "Hybrid MPI/OpenMP Parallel Programming on Clusters of Multi-Core SMP Nodes." Parallel, Distributed and Network-based Processing, 2009 17th Euromicro International Conference, 427–436, 18–20 Feb. 2009. DOI: 10.1109/PDP.2009.43.

8. Wu, C.-C., Lai, L.-F., Yang, C.-T., and Chiu, P.-H. "Using Hybrid MPI and OpenMP Programming to Optimize Communications in Parallel Loop Self-Scheduling Schemes for Multicore PC Clusters." *Journal of Supercomputing*, The Netherlands: Springer, 2010. DOI: 10.1007/s11227-009-0271-z, Feb. 2009.

9. Bach, M., Kretz, M., Lindenstruth, V., and Rohr, D. "Optimized HPL for AMD GPU and Multi-Core CPU Usage." *Computer Science* 26, no. 3–4 (June 2011): 153–164. DOI: 10.1007/s00450-011-0161-5.

10. Chan, E., Geijn, R.v.d., and Chapman, A. "Managing the Complexity of Lookahead for LU Factorization with Pivoting." In Proceedings of the 22nd ACM symposium on Parallelism in algorithms and architectures (SPAA '10). ACM, New York, NY, USA, 200–208. DOI: 10.1145/1810479.1810520.

11. The Top 500 List of Supercomputer Sites. Available from: http://www.top500.org/lists.

12. Marjanovic, V., Labarta, J., Ayguade, E., and Valero, M. "Overlapping Communication and Computation by Using a Hybrid MPI/SMPSs Approach." In Proceedings of the 24th ACM International Conference on Supercomputing (ICS '10). ACM, New York, 5–16. DOI: 10.1145/1810085.1810091.

13. Bach, M., Kretz, M., Lindenstruth, V., and Rohr, D. "Optimized HPL for AMD GPU and Multi-Core CPU Usage." *Computer Science* 26, 3–4 (June 2011), 153–164. DOI: 10.1007/s00450-011-0161-5. http://dx.doi.org/10.1007/s00450-011-0161-5.

14. Rocks cluster distribution. Available from: http://www.rocksclusters.org.

15. CentOS. Available from: http://www.centos.org.

16. Thomadakis, M. E. "The Architecture of the Nehalem Processor and Nehalem-EP smp Platforms." Technical report, December 2010. Available at: http://sc.tamu.edu/systems/eos/nehalem.pdf.

17. Wittmann, M., Hager, G., and Wellein, G. "Multicore-Aware Parallel Temporal Blocking of Stencil Codes for Shared and Distributed Memory." 2010 IEEE International Symposium on Parallel and Distributed Processing, Workshops and Ph.D. Forum (IPDPSW), 1–7, 19–23 April 2010. DOI: 10.1109/IPDPSW.2010.5470813.

18. Jin, H., Jespersen, D., Mehrotra, P., Biswas, R., Huang, L., and Chapman, B. "High Performance Computing Using MPI and OpenMP on Multi-Core Parallel Systems." *Parallel Computing* 37, no. 9, Emerging Programming Paradigms for Large-Scale Scientific Computing, September 2011, 562–575. DOI: 10.1016/j.parco.2011.02.002.

19. Goglin, B. "High-Performance Message-Passing Over Generic Ethernet Hardware with Open-MX." *Parallel Computing* 37, no. 2, February 2011, 85–100. DOI: 10.1016/j.parco.2010.11.001.

20. Abdelgadir, A. T., Pathan, A.-S. K., and Ahmed, M. "On the Performance of MPI–OpenMP on a 12-Nodes Multi-core Cluster." Proceedings of ICA3PP 2011 Workshops (IDCS Workshop), October 24–26, 2011, Melbourne, Australia, (Y. Xiang et al. Eds.): ICA3PP 2011, Part II, *Lecture Notes in Computer Science (LNCS)* 7017, Springer-Verlag 2011, 225–234.

Chapter 18

Smarter Health-Care Collaborative Network

Qurban A. Memon

Contents

18.1 Background

A hospital is an institution for treatment with specialized staff and equipment and often, but not always, providing for longer-term patient stays. Access to health care varies from country to country and across groups largely influenced by socioeconomic conditions and the health policies of governments. In some countries, health-care planning is distributed among market participants, whereas in others, planning is done more centrally by a government or other coordinating body. The health-care industry incorporates several sectors that are dedicated to providing health-care services and products. The management and administration of health care is another sector vital to the delivery of health-care services. In particular, the practice of health professionals and the operation of health-care institutions are typically regulated by national or state/provincial authorities through appropriate regulatory bodies for purposes of quality assurance.

Each hospital is composed of a wide range of services and functional units. These include bed-related inpatient functions, outpatient-related functions, diagnostic and treatment functions, administrative functions, service functions (food, supply), research and teaching functions, etc. This diversity is reflected in the breadth and specificity of regulations, codes, and oversight that govern hospital construction and operations. The functions of a hospital also include complicated mechanical, electrical, and telecommunications systems. Additionally, the hospitals must serve and support many different users and stakeholders. Regardless of location, size, or budget, all hospitals have certain common attributes:

- Staff efficiency and cost-effectiveness within the hospital environment
- Flexibility and expandability to accommodate diversified needs and modes of treatment
- Therapeutic environments, using, for example, familiar and culturally relevant materials, cheerful and varied colors and textures, ample natural light wherever feasible, etc.
- Cleanliness and sanitation
- Accessibility to meet minimum standards set by well-known accessibility standards organizations
- Controlled circulation to handle interrelated hospital functions
- Aesthetics to enhance the hospital's public image and to contribute to better staff morale and patient care
- Security and safety of, for example, hospital property and assets, protection of patients, etc.
- Sustainability in design of hospital infrastructure

Because of strict operational guidelines for operations and governing regulations, it is typically observed that the patients face waiting in lines, paperwork, the cost of such paperwork, transaction times, etc. Typical automation within a hospital addresses these problems but does not solve them altogether. The automated hospital management system typically includes automation of patient-related information in order to reduce operational time for the hospital. Equivalently, this may include providing patients with a user-friendly technology, facilitating transactions to reduce time, enabling quicker reception of medicine from the pharmacy, enabling easy booking and attending of the appointments, enabling easy query of results information, preventing crowded lines in front of reception to ensure medical privacy, reducing some unnecessary duties for the transactions, etc. Regarding the hospital staff requirements for an automated hospital system, it is typically suggested that hospital staff are to be provided with a huge database of patient information to track the patient-related information in order to enable the system to avoid cross work as much as possible and to give the best service to the patient, etc. Thus, it can safely be said that the operational view of an evolving hospital is very complex, and this requires careful study for innovation in services and benefits to be provided by the hospital.

In order to understand the status of public health care, a study was undertaken by the author to investigate the level of health care provided to residents of a typical country in the Gulf Cooperation Council (GCC), such as the United Arab Emirates (UAE). The study revealed that the health-care scenario of the UAE is expected to take a quantum leap in coming years. In the UAE, health-care services are internationally recognized to be of good quality and comparable to other developed countries. In addition to government hospitals, public health care is also shared by private health-care providers. The hospital or patient information and management systems deployed at various government or private hospitals in the UAE look different but are functionally similar, as each one targets the benefits of the streamlining of operations, enhanced administration and control, improved responses, cost control, improved profitability, all largely customized to the requirements of any hospital. Though these hospital or patient information and management systems help provide better care to patients and the streamlining of operations, patient waiting time in hospitals has typically increased.

Radio frequency identification (RFID)-based patient management systems help in tracking the movement of the patient in the hospital and manage the waiting list of patients at a hospital. If a patient is being tracked in the hospital for presence, this means that his or her medical records in the database can be pulled up by using his or her location or presence in the hospital. This RFID-based tracking can be used in various ways to streamline the operations of the hospital and to the benefit of the patient in reducing his or her waiting time in the hospital.

Let us look at a regional view of health care in a typical country. The hospital hierarchy (in the case of government-run hospitals) is usually structured as shown in Figure 18.1. Depending upon population size, a number of similar units may be operating in that town or city. Though operating under a common government code and operating procedure, each hospital operates independently, maintains its patient database, etc. If examined closely, Figure 18.1 also shows that private hospital(s), government hospital(s), and basic health unit(s) may form a collaborative health-care group or can form such a collaboration with similar units in different town or city. The benefits of such collaboration could be to share research-related medical information, medical data records, etc., under a set of constraints. These constraints include privacy of patient data, hospital business information, staff and salary information, etc.

In order to enable smartness in hospital operations and link its information database for collaboration within a health-care network, key technology needs are to be identified and investigated, which can help hospitals achieve the following:

Patient care: A smart identity improves the security of medical records. It is reliable and thus difficult to forge. The smart identity as a digital signature serves as a guarantee and thus provides a barrier to identity theft.

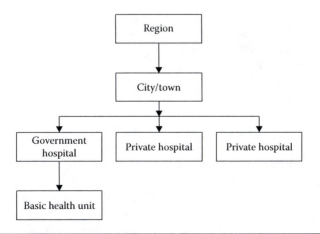

Figure 18.1 Hospital hierarchy in a typical country.

Administrative efficiency: Use of a smart technology helps reduce patient admission time and thus saves the resources of the hospital; it reduces data entry errors as well. It also helps in the streamlining of clinical operations, which, in turn, reduces operational cost and improves the patient experience.

Hospital security: The use of smart cards limits access to those buildings and areas in the hospital that are appropriate for the presence of patients and/or employees.

Medical record management: Linking medical records with a smart ID helps reduce errors in matching records and the creation of duplication and the associated cost. This helps in billing, registration, and continuity of care.

Quality of care: The lack of patient information often causes medical errors and events. These can be reduced by accurately linking a patient's information on smart cards to an institution's medical records. Once in readable range of a reader, the card provides vital medical data at the scene or en route to the hospital. The card can effectively be used to enable the creation of and access to medical records available on a health-care providers' network.

Privacy, security, and confidentiality: The standard and robust cryptography methods, which have been proven to be extremely secure, can be employed to protect information on smart cards. As cards are to be physically held by the patients themselves, patient information on cards is secure and private.

The chapter is organized as follows: In the next section, a review of related work is presented to highlight existing approaches related to intelligent hospital information systems (HISs) with examples taken from field deployment of RFID. Section 18.3 presents a framework to integrate RFID into HISs and discusses RFID deployment considerations in the same perspective. In Section 18.4, a collaborative health-care region is conceived to investigate architectural challenges and issues emerging from such collaboration. The access-control issues are also investigated. Section 18.5 discusses the motivation and benefits for RFID deployment inside a hospital and beyond a hospital's information boundary with practical considerations. A set of relevant standards is highlighted in Section 18.6 that sets the direction for a common approach toward a smart hospital. Section 18.7 presents discussions and emerging trends.

18.2 Related Work

In the literature, various methods and tools have been studied to automate and optimize the functionalities and operations within hospitals. Lina and Yang [1] present a three-layer electronic patient record management system that is expandable, maintainable, secure, and cost-effective. Using various technologies, dataflow, and activities at a hospital and a functional and operational analysis of the clinical departments in a hospital, the database structure has been designed by Lina and Yang [1]. In [2], Hu et al. discuss that hospitals in China are in a competitive environment. The hospital services and management system is discussed by Hu et al. [2] to analyze the contents of health-care services management and functions. They [2] designed the framework and studied the implementation technology to discuss how to integrate a hospital services and management system into a HIS. Similarly, Zhao [3] views that electronic health in China has a broad space of application and promotion. Zhao discusses the role of cooperating health-care systems at the regional level for sustainable electronic health care.

Compared to bar code and electronic cards used for applications in businesses, the RFID has come out as a successful competitor for object identification and automation. The potential of RFID and other sensing devices has been investigated by many researchers [4–7]. In [8], Vanany and Shaharoun have developed and tested a framework to structure the complexity of RFID justification. In fact, the phases of its implementation are used to justify RFID investment, and the deployment is claimed to be applicable in the health-care sector. In terms of automation and efficiency, Jeon et al. [9] propose to use an RFID system for medical-chart management. Once charts are tagged, two issues are addressed in this work: the first one is how to identify more tags simultaneously and correctly by optimizing reader allocation methods; the second is to reduce read time. Along the same lines, RFID has been investigated in [10] to enhance the operation room–management information system for the efficient identification of patients and health-care providers, tracking of critical surgical procedures, and control of medications and medical materials to prevent medical errors. A number of case studies have been initiated to engineer RFID within hospital automation. Lai et al. [11] propose a framework using RFID, integrating it with the hospital's information system and reengineering the inpatient medication processes to improve patient safety and reduce serious medical errors. This framework is implemented in the Taichung Hospital to improve the efficiency of hospital management and patient safety. The RFID technology has also been investigated in [12] for a HIS to monitor patients at the room level. Kim et al. have performed interrupted time-series analysis of the mean waiting time for patients and claim that the mean waiting time of patients significantly reduces (from 5.4 to 4.3 min, a 20% decrease). In another work, Wang et al. [13] investigate RFID as part of the IT infrastructure in the health-care environment of a Taiwan hospital and conclude that RFID deployment is likely to revolutionize hospital medical practices.

Once deployed within an operational domain, the RFID device brings its architectural challenges. The efforts have been exercised to an extent to address relevant issues and constraints. For example, to query physical objects, a temporal RFID model has been investigated in [14] to construct most complex applications in the real world. The respective authors claim that the proposed model provides powerful support on querying physical objects in RFID-based applications. Carbunar et al. [15] have presented a set of algorithms to address three problems associated with tag detection in RFID systems, namely, accurately detecting tags in the presence of interference, eliminating redundant tag reports by multiple readers, and minimizing redundant reports from multiple readers. Similarly, Cao et al. [16] discuss architectural challenges when RFID is deployed in a scalable and distributed environment, such as the supply chain, for the purpose of tracking and monitoring. In that, the centralized and distributed RFID data warehouse concepts are examined and compared for their usefulness to minimize storage of RFID-related data and processing. As a case study, Welbourne et al. [17] present the challenges encountered during a smaller-scale pilot project, where hundreds of antennas and thousands of RFID tags are deployed to uncover the issues in pervasive RFID deployments. Welbourne et al. finally conclude by highlighting the security and privacy of data and overall reliability of the system. In another effort, Chen et al. [18] address security and privacy concerns in enterprise-level RFID deployment using role-based key management. The authors claim that this approach helps eliminate concerns that arise as a result of a leak of RFID code content within cooperative business transactions. As an enabling technology, RFID middleware has been proposed in [19] to provide a seamless environment for moving data from the point of transaction to the enterprise systems.

For the enterprise level, a collaborative network involving RFID study has been carried out in [20] to automatically pay for services in car-parking areas, tolls, and gas stations. For cooperation among health institutions at the regional level, Winter [21] discusses cooperation requirements, such as communication links, connecting heterogeneous software components of different vendors and with different database schemata, etc. For this, Winter proposes 3LGM as a meta-model for the modeling of information systems. Winter defines 3LGM using the unified modeling language

(UML). The tool provides a means for analyzing the information system model to assess its quality. GS1 standard [22] is a common language for health-care professionals to uniquely identify locations, goods, products, patients, services, etc. This system has three distinct components: unique identification numbers, data carriers, and messaging standards for electronic data interchange. The traceability in this whole system, using electronic tracking and identification, is the key for its useful deployment. The RFID has turned out to be a strong candidate for building a smart hospital [23].

In summary, various tools and methodologies have been reported in the literature at the design level or the implementation level using RFID to automate a set of services within a hospital. However, there is room for investigation of RFID to be used beyond a typical hospital information boundary. This area of linking RFID data to an external data warehouse has not been investigated before with regard to cooperating health-care units. This topic will be examined further in Section 18.4. In the next section, the key components of an RFID-based information and management system of a smart hospital are discussed.

18.3 RFID-Based Deployment Considerations

Based on discussions in previous sections, a framework can be developed as shown in Figure 18.2. It describes various components to be integrated into a typical HIS. The data manager includes

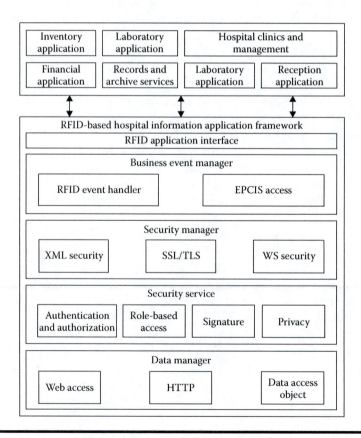

Figure 18.2 RFID-based hospital information application framework.

functions to external systems using protocols, such as web service, socket, etc., and a data access module for input and output processing of data using databases. The security manager provides authentication to distributed services in the network. The business manager connects events from the middleware to high-end application databases. This framework may be used for implementation of RFID-based information system architecture. During the development of this framework, it was assumed that

- There are no assumptions made on the number of tags and readers or on the tag distribution or reader deployment.
- Based on the lower cost, the passive tags are assumed as they also have limited memory.

The prototype development first requires tracking technology to be selected. Three tracking technologies went through consideration and were compared to come up with the most suitable one that fits the requirements. The following is the overview of various components selected and a subsystem development for the prototype.

18.3.1 Technology Selection

Three famous technologies, that is, bar code, contact smart card, and contactless smart card [24], offer the same services, that is, identification and tracking, and database management flexibility. Bar codes can be read by optical scanners, and the patient should show his or her bar code to the scanner in each spot in order to be detected. The case of a smart card, chip card, or integrated circuit card is defined as any pocket-sized card with embedded, integrated circuits that can process information. With a smart card, when it is inserted into a reader, the chip makes contact with electrical connectors that can read information from the chip and write information back. The contactless smart card is the other type in which the chip communicates with the card reader through RFID induction technology. Contactless smart cards can be used without even removing them from a wallet. Thus, it was concluded that the contactless smart card technology based on RFID is better suited for development because of two reasons:

- It provides wide range detection, which can reach greater than 10 m.
- Mobility detection helps in detecting the patients without asking them to stop in front of the reader while their card is in their wallet.

18.3.2 Specifications

The RFID packaging allows the RFID tag to be attached to an object to be tracked. There exist different types of RFID tags, namely, passive tags powered solely by the RFID interrogator within range of the RF field; active tags with their own power source in order to receive a weaker signal from the interrogator and boost the return signal and semipassive tags with a power source for on-tag sensing but not to boost range [24]. Passive tags were selected because of the fact that they are thin and small, they do not consume a lot of power, and they do not need a battery. As regards to whether a read-only, a write-once-and-read-many, or a read/write-passive tag is to be used, the read-only option was selected as it is cheap and because it has the least memory requirement. Because passive tags are selected, a check was done to select the frequencies for passive tags that can target tag readers with a distance of at least 8 to 10 m. For typical hospital-based tracking requirements, passive tags with high-frequency (HF) band were selected as these do not interfere with some known medical equipment [25].

18.3.2.1 Reader Considerations

An RFID reader is simply a radio; the only difference is that the RFID reader picks up analog signals, not hip-hop. The reader not only generates the signal that goes out through the antenna into space but also listens for a response from the tag. Both the tags and the readers operate over a specific frequency. The lower the frequency, the shorter the read distance for an equal tag size. Because the area is wider than a couple of feet, an HF band was used. As regards to type, a read-only reader was selected. Typically, one to four antennas may be attached to a single reader. In the prototype, two antennas attached to opposite sides of a metal gate installed at the entrance gate of a hospital were considered sufficient. The number would differ based on the area or location of the detection range.

18.3.2.2 Management Considerations

Considerations such as what approximate number of patients can typically visit the hospital and how each one of the patients will be able to receive a unique ID number are very critical to the management of the hospital. This is considered when calculating the total number of unique IDs a number system can generate, vis-à-vis a numbering system in a hospital.

Reader position: Considering a typical hospital clinics map, a number of cases were studied, and finally, one of them, based on the number of readers, was selected to be included in the implementation process. In this case, a reader is going to be used on the main door and at the exit of each section so that it will detect if the patient enters the section or not. Based on the number of sections to be covered for tracking patients on a multiple-floor hospital building together with a pharmacy, a main gate, and separate male and female entrances of a typical hospital, a total number of 38 readers were to be spread all over the, say, three floors.

18.3.2.3 Patient Considerations

The health card size typically available in hospitals is that with a length of 8.5 cm, a width of 5.3 cm, and a thickness of 1 mm. This size seems comfortable, and the only thing left is to tag it with RFID on the back of the card with the requirement that the patient must hold the card when he or she enters the hospital. In case of any technical malfunction of the system, the information (which is currently shown on cards), such as name, dates of issue and expiry of the card, card number, blood group, health card issuer phone number, and insurance information, was also agreed to be left printed on the card in order for the hospital management to treat patients during emergencies.

18.3.3 Process Flowchart

Based on the above considerations, the technology that suited the typical requirements was HF RFID technology, and the appropriate product in the market chosen as a prototype to fit the requirements was the Alien ALR 8800 RFID kit [26]. The flow of process for the patient entering the hospital is depicted as shown in Figure 18.3.

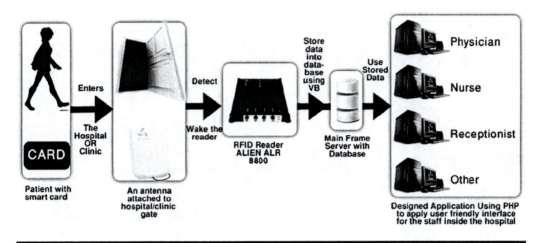

Figure 18.3 Process flowchart.

18.3.4 *Software Development and Database*

The main concentration for the software development remained how to build software that connects to the reader and connects to a MySQL database [27] and then stores in the database table if the patient had entered the clinic or left. Alien RFID Gateway software is the one provided by the manufacturer for the Alien ALR-8800 RFID developer kit [26]. Its software utilities were used to define properties for the reader, for tag detection, and for knowledge of power of signal strength and tag electronic product code (EPC) number.

phpMyAdmin [27] is a tool written in PHP (hypertext processor) and is intended to handle the administration of MySQL over the World Wide Web. Currently, it can create and drop databases; create, drop, or alter tables; delete, edit, or add fields; execute any SQL statement; manage keys on fields; manage privileges; and export data into various formats and is available in 55 languages. It was also found that some sort of visual basic (VB) program was to be written to connect to the device, listen to the port, and then send the information to the database being used. For illustration purposes, the part of VB code written to connect and listen to the port is shown in Figure 18.4.

A PHP code was written to input and output the data from the database to a web page. As an illustration, "config" code, "open db" code, and "close db" code for the pharmacy table are shown in Figure 18.5. The similar programming was done for various forms of pages, such as nurse, patient, doctor, receptionist, administration, etc. Each of the respective staff was perceived to have access to their own data and forms developed to realize data privacy regulations, common to all types of hospitals and clinics nowadays.

For better signal quality from the card to the reader, the control attenuation factor in the reader was set using a hit and trial method. It was found that the attenuation factor needs to be set at 100 on the scale from 0 to 150 allowable in this reader, and antennas need to be placed at a position 109.5 cm off the ground for best results. After finishing the testing procedure, five tags were implemented in different types of materials (such as plastic, iron, etc.). It was found that the iron one was not working when passing through the gate. Another challenge that was encountered was to determine the direction of the patient to know if he or she is going in or coming out of the location. In both cases, it was determined that a memory should be involved in order to know the

```
Imports System
Imports System.IO.Ports
Public Class Form1
    Dim WithEvents port As SerialPort = New System.IO.Ports.SerialPort("COMP9".57600, Parity.None, 8, StopBits.One)
    Private Sub Form1_Load(ByVal sender As Object, ByVal e As System.EventArgs) Handles Me.Load
        CheckForIllegalCrossThreadCalls = False
        If port.IsOpen = False Then port.Open()
    End Sub
    Private Sub port_DataReceived(ByVal sender As Object, ByVal e As System.IO.Ports.SerialDataReceivedEventArgs) Handles port.DataReceived
        TextBox1.Text = port.ReadLine
        If port.ReadExisting.Length = 0 Then
            ListBox1.Items.Add(TextBox1.Text)
            TextBox1.Text = ""
        End If
    End Sub
    Private Sub TextBox1_TextChanged(ByVal sender As System.Object, ByVal e As System.EventArgs) Handles TextBox1.TextChanged
    End Sub
    Private Sub ListBox1_SelectedIndexChanged(ByVal sender As System.Object, ByVal e As System.EventArgs) Handles
ListBox1.SelectedIndexChanged
    End Sub End Clas
```

Figure 18.4   Sample VB code for connecting and listening to the port.

Config code:	Open db code	Close db code

`<?php`
`// This is config.php`
`$dbhost = 'localhost';`
`$dbuser = 'root';`
`$dbpass = '';`
`$dbname = 'gp2_database';`
`?>`
	`<?php`
	`// This is config.php`
	`$dbhost = 'localhost';`
	`$dbuser = 'root';`
	`$dbpass = '';`
	`$dbname = 'gp2_database';`
	`?>`
		`<?php`
		`// an example of closedb.php`
		`// it does nothing but closing`
		`// a mysql database connection`
		`mysql_close($conn);`
		`?>`

Figure 18.5   PHP code for pharmacy table.

previous status. This was effectively done in the database. For a normal human walking with random flow, the speed of the reader to read tags set by the manufacturer turned out to be acceptable, and no read-miss was noted with 15 trials.

18.3.5 Patient Identification

The EPC was conceived by the MIT auto-ID Center [28] as a means to identify physical objects similar in scope to the bar code numbering scheme. It consists of a 96-bit number (header 8 bits, manager 28 bits, object class 24 bits, and serial number 36 bits). The 96-bit code can thus provide unique identifiers for 268 million companies. Each manufacturer can have 16 million object classes and 68 billion serial numbers in each class. The object class may represent a typical hospital adopting this system with a serial number representing the patient ID. Based on this numbering scheme, it can safely be said that this numbering scheme can uniquely represent a patient population within a typical health-care region. The EPC number becomes the primary key to represent a patient's medical record in the hospital.

18.3.6 Database Component

Based on the considerations, the database design was started. In order to enable views on patient data, a typical patient data model is drawn first as shown in Figure 18.6. Figure 18.6 shows a generic patient data model that is used for the patient while visiting a hospital. The "patients"

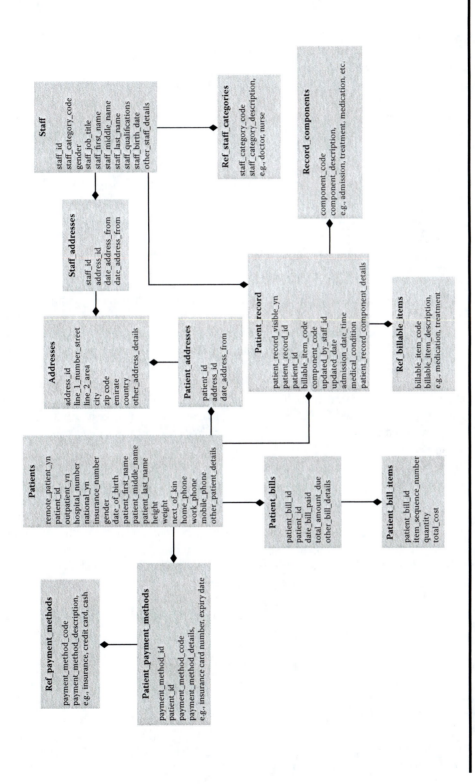

Figure 18.6 Patient's generic data model.

entity in the model contains personal details and is aligned to the "Patient_Record" entity that holds the record of all transactions. For every visit, whether a hospital stay or a routine checkup, a hospital staff member is assigned to the patient, and a "patient_bill_id" is generated. The "Staff" entity in Figure 18.6 contains personal information of all the hospital staff, categorized as per categories stored in the "Ref_Staff_Categories" table. The "Patient_Bills" entity stores all the information regarding the hospital bill, which is itemized as per records of "Patient_Bill_Items." The patient can pay for his or her bills through different methods, such as through insurance, credit card, or cash. Details of such methods are stored in "Patient_Payment_Methods." Itemized bills can be generated through "Patient_Record," "Record_Components," and "Ref_Billable_Items." An additional parameter of "patient_record_visible_yn" has been used to give authority to the doctor to make some of the patient record visible to the network to be viewed by another doctor for reference. It should be noted here that only a doctor can make certain patient records visible or not visible at all.

18.4 Collaborative Health Care and Challenges

When people (as patients) are on the move—visiting either a different area, another city, or relocated to another place for a shorter or longer duration—then usually they undergo similar initial tests and checkups at hospitals at their new location and, probably, a complete reexamination of the health problem. The medical data lying at a hospital at the older location do not contribute in any way to the benefit of the relocating person. The idea here is to make possible access to these medical data within a constrained collaboration of hospitals. In other words, the hospital at the new location (called the remote hospital) uses a collaborative system network and a database to access a patient's medical records from the hospital at the old location (called the local hospital). In order to enable this access, patient RFID can play a role as a key. So at a remote hospital, a patient (with RFID) is detected using a network, and the doctor finds it convenient to access his or her medical records (history) from the local hospital to facilitate medication.

Two main challenges surface in this concept. One is how and which type of medical data are to be shared within a collaborative region. The second is whether RFID can be used to simplify and help streamline the operations within the hospital and be a key in the collaborative healthcare domain. For the first one, the main consideration is how data, such as medical data records, are to be stored and processed among a collaborative network of hospitals. One obvious choice is to store medical records from networked hospitals at a central location in order to enable global viewing of the medical records and, subsequently, simplified archival. The other choice is distributed archival of records at each hospital, and access to patient medical records is allowed by other hospitals on the basis of need. Both of these approaches have advantages as well as disadvantages. In order to simplify analysis, the medical data records may be grouped into two main categories based on frequency of occurrence and data value:

1. Higher-frequency RFID tracking data of patients within the hospital that is practically of no value to a remote hospital.
2. Lower-frequency patient data records based on tests and medication within the hospital, which is of higher value to the doctor at a remote hospital.

If RFID tracking data (typically 30–50 bytes per reading per patient) coming from each deployed reader are also allowed to be stored at a central location, this would increase the volume of data significantly at the central location as each hospital might be handling, typically, thousands of patients per day. Furthermore, as stated previously, this type of data has practically no value to the collaborative region. It is thus viewed that this type of data should be processed and stored locally at each hospital. The analysis of each approach is now described below.

Centralized data processing: The idea in this approach is to employ a centralized architecture where all patient data are sent for stream processing and archived. The clear advantage is simplified stream processing and that the system has a global view of the entire medical data records for a set of regional health-care units. This also helps in medical research on certain diseases as all data are available at one location. Thus, the local servers (at each hospital) need to preprocess the medical data for subsequent archival at the central location. The apparent disadvantage of this approach is the high communication cost of transmitting data streams to the central location and vice versa as network bandwidth costs can be substantial.

Analysis: Consider a collaborative network of five hospitals, where each hospital typically stores 2000 patient cases, and each case contains 20 items (which include patient description, physical checkup, doctor's prescription, tests, medication, pharmacy, etc.), and let each item be of approximately 100 kB. A simple calculation shows that approximately 20 GB per collaborative region are being processed every day for archival at the central location. Because whole data for all five networked hospitals reside at a central location, approximately the same amount of patient data is being accessed by hospital doctors for viewing in order to treat the patient on the follow-up visit. If more persons are allowed to view medical records, the accessed data would be voluminous, and related access delay and communication costs would increase. Furthermore, this will also complicate the privacy of medical records as different hospitals may like to enforce different privacy levels based on patient preferences and business needs.

Distributed data processing: This is an alternative approach to centralized processing, where all patient medical data reside at the local hospital for processing and archival. If patient data need to be viewed by other doctors at a different (remote) hospital, the doctor may log into the system of the local hospital to view the patient's relevant medical records. The clear advantage is that this approach is very natural to the business of the hospital and privacy needs of the patient. Furthermore, the data access and communication costs are expectedly minimal compared to centralized data processing. The apparent disadvantage of this approach is that the system is distributed and may have a different view at each hospital.

Analysis: Consider a collaborative network of five hospitals, where each hospital typically accesses 50 patient cases at a remote hospital and each case contains five items (which include patient description, doctor's prescription notes, tests, medication, etc.), and let each item be of approximately 20 kB (as only a set out of 100 kB may be needed to be viewed by the doctor). A simple calculation shows that approximately 5 MB are being accessed every day by each remote hospital from the local hospital. Because only prescription records may be updated by the doctor in these cases, a few kilobytes of data per patient is to be stored back at the local hospital. The network bandwidth costs are thus minimal, and each hospital is allowed to maintain its privacy needs.

Based on the analysis just presented, the framework of the collaborative health-care region may be drawn as shown in Figure 18.7. For each hospital, the framework remains the same as shown in Figure 18.2 except that here doctors at the remote hospital are allowed access to selected medical records using privileges assigned to them in the collaborative system.

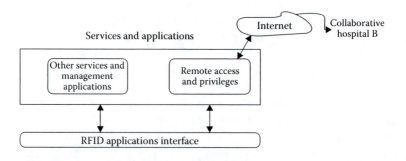

Figure 18.7 Framework for collaborative hospital "A."

A separate service is shown to enforce access to medical records. This server includes access provided to doctors at remote hospitals and thus needs to be linked up to remote access and privileges in services and applications layer—typically through dedicated Internet lines. In order to enforce limited access to medical records, an example development is discussed below, assuming the following:

■ Database design should be simple in order to help integration with other hospital domains for collaboration and should enforce the privacy needs of the hospital.
■ Access to patient medical records (at the remote hospital) should preferably be given to doctors only.

In order to design the database system, it was considered that automated hospitals in a region are willing to collaborate with each other to share patients' medical records. The filtering part is debatable as to whether a hospital is willing to share with another hospital *or* if it is willing, what type of data is being shared. For simplicity, it was considered that all government hospitals in a region were being managed by a similar system (but not necessarily with the same database) and that patients' medical records are being shared with the exception of patients' financial and private information and hospital and doctor information. The hospital where the patient has all his or her information in the database is termed a local hospital, whereas a visited hospital (in the region) is termed a remote hospital. Thus, the database will have multiple views: local as well as remote. The remote view of the patient's medical record may be visualized similar to as is provided by typical web applications to access patient data. The constraints for database were summarized as follows:

■ Privacy of data: Only medical records are to be viewed at a remote hospital.
■ Database independence: Consider that collaborating hospitals do not use the same database technology.
■ Security of medical records: Staff at the local hospital has access to specified pages of the medical record according to assigned privileges, whereas only the doctor at a remote hospital views the medical records from the local hospital.
■ After medical help is provided at the remote (or local) hospital, the database at the local hospital is updated. Storing updates to the database at a remote hospital is optional (upon the patient's discretion).

Based on these constraints, the database design for the remote hospital was started. In order to enable multiple views on patient data, a typical patient data model was considered first, as shown in Figure 18.6.

Figure 18.8 shows the data model for patients when handled by a remote hospital. In this model, the local hospital's staff information is not viewable, and billable items may be recorded differently as per the services available in that remote hospital. The doctor, using his or her privileges, can access patient data through the Internet by logging into the local hospital's system. The doctor may enter his or her prescription and notes, if desired, into the patient medical record. In other words, the treatment received by the patient at the remote hospital will be treated similar to the treatment of the patient at the local hospital. The only difference is that the medical records at the local hospital are now being accessed and updated by the doctor at the remote hospital. The updated medical records of the patient are accessible by doctors and concerned staff members at the local hospital, as well. These multiple views of the database on the Internet can be enabled by various technologies [29–30] to enforce privacy and security of data. This is further investigated in the next subsection.

Database schema tables: A number of tables were designed, for example, for the reader, tag, hospital management tables, etc. For illustration purposes, the database relationship design is shown in Figure 18.9.

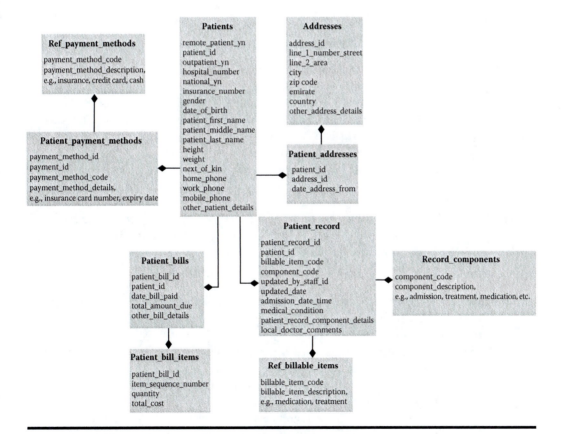

Figure 18.8   Patient's data model for remote hospital.

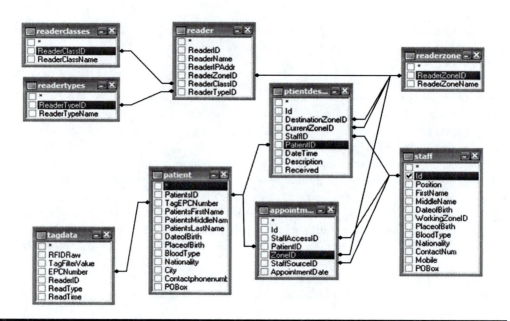

Figure 18.9 Database relationships example.

18.4.1 Access Control on Patient Data

The database application chosen for this implementation is a health-care center and termed a collaborative health-care management system (CHMS). The selection of this model is arbitrary and is chosen as an implementation example, though a similar development can be devised for any group of collaborative health-care units. The data reside in outpatient records, in-hospital records, and family-visit records, and the transactions are executed on a daily basis. The application is used by the collaborative health-care unit manager, doctors, mothers, and the child-care unit manager in charge to perform various transactions. It is also used by nurses, health visitors, the operating theater (OT) in charge, and the office assistant to post data. The accounting manager, accountant, and internal auditor post, generate, and verify accounting data. The idea of data access and its usage is such that users are granted membership into roles based on their responsibilities in the organization and within the framework of collaborative health-care units [31]. The process of developing role-based access control for this application is stated below.

(i) Role definition and functions identification: Based on various categories of participants that are expected to use this network application, the participating roles and functions required for each role in health-care units (hospitals/basic health unit) can be defined as follows:
Hospital roles:
- Doctor (local): Create, modify, or delete inpatient and outpatient records; enter prescriptions; create or modify medical services records for the patient; create, modify, or edit nursing requirements for the patient; enter comments on the records of a patient generated in remote hospitals or basic health units
- Doctor (remote): Modify inpatient and outpatient records, enter prescriptions, modify the medical services record for patient, modify or edit nursing requirements for the patient

- Junior doctor/attached doctor: Modify inpatient and outpatient records; enter comments on prescriptions; comment on medical services; create, modify, or edit nursing requirements for the patient
- Nursing staff: Create or modify nursing requirements of the patient
- Duty nurse: Modify or comment on nursing requirements of the patient
- Finance manager: In addition to accountant functions, modify ledger posting rules
- Billing accountant: Input all hospital transactions, generate general ledger reports, generate patient bills, generate report on receivables
- Medical insurance (corporate client) accountant: Input corporate or insurance-level hospital transactions, generate general ledger reports, generate patient bills, generate report on receivables
- Hospital administration manager: Ability to perform any of the functions of other roles in times of emergency and to view all transactions, account status, and validation flags
- Record section in charge: Input data in patient records, including inpatient and outpatient
- Medical services manager: Ability to perform any of the functions regarding hospital services, view all transactions, account statutes related to medical services of patients
- Laboratory in charge: Input laboratory-related patient data, generate and receive bills for outpatients, generate requirement requisites for material
- Pharmacy in charge: Input pharmacy-related patient data, generate and receive bills for outpatients, generate requirement requisites for material
- Internal auditor: Verify all transactions and ledger posting rules

Basic health unit roles:
- Office assistant: Input data in family folders regarding nutrition and vaccination provided to respective family
- Mother-child unit in charge: Create and delete family folders in addition to tasks defined for office assistant
- Health visitor: Input or modify vaccination information of children under age 10, input and modify mother nutrition chart
- Nurse: Input or modify inpatient record
- Operation theater (OT) in charge: Input or modify OT record
- Doctor: Create, modify, or delete inpatient records; enter prescriptions; create, modify, or delete OT record
- Accountant: Input all health unit transactions and generate general ledger reports
- Accounting manager: In addition to accountant functions, the ability to modify ledger posting rules
- Internal auditor: Verify all transactions and ledger posting rules
- Basic health unit in-charge: Ability to perform any of the functions of other roles in times of emergency and to view all transactions, account statuses, and validation flags

(ii) Role graph: Based on the intended functionality and privilege assignments required for each role, a structural relationship emerges among roles as shown in Figure 18.10. It is clear from the figure that roles higher in the hierarchy accumulate more privileges than the ones lower in the hierarchy. The privileges set for any two roles that are not part of the same chain are disjointed.

(iii) Formulation of constraints:

    a. The maximum number of users that can be assigned to medical services manager, nursing staff coordinator, finance manager, hospital administration manager, internal auditor, and doctor in a local as well as a remote hospital is *one.*

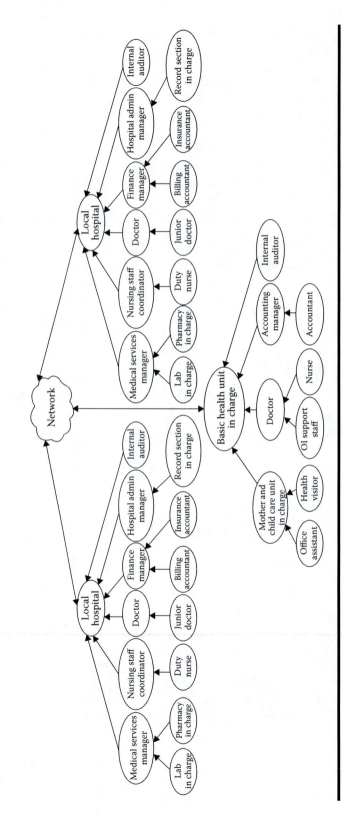

Figure 18.10 Role graph showing structural relationships.

b. The following pair of roles cannot be assigned to the same user [static separation of duty (SSD) or membership mutual exclusively (MME)].

1. Medical services manager and nursing staff coordinator, medical services manager and doctor, medical services manager and finance manager, medical services manager and hospital administration manager, medical services manager and internal auditor, medical services manager and duty nurse, medical services manager and junior doctor, medical services manager and billing accountant, medical services manager and medical insurance accountant, medical services manager and record section in charge

2. Duty nurse and junior doctor, duty nurse and billing accountant, duty nurse and medical insurance accountant, duty nurse and record section in charge

3. Finance manager and hospital administration manager, finance manager and internal auditor

4. Nursing staff coordinator and doctor, nursing staff coordinator and finance manager, nursing staff coordinator and hospital administration manager, nursing staff coordinator and internal auditor, nursing staff coordinator and laboratory in charge, nursing staff coordinator and pharmacy in charge, nursing staff coordinator and junior doctor, nursing staff coordinator and billing accountant, nursing staff coordinator and medical insurance accountant, nursing staff coordinator and record section in charge

5. Doctor and finance manager, doctor and hospital administration manager, doctor and internal auditor, doctor and laboratory in charge, doctor and pharmacy in charge, doctor and duty nurse, doctor and billing accountant, doctor and medical insurance accountant, doctor and record section in charge

6. Finance manager and laboratory in charge, finance manager and pharmacy in charge, finance manager and duty nurse, finance manager and junior doctor, finance manager and record section in charge

7. Hospital administration manager and internal auditor, hospital administration manager and laboratory in charge, hospital administration manager and pharmacy in charge, hospital administration manager and duty nurse, hospital administration manager and junior doctor, hospital administration manager and billing accountant, hospital administration manager and medical insurance accountant

8. Internal auditor and laboratory in charge, internal auditor and pharmacy in charge, internal auditor and duty nurse, internal auditor and junior doctor, internal auditor and billing accountant, internal auditor and medical insurance accountant

9. Laboratory in charge and duty nurse, laboratory in charge and junior doctor, laboratory in charge and billing accountant, laboratory in charge and medical insurance accountant, laboratory in charge and record section in charge

10. Pharmacy in charge and duty nurse, pharmacy in charge and junior doctor, pharmacy in charge and billing accountant, pharmacy in charge and medical insurance accountant, pharmacy in charge and record section in charge

11. Junior doctor and billing accountant; junior doctor and medical insurance accountant; junior doctor and record section in charge

12. Billing accountant and record section in charge

13. Doctor (local) and doctor (remote)

14. Role (local) and role (remote): *equivalents only*

18.5 Motivation and Benefits

As discussed before, a number of critical operational issues are addressed by deploying RFID within collaborative hospitals. Let us analyze this using some cases:

Emergency transfusion system: As an example, while in emergency, any allergic states of the RFID-bearer patient can be verified from his or her records from the local hospital using a PDA while the transfusion is being initiated to the patient. This type of service is not possible to be delivered to the patients in emergencies in typical hospitals where RFID-based IT infrastructure is not used; rather tests are typically done before any such service is delivered. In order to enable this service, the outpatient and emergency transfusion system is to be integrated with the RFID-based IT infrastructure inside the hospital so that medical staff can carry out the emergency service with an unprecedented accuracy. This also helps to monitor the transfusion fluid, the patient, and the medical staff who perform the transfusion.

Front desk initial body checkup: Generally, long lines are visible for initial body checkups within hospitals. In order to decrease the workload of the nurse at the front desk and reduce waiting time for the patient, RFID may successfully be used to automate repetitive procedures. For example, if the patient record is available in the hospital, the patient can directly go to the installed checking kiosk for identification and to initiate payment mode, print checking procedure, and proceed to corresponding testing rooms, and the record is updated in the system. The doctor can then call the patient based on his or her availability in the patient waiting room.

Medical examination by the doctor: In case the patient records are accessible at a remote hospital, patient medical history records can be pulled up by the doctor during the examination to better treat the patient. This service can only be available in collaborative network hospitals.

Examination during travel: Sometimes, a patient requires frequent medical examination on a daily basis and, thus, cannot travel. In regional collaborative health care, the patient may be allowed to travel and continue his or her treatment at network hospitals as his or her medical history can be pulled up by doctors at remote hospitals. Furthermore, in order to better manage time, some medical tests may be done at a local hospital, and detailed examination by a doctor may be completed at a remote hospital.

Drug dispensing system: Normally, the labels are printed and affixed to drug packets before drugs are dispensed. This takes place once the doctor prescribes the medicine for the patient and updates it in the system. With the use of RFIDs in the system, the RFID tags are affixed onto the drug containers and recorded into the system. The patients are also given an additional prescription RFID tag (with relevant drug information) to collect drugs from the dispenser. The dispenser collects the drugs from the shelves ahead of the patient's presence and matches the packet with the prescription tag once it is shown by the patient. Thus, drugs are dispensed with accuracy, and waiting time for the patients is also reduced.

18.6 Standardization Efforts

A number of standards for operation of RFID systems have been developed and are in practice. For the interest of the reader, a few of them are briefly described below.

ISO 18000: A series of international standards for the air–interface protocol used in RFID systems for tagging goods within the supply chain. Along similar lines, medical devices, equipment, and IT infrastructure, etc., can be tagged in the hospital domain.

ISO/IEC 10536 and ISO 15693 (ISO SC17/WG8)—vicinity cards: A series of international standards for the physical interface of contactless smart cards. HF tags and interrogators that comply with the ISO 15693 standard (another subset of the ISO 18000-3 standard) are reported to have been tried in health-care applications.

For health-care operations, the penetration level of RFID still has to be raised to make hospitals really smarter. There are mainly two issues that hinder its penetration on a wider scale: interference with medical devices and privacy concerns. The American National Standards Institute (ANSI, 2009) has recently approved a new standard (ANSI/HIBC 4.0) [32] for using RFID tags to label and track medical products. The standard allows tagging health-care products to prevent RFID interference from interfering with medical devices. The standard recommended that the health-care products be tagged with 13.56-MHz HF coding. In fact, AIM Global, MET Labs, and Georgia Tech have started developing protocols (in 2009) for detecting electromagnetic interference caused by RFID transmissions and their effect on medical devices [33].

In terms of security and privacy, the HIPAA (Health Insurance Portability and Accessibility Act of 1996) regulations address security and privacy of "protected health information." These regulations put emphasis on acoustic and visual privacy and may affect the location and layout of workstations that handle medical records and other patient information, paper and electronic, as well as patient accommodations. One of the key provisions of the HIPAA privacy rule is to assure that an individual's health information is properly protected and that individuals can control how their health information is accessed and used. The HIPAA privacy rule applies to specific covered entities, such as health-care providers (e.g., doctors, dentists, pharmacies, nursing homes), health plans (e.g., HMOs, health insurance companies, company health plans), and health clearinghouses. The use of RFID technology, encryption, and other cryptography measures makes it extremely difficult for unauthorized users to access information, which thus helps to protect patients from identity theft, protects health-care institutions from medical fraud, and helps health-care providers meet HIPAA privacy and security requirements [34]. Thus, the RFID-based cards are an effective tool to facilitate compliance with the HIPAA privacy rule. The American Recovery and Reinvestment Act (ARRA, 2009) establishes a policy committee to examine methods to facilitate secure access by an individual to an individual's protected health information as well as methods, guidelines, and safeguards to facilitate secure access by caregivers, family members, or a guardian. RFID-based patient ID cards address a key ARRA policy concern regarding access to health information. ARRA has expanded the protection beyond the HIPAA rule to include additional entities, such as vendors of personal health records. A major goal of the privacy rule is to define and limit how and when protected health information is used or disclosed by covered entities. Smart cards can help covered entities and ARRA-stipulated entities comply with both the HIPAA privacy rule and the security and privacy mandates under ARRA.

18.7 Discussions and Emerging Trends

In order to simplify implementation, the technologies that were used in the prototype are available in the market and widely accepted worldwide. The foremost issue to be considered is the decision about the sharing of medical records and the trust between hospitals as, nowadays, hospital businesses have become competitive in providing quality health care. If only government and nonprofit hospitals or basic health units are collaborating for better health care, then the sharing of medical records is not serious as a governing body like the government ministry or authority may decide as a policy to implement it among hospitals. This medical record issue is generally considered serious from the perspective of privacy and security because of the vulnerabilities of servers, network intrusions, and the wireless domain. The problems of security and privacy have been addressed successfully in the financial sector [30]. Rather, data privacy may be enhanced as a result of the use of passive RFIDs [31], and security can easily be boosted by the use of appropriate protocols for RFID security [35].

As the costs for smart cards and smart card readers have dropped dramatically, and as the reader infrastructure is replaced or upgraded, it is perceived that RFID-based card technology is expected to revolutionize the markets of financial services, personal identification, and health care—where security, privacy, and information portability are crucial.

Among the many new developments and trends influencing hospital design are the use of hand-held computers and portable diagnostic equipment to allow more mobile, decentralized patient care and a general shift to computerized patient information of all kinds. This might require data ports in corridors outside patient bedrooms, and thus this environment fits well to the deployment of RFID.

Research from the analyst house of Frost and Sullivan [36] has found that the revenue from RFID within health care and pharmaceuticals will rise almost sixfold, from 2004s total of US$370 million to US$2.3 billion in 2011. According to them, the health-care market is likely to see swift uptake of RFID technology because of easily demonstrable benefits beyond the traditional return on investment (ROI)—for example, cutting the risk of drugs being misplaced or given to patients incorrectly. However, bar coding may still be preferred by many of health care's governing bodies, and given the high price of RFID relative to bar codes, bar codes are unlikely to disappear from the market before at least 2015.

RFID has already made its way into health care in both Europe and the United States. For example, the Klinikum Saarbrücken in Germany [37] has conducted pilot studies on the use of RFID to ensure patients are given the correct medication, and the United Kingdom has seen the technology used in drug tracking trials.

References

1. Lina, Y., and Y. Yang. "EPR Management System Development Based on B/S Architecture." *International Seminar on Future BioMedical Information Engineering* 2008, 441–444.
2. Hu, D., W. Xu, H. Shen, and M. Li. "Study on Information System of Health Care Services Management in Hospital." *International Conference on Services Systems and Services Management* 2005, Vol. 2, 1498–1501.
3. Zhao, J. "Electronic Health in China: From Digital Hospital to Regional Collaborative Healthcare." *International Conference on Information Technology and Applications in Biomedicine* 2008, 26.
4. Sarma, S. "Integrating RFID." *ACM Queue* 2, no. 7 (2004): 50–57.
5. Wang, L., and G. Wang. "RFID-driven Global Supply Chain and Management." *International Journal of Computer Applications in Technology* 35, no. 1 (2009): 42–49.
6. Kim, Y., J. Song, J. Shin, B. Yi, and H. Choi. "Development of Power Facility Management Services Using RFID/USN." *International Journal of Computer Applications in Technology* 34, no. 4 (2009): 241–248.

7. Zhong, S., B. Zhang, J. Li, and Q. Li. "High-Precision Localisation Algorithm in Wireless Sensor Networks." *International Journal of Computer Applications in Technology* 41, No. 1/2 (2011): 150–155.

8. Vanany, I., and A. Shaharoun. "The Comprehensive Framework for RFID Justification in Healthcare." *International Business Management* 5, no. 2 (2011): 76–84.

9. Jeon, S., J. Kim, D. Kim, and J. Park. "RFID System for Medical Charts Management and Its Enhancement of Recognition Property." *Second International Conference on Future Generation Communication and Networking* 2008, 72–75.

10. Chen, P., Y. Chen, S. Chai, and Y. Huang. "Implementation of an RFID-Based Management System for Operation Room." *International Conference on Machine Learning and Cybernetics* 2009, 2933–2938.

11. Lai, C., S. Chien, L. Chang, S. Chen, and K. Fang. "Enhancing Medication Safety and Healthcare for Inpatients Using RFID." *Portland International Center for Management of Engineering and Technology* 2007, 2783–2790.

12. Kim, J., H. Lee, N. Byeon, H. Kim, K. Ha, and C. Chung. "Development and Impact of Radio-frequency Identification-Based Workflow Management in Health Promotion Center: Using Interrupted Time-Series Analysis." *IEEE Transactions on Information Technology in Biomedicine* 14, no. 4 (2010): 935–940.

13. Wang, S., W. Chen, C. Ong, L. Liu, and Y. Chuang. "RFID Applications in Hospitals: A Case Study on a Demonstration RFID Project in a Taiwan Hospital." Proceedings of the *39th IEEE Hawaii International Conference on Systems Sciences*, 2006.

14. Wanga, F., S. Liu, and P. Liu. "A Temporal RFID Data Model for Querying Physical Objects." *Pervasive and Mobile Computing* 6, October 2010, 382–397.

15. Carbunar, B., M. Ramanathan, M. Koyuturk, S. Jaganathan, and A. Grama. "Efficient Tag Detection in RFID Systems." *Journal of Parallel and Distributed Computing* 69, (2009): 180–196.

16. Cao, Z., Y. Diao, and P. Shenoy. "Architectural Considerations for Distributed RFID Tracking and Monitoring." *Fifth ACM International Workshop on Networking Meets Databases*, October 2009, MT, US.

17. Welbourne, E., M. Balazinska, G. Borriello, and W. Brunette. "Challenges for Pervasive RFID-Based Infrastructures." *Fifth Annual IEEE International Conference on Pervasive Computing and Communications*, March 2007, 388–394.

18. Chen, C., K. Lee, Y. Wu, and K. Lin. "Construction of the Enterprise-Level RFID Security and Privacy Management Using Role-Based Key Management." *IEEE International Conference on Systems, Man and Cybernetics* 4, October 2006, 3310.

19. Ooi, W., M. Chan, A. Ananda, and R. Shorey. "WinRFID: A Middleware for the Enablement of Radiofrequency Identification (RFID) Based Applications." In *Mobile, Wireless, and Sensor Networks: Technology, Applications, and Future Directions*, IEEE Press, USA, 2006, 313–336.

20. Osório, A., L. Camarinha-Matos, and J. Gomes. "A Collaborative Networks Case Study: The Extended "ViaVerde" Toll Payment System." *Sixth IFIP Working Conference on Virtual Enterprises*, Valencia, Spain, September, 2005.

21. Winter, A. "The 3LGM2-Tool to Support Information Management in Health Care." *Fourth International Conference on Information and Communications Technology*, 2006, 1–2.

22. http://www.gs1.org/healthcare/standards, accessed online on June 29, 2011.

23. Fuhrer, P., and D. Guinard, "Building a Smart Hospital Using RFID." *Lecture Notes in Informatics*, Vol. P-9, 2006.

24. http://en.wikipedia.org/wiki/RFID, accessed online on June 29, 2011.

25. Ohashi, K., S. Ota, L. Ohno, and H. Tanaka. "Comparison of RFID Systems for Tracking Clinical Interventions at the Bedside." *Proceedings of AMIA Annual Symposium*, 2008, 525–529.

26. http://www.barco.cz/en/images/RFID_ALR-9800_Dev_Kit_300px.jpg, accessed online on June 29, 2011.

27. http://en.wikipedia.org/wiki/Web_content_management_system/, accessed online on June 29, 2011.

28. http://autoid.mit.edu/cs/, accessed online on June 29, 2011.

29. Guennoun, M., and K. El-Khatib. "Securing Medical Data in Smart Homes." *IEEE International Workshop on Medical Measurements and Applications*, 2009, 104–107.

30. Callegati, F., W. Cerroni and M. Ramilli. "Man in the Middle Attack to the HTTPS Protocol." *IEEE Security and Privacy* 7, no. 1 (2009): 78–81.

31. Tan, C., L. Qun, and X. Lei. "Privacy Protection for RFID-based Tracking Systems." *IEEE International Conference on RFID*, 2010, 53–60.

32. http://www.hibcc.org/Front%20Page%20Attachments/HIBCC%20RFID%20Standard%20 4.0.pdf, accessed online on June 29, 2011.

33. http://www.rfidjournal.com/article/view/4936/1, last accessed on November 20, 2011.

34. http://www.hipaasecurityandprivacy.com/, last accessed on November 10, 2011.

35. Peris-Lopez, P., J. Hernandez-Castro, J. Tapiador, E. Palomar, and J. van der Lubbe. "Cryptographic Puzzles and Distance-Bounding Protocols: Practical Tools for RFID Security." *IEEE International Conference on RFID*, 2010, 45–52.

36. http://www.doublecode.com/rfid/, last accessed on November 10, 2011.

37. http://ec.europa.eu/information_society/activities/health/docs/studies/rfid/rfid-ehealth1 .pdf, last accessed on November 5, 2011.

WIRELESS NETWORKING V

Chapter 19

Cooperative Services in Next-Generation Wireless Networks: Internet of Things Paradigm

Andreas P. Fatouros, Ioannis C. Fousekis,
Dimitris E. Charilas, and Athanasios D. Panagopoulos

Contents

19.1 Introduction

Mobile networks systems are reaching a peak from the offered capabilities of a single technology. With the use of the same selfish networks, we may not be able to provide further improvement to meet the needs of the users. However, the general idea of cooperation is now on the table. The idea of creating a huge network where all different access technologies can be a part of it is now possible thanks to cooperation without interference. Nevertheless, the development of such a network meets lots of problems along with limitations from the existing technologies. Moreover, end users may refuse to embrace these technologies because of seemingly increased costs. The main focus of this chapter is to highlight the cooperation services between different network technologies. More specifically, the chapter consists of three sections.

In the first section, a generic presentation about the Internet of things (IoT) is made along with the positive results that could succeed in people's everyday lives. The general idea of IoT is introduced along with the already existing related technologies. Furthermore, the difficulties and limitations of IoT are discussed. Finally, an example of a model architecture of a smart home environment network with smart objects is introduced as well as several examples of IoT applications.

The second section of the chapter is devoted to cooperative services in next-generation wireless networks. With the cooperation of different access technologies, significantly enhanced services can be provided to end users. In addition, the already existing services can be supported even faster, thus creating an environment where higher QoS with lower cost can be offered. Furthermore, cooperative services scenarios are discussed along with ideas about implementation in the real world and limitations and problems of cooperative networks. Lastly, motives for users and service providers are proposed based on reputation mechanisms.

The chapter emphasizes on how cooperation can be implemented in the IoT paradigm. Because the development of IoT demands a ubiquitous network where everything can connect and communicate at any time, the cooperation of the existing networks is the only cheap way to make it possible. In addition, to support the idea of cooperation within the frame of this study, a simulation project is developed. In the third section of the chapter, we describe the scenarios that have been simulated and discuss results. The latter demonstrate how significantly a cooperative services scenario can improve the quality of service (QoS) offered to end users.

19.2 Internet of Things

19.2.1 The Concept of IoT

Nowadays, the majority of worldwide connections are made between humans, especially between humans connected through computers or mobile handsets. Consequently, the main communication is from human to human. In the concept of the IoT, every machine or thing

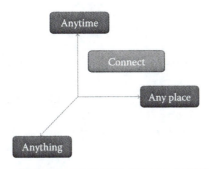

Figure 19.1 Three-dimensional connectivity model.

must be able to connect online to the Internet and communicate with other machines or things without the presence but on behalf of human beings. In addition, there is not a standard definition of the IoT, but the general idea behind IoT is the necessity of giving to every entity the ability to communicate. Perhaps the best idea to describe the IoT is by adding a new dimension to the world of information and communication technologies and transforming the present connectivity at any time, from any place, and for anyone to connectivity for anything as illustrated in Figure 19.1 [1].

The next step is to specify the way these smart things can communicate no matter if the human is present or not. Perhaps this is the essential part of the IoT project in which every "stupid" machine has to become intelligent to be able to communicate. In the IoT, these things or machines are named as "smart" things. With the theoretical approach that basic function can be modeled by expanding the existing form into a machine-to-machine (M2M) communication or a thing-to-thing (T2T) communication [2,3]. Imagine that it is like an evolved form of the peer-to-peer (P2P) logic [4,5]. On the contrary, implementing that extension and creating a stable and trustworthy network is very difficult because of various limitations as will be discussed later.

19.2.2 Related Technologies

As was previously mentioned, all things that are able to connect online and communicate are named as smart things. To make an everyday thing smart, a whole set of algorithms and microprocessors have to be used in order to achieve the desirable result. Furthermore, one detail should be mentioned: the size of these things varies from huge things to tiny things. These tiny things cannot be fully equipped with all these microprocessors and sensors because of their size. In that case, radio frequency identification (RFID) tags can be used, which are small microchips designed for wireless data transmission and are used to help identify objects. On the other hand, in the IoT scenario, things that are equipped with well-known commercial wireless and mobile cards are able to participate.

Generally, RFID tags are attached to an antenna in a package that resembles an ordinary adhesive sticker and transmits data over the air in response to interrogation by an RFID reader. Their cost and size make them suitable for IoT's smart things project. In addition, unique identification and automation are two characteristics that make RFID tags appropriate for object tagging. An RFID tag emits a unique serial number that distinguishes it among many millions of identically manufactured objects. These unique identifiers in the RFID tags can act as pointers to database entries containing rich transaction histories for individual items. Moreover, RFID tags are readable without line-of-sight contact and without precise positioning. RFID readers can scan tags at rates of hundreds per second.

There are three types of RFID tags. In general, the small and inexpensive RFID tags are passive. Especially passive tags do not have any on-board power source. Instead, they derive their transmission power from the signal of an interrogating reader. Some RFID tags contain batteries. There are two such types: semipassive tags, whose batteries power their circuitry when they are interrogated, and active tags, whose batteries power their transmissions. Note that active tags can initiate communication and have read ranges of more than 100 m [6].

19.2.3 Applications of IoT in Everyday Life

The best way to understand the IoT project is by giving a general idea of IoT's application development in people's everyday lives. Because it is difficult to provide real evidence and proof of the definite success of the IoT project, a few examples are provided to support the idea. With the use of RFID tags and RFID readers, a whole set of functions and services can be used to help people. In addition, cooperating entities can support better capabilities and better communication between smart objects. To emphasize, cooperation is the only way to create a safer and more efficient network where RFID-tagged smart objects can communicate with each other, providing accurate results. For example, RFID tags on drugs and food can help people to identify the origin and the quality of them. In addition, with an Internet connection and a database, RFID tags and readers can support a full tracking system [7].

Furthermore, cars, trains, and buses—along with the roads and the rails—equipped with sensors, actuators, and RFID tags may provide important information to the driver and passengers about navigation and safety. Moreover, collision avoidance systems and monitoring of transportation may protect drivers from hazardous materials transportation and other dangers [8]. Of course, the development of such a system is difficult because of different environmental factors, even though research is done to create a common architecture model in which every entity can communicate with every other one [9].

The best example showing the importance of the IoT project is a smart home environment project built by the Korea Institute of Industrial Technology (KITECH) to demonstrate the practicability of a robot-assisted future home environment. This environment consists of smart objects with RFID tags and smart appliances with sensor network functionality. The home server that connects these smart devices maintains information for reliable services, and the service robots perform tasks in collaboration with the environment. Furthermore, an interesting architecture of such a network is proposed, one that shows the way a computer network would seem after the wide use of the RFID tags in smart object appliances as shown in Figure 19.2.

Examining the same smart home environment, we describe a scenario that intends to identify and fetch a requested object that is included in the environment in order to help the owner of the house. The target object is smart, which means it is RFID tagged. Using smart devices, such as a smart table, smart shelf, and smart bookcase, the detection of the presence of the requested item is possible. Once the RFID code of the smart item is detected, it is transferred to the home server through wireless networks. If the user of the smart network wants, for example, the position of a cup, he or she sends a command to the home server. The home server then searches for the status information of the device and sends the position data to the robot. After downloading the data from the home server, the robot moves to the place where the target object is and brings it to the user [10].

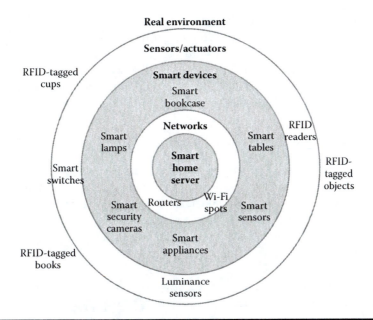

Figure 19.2 Architecture of a smart home environment network with smart objects.

19.2.4 *Limitations and Problems of IoT*

Of course, the IoT project does not have only benefits and positive outcomes in people's everyday lives. Particularly, there are some problems and limitations in the development of the IoT and with the results of the installation as well. First of all, one of the major problems that must be solved is the communication range of the wireless networks. Indoors, this can be solved with the use of a router or a set of routers with wireless connectivity. But outdoors, this is much more difficult because of the time varying fading phenomena. At this point, the only solution that could be applied is the cooperation of the existing networks, which will create a ubiquitous cluster of wireless networks [8,11]. At the same time, another limitation of the IoT's development is an increased demand for power (energy). As a result of the use of wireless connections, two side effects occur mostly as a result of multipath fading: first, the power consumption of smart things increases because of retransmissions, and second, the lifetime of the batteries of smart objects is reduced. Consequently, the solution comes again from the possible cooperation of things in these networks by reducing the retransmitted data and with the smart management of them.

The second part of the IoT's problems involves security, privacy, and identification concerns. The deployment of automatic communication networks of smart objects could possibly represent a danger for society. Embedded RFID tags in every object may result in intrusion into people's personal lives. RFID tags can unknowingly be triggered to reply with their ID and other information without the permission of the users; therefore, security measures must be taken in order to prevent eavesdropping. One of the measures that can be taken is relabeling. According to this mode, users can physically alter tags to limit their data emission and obtain physical confirmation of their changed state. Nevertheless, the use of unique identifiers does not eliminate the threat of clandestine inventorying or the threat of tracking. In addition, even if relabeling can be applied in smart objects with active RFID chips, it cannot be applied in passive RFID tags. This is why the use of a minimalistic cryptography model is better. According to this model, every tag contains a small collection of pseudonyms, which rotates by releasing a different one on each reader query. An authorized reader can store the full pseudonym set for a tag in advance and identify the tag consistently. On the other hand, an unauthorized reader is unable to correlate the different appearances of the same tag. The minimalist scheme can offer some resistance to corporate espionage, such as clandestine scanning of product stocks in retail environments [6]. Finally, cooperative security protocols that have been proposed in wireless sensor networks may find application in IoT configurations.

19.2.5 Benefits of Employment of IoT Technology

The benefits of IoT can be visible in every aspect of everyday life. An especially huge number of applications are possible for development, but only a small part of them is currently available. The domains and environments in which new applications could improve the quality of people's lives are many: from home to work, from gym to traveling, at the hospital, anywhere it is possible to use the capabilities of smart objects. These environments are now equipped with objects with only primitive intelligence, often times without any communication capabilities. Giving these objects the possibility to communicate with each other and to elaborate on the information perceived from their surroundings makes them more effective by making people's lives convenient [8].

Another part of the IoT project concerns the advantages that are created from the cooperation of heterogeneous networks. The whole point of the cooperative networks is to create a ubiquitous network, so everything can be able to communicate at any time. This whole process has the effect of the expansion of the existing network coverage and the increased capabilities of connectivity support. Because of this characteristic, smart objects can easily access the network. As a result, less energy is consumed for the communication of the smart objects, and thus, the lifetime maximization of their batteries is possible. Furthermore, the difficult process to support an efficient security model can be possible with the proportion of the security steps in different entities of the smart network.

19.3 Cooperative Services in Next-Generation Wireless Networks

19.3.1 Cooperative Networks: General Concept

The meaning of the word "cooperative" derives from the very beginning of our society itself and can also be found in terms of biology. For example, people help each other in order to accomplish their tasks faster and in a more productive way. Cooperation's outcome in computer or mobile networks is getting better results in a faster way, wasting much less energy, time, thought, and money. The latest research approaches make great efforts to introduce the concept of cooperation into telecommunications in order to create the next-generation 4G networks. It is envisioned that these evolved networks will offer many advantages to users of mobile communications by augmenting and optimizing the perceived QoS, the number of provided services to end users, and being friendlier to the environment. The demand for higher throughput capabilities, higher reliability, and high-speed mobility, as it is delineated by International Telecommunication Union (ITU) standards for 4G networks, requires great complexity, energy consumption, and expenditure from mobile devices and telecommunication service providers. Some of these standards are data rates up to 1 Gbps, compatibility with former generation networks, and better quality of multimedia services (e.g., video on demand, mobile TV, and high-definition TV). In addition, 4G networks use packet switching and the IP instead of the previous ones.

The envisioning of 4G networks is the idea of cooperative networks. The basic technical characteristics and objectives of cooperative 4G networks are tabulated in Table 19.1. The state-of-the-art for cooperative networks is the coexistence of many heterogeneous networks with different data rates, mobility capabilities, and different technical characteristics [12]. In general, there are two types of wireless mobile networks: wide and local access wireless networks. Wide access wireless networks (WAN), such as general packet radio service (GPRS), global system for mobile communications (GSM), and universal mobile telecommunications system (UMTS) (or 3G), offer lower data rates than the local ones, but they support higher coverage and mobility to users. On the contrary, local access wireless networks (LANs), such as Bluetooth and WLANs, support much higher data rates with much less energy consumption comparatively, but they suffer in mobility and coverage. Thus, the idea of combining these two types of wireless networks in order to achieve the maximum gain regarding data rates, energy consumption, mobility, and coverage has emerged. The tradeoff between coverage and data rates can be minimized in that way. The different networks that are used all over the world will be converged into a wider network [12].

However, many difficulties have to be faced looking toward the implementation of 4G networks. The always-on Internet, the data transfer, the spectrum management, the very high data rates that have been prescribed by ITU, the complexity that will be introduced into the whole system, and the deficiency of IP addresses are some of these problems that scientists and network engineers have to confront and overcome. Some of these characteristics will increase energy consumption, which is one of the main factors that ITU has defined as of great importance for the next-generation networks. This fact adds additional difficulty in the design of network and mobile devices as well.

Table 19.1   Characteristics of 4G Cooperative Networks

Data transfer capability	• 100 Mbps (wide coverage) • 1 Gbps (local area)
Networking	• All-IP network (access and core networks) • Plug and access network architecture • An equal-opportunity network of networks
Connectivity	• Ubiquitous, mobile, continuous
Network capacity	• Tenfold that of 3G
Latency	• Connection delay < 500 ms • Transmission delay < 50 ms
Cost	• Cost per bit: 1/10 to 1/100 lower than that of 3G • Infrastructure cost: 1/10 lower than that of 3G
Connected entities	• Anything to anything
4G networks key objectives	• Heterogeneity and convergence of networks, terminals, and services • Harmonious wireless ecosystem • Perceptible simplicity, hidden complexity • Cooperation as one of its underlying principles

In cooperative networks, the concept of "term node" is introduced. The P2P communication between mobile devices and a base station (BS) is no longer applicable because the node (user or mobile device) can act as a relay. In such a scenario, the destination (D) receives data from both the BS/source (S) via a direct link and the relay (R) via a two-hop transmission as shown in Figure 19.3. This also means that the relay node has the ability not only to receive data from the BS for personal use but also to forward them to other nodes.

Generally speaking, the BS can send data to users through a direct link. However, this link may suffer from noise, which makes the communication difficult or impossible. Fading, long distance between user and BS, high signal-to-noise ratio (SNR), and high packet loss rate are some of the problems that burden the communication. The introduction of the user node may help to overcome these problems as the link between BS and node or between node and destination user. That would make the data transfer more feasible to the destination and could improve the quality of provided services. In such a scenario, the BS can send data simultaneously both to the node and destination user, thus achieving a faster and more reliable transfer of data. This simple model can be applied to more than one cooperative terminal or network. The cooperation, however, between terminals of many heterogeneous networks may demand the development of new network protocols and the introduction of some more bits in the header of data messages that terminals exchange, which may reduce the achieved throughput. Nevertheless, the overall gain through cooperation in data rates, energy consumption, and coverage makes this small decrease in throughput negligible. The benefits of cooperative networks can be summarized in the following aspects:

- *Data rates:* Cooperation between neighboring nodes will significantly increase the data rate because devices will receive data from both the cellular and short-range links. In some situations, the augmentation of the data rate has been double or even more [13,14].
- *Transmission time:* As the overall data rate increases, it is corollary that the needed time to transmit all the data requested by a user would be significantly decreased. Transmission time can be decreased by more than 50% [15].
- *Capacity:* A direct consequence of the last advantage is the increase in the system capacity. In cooperative networks, users tend to download only a part of the overall data information through cellular networks. As a result, they occupy this link much less, they waste far fewer resources from the system, and the overall capacity of the system can be increased.
- *Coverage:* Through the combination and cooperation of networks and mobile devices, the coverage of the system can be increased. Additionally, short-range links can transmit the information further [13,14].

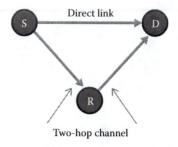

Figure 19.3   Basic cooperative transmission configuration.

- *Energy consumption:* As it is widely known, energy consumption in short-range links is extremely low compared to cellular links. In cooperative networks, users receive a great amount of information through short-range links, decreasing the overall amount of energy spent by users and BSs. However, it must be mentioned that cooperating nodes will spend energy for receiving data from BSs and sending them to other users. This is one of the major problems in the design of next-generation networks and can only be resolved by giving these nodes some motivation in order to cooperate with each other [13,14,16].
- *QoS:* All the above, in combination with the improvement of the SNR and the signal-to-interference ratio (SIR) that short-range links can offer, contribute to the decrease in the bit error ratio and probability (BER and BEP) and, consequently, in a spectacular improvement of QoS. This fact can lead to import new very demanding services for the users.
- *Cost of cooperative services:* Taking as reference either the time terminals spent to receive the data or the amount of data terminals received through cellular links, cooperative networks overmatch significantly compared to the former generations of networks. As a result, the cost is estimated to be much lower [15].

There are two possible ways to transfer data between nodes that are differentiated in the way the cooperating users download the data from the server. In the *static server-aware* approach, it is assumed that users can send or receive data simultaneously through the short-range link or cellular link. In this situation, the server must be aware of the amount of data that has to be sent to every user; consequently, mobile devices must send some information to the server before the procedure begins. Let us assume the existence of two cooperating users and two files in the server side, which are the parts of the whole file. The destination node informs the cooperating node to download the second part of the file while it downloads the first one, and then they exchange data through the short-range link. While installing the short-range link, every cooperating node becomes aware of what it has to download from the server. This scenario can be easily expanded to more than two terminal nodes.

The second scenario relies on a *dynamic server-unaware* approach that takes advantage of the HTTP/1.1 possibilities range request. In that situation, every terminal has the ability to decide which bytes of a file will download from the server through the HTTP GET requests. During the installation of the short-range link, the cooperating terminals decide which bytes each terminal will download, and then an HTTP GET request is sent to the server with the preagreed range request into the GET request. One of the advantages of this scenario is the flexibility that the server offers to users. Users can decide to download data through the best cellular links with the maximum data rate and the minimum noise and fading. Even if the short-range link becomes unavailable for some reason, the terminal will send another HTTP GET request to the server in order to receive the rest of the file [15].

19.3.2 Cooperative Service Scenarios

Observing the 4G networks from a user's point of view, a multitude of new applications and services is supposed to emerge. The new communication system, although cellular, is not completely based on the BS–terminal communication, but is highly dependent on short-range communications. The cooperation between nodes improves the reliability of the offered services, proliferating simultaneously the coverage and data rate and decreasing the energy consumption.

The major drawback of the third-generation networks is the inefficiency to supply users with new investments. This generation of networks has been characterized as an extension to second-generation networks. The third-generation partnership project (3GPP) has deficiency in the ability to add new services to the existent ones and satisfy millions of people who use this network. Recently, 3GPP has developed new services, such as the merging of two high-demanding services, multimedia broadcast and multicast service (MBMS) center in combination with an IP multimedia system (IMS). However, the latter did not appeal to users, and these small improvements did not encourage customers to change their equipment. Thus, these difficulties have brought communication industries and engineers to the foot of 4G networks, an investment that offers to every user multiple new services with high QoS and at a significantly lower price that could be affordable all over the world. In addition, next-generation networks have to be backward compatible and will have instead a user-centric approach as they will be focused on personalized services [16].

As an example of a cooperative service scenario, imagine a middle-sized area where communication through short-range links is possible. This scenario is usual in femtocell networks. Hypothetically speaking, if a number of existing users are trying to access the same service or data from the server, cooperation is possible. First of all, an inquiry through a short-range link is attempted, looking for the desired data in neighboring nodes. If these data exist in one of these nodes, the trade between them can begin, and the destination can get the desired information using a short-range link. Otherwise, if the desired data do not exist in one of the cooperating nodes, then a cooperating download can be applied, that is, users can preagree to download different parts of the desired information and then exchange them through short-range links. In either scenario, terminals make limited use (or none at all) of the cellular link, using a local retransmissions scheme, that way making the reception of data faster and in higher data rates, more accurate, and less energy consuming. In addition, the low use of cellular links greatly facilitates spectrum management as terminals make use of the specific frequencies for less time. Consequently, the spectrum is free most of the time, significantly improving that way the total system capacity. Moreover, short-range communication systems use another part of the spectrum that is not interfering with the cellular one. This service is being managed from the lower OSI layers (physical and link layer), making it less complicated and much faster.

Because of the limitations that cellular links have, terminals cannot process high data rates alone. By distributing the downloading, cooperating nodes receive different parts of the file and then, through local retransmissions, every part arrives to the destination node. Enhanced QoS can be achieved that way. An example using multiple description coding (MDC) is demonstrated, but the same framework can be generalized to more types of services. Thus, the bigger the accumulation of cooperating users becomes, the faster the data receive to destination and the better quality can be provided to users.

The above services can be seen in implemented applications for cooperative networks. Consider, for example, a bit torrent application for wireless networks that has been implemented on a Symbian OS platform [12]. For the cellular link, GPRS is used, and for the short-range link between terminals, Bluetooth has been preferred. Two terminals want to download the same file from server. Next-generation networks give users the possibility to cooperate and download this file in a cooperative way instead of everyone for itself. Through cooperation, every user downloads a part of the overall file and shares it with the others through Bluetooth. Statistics show that with two terminals the transfer time is half that compared with the noncooperative scenario (in which every user downloads the whole), whereas an energy consumption savings by 44% has been noticed. This behavior lies in the low energy per bit and the higher data rates of Bluetooth [12].

Nowadays, another application that is very popular is video streaming. This application has high functional requirements as it requires extremely low delay and packet loss and binds much of the bandwidth. Additionally, it is very demanding concerning energy consumption, limiting in that way the battery life. In order to fulfill these requirements, some video coding schemes, such as MDC and scalable video coding (SVC) have been introduced. Through MDC, the total stream is divided into small substreams. The quality of the video depends on the number of correct substreams users receive, and different substreams have the ability to be decoded separately. SVC, on the other hand, divides the video into a base layer, which is responsible for the quality of the video, and many enhancement layers. These two techniques can be used in cooperative networks when different users download different substreams from the server and exchange them through short-range links. Whatever type of cellular and short-range links have been used, a significantly high energy and transfer time gain has been noticed [17].

Web browsing is another service that will be reinvented by cooperative networks. The demand for the Internet in mobile phones is growing very fast. Low data rates and high prices have kept this service static for years, disinclining users from using it. 4G networks promise high data rates for web browsing at a significantly low price [18]. Web browsing through cooperation can be extremely fast. There are three phases in the web browsing of a user. The first consists of an inquiry from a user to a server for a specific web page, the second is the download of the necessary data information for this page, and the last phase is the processing of that information. Examining the simple situation where only two terminals exist in a small area (where communication through short-range link, usually Bluetooth, is feasible), the user (master) who wants to download a page makes an inquiry to the second user (slave), so the last one downloads a part of the overall information from the server. The slave sends the retrieved data to the master through a high-speed short-range link. The capacity of the system improves, and the transfer time decreases significantly that way. In this scenario, it is necessary that the slave terminal does not have any other web browsing activity [18]. The energy, data rate, transfer time, capacity of the system, and spectrum usage gain can be easily visible through statistics and metrics.

All of the above applications and services can be generalized to support more than two terminals, improving the overall gain in all the above sections. As a conclusion, we can say that cooperative networks can offer great benefits to users and communication industries by introducing new services or by using already existing ones in a more effective way.

In the next scenario, the simple situation of a file download is being examined. Every terminal can support two air interfaces: the cellular and the short-range links. The first one is being used for the communication with the BS, and the second one is used for the communication between terminals. For the first air interface, UMTS communication protocol is being used while, for the second one, terminals use Bluetooth protocol. It is supposed that these interfaces can be used by the terminals simultaneously, meaning that every terminal can receive data from the BS while sending or receiving data through Bluetooth. As it can be seen in the example given in Figure 19.4, MT1 downloads a medium-size file from the BS, while in parallel, it simultaneously can use terminals MT2 and MT3 in order to download the file faster.

In a situation of two neighboring terminals, if the first one wants to download the file, then it makes an inquiry (as the master terminal) to the other terminal (slave) in order to cooperate during the downloading phase. With the assumption that the two terminals are located in the Bluetooth coverage area, three different scenarios are examined in the frame of this study. In the first one, it is assumed that the slave has free resources; in the second one, both terminals download the same file simultaneously, and in the third one, the slave does not have free resources.

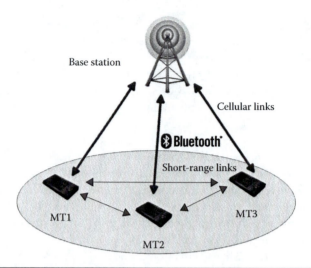

Figure 19.4 Example of three cooperating mobile terminals (MTs).

In the first situation, the slave decides if it will cooperate and give its resources to the master. In the second one, terminals decide either to cooperate (and thus form a cluster, where no distinction between master and slave exists) or not to cooperate (and thus download the whole file by themselves). In both the above situations, if cooperation succeeds, then users can enjoy the benefits mentioned before. In fact, every terminal downloads part of the file and simultaneously sends it to the other members of the cluster [14]. In the last situation, the master terminal can neither use the resources of the slave or form a cluster; thus, it downloads the whole file through a cellular link. In the case that the slave transitions to a pending situation, it can give its resources to the master terminal so as to cooperate [13]. At this point, it must be noted that, as it has been explained in the above scenarios, the basic part of cooperation is simply the existence of many terminals in an area forming clusters that can cooperate in a dynamic way by choosing a master and slaves. This procedure can take place particularly in overcrowded areas where short-range links are feasible.

19.3.3 *Limitations and Problems of Cooperative Services*

In order to render possible the implementation of cooperative networks, specific problems must be solved, and at the same time, motives should be given to users in order for them to cooperate. First of all, in the above-mentioned technologies, the cooperation of intermediary nodes in certain applications (network coding, promotion of packages and data, etc.) was taken for granted; however, this cannot be ensured in real-life networks and applications. More concretely, in many cases, the intermediary node (mobile terminal relay) does not acquire direct profits from cooperation, but only wastes the energy of its battery or even delays the dispatch of its own data if it is already an intermediary node in a call or transport of data. Furthermore, it may even be impossible for the relay to receive packets because all the channels in which it "listens" are occupied.

Another problem that concerns the standardization of cooperative networks is the complexity of algorithms that should be developed. Certain algorithms have already been proposed for various fields, such as the finding of the nearest node and the recognition of the topology of the networks for network coding [19]. However, it is also essential to create algorithms and protocols in the mobile terminals, so they can collaborate efficiently. All of the above increase the complexity needed to design the 4G networks and the corresponding mobile terminals.

Another problem to be solved is the extension of the existing communication protocols and the access techniques so as to cover all different cooperative scenarios. As an example, at Bluetooth, only one master and seven slaves can exist at the same time in each communication. Moreover, slaves cannot communicate with each other; thus, they cannot meet the specifications for cooperative networks [12].

Practically, these three problems reveal the existing complexity and difficulty in order to make all the different networks communicate with each other simultaneously. On the contrary, however, despite these complex problems, the results from the collaboration of networks can lead to enormous growth of the branch of networks and communications with all the benefits discussed so far.

19.3.4 *Motivations for Cooperation*

According to the above, it is essential to provide certain motivations to the intermediary nodes so that they grant their resources to other terminals on demand. A lot of proposals have been made about what these motivations should be. The basic proposal, which is also examined below, is the creation of a mechanism to memorize the nodes that cooperate and the ones that do not.

This mechanism is based on the "reputation" of each terminal, namely, the degree of help that it offers (reputation-based mechanism) [20]. According to this mechanism, the possibility of interaction with each terminal depends on the reputation that it has. The reputation of each terminal can be managed in two ways: centrally or distributed. In a central reputation system, the institution that checks the reputation of the terminals collects elements for them and presents them at the network. Therefore, each user is given the possibility of accessing information relative to the reputation of the other users. In a distributed system, each user stores information on the neighboring terminals. He or she has therefore his or her own database, and he or she compares it with the other users. The basic idea of this particular mechanism is that as it concerns users that collaborate and have a good reputation, it must give them the possibility of acquiring certain profits [21]. As an example, if one user helped another in the transmission of data acting as a relay, it should be possible in the future to use the other user's resources for its own profit [13,22].

A second proposal is the use of concrete and more complicated algorithms, aiming at the localization of users, both of those who collaborate and those who do not, and taking corresponding decisions [13,21–23]. Moreover, there are also other mechanisms for the control and the benefit of motivations in the synergic networks, which are less popular, such as the mechanism of wage (remuneration mechanism). It should be stressed that these mechanisms import overhead into the system as they require mechanisms of coding for the communication between nodes. Accordingly, the traffic is increased and the capacity of the network is decreased, leading to a degradation of throughput [21,24].

Furthermore, certain motives independent from the mechanism can be used for cooperation, which can be given from each provider of mobile telephony. More specifically, the latter can provide lower debits or even the benefit of better services as higher data rates for the "good" users, that is, for the users that cooperate and provide their resources to support other users or, otherwise, for users with good reputations.

19.4 Simulations and Discussion of Cooperative Network Scenarios

19.4.1 Simulation Scenarios

In this paragraph, we describe how the aforementioned cooperative framework has been simulated. In order to quantify the effectiveness of 4G cooperative networks, the following three scenarios have been modeled:

- *Scenario 1 (baseline scenario):* First, a UMTS network that serves some users that make use of the file transfer protocol (FTP) service in order to download a file of variable size has been simulated. In that situation, cooperation does not take place, and every user operates autonomously in order to download the specific file from the nearest BS.
- *Scenario 2 (remote node assistance scenario):* In the second scenario, cooperation is supported by introducing Bluetooth links between a user (master) and all the others that are within the radius of Bluetooth coverage, that is, in a distance less than or equal to 10 m. This simulation corresponds to the ability to provide services at a terminal that is not able to communicate with the BS. Thus, nearby nodes or terminals offer their help by providing their resources to the master terminal in order to download parts of the desired files from the BS. These nodes receive data from the BS through cellular links and forward them to the master terminal via Bluetooth. The rest of the parts of the file are downloaded via a cellular link (UMTS).
- *Scenario 3 (cooperative broadcast scenario):* In the last simulated scenario, the effectiveness of cooperative networks is considered. A single user's mobile device that happens to achieve better data rates from its neighboring nodes through the UMTS protocol becomes automatically the master of the cluster and installs Bluetooth links with terminals within the radius of Bluetooth coverage. The master terminal uses its resources in order to help other users download the desired file faster and with significantly lower power consumption.

The multiple variables of the simulation give the ability for an effective examination of the above scenarios in order to obtain the desired results. Some of these parameters are presented below.

- *Number of users:* The maximum number of users that can participate is limited by the Bluetooth protocol that is used in short-range links to a total of eight mobile devices, namely, one master and seven slaves. The number of participating users plays an important role in cooperative networks and especially in scenarios 2 and 3, as this number significantly affects the results.
- *Dimension of topology:* The BS for the cellular network and the mobile terminals have been placed in a three-dimensional coordinate system covering an area of 3375 m^3 at random positions. Bluetooth coverage is 10 m, whereas the BS is able to communicate with every terminal independently of their distance.
- *File size:* The file size in our study varies from 10 kB to 1 MB.

19.4.2 Network Model

The main file of the simulation is divided into IP packets and protocol data units (PDUs). The number of these packets depends on the file size that has been defined for the FTP service and the size of the IP and PDU packets. In the frame of this work, the IP packet size is equal to 1500 bytes (40 bytes header and 1460 bytes data). The PDU unit size is defined as being equal to 40 bytes, as it has been determined in [25], and a 2-byte header that can be considered negligible.

In a situation of cooperative networks, Bluetooth packets are also transmitted. In that case, the PDUs that master or slave terminals receive through cellular links are encapsulated into Bluetooth packets. Bluetooth packet size varies in every network. In our simulation, considering a low BER (10^{-4}), DH5 packets have been used [26]. Their size is estimated as 339 bytes per packet (15–16 bytes header) and the Bluetooth data rate at 723 kbps [26]. The creation of data packets is different every time, depending on the simulated scenario. In the case of the baseline scenario, the whole file is split into IP packets and then into PDUs. In the case of the remote node assistance scenario, after the inquiry for cooperating users, every cooperating user assumes to download 10% of the whole file. Every time the user downloads a certain number of PDUs, the latter are encapsulated into Bluetooth packets and sent to the master terminal. In the last scenario, every time the master terminal receives some PDUs from the BS, it encapsulates them into Bluetooth packets and sends them to cooperating users. More specifically, every Bluetooth packet consists of eight or fewer PDUs.

Data rates and BEP are considered constant at every link. Data rates have a peak of 384 kbps for the cellular link (UMTS) and 723 kbps for Bluetooth. The above data rates are the maximum feasible rates for every link; however, they are rarely achievable for many reasons. BEP highly affects the data rates as errors in transmission lead to retransmitted packets, that is, lower throughput. Consequently, the higher the BEP, the lower the QoS experienced by end users. The time and manner of retransmission is determined by the TCP. The BEP is set to 10^{-3} for the cellular links and 10^{-4} for the short-range links.

Furthermore, the TCP, which is the most popular protocol for the transport layer, is responsible for the installation of a point-to-point reliable link over the nonreliable IP. This protocol has mechanisms for the control of data flow in order to avoid congestion and has also mechanisms for the packets' retransmission in order to achieve reliability [27]. In the current study, we do not examine the initial condition for the link installation (such as triple handset); on the contrary, we focus on retransmission mechanisms. The BS, after sending PDUs to users, is waiting for the corresponding ACKs (packets for acknowledgment) from the user. In a situation of negative ACK (NACK) or of a lack of the corresponding ACK, the BS has to retransmit the file. In every transmission, a copy of the sent packet goes into a unique buffer (transmission buffer). In case of ACK, the copy of the packet is deleted from this buffer, whereas in case of NACK, it is moved to the retransmission buffer in order to be dispatched again [27,28].

Another thing that should be taken into account is the flow control in which the sliding window mechanism has been used. With this method, the simultaneous dispatch of more than one packet can be achieved, highly increasing the spectrum usage. The packet transmission starts with a small-size window, one to two PDUs, but when BS receives an ACK, this window becomes larger by one until a maximum threshold, which is defined by the bandwidth delay product (BDP), is being reached. In this situation, the window remains stable unless BS receives a NACK, turning the sliding window into half of the current one. For UMTS, BDP varies from 7200 to 64,000 bytes [28,29]. Considered the minimum BDP, BS can send five full-size IP packets, that is, 180 PDUs. This fact could limit the link capacity, however, because of retransmitted packets; this limit is rarely achieved.

Finally, for UMTS, over the physical layer, there is an additional layer, which is divided into the radio link control (RLC) and medium access control (MAC) layers. The RLC layer is responsible for segmentation and reassembly of packets to the lower and upper layers, whereas MAC is responsible for the medium access [28,30]. The timeout during the forwarding of packets from MAC to the physical layer, known as TTI, is usually set at 10 ms.

19.4.3 Energy Consumption Estimation

For the study of energy consumption during the simulation, the energy that is consumed by the mobile terminals for their operation must be calculated. Below, we present the model that we use for the assessment of energy both in cellular and short-range Bluetooth links.

19.4.3.1 Cellular Link (UMTS)

The cellular link is the main link for the communication with the BS. Because of the distance between the terminal and the BS, a large amount of energy is consumed from the MT. The model of a cellular system that could calculate with precision the energy that is consumed at this type of connection is difficult to develop. Practically, the biggest percentage of energy for the file download from the BS concerns the time that the mobile remains in the situation of high state, in which it remains so that it can receive all the PDUs in order to reconstruct the entire file [31]. A constant rate of the consuming energy per kilobit can be taken as a result of simulations, presented in the bibliography, namely, 35,121 J/MB or 0.0044 J/kb [32]. In the framework of this study, we estimate, thus, the consumed energy per PDU as 0.0014 J/PDU.

19.4.3.2 Short-Range Link (Bluetooth)

The short-range link (Bluetooth) concerns the communication between MTs. For the purposes of the simulation, it is assumed that the longest distance that the Bluetooth protocol can uninterruptedly run between MTs is 10 m. Moreover, in order to create a better measurement model of the energy consumption of MTs, we estimated separately the energy consumed by the master and slave MTs. Based on the literature, the energy that is consumed by the mobile terminals per kilobit and per category (master or slave) can be calculated as follows [33]:

■ *From 0 to 5 m.* A stability is observed concerning the distance, so constant values of consumed energy are considered:

$$\text{Master: energy} = 0.001 \text{ J/kb} = 0.0029 \text{ J/Bluetooth Packet}$$

$$\text{Slave: energy} = 0.0008 \text{ J/kb} = 0.0023 \text{ J/Bluetooth Packet}$$

■ *From 5 to 10 m.* A linear dependence of energy from the distance is observed. Moreover, the bents of interrelations for master and slave MTs can be considered similar [33]. Energy per Bluetooth packets is estimated thus as follows:

$$\text{Master: energy} = 5.74 \times 10^{-5} \times d[\text{m}] + 2.6 \times 10^{-3}, \text{ for } 5 \leq d \leq 10$$

$$\text{Slave: energy} = 5.74 \times 10^{-5} \times d[\text{m}] + 2.10^{-3}, \text{ for } 5 \leq d \leq 10$$

19.4.4 Simulation Results

In this section, the results from the extensive simulations of the different scenarios are presented in order to show the possible profits from the future cooperative networks. These results are based on measurements that took place during a large series of simulations. The analysis of the results can be divided into two parts: the first part concerns the comparison of the two first scenarios, whereas in the second part, the comparison is between the first and the third scenarios. Data rate, transfer time, and the energy consumption are metrics (figures of merit) that we consider in order to perform the comparisons in this study.

19.4.4.1 Comparison between Remote Node Assistance and Baseline Scenarios

It is expected that as more users contribute in downloading, the greater the gain we have in data rate and transfer time for the master terminal. For this scenario, the file size is 500 kB. As it can be seen in Figure 19.5, the master terminal improves its data rate proportionally to the number of cooperating users. The number of cooperating terminals can be less than eight as Bluetooth protocol dictates. In the case of seven contributing nodes, the data rate gain can be more than 100%. The rest of the users do not enjoy any benefit from this cooperation. For these users, as explained before, some motivations have to be given from the service provider in order to cooperate. In addition, transfer delivery time has a remarkable decrease. With cooperation, the master terminal receives parts of the file from other nodes decreasing this way with the transfer time.

Concerning the energy consumption as it is explained in Section 19.4.3, it would be expected to be less than in a single UMTS network, and especially for the master terminal because it receives part from the whole file via Bluetooth links. However, the master terminal wastes 9%–11% more energy compared with a UMTS network. This paradox lies in the fact that the master terminal wastes more energy than slaves to download the file from short-range links. Nevertheless, the overall energy consumption is similar to a UMTS network. Results are depicted in Figure 19.6. As a conclusion, we could claim that, in this specific scenario, we could achieve better data rates and transfer time, wasting the same energy in order to help the master terminal to download the desired file.

Except for the number of cooperating users, the aforementioned metrics depend also on the file size that the master terminal expects to download. In this study, the file size varies from 100 kB to 1 MB, and the number of cooperating users takes discrete values. It is important to mention that the bigger the file is, the greater the data rates are expected, as some time is needed

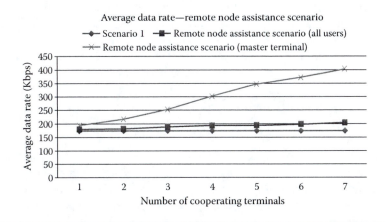

Figure 19.5 Average data rate versus number of cooperating terminals (scenario 2).

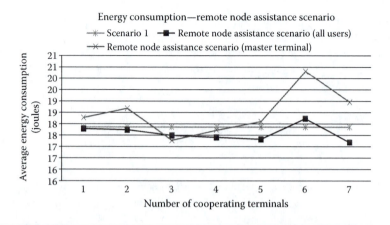

Figure 19.6   Energy consumption versus number of cooperating terminals (scenario 2).

to reach a peak. More specifically, data rates have a good improvement analogically to the file size. The data rate gain reaches 35%–40%. Generally, the data rate has a stable greater value compared to a UMTS network, and this fact affects greatly the transfer delivery time, which is reduced. Last, but not least, as it is expected, higher file sizes lead to higher energy consumption from terminals.

19.4.4.2 Comparison between Cooperative Broadcast and Baseline Scenarios

In the third simulated scenario, one user who has a better cellular link from his or her neighboring nodes receives inquiries in order to help the transmission and installs Bluetooth links with them. The master terminal downloads parts of the whole file and sends them to other nodes that want this file. Imagine a very popular file, such as a video, that many users want to download in a small area. The transfer of Bluetooth packets does not stop until the whole file is received from users, and this transfer is dynamic. Assuming, however, that all users start simultaneously to use the FTP service, the percentage of the file they receive via Bluetooth is 35%–40%.

In that case, as in the previous scenario, a file of 500 kB is used for the FTP service, whereas the number of cooperating users is variable from one to seven. As depicted in Figure 19.7, the cooperating users have gains bigger than 120% compared to the UMTS network. This gain is independent from the number of cooperating users as the master terminal distributes bandwidth evenly at the slaves. In that way, all cooperating users have the same gain.

The huge gain in data rate leads to an equally great gain in transfer time. This gain is constant and does not depend on the number of cooperating terminals because the gain in data rate is stable, and the transfer time is reduced by approximately 50%–60% compared to the UMTS network (baseline scenario).

Because slaves have greater data rates as a result of cooperation and download their files much faster from noncooperative users, there is a great reduction in energy consumption. As before, the energy gain is stable and significantly greater compared to a UMTS network. As can be seen in Figure 19.8, the energy benefit is approximately 50%, whereas if we examine the average energy consumption of all users of the system (whether they participate in cooperation or not), the energy gain varies from 0% to 50%. Thus, there is a great energy saving in mobile devices, leading to

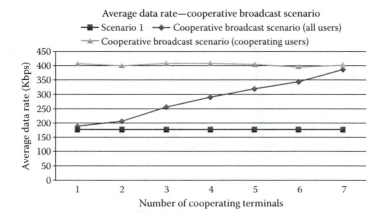

Figure 19.7   Average data rate versus number of cooperating terminals (scenario 3).

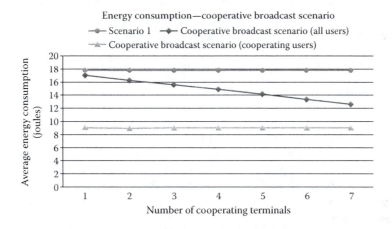

Figure 19.8   Energy consumption versus number of cooperating terminals (scenario 3).

improvement in battery life and, simultaneously, an equal reduction in the rates of electromagnetic energy that may be harmful.

In addition, data rate, transfer time, and energy consumption have also been examined in relation to the file size. As long as data rate transfer time and energy gain are stable for this scenario, the average gain for cooperative users and all users is examined. As was discussed before, the number of cooperating users does not affect the metrics in this scenario. Lastly, concerning the energy consumption, a growing gain has been noticed that depends on file size and varies from 0% to 50%. For larger files, the energy gain could be even larger. This derives from the fact that the larger the file is, the faster the simulation time increases, and the master terminal sends a greater percentage of the total file via Bluetooth links.

19.4.5 Discussion

Based on the simulation's results and the figures that have been presented in the previous section, the superiority of an integrated system with cooperation compared to a single mobile network without cooperation, such as the UMTS network, is evident. Particularly in the remote node assistance scenario, data rates are doubled, and the download time is reduced by half. This is the scenario in which only one user has benefits. Moreover, the results on data rates are impressive, but they are dependent on the number of cooperative terminals.

In the cooperative broadcast scenario, the energy gain is increased along with the size of the downloaded file. In addition, there is no dependence on the number of cooperative terminals. The only dependence that exists is that of the master terminal. The faster the master terminal downloads the file, the faster it can seed the parts of the file to the slaves. In other words, the speed of the Bluetooth protocol is constrained by the lower speed of the UMTS protocol. That problem is caused because the master terminal should download the parts of the file before it can seed them. The most encouraging part of this simulation scenario is that as long as the slaves download the most parts of the file from the master, better data rates are observed, approaching the theoretical value of 723 kbps because they reach the average speed of 600–650 kbps. Nevertheless, the gain from this scenario is very important.

As a final remark, simulation results give a sense of optimism that 4G networks could support data rates over 1 Mbps, and their energy consumption needs can be even less if user nodes cooperate. Of course, a lot of obstacles have to be overcome before these simulated scenarios can be deployed in real-life networks, and similar results can be achieved.

19.5 Future Research Directions

One of the issues that should be taken into account is that the protocol that is used for the secondary air interface can make a difference. Considering the case of using a Wi-Fi Internet connection, we can get a result of a tremendous change of time and data rate in a positive way. Based on the results of the simulation, it can be ensured that researching the field of cooperation services and networks will not be a waste of time. More research can be done in many cooperative communication fields, but both mobile users and mobile telecommunication providers have to understand the gain of cooperative services.

In addition, the evolution of a mobile terminal's software that can support cooperative services is one of the most interesting aspects. The adjustment of the protocols that can ensure a communication without fading and interferences between the objects can create a stable environment where different protocols work undisturbed, providing even better data rates and even less energy consumption. Because the right trade between mobile users is vital to make users cooperate, it is essential to create the right software, which tracks down the use of cooperative services and, of course, merit the cooperative users. Novel software development theories and supporting technologies need to be introduced, given the IoT software system context awareness and dynamic deployment capability in order to realize the environmental adaptability of IoT services. IoT software should be able to adapt to the dynamic environment flexibly to provide intelligent services.

19.6 Conclusions

The main goal of this chapter was to discuss the benefits from the development of cooperative networks. First, an analysis of the IoT was made and the possible benefits from the cooperation strategies on the IoT concept were discussed. In the second section of this chapter, we described the cooperative network characteristics, and we presented the benefits from their use. Finally, we presented some simulation results of a cooperative service, demonstrating the advantages of cooperative networks in energy consumption minimization and the increase in data transfer speed. Results show that many benefits can be derived from node cooperation. Nevertheless, many technical challenges of cooperative networks are still open for future research, such as interoperability issues or motivations for cooperation. Even though theoretical cooperative scenarios can be well formulated and simulated, in real life, incentives have to be provided to rational nodes in order for them to share their valuable resources.

The clearly emerging vision of future networks is that the coexisting heterogeneous wireless networks and mobile terminals will evolve so as to cooperate with each other in order to facilitate network traffic and guarantee QoS requirements to even the most demanding users. Large-scale applications of IoT for surveillance (e.g., for forests) currently under development integrate wireless-sensor networks with vehicle-carrying networks, intelligent power-usage networks, 3G mobile networks, and the Internet in order to establish multiapplication verification platforms for IoT. Access to large-scale heterogeneous network elements and massive data exchange among them are the important novel features associated with the wide application of IoT. At the same time, network elements in each local region must be able to self-organize dynamically and realize interconnection in an interoperable manner. Consequently, one of the most important challenges related to IoT is the support of data exchange in large-scale heterogeneous network elements with simultaneous local dynamic autonomy.

References

1. Tan, L., and Neng Wang. "Future Internet: The Internet of Things." 3rd International Conference on Advanced Computer Theory and Engineering (ICACTE), 2010, 376–380.
2. Puyolle, G. "An Autonomic-Oriented Architecture for the Internet of Things." Modern Computing, 2006. JVA '06. IEEE John Vincent Atanasoff 2006 International Symposium, Sofia, 3–6 October 2006, 163–168.
3. Pereira, J. "From Autonomous to Cooperative Distributed Monitoring and Control: Towards the Internet of Smart Things." at ERCIM Workshop on eMobility, Tampere, 30 May 2008.
4. Takaragi, K., M. Usami, R. Imura, R. Itsuki, and T. Satoh. "An Ultra Small Individual Recognition Security Chip." *IEEE Micro* 21, no. 6 (2001): 43–49.
5. Kortuem, G., D. Fitton, F. Kawsar, and V. Sundramoorthy. "Smart Objects as Building Blocks for the Internet of Things." *IEEE Internet Computing*, (January/February 2010): 44–51.
6. Juels, A. "RFID Security and Privacy: A Research Survey." *IEEE Journal on Selected Areas in Communications* 24, no. 2, (2006).
7. Koshizuka, N., and K. Sakamura. "Ubiquitous ID: Standards for Ubiquitous Computing and the Internet of Things." *Pervasive Computing* 9, no. 4, (October–December 2010): 98–101.
8. Atzori, L., A. Iera, and G. Morabito. "The Internet of Things: A survey." Computer Networks, 31 May 2010.
9. Terziyan, V., O. Kaykova, and D. Zhovtobryukh. "UbiRoad: Semantic Middleware for Cooperative Traffic Systems and Services." *International Journal on Advances in Intelligent Systems* 3, nos. 3 and 4 (2010): 286–302.
10. Baeg, S. H., J. H. Park, J. Koh, K. W. Park, and M. H. Baeg. "Building a Smart Home Environment for Service Robots Based on RFID and Sensor." Control, Automation and Systems, 2007, ICCAS '07, International Conference, 17–20 Oct. 2007, 1078–1082.
11. Prasad, N. R. "Secure Cooperative Communication and IoT: Towards Greener Reality." CTIF Workshop 2010, May 31–June 1, 2010.
12. Zhang, Q., F. H. P. Fitzek, and M. Katz. "Evolution of Heterogeneous Wireless Networks: Towards Cooperative Networks." Third International Conference of the Center for Information and Communication Technologies (CICT)—Mobile and Wireless Content, Services and Networks—Short-Term and Long-Term Development Trends, Copenhagen, Denmark, November 2006.

13. Fitzek, F. H. P., and M. Katz. *Cooperation in Wireless Networks: Principles and Applications.* Springer, 2006.

14. Kristensen, J. M., and F. H. P. Fitzek. "The Application of Software Defined Radio in a Cooperative Wireless Network." Software Defined Radio Technical Conference. SDR Forum. Orlando, Florida, USA, 2006.

15. Militano, L., F. H. P. Fitzek, A. Iera, and A. Molinaro. "On the Beneficial Effects of Cooperative Wireless Peer to Peer Networking." Tyrrhenian International Workshop on Digital Communications 2007 (TIWDC 2007). Ischia Island, Naples, Italy, 2007.

16. Frattasi, S., B. Can, F. Fitzek, and R. Prasad. "Cooperative Services for 4G." 14th IST Mobile and Wireless Communications Summit, Dresden, Germany, 2005.

17. Albiero, F., M. Katz, and F. H. P. Fitzek. "Energy-Efficient Cooperative Techniques for Multimedia Services over Future Wireless Networks." IEEE International Conference on Communications (ICC 2008), 2008.

18. Perrucci, G. P., F. H. P. Fitzek, A. Boudali, M. Canovas Mateos, P. Nejsum, and S. Studstrup. "Cooperative Web Browsing for Mobile Phones." International Symposium on Wireless Personal Multimedia Communications (WPMC'07), India, 2007.

19. Zhang, J., and Q. Zhang. "Cooperative Network Coding-Aware Routing for Multi-Rate Wireless Networks." IEEE INFOCOM 2009, Rio de Janeiro, Brazil, 181–189.

20. Charilas, D. E., S. G. Vassaki, A. D. Panagopoulos, and P. Constantinou. "Cooperation Incentives in 4G Networks." In *Game Theory for Wireless Communications and Networking*, CRC Press, Boca Raton, FL, 2011, 295–314.

21. Oualha, N., and Y. Roudier. "Cooperation Incentive Schemes." Rapport de recherche RR-06-176, France, 2006.

22. Hales, D. "From Selfish Nodes to Cooperative Networks—Emergent Link-Based Incentives in Peer-to-Peer Networks." The Fourth IEEE International Conference on Peer-to-Peer Computing, Zurich, Switzerland, 25–27 August 2004.

23. Sun, Q., and H. Garcia-Molina. "SLIC: A Selfish Link-Based Incentive Mechanism for Unstructured Peer-to-Peer Networks." Hector, Stanford, 2003.

24. Frattasi, S., F. H. P. Fitzek, and R. Prasad. "A Look into the 4G Crystal Ball." IFIP International Federation for Information Processing, 2006, 281–290.

25. Lo, A., G. Heijenk, and C. Bruma. "Performance of TCP over UMTS Common and Dedicated Channels." IST Mobile and Wireless Communications Summit 2003, Aveiro, Portugal, 15–18 June 2003, 138–142.

26. Kim, J., Y. Lim, Y. Kim, and J. S. Ma. "An Adaptive Segmentation Scheme for the Bluetooth-Based Wireless Channel." Tenth International Conference on Computer Communications and Networks, Scottsdale, AZ, USA, 15–17 Oct. 2001, 440–445.

27. Transmission Control Protocol—Wikipedia, the free encyclopedia, http://en.wikipedia.org/wiki/Transmission_Control_Protocol.

28. Teyeb, O. M. "Quality of Packet Services in UMTS and Heterogeneous Networks, Objective and Subjective Evaluation." 2006.

29. TCP over Second (2.5G) and Third (3G) Generation Wireless Networks [RFC-Ref], http://rfc-ref.org/RFC-TEXTS/3481/chapter4.html#d4e443021.

30. Bestak, R., P. Godlewski, and P. Martins. "RLC Buffer Occupancy When Using a TCP Connection Over UMTS." 2002 13th IEEE International Symposium on Personal, Indoor and Mobile Communications Vol. 3, 15–18 Sept. 2002, 1161–1165.

31. Balasubramanian, N., A. Balasubramanian, and A. Venkataramani. "Energy Consumption in Mobile Phones: A Measurement Study and Implications for Network Applications."

32. Perrucciy, G. P., F. H. P. Fitzeky, G. Sassoy, W. Kellererx, and J. Widmer. "On the Impact of 2G and 3G Network Usage for Mobile Phones' Battery Life." European Wireless, 2009.

33. Militano, L., A. Iera, A. Molinaro, and F. H. P. Fitzek. "Wireless Peer-To-Peer Cooperation: When Is It Worth Adopting This Paradigm?" The 11th International Symposium on Wireless Personal Multimedia Communications (WPMC'08).

Chapter 20

Schedule-Based Multichannel MAC Protocol for Wireless Sensor Networks

Md. Abdul Hamid and M. Abdullah-Al-Wadud

Contents

20.1 Introduction

Designing a multichannel media access control (MAC) protocol attracts the interest of many researchers as a cost-effective solution to meet the higher bandwidth demand for the limited bandwidth in a wireless sensor network (WSN). Because of rapid technological advances, a certain geographical location can be visualized as a fully connected information space using fine granularity processing, which can be implemented using sensor technology. Sensor nodes may be regarded as atomic computing particles, which can be deployed to geographical locations for capturing and processing the data of their surroundings. The expected achievement of such sensor networks is to produce, over an extended period of time, global information from local data sensed by individual sensors. Harmonizing sensor nodes into a sophisticated computation and communication infrastructure, called a WSN, may have a strong impact on a wide variety of sensitive applications [1–4], such as military, scientific, industrial, health, and home networks. However, because of the half-duplex property of the sensor radio and the broadcast nature of the wireless medium, limited bandwidth remains a pressing issue for WSNs. The bandwidth problem is more serious for multihop WSNs because of interference between successive hops on the same path as well as that between neighboring paths. As a result, conventional single channel MACs cannot adequately support the bandwidth requirements.

In the state-of-the-art research, significant attention has been put forward to design throughput maximizing MAC protocols [5–11] that work well when one physical channel is used. However, because of the limited radio bandwidth in WSNs (e.g., 19.2 kbps in MICA2 [12], 250 kbps in MICAz [13] and Telos [14]), single channel MAC protocols further limit the higher demand for the bandwidth. Radio transceivers for WSNs are typically cheap devices offering low bandwidth communication only. When physical events in the real world trigger spontaneous communication in many nodes, the single communication channel is under heavy load and many messages are lost as a result of collisions. Carrier sense multiple access/collision avoidance (CSMA/CA) schemes are well suited to spontaneous communication but do not provide high channel utilization under heavy load. Therefore, another cost-effective solution has drawn attention with the possibility of using multiple channels. The solution works for parallel data transmission based on current WSN hardware, such as MICAz and Telos, that provide multiple channels with a single radio.

A number of multichannel MAC protocols have been developed for general wireless networks [15–17] with a single radio. Considering typical applications and the capability of WSNs, these protocols are not suitable. Because of the small MAC layer packet size in WSN compared to general wireless networks, protocols such as [15–17] designed with request to send/clear to send (RTS/CTS) or a three-way handshake for channel-time negotiation provide significant control overhead for the constrained sensor nodes. Therefore, a multichannel MAC protocol for WSN should consider the minimum control overhead possible in negotiating the time-channel selection. Researchers have proposed a few multichannel MAC protocols [18–20,21] that exploit multiple channels to increase the network throughput in WSNs. However, these protocols suffer from high control overhead.

In this chapter, we discuss various multichannel MAC protocols for wireless ad hoc and sensor networks along with their pros and cons. Then, we describe a new approach, a group-wise schedule-based multichannel MAC protocol (GS-MAC) for static WSNs [22]. The approach targets the improvement of the network throughput using conflict-free, multichannel scheduling by grouping the neighboring nodes. The approach is fully decentralized and efficient within the sensors' localized scope. The scheme has been simulated to evaluate the effectiveness in terms of aggregate throughput, packet delivery ratio, end-to-end delay, and energy consumption.

The rest of the chapter is organized as follows: In Section 20.2, we introduce multichannel MAC in WSNs and review the related works of existing MAC protocols in wireless networks with particular emphasis on ad hoc and sensor networks. In Section 20.3, we discuss the schedule-based, multichannel MAC for WSNs along with the algorithms that schedule a conflict-free time slot/channel assignment for the sensor nodes. In Section 20.4, we evaluate and discuss the performance of the approach through extensive simulations. In Section 20.5, we outline future research directions about how this work might be augmented to develop a more efficient multichannel

MAC protocol applicable to resource-constrained sensor networks. Finally, we conclude the chapter in Section 20.6 with a brief summary and the avenues of potential research directions.

20.2 Background

In the context of WSNs, there exist recent proposals that use the concept of multichannel media access techniques to improve the network performance. Zhou et al. [18] have recently introduced the multifrequency media access control for wireless sensor networks (MMSN) multifrequency MAC protocol especially designed for WSN. It is a slotted CSMA protocol, and at the beginning of each time slot, nodes need to contend for the medium before they can transmit. MMSN assigns channels to the receivers. When a node intends to transmit a packet, it has to listen for the incoming packets both on its own frequency and the destination's frequency. A snooping mechanism is used to detect the packets on different frequencies, which makes the nodes switch between channels frequently. MMSN uses a special broadcast channel for the broadcast traffic, and the beginning of each time slot is reserved for broadcasts. MMSN requires a dedicated broadcast channel.

Multichannel lightweight medium access control (MC-LMAC) protocol [21] used a scheduled access, where each node is granted a time slot beforehand and uses this time slot without contention. At the start of each time slot, all the nodes are required to listen on a common channel in order to exchange control information. However, the protocol overhead is significantly high. Tree-based multi-channel protocol TMCP [23] is a tree-based, multichannel protocol for data collection applications. The goal is to partition the network into multiple subtrees while minimizing the intratree interference. The protocol partitions the network into subtrees and assigns different channels to the nodes residing on different trees. TMCP is designed to support convergecast traffic, and it is difficult to have successful broadcasts because of the partitions. Contention inside the branches is not resolved because the nodes communicate on the same channel.

There are many MAC protocol proposals that consider single-channel communication [5,7,8,10] in the domain of WSNs. These protocols perform to be good in single-channel scenarios where the primary design goal is energy efficiency [24], scalability, and adaptability to changes [25].

There are single-channel MAC protocols that aim to provide high throughput especially with scheduled communication, such as Z-MAC [26] and Burst-MAC [27]. While these protocols perform well in single-channel scenarios, parallel transmissions over multiple channels can further improve the throughput by eliminating the contention and interference on a single channel.

Besides multichannel communications, there exist other methods to reduce the impact of interference, such as transmission power control [28], creating minimum interference sink trees [29]. In a previous work, Incel and Krishnamachari [30] have investigated the impact of transmission power control on the network's performance with a realistic setting and found that discrete and finite levels of adjustable transmission power on the radios may not completely eliminate the impact of interference.

In a GS-MAC mechanism, data transmission and reception scheduling, as well as actual data transmission, are performed in a collision-free manner. Unlike the existing protocols, in GS-MAC, a cycle's time is split into three parts. In the first part, the beginning of a cycle, each node simply acquires the order in which each node will announce the data transmission-reception schedule. In the second part, nodes broadcast their schedule according to this order. Each node broadcasts its transmission schedule along with the scheduling information of its neighbor nodes only once. And finally, each node actually transmits the data packets according to the schedule announced in the second part of the cycle. Because each node uses one broadcast to announce its scheduling information, some of the two-hop nodes may not receive the schedule. GS-MAC overcomes this problem by splitting the data transmission time slots into different groups and introducing safety spaces, where each node transmits and receives actual data packets in one of the groups. Each node calculates its group using the number of groups and the order. This makes the scheduling collision free.

20.3 Multichannel MAC Protocol: Schedule-Based Approach

20.3.1 Network Model and Assumption

We consider a WSN that monitors a vast terrain of interest via a large number of static sensor nodes and a data collection point called the sink/base station (BS). This WSN can be represented by an undirected graph $G = (V,E)$, where V represents the set of all sensors in the network, and $E \subset V \times V$ represents the set of communication links between any pair of nodes. There is one data collection point called the BS in V. All traffics generated at sensors are destined for the BS. Such a network is called a many-to-one sensor network. The distance $d(i, j)$ between nodes i and j is defined as the minimum number of edges needed to traverse from one to the other. From this definition, the topology of the sensor network can be described by an $N \times N$ symmetric adjacency matrix C, which is defined as $C_{ij} = 1$, if $d(i, j) = 1$; or else $C_{ij} = 0$.

We assume that every sensor node has a unique ID. Each node is equipped with a half-duplex transceiver; a node can either transmit or listen but cannot do both simultaneously. A transceiver can be tuned to different channels (nonoverlapping frequencies), and all channels have the same bandwidth. The sink (or BS) is a data-collection center equipped with sufficient computation and storage capabilities, while the sensors are battery-operated and are empowered with limited data-processing engines. Nodes are time-synchronized [31] to provide efficient broadcast support. The task of the sensors is to dynamically serve the need of the data from the target area to the sink.

20.3.2 Problem Statement

The conflict relationship in the network can be described by an interference matrix $I_{N \times N}$, where if $d(i, j) \le 2$, $I_{ij} = 1$, or else $I_{ij} = 0$. This conflict relationship (because of interference) leads to two conditions for parallel transmission to be successful: (1) nodes i and j can transmit data on the same channel at the same time if the communication distance $d(i, j)$ is larger than 2, and (2) if the communication distance $d(i, j)$ is less than or equal to 2, nodes i and j can transmit data at the same time on different channels.

To design a multichannel MAC, usually a period of time is split into some equal intervals called time slots. Each time slot is designed to accommodate one or more packets to be transmitted and received between pairs of nodes in the network. Hence, the allocation of time slots directly influences the network's performance. Furthermore, proper channel/time slot allocation also ensures collision-free communication when several transmissions run simultaneously. So an efficient way of scheduling the channels/time slots is required to maximize the network throughput and improve other performance issues, such as delay, energy consumption, etc.

20.3.3 Description of Protocol

In this section, we discuss the GS-MAC approach in detail. The goal is to devise an efficient multichannel MAC protocol that carefully schedules message transmissions so as to avoid collisions at the MAC layer and, thereby, to utilize multiple channels to maximize parallel transmission among neighboring nodes. The media access design of GS-MAC is fully distributed and avoids a multi-channel hidden terminal problem [16].

The main concern here is to devise a methodology so as to avoid the collisions among transmissions of different sensor nodes. And the key reason behind the collisions is the so-called hidden terminal problem, which is caused when a sender is not aware of the transmission of the other sender. Moreover, in many cases, senders may not even notice the collisions if they are out of the interference range. It is the receiver who actually faces the problems in receiving as a result of the collisions. Keeping this in mind, channel assignment in GS-MAC protocol is made receiver-based. During the network initialization, receiving channels are assigned to the nodes for data reception, and each node broadcasts its receiving channel to its neighbors. When a node wants to transmit data, it needs to switch to the receiver's receiving channel.

In GS-MAC protocol, different time slots are assigned to different sender–receiver pairs, and the use of multiple channels assures parallel transmissions between different sender–receiver pairs in the same time slot over different channels. The data transmission schedules in different channels in different time slots are done carefully, avoiding collisions. When a receiver selects a channel as well as a data-reception slot, it is aware of the other schedules that are already chosen by the other nodes within its interference range (typically the nodes within its two-hop distance).

In GS-MAC protocol, a cycle (time duration) consists of three parts: (1) the contention period (CP) to provide an ordering to the nodes, (2) the control slot window (CSW) to perform the data transmission scheduling algorithm, and (3) the data transfer window (DTW) where the actual data transmissions take place. However, once the data transmission slots are chosen (during CP and CSW), the nodes can use the schedule (repeating the DTW only) until any change is necessary as a result of topology changes (e.g., node failure, etc.).

We describe GS-MAC protocol in the following subsections in detail.

20.3.3.1 Cycle Structure

The structure of a cycle is shown in Figure 20.1. As stated earlier, one cycle (time duration) is divided into three parts, namely, the CP, a CSW, and a DTW. Both the CSW and DTW are contention-free periods (CFPs). The CSW is divided into m $(0,1,2,\ldots, m - 1)$ equal-sized slots.

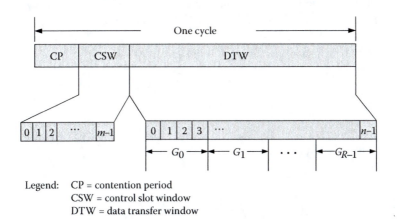

Legend: CP = contention period
         CSW = control slot window
         DTW = data transfer window

Figure 20.1 Cycle structure: cycle is divided into CP, CSW, and DTW. DTW is divided into R groups.

The duration of a slot in CSW is set to the time for picking up the desired reception slots plus the transmission-reception time of a control message. Similarly, the DTW is divided into n (0,1,2,…, $n - 1$) time slots of equal length, and the duration of a slot in the DTW is set to the transmission-reception time of one or more data packets along with the ACKs. Time slots in the DTW are further categorized into R (G_0, G_1, G_2,…, G_{R-1}) groups with equal numbers of time slots. The selection of the parameters m, n, and R is discussed in the later sections.

20.3.3.2 Scheduling Transmission

For communication during the CP at the beginning of a cycle, nodes use the broadcast mechanism used in 802.11 CSMA/CA. During this period, each of the nodes stays in a common channel and contends for a slot in the CSW. Each node obtains an order s ($0 \le s < m$), that is, the slot in the CSW, by contending each other during the CP (the parameter m is set to the maximum number of nodes that may fall within the one-hop neighborhood of a node, including itself). By doing so, Theorem 20.1 guarantees that no other node within the two-hop communication distance of a node will have the same order.

Theorem 20.1

If every node within the one-hop neighbors of a node selects a distinct slot in the CSW, it ensures that no other node within its two-hop neighborhood will select the same slot.   ■

Proof

Suppose that two nodes A and B fall within the two-hop neighborhood of each other, and they have selected the same slot s in the CSW. There must be another node C, which has node A and node B within its one-hop neighborhood. This contradicts with the proposition.   ■

During the CSW, all nodes select time slots/channels to be used during the following DTW. The nodes also tune to a common channel to broadcast their selections to others. The CSW is divided into m slots, which are allocated to the nodes according to the order s that they get during the CP. During its assigned slot in the CSW, every node selects some empty slots (in the DTW) in its receiving channel for data reception from the nodes that will send data to it. Allocation of such slots is done by every node that is likely to receive data from some other nodes. The selection can be done in a distributed manner according to Algorithm 20.1. Every node $node_r$ follows Algorithm 20.1 during its slot s in the CSW.

The data structures that are used in GS-MAC algorithms are listed here.

Channel: stores transmission schedule of n slots in a channel

■ Sender [n]
■ Receiver [n]

Node: stores information of a node

■ *recvChannel* /*receiving channel */
■ *s* /* order in the CSW */
■ *channel[nc]* /*nc = no of channels*/

In Algorithm 20.1, all the nodes having the same order s allocates their time slots in the DTW simultaneously. As the orders are two-hop aware, there will be no chance of collisions among the simultaneous transmissions that are received by these nodes. After selecting the slots, a node updates and broadcasts its *channel* information containing its schedule along with that of the others available to it. Nodes within the transmission range update this information by overhearing this broadcast message. Upon receiving the message from a node *node_r*, a transmitting node can know at which slot in the DTW it should transmit data to *node_r* and what channel to use.

ALGORITHM 20.1 **AssignTransmissionSlots**

$y = s \bmod R;$ /*select the group*/
$f = \min(0 \le i < m)$, where $i \bmod R = y$
$Pos = s/R;$ /*Position in group*/
for $k = Pos\text{-}1$ **down to** f **do**
 if *node_r* does not have slot allocation information of any node having order k in its 2-hop neighborhood
 then
  break;
 end if
end for
$NoInfo = k - f - 1;$
$AssignSlots(node_i, G_y, NoInfo);$
Broadcast $node_i.Channel$ to the one hop neighbors.

A node *node_r* picks a time slot *slot* for a transmitting node *node_i* in its receiving channel *ch* following Algorithm 20.2. Prior to selecting the time slot, *node_r* checks if either *node_i* or *node_r* preexists in *slot* or if *ch* is occupied by another transmission during *slot*. This confirms collision-free scheduling.

In some cases, it may happen that a node *node_r* does not have the information about the scheduling done by some of the other nodes in its two-hop neighborhood having lower control slot orders in the CSW. In such cases, it reserves some *SafetySpace*s in the beginning of the DTW from which those nodes may have selected their *slot*s; *node_r* tries to allocate from the other slots for itself. However, this may lead to reservation of a large number of slots as *Safetyspace*, which may not be always feasible. To minimize this, GS-MAC protocol partitions the DTW in R groups, G_y ($0 \le y \le R - 1$). A node can select its time slot(s) from one of the R groups according to Equation 20.1:

$$y = s \bmod R \qquad (20.1)$$

This grouping has two advantages. First, because every node has to be aware only of the nodes in its respective group, the information overhead for a node is minimized. Second, it minimizes the amount of unavailable information necessary for a node in selecting channels as well as time slots. This reduces the time slot(s) to be reserved as *safetyspace*(s). Consequently, the reduction in *safetyspace*(s) reduces the required number of time slots n in the DTW, which, in turn, increases the throughput.

The DTW is grouped according to the ratio, k, of the interference and transmission range. In this chapter, we consider the value $k = 2$, and the DTW is grouped into $R = k + 1$.

ALGORITHM 20.2   **AssignSlots(*Node_i*, *G_y*, *NoInfo*)**

start = group*Gsize;   /* *Gsize = total slots in a group* */
SafetySpace = NoInfo*ns;   /* *ns = maximum number of*
                                                    intended senders of a node */

start = start + SafetySpace;
end = start + Gsize − 1;
for each receiving node *node_r* of *node_i* **do**
    for *slot = start* **to** *end* **do**
        if neither *node_i* nor *node_r* has a previous entry in *slot* **then**
            ch = node_r.recvChannel;
            if there is no entry for channel *ch* in *slot* **then**
                node_i.Channel[ch].Sender[slot] = i;
                node_i.Channel[ch]. Receiver [slot] = r;
                break;   /* *Try the next receiving neighbor* */
            end if
        end if
    end for
end for

To ensure that every node receives at least one time slot in the DTW, the number of time slots n in the DTW should be $ns\left(\dfrac{m}{R}-1\right)+1$, where ns is the maximum number of senders of a receiving node, and the product term defines the maximum number of possible *safetyspaces* that a receiving node may have to reserve for the other nodes in its group (in the worst case). Note that if such a maximum number is used for n, it is much more likely to have some empty (unused) time slots in the DTW. However, as mentioned earlier, the need for leaving *safetyspaces* is not very high because the necessary information of the other nodes in the group is likely to be available in most cases. Hence, some compromises can also be made in determining n, which may cause some nodes to receive no packets in a DTW. However, because the packets are transmitted to all possible receivers, paths to the sink are much likely to exist even if some nodes are avoided.

Unlike the RTS/CTS (two-way handshake) or request–response–reservation (three-way handshake), the GS-MAC scheme involves only one broadcast message needed for each node to schedule their transmissions.

20.3.3.3 Scheduling Example

We describe the transmission-scheduling algorithm with an example shown in Figure 20.2. Consider Figure 20.2a where each node is represented with its node ID and receiving channels, and each arrow shows its intended receivers to which it may transmit its data.

This example considers a small snapshot of nine nodes (S2, S3, S5, S6, S8, S9, S10, S11, and S12) from a large network that fall within a one-hop communication distance of S6. Assume that these nine nodes have got the slots in the CSW according to the order shown in Figure 20.2b after contending during the CP. As can be seen from Figure 20.2b, S11 is the first node ($s = 0$) to select and announce its reception schedule, and it has three intended senders: S3, S6, and S9. To determine the group, y, from which it selects its time slots to receive data from its intended senders, S11

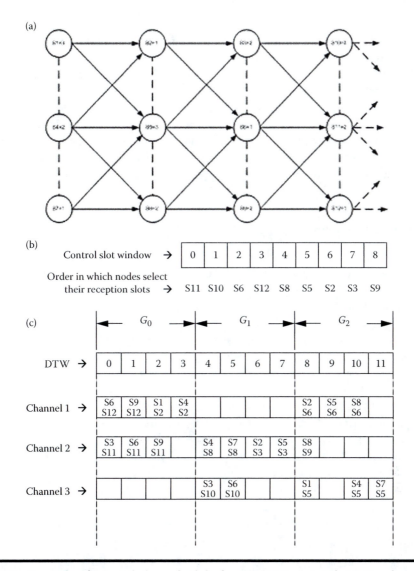

Figure 20.2 **Example of transmission schedule for GS-MAC protocol. (a) Topology with the node ID and receiving channel. (b) CSW and the order in which each node selects its reception schedule. (c) Allocated slots with three channels in a DTW where 12 time slots are divided into three groups.**

gets $y = 0$ according to Equation 20.1, which is the first group in the DTW. Because it has three intended senders, it can select three slots from channel 2 from the beginning of group 0 as shown in Figure 20.2c. Then, it broadcasts its schedule so that its immediate neighbors can update this information.

In a similar manner, other nodes may select their time slots/channels. Consider node S5 with control slot order 5. It will select the time slots from group 2 according to Equation 20.1. Before selecting the slots, S5 checks the schedule of S6, which should have already chosen its slots. It knows, from the broadcast message of S6, which slots are selected from group 0 by S6, and accordingly, S5 selects the slots 0 in channel 3 as shown in Figure 20.2(c).

In a case when two nodes fall into the same group, but one node does not know the previous node's schedule (i.e., two nodes are not immediate neighbors, and a node has not yet received the previous one's broadcasted message via other nodes), the node can leave the time slots (safetyspace) from the beginning of that group (because the previous node has already taken the slots from that group) and select the time slots accordingly.

Nodes that are beyond two hops may get the same pattern of control slot order in the CSW. For example, node S1, with same control slot order, may select its schedule at the same time as S10 and so on. In this way, only one node is receiving in a channel during a particular slot within two hops. Also, the schedule allows parallel transmission within two hops with disjointed sets of source–destination pairs with different channels.

20.4 Simulated Performance

To judge the effectiveness of this multichannel MAC protocol, we have considered three important performance metrics: aggregate throughput, packet delivery rate, and average end-to-end packet delay as a function of the number of channels. We compare GS-MAC with the CSMA, MMSN, and MC-LMAC protocols.

In our simulation model, we assume a multihop network environment, where 100 nodes are uniformly randomly distributed over a square-shaped terrain. The sink is positioned on the midpoint of a boundary. We assume the network topology to be static, and the radio range of all nodes are the same. A free space propagation channel model is assumed with the capacity set to 250 kbps. Packet lengths are 32 bytes for data packets. The maximum transmission range for a sensor node is assumed to be 40 m. The number of channels is varied from 1 to 10. Each node has a maximum of three intended receivers (immediate neighbors) toward the sink node.

Figure 20.3 presents the results in terms of aggregate throughput. The aggregate throughput is shown (in number of bytes per second received by the sink node) as a function of the number of channels. With the GS-MAC protocol, the maximum aggregate throughput at the sink node

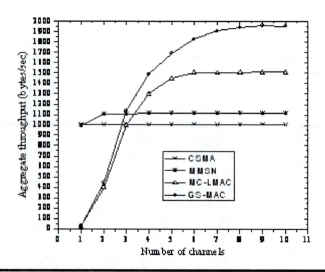

Figure 20.3 Aggregate throughput with different number of channels.

is 1963 bytes/s. The results from CSMA, MMSN, and MC-LMAC protocols are also presented to compare the performance. Figure 20.3 shows that the aggregate throughput increases as the number of channels increases from 1 to 10 except CSMA, where the number of channels is fixed to 1. A significant improvement is achieved using GS-MAC protocol as compared to the CSMA, MMSN, and MC-LMAC protocols. On average, single-channel CSMA achieves an aggregate throughput less than all the other protocols. Because of the high contention, the protocol fails to successfully allocate the medium to the nodes. Aggregate throughput with MMSN is observed to be limited and does not increase after six channels. This is a result of the failure of the nodes around the sink to successfully sense the channel and prevent the collisions. MC-LMAC suffers from clashes that occur during the selection of the free time slot(s) within two-hop nodes.

Figure 20.4 presents the results in terms of a packet delivery rate, which is the ratio between the number of packets received by the sink and the total number of packets generated by the nodes. The performance is better than both protocols and achieves delivery of more than 99% of the packets.

Figure 20.5 presents the average end-to-end packet delay, which is the time between the transmission of a packet at the source node and reception at the sink node. GS-MAC protocol achieves much lower delay than the MC-LMAC protocol. Unlike the MC-LMAC, GS-MAC protocol has decreasing end-to-end delay as the number of channels increases. This is because the average delay from the source to the sink is influenced by the size of a frame in MC-LMAC protocol. Furthermore, decreasing the frame size would not reduce the delay because the number of packets that can be delivered per time slot will also decrease, and the packets will be buffered to be transmitted later. CSMA experiences a higher delay than all the protocols because of the exponential and higher number of backoffs resulting from the high contention.

Figure 20.6 shows the results in terms of energy efficiency per successfully delivered packet. We consider both the energy spent to receive and transmit as well as the energy spent for relaying the packet toward the sink node. Energy spent per delivered packet is quite high with MC-LMAC when there is only a single channel. This is because of the very low delivery rate. As the number of channels increases, all the three protocols, MC-LMAC, MMSN, and the GS-MAC, spend much

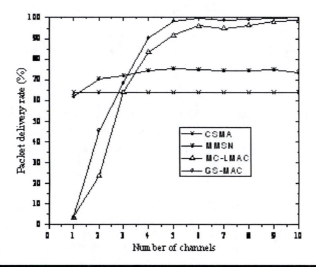

Figure 20.4 Packet delivery rate with different number of channels.

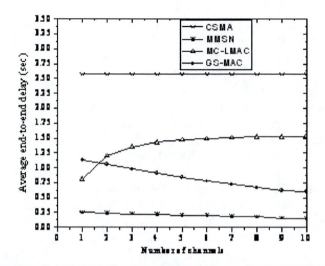

Figure 20.5   End-to-end packet delay with different number of channels.

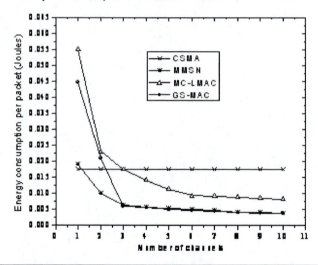

Figure 20.6   Energy consumption per successfully delivered packet with different number of channels.

less energy than CSMA. Although MMSN has much less energy consumption as compared to GS-MAC in cases of one and two channels, GS-MAC protocol is more energy efficient than all the protocols. This is because GS-MAC has much fewer collisions as compared to the existing ones.

20.5 Future Research Directions

As for future works, it would be interesting to run the experiment with different system loads and with different node densities. Nevertheless, more research interest to delve into the performance issues for the mobile sink and multiple sinks might further be judged to quantify the suitability of the GS-MAC approach for WSNs.

20.6 Conclusions

In this chapter, we have discussed multichannel MAC protocols and described a new schedule-based multichannel MAC protocol for WSNs. The new protocol consists of a CP to provide an ordering to the nodes, following which, in a CSW, every receiving node selects some time slots/channels from a DTW. The approach is fully decentralized and efficient within the sensors' localized scope. The GS-MAC protocol takes the advantage of the control slot order along with the groups of data transmission and reception windows to maximize the parallel transmission in a collision-free manner. Also, each node needs only one broadcast message to advertise the schedule information that it has. Through simulations, it was shown that the GS-MAC mechanism provides significant performance improvements in terms of aggregate throughput, packet delivery ratio, average delay, and energy consumption.

References

1. Akyildiz, I. F., W. Su, Y. Sankarasubramaniam, and E. Cayirci. "Wireless sensor networks: A survey." *Computer Networks* 38, (2002): 393–422.
2. Akyildiz, I. F., T. Melodia, and K. R. Chowdhury. "A survey on wireless multimedia sensor networks." *Computer Networks* 51, (2007): 921–960.
3. Gurses, E., and O. B. Akan. "Multimedia communication in wireless sensor networks." *Annals of Telecommunications* 60, no. 7–8 (2005): 799–827.
4. Culler, D., D. Estrin, and M. Srivastava. "Overview of sensor networks." *IEEE Computer*, Special Issue on Sensor Networks, 2004.
5. Polastre, J., J. Hill, and D. Culler. "Versatile Low Power Median Access for Wireless Sensor Networks." In *ACM SenSys*, 2004.
6. El-Hoiyi, A., J. D. Decotignie, and J. Hernandez. "Low Power MAC Protocols for Infrastructure Wireless Sensor Networks." In *The Fifth European Wireless Conference*, 2004.
7. Ye, W., J. Heidemann, and D. Estrin. "An Energy-Efficient MAC Protocol for Wireless Sensor Networks." In *IEEE INFOCOM*, 2002, 1567–1576.
8. Dam, T., and K. Langendoen. "An Adaptive Energy-Efficient MAC Protocol for Wireless Sensor Networks." In *ACM SenSys*, Los Angeles, CA, 2003, 171–180.
9. Woo, A., and D. Culler. "A Transmission Control Scheme for Media Access in Sensor Networks." In *ACM MobiCom*, 2001.
10. Rajendran, V., K. Obraczka, and J. J. Garcia-Luna-Aceves. "Energy-Efficient, Collision-Free Medium Access Control for Wireless Sensor Networks." In *ACM SenSys*, 2003.
11. Van Hoesel, L., T. Nieberg, J. Wu, and P. Havinga. "Prolonging the lifetime of wireless sensor networks by cross layer interaction." *IEEE Wireless Communication Magazine* 11, (2005): 78–86.
12. Hill, J., R. Szewczyk, A. Woo, S. Hollar, D. Culler, and K. Pister. "System Architecture Directions for Networked Sensors." In *The Ninth International Conference on Architectural Support for Programming Languages and Operating Systems*, 2000, 93–104.
13. "XBOW MICA2 Mote Specifications. http://www.xbow.com.
14. Polastre, J., R. Szewczyk, and D. Culler. "Telos: Enabling Ultra-Low Power Wireless Research." In *ACM/IEEE IPSN/SPOTS*, 2005.
15. Miller, M. J., and N. H. Vaidya. "A MAC protocol to reduce sensor network energy consumption using a wakeup radio." *IEEE Transactions on Mobile Computing* 4, no. 3 (2005): 228–242.
16. So, J., and N. Vaidya. "Multi-Channel MAC for Ad-Hoc Networks: Handling Multi-Channel Hidden Terminal Using a Single Transceiver." In *ACM MobiHoc*, 2004.
17. Tzamaloukas, A., and J. J. Garcia-Luna-Aceves. "A Receiver-Initiated Collision-Avoidance Protocol for Multi-Channel Networks." In *IEEE INFOCOM*, 2001.
18. Zhou, G., C. Huang, T. Yan, T. He, J. Stankovic, and T. Abdelzaher. "MMSN: Multi-Frequency Media Access Control for Wireless Sensor Networks." In *IEEE Infocom*, 2006.

19. Incel, O. D., S. Dulman, and P. Jansen. "Multi-Channel Support for Dense Wireless Sensor Networking." In *EUROSSC*, 2006, LNCS 4272, 1–14.
20. Chen, X., P. Han, Q. S. He, S. Tu, and Z. L. Chen. "A Multi-Channel MAC Protocol for Wireless Sensor Networks." In *Proceedings of The Sixth IEEE International Conference on Computer and Information Technology*, 2006.
21. Incel, O. D., P. G. Jansen, and S. J. Mullender. "MC-LMAC: A Multi-Channel MAC Protocol for Wireless Sensor Networks." *Technical Report TR-CTIT-08-61*, Centre for Telematics and Information Technology, University of Twente, Enschede, 2008.
22. Hamid, M. A., M. A. Wadud, and I. Chong. "A schedule-based multi-channel MAC protocol for wireless sensor networks." *Sensors* 10, October 21, 2010, 9466–9480.
23. Wu, Y., J. Stankovic, T. He, and S. Lin. Realistic and Efficient Multichannel Communications in Wireless Sensor Networks. In *Proceedings of IEEE INFOCOM*, 2008, 1193–1201.
24. Langendoen, K., and G. Halkes. "Energy-Efficient Medium Access Control." In *Embedded Systems Handbook*, R. Zurawski, Ed. CRC Press, Boca Raton, FL, 2005.
25. Demirkol, I., C. Ersoy, and F. Alagoz. "MAC protocols for wireless sensor networks: A survey." *IEEE Communications Magazine* 44, no. 4 (2006): 115–121.
26. Rhee, I., A. Warrier, M. Aia, and J. Min. "Z-MAC: A Hybrid MAC for Wireless Sensor Networks." In *Proceedings of The 3rd International Conference on Embedded Networked Sensor Systems (SenSys)*. ACM, New York, 2005, 90–101.
27. Ringwald, M., and K. Römer. "BurstMAC—A MAC Protocol with Low Idle Overhead and High Throughput." In *Adjunct Proceedings of The 4th IEEE/ACM International Conference on Distributed Computing in Sensor Systems (DCOSS)*, Santorini Island, Greece, 2008.
28. El Batt, T. A., and A. Ephremides. "Joint Scheduling and Power Control for Wireless Ad-Hoc Networks." In Proceedings of *IEEE INFOCOM*, 2002; vol. 2, 976–984.
29. Fussen, M., R. Wattenhofer, and A. Zollinger. "Interference Arises at the Receiver." In *Proceedings of the International Conference on Wireless Networks, Communications and Mobile Computing*, 2005; vol. 1, 427–432.
30. Incel, O. D., and B. Krishnamachari. "Enhancing the Data Collection Rate of Tree-Based Aggregation in Wireless Sensor Networks." In *Proceedings of SECON*, 2008, 569–577.
31. Maróti, M., B. Kusy, G. Simon, and A. Lédeczi. "The Flooding Time Synchronization Protocol." In *ACM SenSys*, 2004.

Chapter 21

Mobile IPv6-Based Autonomous Routing Protocol for Wireless Sensor Networks

Kashif Saleem, Zahid Farid, and Mohammad Ghulam Rahman

Contents

21.1 Introduction

In the fields of wireless communication, electronics, and IC fabrication, there have been tremendous advances over the last few years. The advancement in deployment of networks of low cost, low power, and multifunctional sensors has received much attention. Sensors are generally equipped with data-processing and communication capabilities. A wireless sensor network (WSN) is designed to link the physical world to the digital world by capturing and revealing real-time activities and converting these into that form after which it can be processed and stored, and then action is taken on it. The WSN serves an extremely valuable position in sensing and monitoring systems. The monitoring systems include military and civil applications, such as target field imaging, intrusion detection, weather monitoring, security and tactical surveillance, distributed computing, and detecting ambient conditions.

One of the most famous initiatives consolidating the possible deployment of WSN was the IEEE 802.15.4 [1]. IEEE 802.15.4 offers simple, energy-efficient, and inexpensive solutions to a wide variety of applications in WSNs. IEEE standard 802.15.4 specifies a physical (PHY) and a medium access control (MAC) layer dedicated for a low-rate wireless personal area network (LR-WPAN). The main motivation of IEEE 802.15.4 is to develop a dedicated standard and not to rely on existing technologies, such as Bluetooth or WLAN, and to ensure low-complexity, energy-efficient implementations.

WSN consists of wireless sensor nodes that are small in size and are able to sense, process data, and communicate with each other, typically over a radio frequency (RF) channel. Sensor nodes are devices that contain the four main components, as shown in Figure 21.1. Broadly speaking, WSN is designed for detection of events and phenomena, to collect and process data, and then to transmit sensed data to an interested user. Because the sensor nodes have limited memory and are typically deployed in difficult-to-access locations, a radio is implemented for wireless communication to transfer the data to a base station (e.g., a laptop, a personal handheld

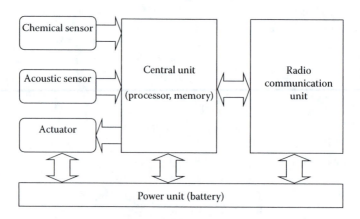

Figure 21.1 Components of wireless sensor nodes.

device, or an access point on a fixed infrastructure). Limitation in energy, transmit power, memory, computing power, self-organization, short-range broadcast communication, multi-hop, dense deployment, changing topology frequently, and node failure are the basic features of the sensor network [2,3].

The most challenging issues in a sensor network are limited and unrechargeable energy provision and low power and memory, so many researchers are focusing their efforts to improve the energy efficiency from different angles. In sensor networks, energy is consumed mainly for three purposes [4]: data transmission, signal processing, and hardware operation. WSNs send sensor data to the sink node, which is one of the objectives of sensor nodes. Though the dimensions of WSN are increasing day by day, the majority of sensor nodes do not have enough transmitting power to reach the sink node directly. To overcome this problem, hierarchical or mesh-based routing is often used. In recent years, many researchers have been trying to improve transport and network protocols for WSN considering the most important metric: energy efficiency.

Work on new protocols was conducted by keeping WSN challenges in mind; Transmission Control Protocol/Internet Protocol (TCP/IP) and other recent solutions were considered inappropriate for WSNs [5]. As most WSN applications are required to perform monitoring or detection of phenomena, for such applications, the network cannot operate in complete isolation; there must be a way to guarantee the data acquisition. For example, when the WSN communicates directly to an existing network infrastructure, such as the global Internet, a local area network (LAN), or private intranet, remote access to the WSN can be achieved [4]. A WSN with Internet Protocol (IP)-based Internet connectivity can make ubiquitous computing realistic.

21.1.1 Constraints on Current IP-Based WSNs

The TCP/IP is the de facto standard for Internet communication and not only for the Internet but also for local area networks. The desired WSN Internet connectivity must use the TCP/IP suite at some point, by means of a gateway, bridge, or router, enabling protocol transformation or directly at the sensor level. In WSNs, the use of IP has always been considered inadequate and in contradiction to the needs of wireless sensor networking.

IP was not designed for energy-restricted, low-memory, and low-processing power devices, and hence, IP was not suitable for WSNs. Despite these issues, researchers are looking into the use of IP in WSNs because of the potential it has. The problems of IP connectivity with WSNs include larger header overhead, the need for a global address scheme, limited bandwidth, limited energy of the node, architectural model, data flow pattern, implementation challenges, and also the TCP/IP transport protocol [5].

After the evolution of Internet protocol version 6 (IPv6), it introduces a large unified addressing structure of 128 bits to alleviate the IPv4 address shortage. Autoconfiguration (autoconf) allows hosts to autoconf IP addresses with improved security (security extension headers, integrated data integrity) and better performance (aggregation, neighbor discovery instead of Address Resolution Protocol (ARP) broadcasts, no fragmentation, no header checksum, flow, priority, integrated Quality of Service (QoS)) [6,7]. IPv6 also supports a richer set of communication paradigms, including a scoped addressing architecture and multicast into the core design. The targets for IP networking for low-power radio communication are the applications that need wireless Internet connectivity at lower data rates for devices with a very limited form factor.

Recently, the Internet task force (IETF) ROLL working group (routing over low-power and lossy network) with the aim of specifying a routing solution for low-power and lossy networks

(LNNs) is supporting a variety of link layers, low bandwidth, lossy, and low power and standardized an IPv6-based routing solution for an IP-based smart object network. The result of this working group was the "ripple" routing protocol (RPL) [8]. Moreover, to allow better estimations, RPL needs to be improved with the link quality valuations [9].

The advancement of IPv6 is required. By using optimization mechanisms, it is possible to achieve performance in terms of energy and throughput. IPv6-based currently available routing protocols such as 6LoWPAN [10], RPL [8], etc., do not handle or support mobility. There are lots of areas in available routing protocols to perform research, such as node discovery, routing protocol, security deployment, and power management. Additionally, besides all WSN constraints, network lifetime or wireless sensor node energy still remains as the main issue in WSNs.

21.1.2 Proposed Solutions and Objectives

To address the problems and challenges mentioned above, we have proposed a novel architecture of mobile IPv6-based autonomous routing protocol (MARP) that can efficiently forward data packets from mobile source nodes to the required destinations.

A MARP comes up with a high packet-delivery ratio while minimizing packet overhead and efficient power consumption for routing in WSNs. Data packets in the proposed routing protocol should be delivered within a given time to live (TTL) or deadline in order to satisfy routing features. A selective optimal forwarding node may result in prolonging the WSN lifetime. RPL is enhanced with a packet reception rate (PRR) that is based on end-to-end delay and energy consumption metrics. Additional metrics help the network to maximize the throughput while minimizing delay and packet loss.

Topological node mobility is the result of physical movement and/or of a changing radio environment. Hence, mobility needs to be handled even in a network with physically static nodes. We have tackled the mobility factor under MARP by enhancing RPL with the self-optimized routing mechanism biological-inspired autonomous routing protocol (BIOARP) [11–13]. The proposed MARP can handle the highly dynamic changes in the environment and mobility in an autonomous manner.

21.1.3 Organization of the Chapters

The next section reviews the related research works on IPv4, IPv6, and mobile-based IPv6-routing approaches. Section 21.3 describes the way to implement the mobile IPv6 autonomous routing mechanism. Section 21.4 shows the work and results obtained through the work done up until this time. The future directions for research and conclusion are stated under Section 21.5.

21.2 Background

WSNs consist of a very large number of sensor nodes, which are deployed in the target field, and they collaborate to form an ad hoc network capable of reporting the phenomenon to a data collection point called a sink or base station. An ad hoc network is a peer-to-peer wireless network, which consists of nodes that are connected to each other without infrastructure. Because the nodes in a network can serve as routers and hosts, they can forward packets on behalf of the other

Table 21.1   Comparison between Traditional Network and WSNs

Traditional Networks	WSNs
General-purpose design; serving many applications	Single-purpose design; serving one specific application
Typical primary design concerns are network performance and latencies; energy is not a primary concern	Energy is the main constraint in the design of all node and network components
Networks are designed and engineered according to plans	Deployment, network structure, and resource use are often ad hoc in nature (without planning)
Devices and networks operate in controlled environments	Sensor networks often operate in environments with harsh conditions
Maintenance and repair are common and networks are typically easy to access	Physical access to sensor nodes is often difficult or even impossible
Component failure is addressed through maintenance and repair	Component failure is expected and addressed in the design of the network
Obtaining global network knowledge is typically feasible; centralized management is possible	Most decisions are made locally without the support of a central manager

nodes and run user applications [14]. A comparison has been shown in Table 21.1 between traditional networks and WSNs [15].

21.2.1 Operating System for WSNs

The operating system (OS) plays an important role in the hardware management. The performance of the hardware or system on a machine mainly depends on the OS reliability. This is very relevant especially in WSNs, where wireless sensor nodes are designed to operate with limited resources. In a WSN, the most common constraint is the lifetime of the entire network, which depends on wireless sensor nodes' battery power. Second, the memory and operational capabilities as sensing are less demanding of resources than computation in a conventional OS. WSNs are often designed for reliable real-time services. A variety of OS solutions are given for WSNs; the most popular of them are illustrated below.

21.2.1.1 TinyOS

TinyOS [16] is a tiny (fewer than 400 bytes), flexible OS built from a set of reusable components that are assembled into an application-specific system. TinyOS was developed and is maintained by the University of Berkeley, California. A large number of manufacturing companies making wireless sensor nodes employed TinyOS. The current version of TinyOS is 2.1.1. TinyOS is component-based programming and coded by NesC language (networked embedded system C), a dialect of C [17]. TinyOS is not an OS in the traditional sense. It is a programming framework for an embedded system and set of components that enables building an application-specific OS into each application. A typical application is about 15 kB in size, of which the base OS is 400 bytes. The largest application is a database-like query system that is about 64 kB [18].

The OS for 6LoWPAN, called blip, is based on TinyOS and was implemented by the University of Berkeley [16]. It uses 6LoWPAN/HC-01 header compression and includes IPv6 features, such as neighbor discovery, default route selection, point-to-point routing, and network programming support. Standard tools, such as ping6, tracert6, and nc6, can be used to interact with and trouble-shoot a network of blip devices; PC-side code was written using the standard Berkeley sockets (BSD sockets) application programming interface (API) (or any other kernel-provided networking interface). A sensor network can also be easily mapped into the public subnet to provide global connectivity.

21.2.1.2 Contiki

Contiki is the open-source OS for the next billion connected devices called the Internet of things (IoT) [19,20]. Contiki is a highly portable, multitasking OS for memory-constrained networked embedded systems and WSNs. It is developed and maintained by a group of developers from industry and academia led by Adam Dunkels from the Swedish Institute of Computer Science. Contiki has been used in a variety of projects, such as road tunnel fire monitoring, intrusion detection, wildlife monitoring, surveillance networks, etc. It is designed for microcontrollers with small amounts of memory.

A typical Contiki configuration is 2 kB of Random Access Memory (RAM) and 40 kB of Read only Memory (ROM). Contiki runs on a variety of platforms, ranging from embedded microcontrollers such as the Texas Instruments (TI) MSP430 and the Atmel AVR to old home computers. Code footprint is on the order of kilobytes, and memory usage can be configured to be as low as tens of bytes. The OS is written in the C programming language and consists of an event-driven kernel on top of which application programs can be dynamically loaded and unloaded at run-time.

Furthermore, Contiki uses lightweight protothreads that provide a linear, threadlike programming style on top of the event-driven kernel. In addition to protothreads, Contiki also supports per-process optional multithreading and interprocess communication using message passing. Contiki provides IP communication, both for IPv4 and IPv6. The latest version of Contiki is 2.5 (released September 9, 2011). ContikiRPL is a new implementation of the proposed IETF standard RPL protocol for low-power IPv6 routing. It is now the default IPv6 routing mechanism in Contiki.

21.2.2 Network Simulators

Network simulators (NSs) on top of the OS give the programmer appropriate results and outputs. With the help of NSs, the coded program output for the required network scenario is checked and rectified. The logical implementation and investigation through NSs saves a lot of resources that are wasted in reflashing the real-time devices by a written program and in deployment of these programmed devices. Mostly as a result of the reflashing/reprogramming, real-time devices (wireless sensor nodes) get physically damaged. The strength of logical output and analysis absolutely depends on NS efficiency.

Some of the open-source simulators are compared with the help of Table 21.2 [8].

Table 21.2   Open-Source Simulation Comparison

Simulator	NS2	Castalia OMNet++	TOSSIM	Cooja/ MPSim	WSim/WSNet
Level of details	Generic	Generic	Code level	All levels	All levels
Timing	Discrete event	Discrete event	Discrete event	Discrete event	Discrete event
Simulator Platforms	FreeBSD, Linux, SunOS, Solaris, Windows (Cygwin)	Linus, Unix, Windows (Cygwin)	Linux, Windows (Cygwin)	Linux	Linux, Windows (Cygwin)
WSN Platforms	Zigbee	n/a	Micaz	Tmote Sky, ESB, Micaz	Micaz, Mica2, TelosB, CSEM, Wisenode, ICL, BSN nodes
GUI Support	Monitoring of simulation flow	Monitoring of simulation flow, C++ development	None	Yes	None
Physical [21–23]	Lucent Wave-Lan DSSS	CC1100, CC2420	CC2420	CC2420, TR 1001	CC1100, CC1101, CC2500, CC2420
MAC	802.11, preamble-based TDMA	TMAC, SMAC	Slandered, TinyOS 2.0 CC2420 Stack	CSMA/CA, TDMA, X-MAC, LPP, Contiki MAC	DCF, BMAC, ideal MAC
Network	DSDV, DSR, TORA, AODV	Simple tree, multipath rings	No data	RPL, AODV	Greedy geographic
Transport	UDP, TCP	None	No data	UDP, TCP	None
Energy Consumption Model	Yes	Yes	With power TOSSIM add-on	Yes	Yes

21.2.3 Literature Review

Research has been carried out for interconnection of WSNs with the Internet. For this, power limitation is the major constraint and is considered as the restricting factor of the interconnection functionality. We categorize the research of interconnection of WSNs with the Internet into two approaches: TCP/IP (v4) and IPv6, which are discussed below.

21.2.3.1 Interconnection of TCP/IP (Version 4) with WSNs

WSN interconnection with an external network is presented in [5]. Liutkevicius et al. presented two approaches to connect a WSN to an external network: the gateway approach and the overlay approach. The gateway approach is better in terms of power efficiency, and the integration with the external network is simple. Liutkevicius et al. [5] specifically pointed out that cluster-based architecture (topology) is more power efficient and scalable, as shown in Figure 21.2 [5].

In [24], Neves and Rodrigues present the advantages and challenges of IP on sensor networks [24]. They survey the state of the art with some implementation examples and point out further research topics in this area. They also present some concepts about WSNs, routing protocols, IPv4, and body sensor networks and mention further research in the areas of interconnection of WSNs with IP.

In [25], Dunkels et al. present the connection of wireless sensornets with TCP/IP networks. In this paper, they discuss three different ways to connect sensor networks with TCP/IP networks: proxy architectures, Delay-Tolerant Networking (DTN) overlay, and TCP/IP for sensor networks. Additionally, they present three architectures that are in some respect orthogonal and can be best used by making combinations, such as a partially TCP/IP-based sensor network with a DTN overlay connected to the global Internet using a front-end proxy.

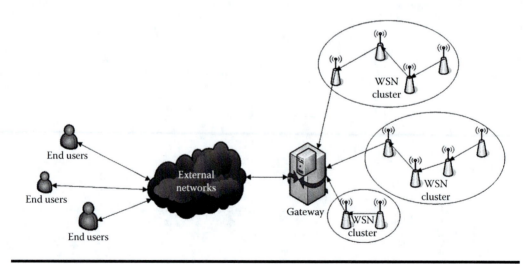

Figure 21.2   Multihop WSN.

21.2.3.2 Interconnection of IPv6 with WSNs

As noted earlier, despite the issues found in WSNs, researchers are looking into the use of IP in sensor networks because of the potential it has [24]. A huge collection of protocols have been invented and evaluated. In the mean time, the Internet has evolved as well. In 1998, request for comments (RFC) 2460 defined IPv6 [26]. IPv6 is the designated successor of IPv4. An increase in IP hosts and scalability are the primary goals of IPv6. IPv6 is better suited to the needs of WSNs than IPv4 in every respect.

A new innovation in Internet protocol technology, called 6LoWPAN, is making the IoT a reality. 6LoWPAN [27] is a standard from the IETF published in 2007, which optimizes IPv6 for use with low-power, low-bandwidth communication technology, such as IEEE 802.15.4. The IETF 6LoWPAN group was formed to standardize framing and header compression for the transmission of IPv6 packets over IEEE 802.15.4 [27]. The document describes the frame format for transmission of IPv6 packets and the method of forming IPv6 link-local addresses and stateless autoconfigured addresses on IEEE 802.15.4 networks.

6LoWPAN works by compressing 60 bytes of a standard IPv6 header down to just 7 bytes and optimizing the mechanism for wireless embedded networking. Additional specifications include a simple header compression scheme using shared context and provisions for packet delivery in IEEE 802.15.4 meshes. The IEEE 802.15.4 link layer has severe constraints on size as packet size cannot be larger than 127 bytes. For this reason, an IPv6 packet may need to be fragmented into multiple link-layer frames. Furthermore, to make efficient use of available bandwidth, the IPv6 packet header needs to be compressed.

6LoWPAN opens a great opportunity for WSNs to be operated remotely. In [28], a gateway construction is designed for interconnection of 6LoWPAN WSNs with IPv6 and IPv4 clients, enabling them to receive sensor node reading or send a command to 6LoWPAN WSNs at any time through direct communication with the sensor node. The gateway acts like a bridge between IPv4 hosts and the WSN, receiving a client's request from the IPv4 network, sending a corresponding compressed IPv6 message for the WSN, receiving a response from the WSN, and finally returning the result to IPv4 clients through the IPv4 network, as shown in Figure 21.3. A laboratory test bed was created for the validation of the proposed architecture.

A gateway-based framework (the VIP bridge approach) for sensor network over IP network interconnection is presented in [29]. The VIP bridge approach virtually assigns an IPv6 address to the sensor node and uses a lower-level gateway (bridge) for interconnection. The frame entities are the sensor node, the IP host, and the gateway node. The framework is further divided into two

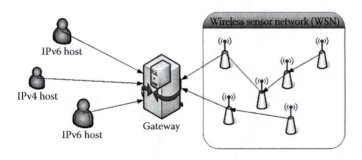

Figure 21.3   Interaction between IPv6 and IPv4 hosts and the gateway.

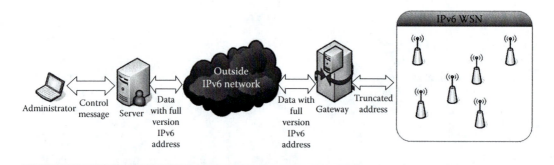

Figure 21.4   Data flow between WSN and external networks.

main components: virtual address assignment and packet translation. Two mapping processes are performed by the gateway by assigning an IP virtual address to the gateway. The gateway maps the WSN internal IP virtual address into a global address and vice versa. It supports both address-centric and data-centric WSN protocol. Emara et al. [29] mentioned the drawbacks of a VIP bridge, that is, a single point of failure, a bottleneck problem, and a major limitation in terms of routing protocol.

In [30], an interconnection method was proposed that connects IPv6 with a large-scale WSN. Juan et al. proposed a data-centric method to make the WSN access the Internet under the IPv6 protocol. In this method, the WSN was divided into grid networks and assigned an IPv6 address for each grid map. A clockwise grid routing protocol (CGRP) was proposed that bypasses the routing holes-based boundary effects and also achieves a high success rate of packet transmission. Implementation of the mechanism between the WSN and the external network was performed in OMNET++.

In [31], Ludovici et al. discussed forwarding techniques for IP fragmented packets in a real 6LoWPAN. Real 6LoWPAN implementation is done to achieve the results. The authors presented a new routing scheme based on mesh, which improves the critical aspects of the mesh forwarding.

An IPv6-based address allocation for WSNs is presented in [32]. An IPv6 address is too long, that is, 128 bits, so it is difficult to implement in a WSN, as energy cost in a WSN is a key constraint for using an IPv6 full address. This work proposed an improved, self-organized address scheme with a simplified IPv6 address format that is used in between gateway and the WSN, as shown in Figure 21.4.

21.2.3.3 Mobility Enabled Intelligent Routing Protocols for WSNs

Lots of research has been performed on WSNs, especially in the area of developing routing protocols despite key problems, such as energy efficiency, mobility, and traffic pattern, still remaining as open areas of research. IP-based WSNs have many applications, such as home automation, industrial control, health care, and agriculture monitoring. The research community takes interest in layer-three routing to solve problems, such as reliability and end-to-end delay communication enhancement.

A global addressing scheme-based routing approach that focuses on the above-mentioned issues is presented in [33]. Islam and Huh proposed a sensor proxy mobile IPv6 routing protocol, which is a network layer protocol. They also proposed a hierarchical addressing by which individual IP WSN nodes will be identified by unique global IPv6 addresses.

Self-optimization of the WSNs in terms of energy efficiency has attracted great attention. In BIOARP [12,15], the enhanced ant colony optimization (ACO) mechanism is adapted to perform self-optimized routing in WSNs. The routing protocol depends on three modules: routing management, neighborhood management, and power management. These functions cooperate and coordinate with each other to provide self-optimized routing capability. BIOARP is a real-time application that reduces delay, packet loss, packet overhead, and battery power consumption.

The self-organization is a major factor that is necessarily required by WSNs to tackle the uncertain environmental behavior. Therefore, much concentration is developed among researchers in terms of self-organization of WSNs. The communication in Mobile Wireless Sensor Network (MWSN) is self-organized by involving biological-inspired techniques and algorithms [34].

21.3 Research Methodology

Research methodology contains the initial design and architecture of the proposed MARP mechanism.

21.3.1 Design Concept of MARP

A WSN is intended to operate for months or years without battery replacement or human intervention. Applications such as medical care, battlefield surveillance, fire detection, and structural and environmental monitoring can benefit from wireless sensor nodes that communicate in a multihop manner to collect and transfer the required data to their destination. Several WSN applications require MIPv6 connectivity to ensure universal interaction, especially in a disaster situation. The MARP is proposed primarily to overcome problems such as battery consumption, traffic congestion, mobility, and packet delay to maximize traffic throughput while maintaining the QoS in terms of packet reception ratio (PRR), packet overhead, and power consumption in WSNs. MARP is an extension of RPL and BIOARP; RPL provides an IPv6 platform, and by enhancing it with BIOARP, the autonomous routing capability is enabled.

21.3.2 MARP Design Approach

The design of the proposed MARP is shown in Figure 21.5. The process begins when the data packet needs to be transferred or forwarded. First, the algorithm checks whether the neighboring table at every hop contains the pheromone value that determines the next best hop. If the pheromone value is found, then the predetermined path is followed to forward the packets. Otherwise, RPL discovery process is invoked to perform neighbor discovery and determine the next best hop. At the transfer of first data packet on every hop, the BIOARP-based calculated pheromone value is stored in a routing table. This process continues until the first data packet reaches the destination. However, if problem occurs on the way toward the destination, the RPL discovery process will be reinvoked.

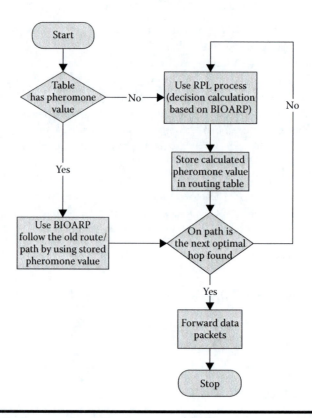

Figure 21.5 **MARP design approach.**

21.3.3 Packet Reception Ratio

A wireless medium does not guarantee reliable data transfer. Performance of WSNs depends on the link quality of the wireless medium [35]. The probability of receiving a packet between two wireless sensor nodes is defined as a PRR [36]. MARP uses the link layer model as derived in [21,37], and the PRR is determined by

$$\text{PRR} = \left[1 - \left(\frac{8}{15} \right) \left(\frac{1}{16} \right) \sum_{j=2}^{16} (-1)^j \binom{16}{j} \exp\left(20\text{SNR}\left(\frac{1}{j} - 1 \right) \right) \right]^m .$$

The signal-to-noise ratio (SNR) is calculated in [1,38] as

$$\text{SNR} = P_t - PL(d) - S_r$$

where P_t means the transmitted power (in dBm; maximum is 0 dBm for Tmote Sky) and is for the receiver's sensitivity (−90 dBm in Tmote Sky). A routing metric, such as the PRR, will be found with the addition of already defined metrics of IPv6 RPL such as end-to-end delay and energy consumption.

21.3.4 Data Structure at Every Wireless Sensor Node

The BIOARP mechanism is used for calculation of the best pheromone value. New metrics will be added to the routing table as shown in Figure 21.6 in existing IPv6-based RPL, that is, PRR and record expiry β as shown in Table 21.3. In order to find the next hop node, MARP uses the best pheromone value-based neighboring node ID, which is marked for the data to be forwarded. In the design of MARP, link quality is considered in order to improve the delivery ratio.

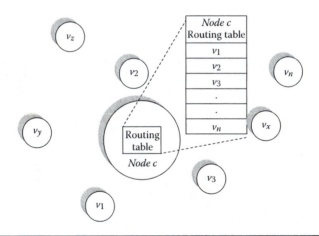

Figure 21.6 Routing table at every node.

Table 21.3 Routing Metrics

			New Metric Added	New Metric Added
	Velocity (End-to-End Delay)	Energy (Battery Remaining)	Link Quality (PRR)	Record Expiry
Node 1	τ^1	η^1	ω^1	β^1
Node 2	τ^2	η^2	ω^2	β^1
.	.	.	.	.
.	.	.	.	.
.	.	.	.	.
.	.	.	.	.
Node n	τ^n	η^n	ω^n	β^1

The routing table records are generated through a neighbor discovery mechanism that involves the request-to-register (RTR) packets and RTR replies. Default static node expiry record time is 180 s [15], which will be reduced for availing the most updated environmental information. Hence, most updated neighboring information will enhance and fulfill the mobility factor [39]. The new neighboring node information that is added in the routing table helps data packets forwarding. If some link error occurs or time value exceeds the expiry time value, the respective neighboring node record will be deleted from the routing table. The process of rediscovery will be reinitiated when the routing table is empty.

21.4 Simulation

The MARP designing and analysis process has been carried out through an NS.

21.4.1 Simulation Tools

The scenario was simulated using the NS Cooja, a Contiki OS-based NS. We have chosen the Cooja NS because the results generated by it are similar to the real-time experiment [40]. Instant

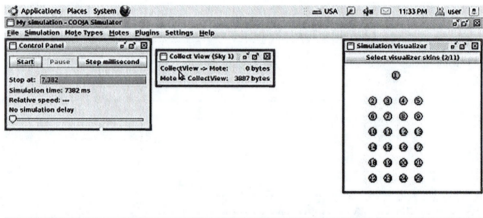

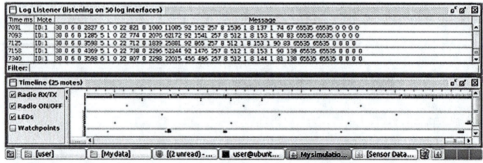

Figure 21.7 NS Cooja, a simulator of Contiki OS.

Contiki OS 2.5 (release date September 12, 2011) was chosen for simulation with Ubuntu 11.4. After installation of the Instant Contiki 2.5, an upgradation was performed for the latest CVS. The IPv6 RPL is implemented under Cooja; the code is written in C++. Twenty-five WSNs were deployed by assigning the distance between nodes manually, that is, the nodes are in the form of a grid and have equal distances between any two nodes. The distance between two nodes is 9 m, as shown in Figure 21.7.

21.4.2 Graphical Animation of the Network

During the animation produced by the Cooja animator as shown in Figure 21.8, we can examine the output of the network. The CBR traffic is generated from all nodes toward node 0. Each node contains a table with the pheromone value as shown previously in Figure 21.6. The pheromone table at each node contains the pheromone value for the next node toward the required destination. In this way, all the ants disperse in as many paths as possible to achieve a load balancing provisioning over a WSN.

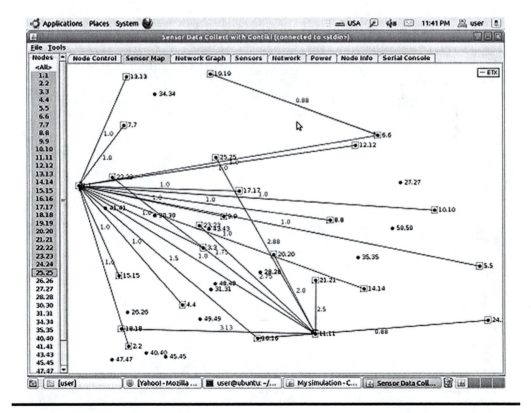

Figure 21.8   Animation produced by Cooja.

21.4.3 Network Model, Performance Parameters, and Preliminary Result

Twenty-five wireless sensor nodes were deployed over a 50 × 50 m² grid as shown in Figure 21.8. A fixed size for one packet is considered in the simulation. Total simulation time was equal to 100 s (100,000 ms) and the number of packets per second was one. In order to avoid cycles and the routing table's freezing, we have enabled every packet with a node ID and packet sequence ID. Currently, while running RPL on the given scenario, we have experienced a packet loss of 2087, as shown in Figures 21.9 and 21.10.

21.5 Conclusion and Future Works

In this chapter, we present a novel MARP for WSNs to enhance RPL with our previous autonomous routing mechanism, the BIOARP. In MARP, the autonomous routing decision depends on end-to-end delay, remaining battery power, and SNR metrics. The proposed MARP has been designed and studied through real-time simulator Cooja. The details show that our MIPv6-based proposed protocol will provide better data delivery while minimizing the power consumption, delay, and packet loss in mobility-enabled WSNs.

MARP is a mobility-enabled autonomous routing protocol for WSNs equipped with all kind of sensors. This eventually leads to a potential area to further enhance MARP for future

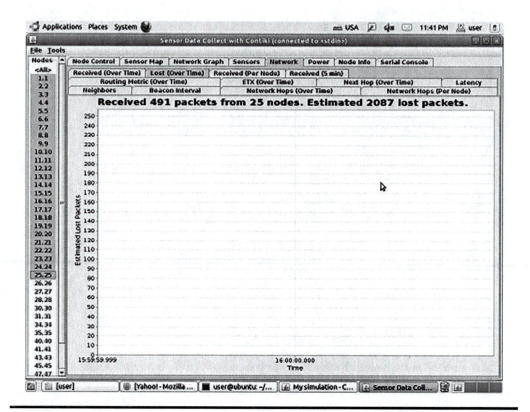

Figure 21.9   Packet loss.

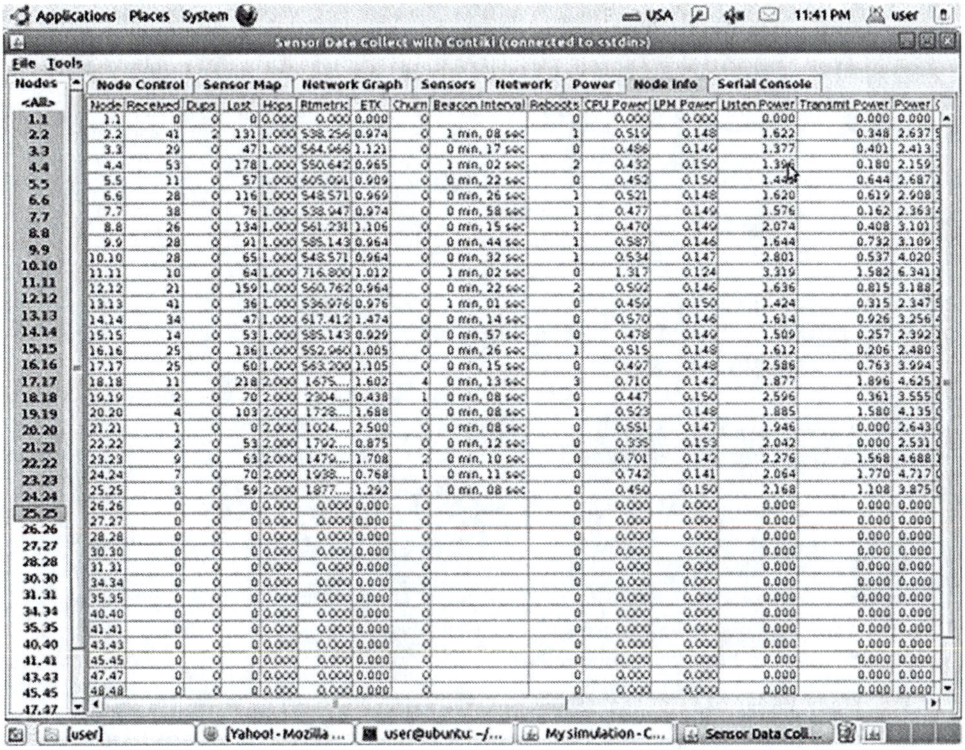

Figure 21.10 Node-by-node network simulation result.

applications. The future applications include a monitoring and surveillance system for indoor and outdoor applications and the like. However, further exploitation and development can increase the performance of the proposed routing protocol. The future efforts may include the following:

- Other variants, such as the cognitive, reinforcement learning max–min ant system, can be considered accordingly for different application needs.
- Including multipath routing capability can improve the performance in terms of the data delivery ratio, but on the other hand, it will consume more battery power.
- The architecture of MARP can be tested in the real test bed, according to different applications/scenarios.
- The performance of IPv6 applications based on this could be investigated.

References

1. IEEE. "IEEE P802.15 Working Group: Draft Coexistence Assurance for IEEE 802.15.4b." 2005.
2. Akylidiz, F., W. Su, Y. Sankarasubramaniam, and E. Cayirc. "A survey on sensor networks." *IEEE Communication Magazine* 40, (2002): 102–114.
3. Yick, J., B. Mukherjee, and D. Ghosal. "Wireless sensor network survey." *Computer Networks* 52, (2008): 2292–2330.

4. Cecilio, J., J. Costa, and P. Furtado. "Survey on Data Routing in Wireless Sensor Networks." In *Wireless Sensor Network Technologies for Information Explosion Era*. vol. 278, edited by T. Hara, 3–46. Springer book series "Studies in Computational Intelligence," 2010.

5. Liutkevicius, A., A. Vrubliauskas, and E. Kazanavicius. "A Survey of Wireless Sensor Network Interconnection to External Networks." In *Novel Algorithms and Techniques in Telecommunications and Networking*, edited by T. Sobh, K. Elleithy, and A. Mahmood, 41–46. The Netherlands: Springer, 2010.

6. Narten, T., E. Nordmark, and W. Simpson. *Neighbor Discovery for IP Version 6 (IPv6)*. RFC Editor, United States. 1998.

7. Thomson, S., T. Narten, and T. Jinmei. *{IPv6} Stateless Address Autoconfiguration*. RFC Editor, United States. 2007.

8. Ben Saad, L., C. Chauvenet, and B. Tourancheau. "Simulation of the RPL Routing Protocol for IPv6 Sensor Networks: Two Cases Studies." In *International Conference on Sensor Technologies and Applications SENSORCOMM 2011*, France, 2011.

9. Strübe, M., S. Böhm, R. Kapitza, and F. Dressler. "RealSim: Real-Time Mapping of Real World Sensor Deployments into Simulation Scenarios." In *Proceedings of the 6th ACM International Workshop on Wireless Network Testbeds, Experimental Evaluation and Characterization*, Las Vegas, Nevada, USA, 2011, 95–96.

10. Moravek, P., D. Komosny, M. Simek, M. Jelinek, D. Girbau, and A. Lazaro. "Investigation of radio channel uncertainty in distance estimation in wireless sensor networks." *Telecommunication Systems*, 1–10.

11. Saleem, K. "Biological Inspired Self-Organized Secure Autonomous Routing Protocol for Wireless Sensor Networks." Doctor of Philosophy (Electrical Engineering), Faculty of Electrical Engineering, Universiti Teknologi Malaysia, Malaysia, 2011.

12. Saleem, K., N. Fisal, M. A. Baharudin, A. A. Ahmed, S. Hafizah, and S. Kamilah. "Ant colony inspired self-optimized routing protocol based on cross layer architecture for wireless sensor networks." *WSEAS Transactions on Communications (WTOC)* 9, (2010): 669–678.

13. Saleem, K., N. Fisal, M. A. Baharudin, A. A. Ahmed, S. Hafizah, and S. Kamilah. "BIOSARP–Bio-Inspired Self-Optimized Routing Algorithm using Ant Colony Optimization for Wireless Sensor Network: Experimental Performance Evaluation." In *Computers and Simulation in Modern Science, Included in ISI/SCI Web of Science and Web of Knowledge*, vol. IV, edited by N. E. Mastorakis, M. Demiralp, and V. M. Mladenov, 165–175. 2011.

14. Frodigh, M., P. Johansson, and P. Larsson. "Wireless ad hoc networking—the art of networking without a network." *Ericsson Review* 4, (2000): 248–263.

15. Saleem, K., N. Fisal, S. Hafizah, and R. Rashid. "An Intelligent Information Security Mechanism for the Network Layer of WSN: BIOSARP." In *Computational Intelligence in Security for Information Systems*. vol. 6694, edited by Á. Herrero and E. Corchado, 118–126. Berlin: Springer, 2011.

16. UC-Berkeley. 2011. *TinyOS Tutorial*, http://docs.tinyos.net/index.php/TinyOS_Tutorials.

17. Gay, D., P. Levis, R. v. Behren, M. Welsh, E. Brewer, and D. Culler. "The Nesc Language: A Holistic Approach to Networked Embedded Systems." In *Conference on Programming Language Design and Implementation, SIGPUN 2003*, San Diego, California, USA, 2003, 1–11.

18. Levis, P., and N. Lee. 2011, *TOSSIM: a simulator for TinyOS Networks, User's Manual*, http://www.cs.berkeley.edu/~pal/research/tossim.html.

19. U. Berkeley. 2011, *Berkeley IP*.

20. Dunkels, A., B. Grönvall, and T. Voigt. "Contiki—A Lightweight and Flexible Operating System for Tiny Networked Sensors." In *Proceedings of the 29th Annual IEEE International Conference on Local Computer Networks*, Tampa, FL, USA, 2004, 455–462.

21. Atmel. "Atmel Extends Trusted Computing Standard To Embedded Systems." 2012.

22. T. Instruments. "MSP430 Data Sheet, Texas Instruments, http://focus.ti.com/mcu/docs/mcuprodoverview.tsp?sectionId=95&tabId=140&familyId=342), 2012.

23. Crossbow. "MPR-MIB Users Manual." June 2007, 2011.

24. Neves, P. A. C. d. S., and J. J. P. C. Rodrigues. *Internet Protocol over Wireless Sensor Networks, from Myth to Reality, Journal of Communications*, vol. 5, no. 3 (2010), 189–196, Mar 2010.

25. Dunkels, A., J. Alonso, T. Voigt, H. Ritter, and J. Schiller. "Connecting Wireless Sensornets with TCP/IP Networks." In *Proceedings of the Second International Conference on Wired/Wireless Internet Communications (WWIC2004)*, 2004, 143–152.

26. Deering, S., and R. Hinden. *Internet Protocol, Version 6 (IPv6) Specification*. RFC Editor, United States. 1998.

27. Kushalnagar, N., G. Montenegro, D. Culler, and J. Hui. "Transmission of IPv6 Packets over IEEE 802.15.4 Networks." RFC Editor 2070–1721, 2007.

28. da Silva Campos, B., J. J. P. C. Rodrigues, L. D. P. Mendes, E. F. Nakamura, and C. M. S. Figueiredo. "Design and Construction of Wireless Sensor Network Gateway with IPv4/IPv6 Support." In *Communications (ICC), 2011 IEEE International Conference*, 2011, 1–5.

29. Emara, K. A., M. Abdeen, and M. Hashem. "A Gateway-Based Framework for Transparent Interconnection between WSN and IP Network." In *EUROCON 2009, EUROCON'09. IEEE*, 2009, 1775–1780.

30. Juan, L., L. Zhen, L. De-xiang, and L. Re-fa. "Wireless Sensor Network Inter Connection Design Based on IPv6 Protocol." In *Wireless Communications, Networking and Mobile Computing, 2009. WiCom '09. 5th International Conference*, 2009, 1–4.

31. Ludovici, A., A. Calveras, and J. Casademont. "Forwarding techniques for IP fragmented packets in a real 6LoWPAN network." *Sensors* 11, (2011): 992–1008.

32. Zhan, J., B. Yang, and A. Men. "Address Allocation Scheme of Wireless Sensor Networks Based on IPv6." In *Broadband Network and Multimedia Technology, 2009. IC-BNMT'09. 2nd IEEE International Conference*, 2009, 597–601.

33. Islam, M. M., and E.-N. Huh. "A novel addressing scheme for PMIPv6 based global IP-WSNs." *Sensors* 11, (2011): 8430–8455.

34. Banitalebi, B., T. Miyaki, H. R. Schmidtke, and M. Beigl. "Self-Optimized Collaborative Data Communication in Wireless Sensor Networks." In *Proceedings of the 2011 Workshop on Organic Computing*, Karlsruhe, Germany, 2011, 23–32.

35. Dunkels, A., O. Schmidt, N. Finne, J. Eriksson, F. Österlind, N. Tsiftes, and M. Durvy. 2012, 16 May. *The Contiki OS,* http://www.contiki-os.org/. Available: http://www.contiki-os.org/.

36. Rodrigues, J. J. P. C., and P. A. C. S. Neves. "A survey on IP-based wireless sensor network solutions." *International Journal of Communication Systems* 23, (2010): 963–981.

37. Levis, P., S. Madden, J. Polastre, R. Szewczyk, K. Whitehouse, A. Woo, D. Gay, J. Hill, M. Welsh, E. Brewer, and D. Culler. "TinyOS: An Operating System for Sensor Networks Ambient Intelligence." In *Ambient Intelligence*, edited by W. Weber, J. Rabaey, and E. Aarts, 115–148. Berlin: Springer, 2005.

38. Sklar, B. *Maximum A Posteriori Decoding of Turbo Codes, In Digital Communications:* Fundamentals and Applications, Second Edition, Prentice-Hall, 2001.

39. Nasser, N., A. Al-Yatama, and K. Saleh. "Mobility and Routing in Wireless Sensor Networks." In *Electrical and Computer Engineering (CCECE), 2011 24th Canadian Conference*, 2011, 573–578.

40. Eriksson, J., F. Österlind, N. Finne, A. Dunkels, N. Tsiftes, and T. Voigt. "Accurate Network-Scale Power Profiling for Sensor Network Simulators Wireless Sensor Networks," Vol. 5432, edited by U. Roedig and C. Sreenan, 312–326. Berlin/Heidelberg: Springer, 2009.

Chapter 22

Taxonomy of QoS-Aware Routing Protocols for MANETs

Chhagan Lal, Vijay Laxmi, and Manoj Singh Gaur

Contents

22.1 Introduction

With the proliferation of inexpensive and infrastructure-less mobile ad hoc networks (MANETs), research focus has shifted to issues related to security and quality of service (QoS) in these networks. MANETs are collections of mobile hosts (also called nodes), which are self-configurable, self-organizing, and self-maintainable. The nodes communicate with each other through wireless channels with no centralized control. With the evolution of wireless prevalence in the last decade, we are witnessing more and more applications moving and adapting to wireless methods to communicate. MANET nodes rely on multihop communication; that is, nodes within each other's transmission range can communicate directly through radio channels, whereas those outside the radio range must rely on intermediate nodes to forward messages toward their destinations. Mobile hosts can move, leave, and join the network whenever they want, and routes need to be updated frequently because of the dynamic network topology. This is illustrated in Figure 22.1. Suppose, node A wants to communicate with node B. At time t_1, the routing path is $A \rightarrow C \rightarrow B$. At time t_2 $(>t_1)$, node C moves out of range of node A. Because of this, the changed route for node B at time t_2 is $A \rightarrow D \rightarrow B$.

In MANETs, one of the important issues is routing, that is, finding a suitable path from a source to a destination. Because of the rapid growth in the use of applications, such as online gaming, audio/video streaming, voice-over IP (VoIP), and other multimedia streaming applications in MANETs. It is mandatory to provide the required level of QoS for reliable delivery of data. Providing the required QoS guarantees in wireless multihop networks is much more challenging than in wireline networks mainly because of its dynamic topology, distributed on-the-fly nature, interference, multihop communication, and contention for channel access. In particular, it is important for routing protocols to provide QoS guarantees in terms of metrics, such as achievable throughput, delay, packet loss ratio, and jitter.

Despite the large number of routing solutions available in MANETs, their practical implementation and use in the real world is still limited. Multimedia and other delay- or error-sensitive applications that attract a mass number of users toward the use of MANETs have led to the realization that best-effort routing protocols are not adequate for them. Because of the dynamic topology and physical characteristics of MANETs, providing guaranteed QoS in terms of achievable throughput, delay, jitter, and packet loss ratio is not practical. So QoS adaptation and soft QoS have been proposed instead [1]. Soft QoS means failure to meet QoS is allowed for certain cases, such as when a route breaks or the network becomes partitioned [1]. If node mobility is too high and topology

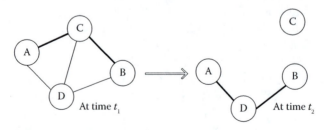

Figure 22.1 Communication in ad hoc networks.

changes very frequently, providing even soft QoS guarantees is not possible. For a routing protocol to function properly in a wireless network where mobility is high, the rate of topology state information propagation must be higher than the rate of topology change. Otherwise, the topology information will always be stale, and inefficient routing will take place, or there may be no routing at all. This applies equally to QoS state and QoS route messages. A network that satisfies the above condition is said to be combinatorially stable [2]. Some routing protocols provide the information about available network resources to the applications so that these can adjust their execution accordingly for achieving their required level of QoS. Other routing approaches may not serve the application directly but try to increase the overall network performance in terms of QoS metrics.

The proposed chapter is organized in the following manner: We start with a precise analysis of the challenges posed by the provision of QoS over a multihop MANET environment. This is followed by an overview of commonly used metrics for QoS protocol performance estimation and a discussion of the factors affecting the performance of QoS protocols. Next, we present an overview of the most important factors and choices involved in the design of QoS support protocols for multihop MANETs. We continue by providing a unique taxonomy that classifies the QoS solutions into different groups in order to organize the many candidate solutions. Following this, we summarize the key features, basic operation, and major pros and cons of a selection of QoS approaches. To obtain a useful and essential subset of QoS protocols from the large array of solutions provided in the literature, we mainly focus on journal articles, transaction papers, and peer-reviewed conferences. We will be presenting an overview and survey of current trends and patterns in this field. We will conclude the chapter with the present state of the art and the future work areas.

22.2 QoS in MANETs

QoS in MANETs is defined as a set of service requirements that should be satisfied by the network when a stream of packets is routed from a source to a destination [3]. A data session can be characterized by a set of measurable requirements, such as maximum delay, minimum bandwidth, minimum packet delivery ratio, and maximum jitter. All the QoS metrics are checked at the time of connection establishment, and once a connection is accepted, the network has to ensure that the QoS requirements of the data session are met throughout the connection duration [4]. In wireless networks, the problem of guaranteeing QoS to a data session is more complex as compared to wired networks. This is because of the characteristics of ad hoc networks, such as unreliable and error-prone wireless media, limited bandwidth (which limits the use of control messages), dynamic topology (i.e., nodes are free to move, join, or leave the network), and low battery power and processing power. Moreover, the protocols at various layers may need to self-tune to adjust to environments, mission changes, and traffic so that the MANETs can retain their efficiency [4].

In the literature, several solutions have been proposed to address the issue of QoS provision in MANETs. These solutions include call admission control (CAC) protocols that admit a flow only if sufficient resources are available, cross-layer design to provide interaction between layers to increase network performance, data rate adaptive protocols that require applications to change their coding scheme to achieve a data rate that can be supported by the network, routing protocols that find paths having sufficient resources to meet the given QoS requirements, multipath routing schemes for fault tolerance, and load balancing and reservation protocols that attempt to reserve the resources along the path. At each layer, a set of metrics is used to evaluate the performance of QoS-aware routing protocols.

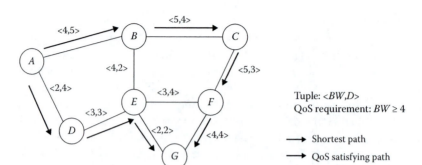

Figure 22.2 QoS-aware routing in MANETs.

The above-mentioned QoS metrics are used by applications to specify their QoS requirements. QoS requirements can be defined in terms of a set of metrics. For example, Figure 22.2 shows a network topology in which an application at node A has certain bandwidth ($BW >= 5$ kbps) and delay ($D <= 5$ ms) requirements. A QoS-aware routing protocol selects ($A \to B \to C \to F \to G$) a route that satisfies the QoS requirements of the application instead of selecting the shortest path ($A \to D \to E \to G$). Providing a multiconstrained QoS aims at optimizing multiple QoS metrics while provisioning QoS over MANETs and is, literally, a complex task.

22.3 Issues and Design Considerations in QoS Provisioning over MANETs

Before describing different types of QoS-aware routing solutions, it is important to discuss the design issues that greatly affect the performance of QoS solutions. An optimal QoS protocol development process is dependent on the design issues under consideration. Routing is used to create and maintain a route for transporting a packet stream from a source to a destination. Early MANET routing protocols focused on finding a feasible route from a source to a destination without considering the stability and reliability of that route. Furthermore, no provisions were made for QoS support. For QoS-aware applications, a path with sufficient resources to meet stringent QoS guarantees should be used. Following are design considerations and challenges for QoS-aware ad hoc networking routing protocols.

22.3.1 Challenges in Provisioning QoS over MANETs

A summary of major challenges faced in MANETs while provisioning QoS support are given below.

1) *Unpredictable physical characteristics:* In wireless networks, data are transmitted over radio channels that are highly unreliable because of atmospheric and operating conditions. Wireless channels are inherently prone to bit errors caused by network conditions, such as interference (transmissions of neighboring nodes), thermal noise, shadowing (reflections from obstacles), and multipath fading (multiple paths to destination) [5]. This leads to unpredictable channel capacity and increases the probability of link failure.

2) *Infrastructure-less and uncentralized network architecture:* The inherently infrastructure-less, inexpensive, and quick-to-deploy nature of MANETs is providing a promise for its use in

diverse domains. Because of the uncentralized nature, only contended protocols are used at the MAC layer. Because of a lack of centralization, each node has to send its QoS state information to all other nodes. This increases the overhead and complexity of QoS-aware protocols, as efficient dissemination of QoS state information is necessary.

3) *Channel contention:* In order to keep up-to-date information about network state, nodes should communicate with each other on a shared channel. This increases collisions resulting from channel contention and interference. Increases in collisions result in increased delay, decreased packet delivery ratio, and low utilization of channel bandwidth and drain the nodes' battery power. One way to avoid these is with the use of a TDMA scheme with the help of global clock synchronization. In TDMA, each node transmits in its predefined slot, so there is no need for channel contention. Contention-free transmission can also be achieved by using a different spreading code (such as CDMA/FDMA) for each transmission. Practical implementations of the above-mentioned methods are difficult to achieve because of dynamic topology change, lack of central authority, and the overhead and complexity involved for channel access synchronization [6].

4) *Multihop communication:* Hosts in MANETs also work as routers to forward packets toward their destinations. To send data from a source to a destination, a path is discovered using intermediate nodes. Every node on the path is equally important for reliable transfer of data as failure of even a single node in the route shall result in MANET communication breakdown.

5) *Limited network resources:* Because of the mobile nature of devices used in MANETs, the size of these devices should be small. This requirement leads to constraints on available battery power, processing speed, and memory space. The wireless channels have low link capacity as compared to wired networks. Effective resource management schemes are required for the best utilization of these scarce network resources. Solutions for QoS routing problems are NP-complete [2] in nature and complex heuristics that may place an undue strain on resource-constrained mobile devices for approximating these solutions.

22.3.2 Design Trade-Offs

The general design trade-offs that may affect the design of QoS-aware routing protocols is as follows:

1) *Reactive versus proactive routing.* Route discovery and maintenance processes are characterized as proactive (table-driven) or reactive (on-demand). Proactive protocols produce low delay during route discovery and route setup. But the overhead caused by proactive protocols to maintain the up-to-date routing topology consumes a significant amount of scarce channel capacity. Providing QoS guarantees becomes more difficult for proactive routing protocols as the network mobility and size increase. A reactive approach avoids the potential wastage of resources by only discovering routes on demand and maintaining only active routes. However, route discovery using a reactive approach causes initial delay when an application needs a route to a destination. Reactive protocols produce low overhead and adapt quickly to dynamic topology changes. Therefore, the first trade-off is selection of the nature of routing protocols, proactive or reactive, for provisioning QoS over MANETs while balancing overhead and delay.

2) *Capacity versus battery consumption and delay.* One way through which the capacity of a network can be increased is by sending more data session packets concurrently via different

neighbors [7]. These neighbors forward the packet to the destination when it comes into their transmission range. This scheme increases network capacity on the cost of delay. On the other hand, if redundant packets are sent on multiple paths, it is observed that the destination receives the packets with low delay [8]. This method decreases the capacity of the network and also increases the battery utilization of each node.

3) *Transmission range: short versus long hops.* Variation in the transmission range of nodes affects the number of hops required to forward a packet to its destination. Using low transmission power increases the number of hops. This reduces energy consumption at intermediate nodes and produces a higher signal-to-interference ratio. On the other hand, long hops produce low routing overhead and route maintenance overhead, decrease the route failure rate, and increase path efficiency and end-to-end reliability [9].

4) *QoS provisioning: global versus individual.* Some QoS-aware routing protocols provide information about available network resources to the applications that can adjust their coding scheme accordingly to achieve the required level of QoS. Other QoS solutions may not serve the application directly but try to increase the overall network performance in terms of QoS metrics.

22.3.3 Factors Affecting Performance of QoS Protocols

When evaluating the performance of QoS-aware routing protocols, many factors affect the results. Most of the factors affecting network performance are directly influenced by the basic characteristics of MANETs. These factors together define the term "scenario," whether in a practical or simulation case, and are summarized as follows:

1) *Traffic source.* Type, number, and data rate of traffic sources in a network greatly affect the performance of QoS protocols. As the number and data rate of a traffic application increase, the number of packets in the network also increases, leading to increased channel contention and interference. An increase in the number of traffic sources results in more route computations that further increases network overhead. Increase in data rates may cause an unbalance between nodes transmission rate and available channel bandwidth.

2) *Node mobility and placement.* Node mobility comprises three parameters: the nodes' mobility pattern, pause time, and speed (maximum and minimum). The mobility pattern defines whether the node moves at a uniform speed or a varying speed and the pattern of movement of the nodes, that is, whether they move independently or in a group. The pause time specifies the time during which nodes remain stationary between each period of movement. This parameter, along with the nodes' maximum and minimum speeds, gives an idea of how frequently the topology of the network changes. Thus, we can determine how often we have to update network state information.

3) *Network size.* As the network size increases, the message overheads caused by hello and topology control packets increase. If QoS support is given, then the overhead for QoS state gathered and disseminated further increases the control overhead, which affects the network performance. Increase in size also causes message update latency.

4) *Node transmission power.* Nodes have the capability to control their transmission power. If transmission power is high, nodes have a higher number of direct neighbors that increase their connectivity. On the other hand, high transmission power causes interference and increased battery drain rate and may result in unidirectional links between nodes.

5) *Channel characteristics.* As discussed earlier about wireless channel characteristics in previous sections, there are many reasons for radio links being unstable and unreliable and having a high bit error rate causing incorrect delivery of data packets. These channel characteristics affect the network's ability to provide the desired level of QoS in MANETs.

22.4 QoS-Aware ad hoc Routing Protocols

There are many existing QoS-aware routing protocols for ad hoc networks emphasizing a variety of implementation scenarios. The basic objective has been to devise a QoS solution that minimizes control overheads, end-to-end delay, and energy consumption while maximizing the throughput and packet delivery ratio. Because these types of networks can be used in a variety of applications (online gaming, audio/video streaming, VoIP, and other multimedia streaming applications), they differ in terms of their QoS requirements and complexities. One possible way is to classify the existing solutions on the basis of QoS metrics used at the time of path evolution and path selection during route discovery. This classification is not feasible because many protocols use several QoS metrics. We classify the QoS-aware routing protocols on the basis of their dependence on the MAC layer. There are two board categories: (1) dependent, that is, protocols that interact with the MAC layer, and (2) independent, that is, protocols that do not interact with the MAC layer.

Figure 22.3 illustrates the classification of protocols based on their interaction with the MAC layer. It shows that all protocols are divided into three board categories:

Contention-free, MAC-based solutions. These protocols rely on accurately measured resource availability and resource reservations and require a contention-free MAC layer, such as TDMA or CDMA. These protocols are able to provide semihard QoS. Providing hard QoS guarantees is only possible in wireline networks where the link is not inherently prone to bit errors, and the nodes are not mobile. Solutions that are based on noncontended MAC

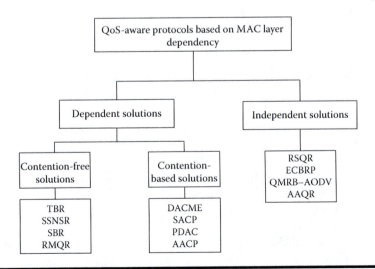

Figure 22.3   Classification of QoS-aware routing protocols based on their dependence on MAC layer.

protocols provide hard QoS only for certain periods of time when the channel or network conditions are stable. The term "semihard" is used to describe the characteristics of such protocols. These protocols require synchronization among network nodes for resource estimation and reservation. This synchronization, however, is not possible in MANETs because of their inherently uncentralized architecture.

Contention-based, MAC-based solutions. The protocols that rely on contended MAC fall in this category. These protocols can estimate the available network resources using statistical methods that are not very accurate. Such protocols try to provide soft QoS guarantees by making implicit resource reservations using CAC schemes. All guarantees are based on current network states and may not be sustained if the network states change frequently. In MANETs, the IEEE 802.11 specification is used at the MAC layer requiring nodes to contend for access to the channel.

MAC-independent solutions. Those that are independent from the MAC layer, that is, that do not require any kind of interaction with the MAC layer, are discussed in this category. Such protocols try to increase the overall network performance by improving the route discovery and route maintenance process. These solutions do not offer any type of soft QoS guarantee that relies implicitly on a certain MAC access scheme. The main purpose of these solutions is to provide better average QoS for all packets according to one or more QoS metrics. This often comes at the cost of increased routing complexity and packet overhead.

22.5 Protocols Relying on MAC Layer

22.5.1 QoS-Aware Routing Solutions Based on Contention-Free MAC Protocols

In this section, we review some popular QoS-aware routing solutions that are based on the IEEE 802.11 MAC protocol. We identify the advantages and disadvantages of each approach to review its strengths and weaknesses.

22.5.1.1 Ticket-Based Probing (TBR)

In [1], Chen and Nahrstedt proposed a distributed, ticket-based multipath QoS routing protocol. This protocol discovers QoS routes equal to the number of tickets with sufficient resources to provide certain throughput and delay guarantees. The dynamic nature of ad hoc networks makes the available network state information inherently imprecise. The main aim of the proposed approach is to develop a routing algorithm that can work well with a certain level of nonprecision in network state information. The novelty of the proposed scheme is the method used for discovering QoS paths. First of all, a proactive routing protocol, such as DSDV [10], is assumed to keep the routing table up to date at each node. The routing table at each node contains information about QoS metrics, such as bandwidth or residual link capacity, delay of link, and cost of link. The bandwidth and delay information of a link are provided by the lower layers. A weighed cost function is used to calculate the current delay/bandwidth state based on the old delay/bandwidth state and new delay/bandwidth state. When a QoS-constrained path is required by an application session, the number of tickets is issued by a source node in the form of probes, which are used to discover and reserve the available resources along the path.

The design of the protocol is based on two observations: 1) QoS routing is done for individual connections, and 2) there exist multiple paths from a source to a destination. The basic idea of TBR is that one ticket gives permission to search one QoS path. The source node issues a number of tickets, depending on the QoS requirements of the application and the state information stored in the routing table. TBR uses two colors of tickets. Yellow tickets are used to find feasible paths that contain low-delay links, and green tickets maximize the probability of finding low-cost paths that may have larger delays. If the QoS requirements of an application data session are tighter, more tickets are issued. Each probe contains one or more tickets and is sent from a source to a destination. At intermediate nodes, probes with more than one ticket may split into multiple ones, each finding a different downstream subpath. The next hop with lower delay and high bandwidth toward the destination is discovered based on the state information of nodes and assigned more tickets. Intermediate or destination nodes can mark tickets invalid if they are not able to find any path that satisfies the given QoS requirements. The route discovery process is terminated when all tickets reach their destination.

The route discovery process of TBR is shown with a simple example in Figure 22.4. Here, S is the source, and T is the destination. Three tickets are issued by S within one probe P_0 and forwarded to A (i.e., the next hop of S), where P_0 is split into two probes: P_1 (with two tickets) and P_2 (with one ticket). P_1 with two tickets is forwarded to B, and it splits into P_3 (through F) and P_4 (through E). When all the tickets reach a destination successfully, they select the path with the highest resources. The destination initiates the resource reservation process by sending the confirmation message back to the source. This message toward the source makes reservations at intermediate nodes on its way and enables intermediate nodes along the route to update its delay/bandwidth estimations and reserve the corresponding resources for the established path. Route breaks are handled locally by the node that detects them. Upon detecting a route break, the node tries to find alternate paths satisfying the ongoing session's QoS requirements. If the node is unable to repair the route locally, it sends a route break message to the source node that initiates the route discovery process again.

This protocol is an example that shows how multiconstrained metrics can be used for route discovery. TBR avoids flooding of *RREQ* messages by limiting them to less than or equal to the number of tickets issued. Furthermore, multiple paths are searched when QoS requirements are tighter. This helps the protocol to tolerate the imprecise state information. Along with the

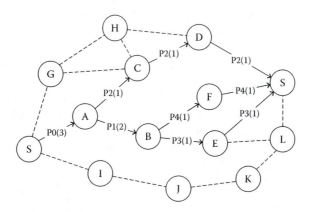

Figure 22.4 Route discovery in TBR.

above-mentioned advantages, TBR has a few disadvantages, too. The protocol uses a proactive approach to collect routing information that requires large maintenance overheads and does not scale well with the size of the network. Second, assuming CDMA over TDMA at the MAC layer for channel accessing requires time to be divided into fixed-size slots, and each group of nodes is assigned a different spreading code. Spreading codes can be assigned statically at the start-up time, or they can be assigned dynamically. Static assignment is not possible in MANETs because of the mobile nature of the nodes. On the other hand, assigning spreading code and synchronizing nodes with the start of the time slot require some global entity. That is not possible in MANETs because of their uncentralized nature. One solution is the use of synchronization signaling, which incurs extra overheads on the network. Either way, authors do not specify how time slot and code allocation is achieved.

22.5.1.2 Synchronous Signaling and Node State-Based Routing (SSNSR)

According to Stine and de Veciana [11], the approach taken by most QoS routing solution developers to adapt the wireline paradigm to an ad hoc network is not adequate. Authors argue with the fact that in a wireline paradigm, communicating nodes are linked with physical entities called links. On the other hand, nodes in wireless networks share geographical space and frequency spectrum with each other. This implies that communicating node pairs are influenced by nodes in their carrier sensing range. To address these problems, synchronous collision resolution (SCR) is used at MAC, and node state routing (NSR) is used for routing in [11].

In NSR, each node maintains a routing table that contains all necessary information about itself and the nodes in its CS range. The basic information at each node consists of its IP address, residual power, and input/output packet queue size. Routing protocols assume that at each node location information is provided by GPS and that the radios have the ability to calculate the path loss to their neighbors. This provides more state information to nodes, such as their relative speed, direction of movement, and the strength and quality of signals. With the above state information, each node builds location and path loss maps.

Each node exchanges its state information periodically with neighbor nodes. They may exchange that information with their neighbors if certain parameter value changes are above some threshold. Routing functions are performed using location and path-loss maps that provide sufficient information to determine link connectivity and network topology. After inferring connectivity using the QoS metrics between pairs of nodes, a node can easily calculate QoS routes using Dijkstra's algorithm. A routing metric assigned to a link is estimated using states of both ends of that link plus the state of their neighbors.

There are many advantages of using node states instead of link states. (1) A node can store multiple routing tables that contain routes as per different QoS requirements. For example, a routing table created using energy conservation metric may be set as default, and when a multimedia session with a delay/bandwidth constraint is admitted, the table that supports the delay/bandwidth constraint is used for routing. (2) There is no need to send many link update massages when a node moves, as in other protocols. Instead, only that node's state table is updated in nodes that come in its CS range. Furthermore, this protocol shows great potential to satisfy multiconstrained data sessions due to the information available in node states.

Despite the above advantages, the protocol has many drawbacks. First and foremost, it relies on accurate location information through GPS that limits its use to devices that are capable of being equipped with these kinds of devices. Second, as described in [1], this protocol also relies on TDMA/CDMA protocols for bandwidth estimation and reservation, which have their own disadvantages in ad hoc networks as mentioned in the previous protocol. Third, routing information is gathered and maintained using a proactive approach with overheads that increase with the size of the network.

22.5.1.3 Signal-to-Interference and Bandwidth Routing (SBR)

SBR [12] is a TDMA-based reactive routing protocol, that is, channel capacity is measured in terms of time slots. SBR explicitly fulfills both throughput and signal-to-noise ratio (SNR) (by minimizing the bit error rates on links) requirements for different multimedia sessions. SBR achieves low bit error rates by assigning adequate power to intermediate nodes between transmitter–receiver pairs. The maximum achievable SIR is limited by limiting the transmission power of nodes. The power assignment schemes used in SBR support searching paths that satisfy SIR requirements and reduce the level of interference. This makes it different from our previous QoS solutions aimed at merely fulfilling a single QoS constraint at a time.

In SBR routing protocol, a node sends an *RREQ* packet to search a route only when it has data to send. This makes SBR an on-demand routing protocol like AODV [13]. The difference in SBR and other QoS routing protocols lies in the quality of routes selected for data transmission. The *RREQ* message contains information about required bandwidth and SIR. Every link on a discovered route has to satisfy these requirements. The destination responds using *RREP* messages after it has received the *RREQ* packets. The *RREP* message contains information about the estimated power so that the correct power can be set in the slots used for data transmission. If it is not possible to find a single path that satisfies the QoS requirements of multimedia users, SBR has a backup capability to find multiple paths that combined together satisfy the given QoS requirements. The backup method of the multiple-path search might decrease the session rejection rates. A simple example of the operation of SIR and bandwidth-guaranteed routing is depicted in Figure 22.5. Assume *S* is the source node, and *T* is the destination node. A section of each node's time slot schedule is shown next to it. Dark shading indicates a slot used for transmission, and light shading indicates a slot for reception. Unshaded slots are used by other data sessions. In this example, the bandwidth requirements for the source for its data session are two time slots. The route discovery and time slot assignment phase is over, and at the source, slots 1 and 2 are assigned for data transmission. However, each of the two possible next hops has only two slots to spare, and one must be used for receiving the source's transmission. The two available routes are used to serve the session's bandwidth requirement cooperatively by dedicating one time slot each to transmission. The labels T_1 and T_2 illustrate the fact that different transmission powers are used in each time slot. As in

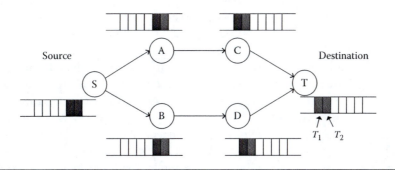

Figure 22.5 Multiple route discovery in SBR.

previous TDMA examples, forwarding nodes must be careful not to transmit in a slot in which their upstream node is receiving.

This protocol presents a good example of a simple, multiconstrained routing protocol. But the assumption that a global clock is present for the synchronization required in time-slotted systems limits the usefulness of the proposal. Furthermore, the signaling process used for a multiple-path search consumes a significant amount of channel bandwidth. The effect of topology change, an inherent MANET property, is not discussed in detail. All the protocols discussed above rely on contention-free MAC protocols for resource estimation and reservation. This limits their use in real-world applications, such as disaster recovery schemes, military operations, etc., that do not support contention-free MAC protocols.

22.5.1.4 Reliable Multipath QoS Routing Protocol

Reliable multipath QoS routing (RMQR) protocol in [14] proposes a mechanism that searches a reliable multipath (or unipath) QoS route. These selected routes satisfy the bandwidth requirement of admitted data sessions. Like SSNSR, RMQR protocol also uses GPS to obtain the location information of the nodes. By continuously updating the location information through GPS, a node can also calculate its average velocity and direction of movement. When a source node requires a path, it sends an *RREQ* packet together with the information about its location, direction, and average velocity. The next hop of the source node will use this extra information to predict the link expiration time (LET) between itself and the source node. The information of link expiration together with the number of hops over a path is used to select a low-latency and high-stability route.

The authors assume that the MAC layer is implemented using the CDMA-over-TDMA channel model same as defined in [15]. In CDMA-over-TDMA, multiple data sessions can share the same TDMA slot using different spreading codes. The same spreading code can be used only if the two nodes have a hop distance greater than two. The link bandwidth in this channel model can easily be defined by simply knowing the number of free slots between two nodes. This link bandwidth is further employed to calculate the path bandwidth.

The route discovery process of RMQR protocol is described in Figure 22.6. We assume that the number over the arrows represents LET and the numbers within the boxes are the available slots between two nodes. Let the bandwidth requirement of the application be two slots. Node S is the source, and node D is the destination. When node S receives an *RREQ* packet, it searches its routing table to find a suitable path for node D. If there is no path for D in node S's routing table, S broadcasts the *RREQ* packet (refer to Figure 22.6a). Upon the reception of the *RREQ* packet, nodes A, C, and E check the signal strength of the packet and drop the packet if the signal strength is less than a predefined threshold value. Finally, when all *RREQ* packets are received, node D calculates three feasible paths: ($S{\rightarrow}A{\rightarrow}B{\rightarrow}D$), ($S{\rightarrow}C{\rightarrow}G{\rightarrow}D$), and ($S{\rightarrow}E{\rightarrow}F{\rightarrow}G{\rightarrow}D$) (refer to Figure 22.6b). The path bandwidth of all the feasible paths is calculated using the information stored in *RREQ* packets. Thus, the path bandwidth of these paths is eight, five, and six time slots, respectively. RMQR uses the signal strength (i.e., the ratio of LET and hop counts) of all the links along a feasible path to calculate the final route between S and D. The path ($S{\rightarrow}C{\rightarrow}G{\rightarrow}D$) has the maximum signal strength and minimum hop counts, so node D sends an *RREP* packet back to source S along this path. To eliminate the problem of hidden nodes, the path with twice (i.e., in this case, it is four times) the required bandwidth is selected in a route discovery process. If the bandwidth requirement is increased and the application needs four free slots, the current route is no longer sufficient to provide the required bandwidth (i.e., eight). To handle these kinds

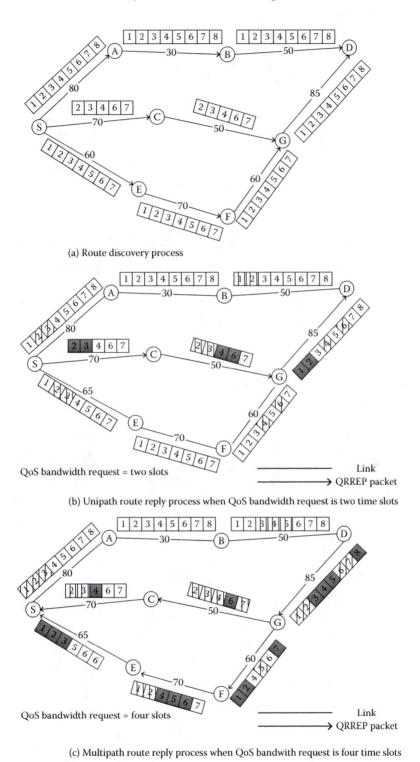

(a) Route discovery process

(b) Unipath route reply process when QoS bandwidth request is two time slots

(c) Multipath route reply process when QoS bandwith request is four time slots

Figure 22.6   Route discovery process of RMQR.

of situations, RMQR uses multiple paths that combined together can satisfy the requested band-width guarantees of the application (refer to Figure 22.6c).

The main advantage of RMQR is its robust multipath (or unipath) route discovery scheme. Furthermore, link breakages are predicted before links go down, and traffic is rerouted over new links. The protocol has several disadvantages: (1) the overhead caused by control packets used for estimating link-expiration time and signal strength is very high, and (2) use of GPS for location information limits the usefulness of the proposed protocol to devices with GPS.

Finally, we can conclude from the fact that MANETs have no centralized control, the key requirement for the use of contention-free MAC protocols. Recently, researchers have shifted their focus from QoS provisioning solutions that rely on contention-free MAC to the approaches that use contended MAC protocols.

22.5.2 QoS-Aware Routing Solutions Based on Contented MAC Protocols

In this section, we present the detailed review of some novel QoS-aware routing solutions pre-sented in the literature, which are based on contended MAC protocols.

22.5.2.1 Distributed Admission Control for MANET Environment

A QoS framework that supports the delay, bandwidth, and jitter requirements of the application using cooperation from different layers is proposed in [16]. The proposed framework is modu-lar and offers great flexibility by allowing plugging in to different protocols on different layers. Optimizations are performed using interactions between layers. For evaluation, the protocol uses H.264/AVC video traces [17] to simulate the video on source nodes. The quality of received video stream is measured in terms of the SNR so that the efficiency of the proposed scheme can be defined in the form of quality of experience (QoE) of end users.

The main component of the proposed QoS framework is distributed admission control *for MANET* environment (*DACME*). It uses an end-to-end probe-based admission control mech-anism to estimate QoS requirements specified by multimedia applications. Figure 22.7 shows

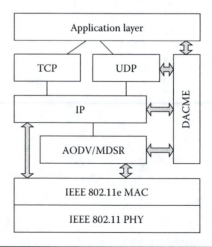

Figure 22.7   Modular QoS framework with cross-layer interactions.

different architectural elements and their interdependencies (represented by solid lines). The basic requirement of the use of the proposed scheme is that the source and destination nodes of a QoS flow must have a DACME agent running. There are two main modules of DACME (refer to Figure 22.8). First, is the *QoS measurement module,* responsible for measuring the QoS metrics on an end-to-end path. Second, is the *packet filter module* that blocks all traffic not accepted into the MANET according to the measurements provided by the QoS measurement module. An application that wants to send data first registers itself with the DACME agent by specifying its QoS requirements. Once the registration is successfully completed, the QoS measurement module is activated and probes are sent toward the destination to measure the available resources along the path. The destination uses the interarrival time of the bandwidth probes for measuring the path bandwidth, and end-to-end delay is calculated by taking half of the round-trip time of a probe.

The DACME uses information about the destination path stores in the node's routing table to send probes. If no path is available for a destination, no probes are sent, and this is notified to the application. The network layer also helps to rediscover the QoS paths when the route to destination is changed in the node's routing table by notifying it to the DACME agent. DACME is independent of the routing scheme used at the network layer, but simulation results show that this framework gives the best results with reactive multipath routing protocols. The protocol uses a multipath dynamic source routing (MDSR) routing protocol to calculate multiple QoS paths toward the destination. The MDSR protocol splits the traffic coming from the data session over at

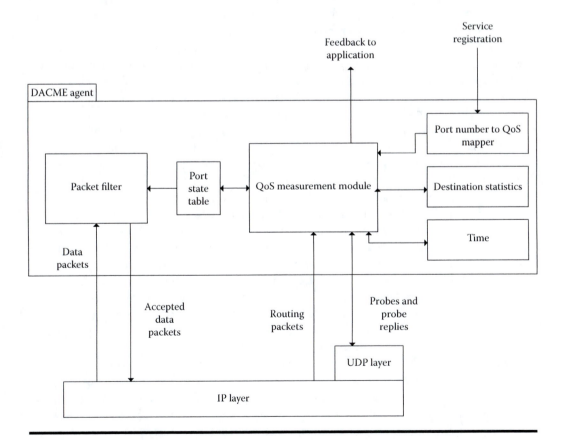

Figure 22.8 Functional block diagram of the DACME agent.

least two routes to increase the robustness. DACME is optimized to work with the IEEE 802.11e MAC specification that maps the TOS values of incoming packets with its access categories. In [18], Chaparro et al. extend the DACME for scalable videos. The protocol shows fairness between video sessions in terms of resource consumption. It also avoids network congestion through efficient admission control and guarantees an acceptable video quality at the receiver.

DACME is a stateless QoS provisioning scheme, that is, no information about ongoing traffic flows is stored at intermediate nodes, saving their memory and processing power. Intermediate nodes need not implement any of the protocol's functionality because DACME agents are running on source and destination nodes only. The biggest disadvantage of being stateless is that reservation of resources is not possible at intermediate nodes as required for multimedia streams. DACME is a good example that satisfies the applications' multiconstrained QoS requirements. This comes at the cost of large overhead caused by multiple probes sent by source nodes to measure end-to-end bandwidth, delay, and jitter. Multiple paths increase the fault tolerance ratio and robustness of the proposed scheme. But as the number of paths to the destination increases, a higher number of probes are required, which decreases network performance.

22.5.2.2 Staggered Admission Control Protocol

Existing QoS-aware admission control-based routing solutions neither take into account the effects of collisions nor do they consider the effect on the collision rate of traffic admission. Staggered admission control protocol (SACP) [19] is a bandwidth-constrained admission control scheme for multihop ad hoc networks like one proposed in [6], based on the concept of staggered admission. SACP tries to avoid congestion in the network, which helps to decrease packet drop ratio because of collisions and intermediate router queue overflows. In [19], congestion in the network is avoided at the first place by controlling the flow admission rate and routes used for QoS transmissions.

The basic idea of the proposal is to admit the QoS-constrained sessions gradually. In SCAP, the traffic session should begin with the lowest possible data rate and gradually increase the data rate to the requested required rate over the period of time. SACP is designed as an extension of QoS-dynamic source routing (DSR) [20]; only the admission control mechanism is modified to minimize the collision rate. Bandwidth is calculated at each node by monitoring the channel idle time ratio (CITR) with the help of MAC protocol.

This protocol performs admission control in three steps. In the first step, a bandwidth-constrained path is discovered by considering the intraroute contention. After receiving route replay messages from the destination, multiple node-disjoint paths are formed and stored in a source routing table. The second step consists of estimating the capacity of nodes that lie in the CS-range of intermediate nodes in the route. Information about the session is also stored at nodes in the CS-range of intermediate nodes via admission control request (ACR) messages. If no nodes lie on the CS-range, reject the session; it reaches the third step when the session is partially admitted. For the next few seconds, the data rate increases gradually; during this time, it may still be possible that the session is rejected. The data rate reaches the required rate in a small fraction of time if no collisions are heard in the network.

Figure 22.9 shows the effect of CS-range nodes on the admission control scheme. The small solid-lined circles are nodes; the medium-sized circles represent the transmission range coverage region, and the dashed largest circle represents node X's average CS-range coverage region. Note that all nodes have the same transmission and CS-ranges. Therefore, node X's transmission causes busy channel states to all nodes within its CS-range and vice versa. When node X broadcasts an ACR message, it is received and forwarded by all its neighbors. As shown in Figure 22.8, there

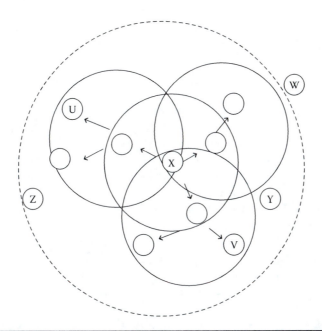

Figure 22.9   Role of CS-range nodes on admission control scheme.

are no relay nodes to reach to node *Y*, and therefore, its capacity cannot be estimated by node *X*. To avoid such situations, one method is to increase the transmission power of node *X*, so it can directly reach all its CS-range nodes. The disadvantage of this approach is a much larger level of interference, which is imposed on all nodes during transmission, possibly to increase collision rate. Also, with a time-to-live field of two hops, the ACR messages are not able to reach all nodes in the CS-range even if there are suitable relay nodes available; for example, see node *Z* in Figure 22.9. When increasing time-to-live to three hops, for example, too many nodes may be reached. This causes a problem like if node *W* had insufficient bandwidth to support the session, it would reject it. This is a wrong decision because node *W* would not actually be significantly affected by node *X*'s transmissions.

In the SACP route, failures are handled in two ways. First, a portion of each node's bandwidth is reserved for ARP messages and DSR route requests. Second, each packet carries a small header extension, which stores the traffic session source's view toward the residual capacity of intermediate nodes. If, because of mobility or route failures, this view of the source changes for any intermediate node, an update message is sent immediately. If a route fails, the source node of the failed session reroutes the traffic on some known route with the highest bottleneck bandwidth calculated from the routing cache built by QoS–DSR during the route discovery process. In [21], SACP is extended to deal with route failures using efficient backup route discovery and maintenance schemes.

There are two advantages of this CAC scheme over other CAC protocols. First, the effect on collision rate in the network when a new session is admitted is considered, which is ignored by other CAC protocols. Second, the efficient backup route discovery and maintenance schemes greatly increase the protocol's fault-tolerance rate and make it more suitable for mobile and large-size networks. However, because of its staggered admission control session, this scheme is not suitable for small sessions. Furthermore, during the third step of admission control, if a temporary burst of overhead causes a drop in a session's throughput, the session is falsely rejected. Finally, the weaknesses of using only two hops for time to live as discussed in Figure 22.8 should be addressed.

22.5.2.3 Priority-Based Distributed Flow Admission Control

In [22], Pei and Ambetkar propose a priority-based, distributed flow admission control (PDAC) protocol over the flow-state extensions of DSR protocol [23]. The protocol provides guaranteed throughput to the admitted sessions using a new DSR option called the "admission control option" and a scalable transmission rate reservation protocol. Source nodes can admit or discard a traffic session based on global knowledge (such as the traffic flow priority) and local knowledge (such as interference and effective transmission rate).

The route toward the destination is discovered on demand using a DSR route-recovery process. The protocol assumes that the information about available bandwidth at each node is provided by the MAC layer using the CITR. PDAC assigns a priority to each admitted session based on its QoS requirements. The proposed protocol works in two phases. The first phase is known as the end-to-end flow establishment phase; in this phase, flow establishment packets (FEPs) are forwarded on the known route toward the destination. A node will only forward FEPs to the next hop if it has sufficient available channel capacity or if it can make sufficient capacity by preempting the already-going sessions whose priority is lower than the current requesting session. The effective transmission rate for a traffic session can be reserved if the FEP reaches the destination node. The second phase of the PDAC scheme is called the "resource reservation phase," and during this phase, the FEP response packet is sent from the destination toward the source. This triggers the preemption of low-priority traffic sessions along the route if required, and the corresponding sources are notified.

The main advantage of PDAC is its simple implementation with the existing DSR protocol and its low control overhead. Another advantage of the proposed scheme is that it allows the coexistence of nodes, which support or do not support the PDAC. But the authors do not account for the interference caused by CS-neighbors of nodes along the QoS path to keep the overhead low. This may result in wrong admissions. Furthermore, it is possible that after admission a session can be forcefully rejected because of some other high-priority session in the middle of its transmission. This limits the applicability of PDAC protocol.

22.5.2.4 Adaptive Admission Control Protocol

The adaptive admission control protocol (AACP) [24] is an accurate, reactive, low-cost admission control protocol, which uses robust and accurate available resource estimations and prediction mechanisms for relevant admission decisions. AACP accurately quantifies the available bandwidth at each node. It avoids interference with ongoing sessions and adapting to QoS violations caused by interflow interference, intraflow interference, and mobility. The basic idea of AACP is threefold. First, through a data aggregation process, an accurate low-cost signaling scheme is used to retrieve the available bandwidth of CS nodes. Second, the contention count for a path is calculated using routing metrics and topology parameters, which help to adapt the path's roughness. Third, to prevent the loss of QoS traffic along a session because of mobility, efficient adaptation mechanisms are used.

Based on the approach in [24], the most accurate estimation of the impacted region (i.e., the area covered by a node's CS-range) is obtained by considering nodes within a three-hop area. AACP uses hello messages, exchanged between nodes, for connectivity awareness to disseminate

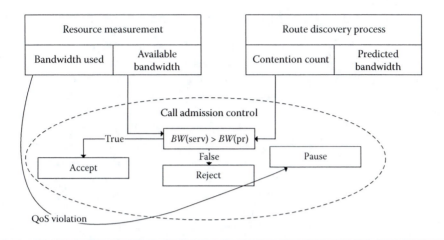

Figure 22.10 Functional block diagram of AACP.

nodes' available bandwidth. The hello messages can travel only one hop, but the residual bandwidth of one-hop neighbors are carried, and therefore, each node is aware of the minimum bandwidth available in its two-hop estimated CS-range. In order to avoid interflow interference, each node estimates its available bandwidth, which is equal to the smallest of the CS nodes' available bandwidth. AACP admission control structure is depicted in Figure 22.10. AACP utilizes a QoS–AODV approach for route discovery with available bandwidth determined as described above.

AACP handles the QoS violations caused by node mobility in the following manner: Whenever QoS packets occupy a significant amount of the packet interface queue, one preselected source node is notified. On receiving this information, the source node pauses the data session that requires the highest bandwidth. In doing so, the largest number of reserved resources are freed, which minimizes the risk of pausing another QoS data session. However, the paused flow might face difficulties to readmittance because of its large bandwidth requirements. In order to avoid QoS data sessions being paused too often because of mobility and congestion, AACP requires that source nodes increase the bandwidth requirements of readmitted sessions right after a loss of QoS guarantees takes place.

The major advantage of AACP is its robust mechanism that deals with congestion caused by route failures that disrupt the minimum number of QoS sessions. Second, a source node causes no delay in testing the available resources of its CS nodes, resulting in lower session admission time as compared to other admission control protocols. It also provides an accurate available bandwidth estimation, which lacks in other QoS protocols. Furthermore, because of the exchange of hello messages, which contain information of up to three-hop neighbors, these hello packets implement a form of proactive routing protocol. This decreases the route discovery and session admission time of the data sessions if the destination is within the range of two hops from the source node. However, this protocol also suffers the problems discussed with Figure 22.9, resulting from inappropriate selection of hop counts to calculate the contention count of a path.

22.6 QoS-Aware Routing Solutions Independent on MAC Layer

Routing protocols proposed for provisioning QoS in MANETs that are independent of the underlying MAC layer are reviewed in this section.

22.6.1 Route Stability-Based QoS Routing Protocol (RSQR)

Because of the inherently dynamic nature of ad hoc networks, it is difficult to ensure a data path will be valid for longer periods of time. The QoS routing approach proposed in [25] is an extension of QoS routing with bandwidth and delay constraints. RSQP is a simple model that ensures route stability by computing link stability along the route using the received signal strength of two consecutive packets received from a neighbor. Route stability-based QoS routing protocol (RSQR) uses an on-demand routing scheme (i.e., enhanced AODV) for route discovery. Route discovery is done by adding some extra fields carrying route stability information in route request/ reply messages. Based on the route stability information, the route discovery process selects a route with higher stability among all possible routes between a source destination pair. Furthermore, CAC is included based on signal strength, which further enhances the routing performance. The protocol also detects QoS violations and uses necessary recovery processes.

RSQR proposes a route stability model (RSM) with the following characteristics: (1) it uses signal strength and node mobility to measure the link stability, (2) the signal strength estimation

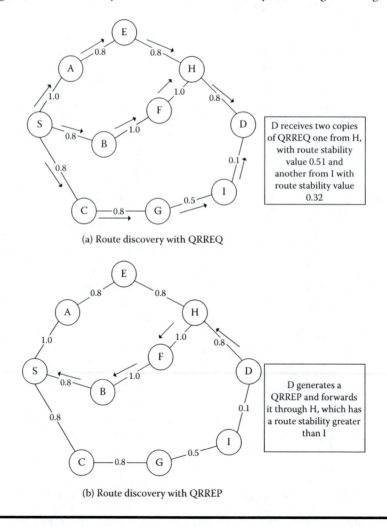

(a) Route discovery with QRREQ

(b) Route discovery with QRREP

Figure 22.11 Route discovery process of RSQR.

is provided by MAC layer protocol, and (3) the signal stability value is within the range of [0,1]. Though it is very difficult to calculate the exact value of link scalability, a scheme to calculate the relative stability using the recent and current values of signal strengths is used. When a data session with desired QoS metrics requests a route, the routing protocol at the source node (in this case AODV) broadcasts a QoS route request (QRREQ) packet. Intermediate nodes, upon receiving the QRREQ, check the received signal strength, and if it is less than some predefined threshold, the QRREQ is dropped. If the received signal strength is good, then the QoS requirements of the session are checked against the available resources to perform admission control. After this processing, the route stability and end-to-end delay fields of QRREQ and the route request forwarded table (RTF) are modified, and the message is broadcast. When the first QRREQ is received by the destination, it waits for some predefined time to receive more QRREQ packets belonging to the same session. Afterward, this destination chooses the route with the maximum stability and sends a QRREP along that route. Figure 22.11 shows the route discovery process of RSQR. *S* needs a QoS route to node *D*, so *S* initiates route discovery by flooding a QRREQ. The values on each link in Figure 22.11 represent its link stability. When node *H* receives a QRREQ message from *B* and *E* nodes, it forwards only one QRREQ based on the route stability value in its RFT. Finally, node *D* receives the two QRREQ packets from *H* and *I* with different route stability values. Node *D* generates a QoS route reply (QRREP) packet and forwards it through *H*, which has a route stability value (0.51) that is higher than the route stability (0.32) for QRREQ through *I*.

RSQR uses a simple approach to discover and reserve the available bandwidth at each node. Less control overhead is caused by the use of extra fields in the route request/reply messages. The routes are calculated based on the link stability of all links over the route. This decreases the probability of link failures in cases of highly mobile environments. On the other hand, the resource estimation method is not discussed in the paper. Furthermore, if an active link breaks, all the packets on that route are dropped because of the reestablishment time, which is very high because of the reactive route recovery method.

22.6.2 Efficient Cluster-Based Routing Protocol

Tao et al. [26] propose an efficient cluster-based routing protocol (ECBRP) to support QoS for real-time multimedia streaming in MANETs. The basic goal of this protocol is three-fold. First, design an efficient clusterhead selection algorithm by taking account of the node mobility and connectivity. This reduces the flooding of *RREQ* packets during the route discovery process. Second, a link-failure detection scheme is proposed, which is able to identify whether the packet loss is a result of congestion or mobility. Third, enhance the routing protocol so that in case of packet loss it can adjust its transmission strategy dynamically as the network condition changes. The protocol uses the above three mechanisms together to provide QoS to real-time multimedia streaming in terms of decodable frame ratio and end-to-end delay.

The traditional CBRP [27] uses only a node's ID to elect clusterheads. If a node with a low ID happens to be highly mobile and possesses weak connectivity to other nodes in the network, this causes severe reclustering periodically when it moves through other clusters frequently and, because of its low connectivity, the number of clusters increases in the network. To resolve the above-mentioned problems, ECBRP uses a node's mobility and connectivity information along with its ID to make clusterheads in the network. The protocol assumes that each node is equipped with GPS. GPS provides the location information, which is used to calculate a node's mobility and connectivity degree. Each node sends hello messages to its neighbors periodically. The format of a hello message with additional fields in a neighbor table is shown in Figure 22.12. With the help

Neighbor ID	Neighbor status	Link status	Time	Connectivity degree	Relative velocity	Mobility degree	Distance

Figure 22.12 Neighbor table with extended fields.

of the additional information provided in the node's neighbor table, only nodes with low mobility and high connectivity are able to make clusters—unlike CBRP, ECBRP is able to distinguish that a packet loss is a result of congestion or node mobility. In ECBRP, when a packet drop occurs at a node, it checks if the next hop is out of range; then the packet loss is a result of mobility or else it is because of congestion. With the help of location information provided by GPS, a node can determine whether its next hop node goes out of its range or not. Finally, to alleviate the congestion, the protocol defines an adaptive packet salvage strategy. A two-bit unused field in an IP packet header is used to differentiate whether the packet belongs to a multimedia application or a cross-traffic application. In case of a broken link resulting from mobility, the packets belonging to cross-traffic and B-frames (which causes the least impact on decoding the video) are dropped first, and the other packets are salvaged if the node can repair the link locally. If the link-broken mechanism detects congestion, it salvages all the packets until a new route is discovered through local repair, and if local repair is not possible, it drops B-frames and continues transmitting other traffic on the current route.

The flooding of *RREQ* messages is limited by allowing nodes to send or forward *RREQ* messages only to clusterheads. This reduces the control overhead during the route discovery and maintenance phases. On the other side, when the network scales and node mobility increases, the overhead of cluster formation/reformation and maintenance is also increased. Furthermore, the use of GPS limits the applications of the proposed protocol to the devices equipped with GPS.

22.6.3 QoS Mobile Routing Backbone over AODV

The basic goal of QoS mobile routing backbone over AODV (QMRB–AODV) [28] is to select a route that can support the QoS requirements of a data session by dynamically distributing the traffic within the network using a mobile routing backbone (MRB). Nodes in the network are considered heterogeneous and have different characteristics. This protocol classifies nodes in MANETs as either QoS routing nodes, simple routing nodes, or transceiver nodes. Only nodes having sufficient resources in terms of remaining battery power and link stability will take part in MRB formation. Once the MRB is created, the routes toward the destination are searched on this MRB only.

The basic aim of the QMRB–AODV is to identify those nodes whose communication capabilities and processing characteristics will allow them to take part in the MRB formation and optimize the routing process. In [28], four metrics that support QoS provisioning in MANETs are used to differentiate nodes in the network. The first metric is known as *static resources capacity* (SRC), which is estimated at each node using metrics, such as the packet queue length, the processing power, the available bandwidth, and the battery power. The second metric used is *dynamic resource availability* (DRA), which is estimated at any time by taking the ratio of the quantity of resources that is currently being used with the SRC. The third metric defined is the *neighborhood quality* (NQ), that is, the number of nodes in the neighborhood that are able to forward packets. The fourth and most important is the *link quality and stability* (LQS). To measure the link stability, each node records the lifetime of the links it encounters. Whenever a node detects a new link, it starts a timer that ends when the link is broken. On the other hand, link quality is measured by estimating the signal strength of received packets from the MAC layer. Using the LQS, the links in the network can be labeled as high-quality links (HQLs), low-quality links (LQLs), or unusable links.

Optional hello messages of AODV are used to disseminate the values of the above-mentioned four QoS metrics in the network. A QoS route discovery is initiated when a node's network layer receives a route request message with specified QoS requirements from the application layer. If the node with an *RREQ* message is the source node, and the destination is present in its neighbor table, then the packet is forwarded to the destination. If it is an intermediate node, then the required bandwidth is reserved, and this information is sent to the destination with the *RREQ* message. However, if the neighbor list of a node does not have an entry for destination, the node selects neighbors from the list with sufficient resources to forward the *RREQ* message. The node also makes a new entry in the routing table, and bandwidth resources are reserved for the data session. On the other hand, if the source node encounters that none of its neighbors can satisfy the bandwidth requirements of the requested flow, it tries again after some time. If the node is an intermediate node, it simply waits for another *RREQ*. The route selection process can be initiated as soon as the destination gets its first *RREQ* message. The destination chooses the route with the lowest number of hops and the highest allocated bandwidth if it is serving a QoS flow.

The advantage of the proposed protocol is twofold. First, the congestion is handled by directing the traffic through less-congested areas of the network that are rich in resources. Second, the flooding of *RREQ* messages is limited to only those nodes that have sufficient resources to handle a QoS session. Apart from the above advantages, there are several drawbacks to proposed protocol. The hello messages used for disseminating QoS metrics incur extra overhead in the network. The initial time required for gathering information about the QoS metrics and its dissemination is very long. Furthermore, the response against route failures is not discussed, which is very important in MANETs because of their dynamic topology.

22.6.4 Application-Aware QoS Routing

A rather unique mechanism is proposed in [29], which satisfies an application's bandwidth and delay guarantees. The approach is unique because instead of using the MAC protocols to obtain information about QoS metrics, it uses a transport layer protocol. Application-aware QoS routing (AAQR) uses a real-time transport (RTP) protocol to estimate the state information of the nodes and provide a stable path based on that state information. RTP control packets are used to calculate delay and jitter between two adjacent nodes. The difference between timestamp values on transmission and receipt of RTP control packets between two nodes is used to statically estimate the delay between those nodes. To estimate the residual capacity of each node, first the raw bandwidth at each node is calculated. Then, the bandwidth used by the ongoing RTP sessions is subtracted from the nodes' raw bandwidth to calculate the residual bandwidth on that node.

The routes are discovered reactively, although the authors have not discussed the route discovery process in detail. A subset of routes that satisfies the delay requirements of the multimedia application is formed. From this subset, a sub-subset is selected, which consists of the routes that satisfy the bandwidth as well as the delay requirements. Finally, from this sub-subset, a path with the highest bandwidth and lowest delay is selected for routing that satisfies the QoS session requirements. In AAQR, the delay is considered the most important QoS constraint. If there is insufficient available bandwidth for a QoS data session, the protocol can select a quasi QoS route that satisfies the delay requirements only.

The major advantage of the AAQR is that no extra overhead is incurred for finding the QoS metric-satisfied routes. The existing transport layer RTP packets are used to estimate the QoS metrics along a path. Additionally, the protocol provides multiconstrained routing paths. However, the use of RTP at the transport layer limits its usefulness over the range of application scenarios.

22.7 Future Work

Following the work summarized in this chapter, we still believe that there are unresolved issues to be addressed to improve the performance of QoS-aware routing protocols. Even in low-mobility scenarios, there is a way to go in the area of QoS-constrained routing before perfect CAC is performed. Protocols proposed in [18,19] place great emphasis on the CAC mechanism, which is definitely very important. Also, CAC protocols [30] that provide security with the required QoS guarantees are preferred as security is an important issue in MANETs. In contrast, existing solutions often ignore or downplay the importance of session maintenance and completion, which is maintaining the QoS routes as long as the application data session requires. Based on a survey of previous research articles, we have identified some of the open research problems as listed below:

- From the point of view of an end user, session completion is more important than session admission. To increase the probability of session completion, it is required to make appropriate initial admission decisions. This decision-making is still vague and needs to be application-dependent. The accuracy of initial admission decisions depends on how close the admitting nodes' view of network resources matches reality. This also depends on how quickly and efficiently the QoS-aware solution has adapted to the most recent resource and/ or topology changes in the network. Furthermore, fast local route repair without disturbing the ongoing QoS sessions requires additional investigation because it can improve the session completion rate and protocols' robustness against mobility.
- A previous study reveals that one of the major challenges in QoS provisioning over MANETs is the unpredictable radio channel. Most of the QoS solutions consider that the underlying physical channel is perfect, ignoring the effects of shadowing and multipath fading. Furthermore, most protocols assume that transmission, collision, and the CS-range of nodes are fixed, not realistic if shadow-fading and multipath fading are considered. When real-world test beds and network emulation solutions are used to investigate the performance of proposed solutions, these modeling inaccuracies come into the light. Therefore, studying and analyzing the impact of a more realistic physical layer model [31] on QoS routing solutions performance constitute another area of future work.
- The type of traffic application used for traffic generation greatly affects the admission control process of QoS-aware solutions. With the exception of a few QoS routing solutions that utilized the real-time video traces produced from different video CODECs, such as H.264 and MPEG-4, most of the QoS solutions evaluate their performance with constant-bit-rate (CBR) data sessions. With CBR data sessions, the admission decision-making features of a protocol to be tested under various loads still may not represent the performance that would be achieved with real-time traffic [32]. More work that models real-time traffic is required. The random movement-based mobility models, such as the random waypoint model used in most of the proposed protocols, do not accurately represent mobility patterns for many wireless networks. Again, QoS solutions with more realistic mobility models for various scenarios would be useful.

- We have seen that there are many solutions available in the literature that satisfy the multi-constraint QoS requirements of data sessions. Such methods have limited usability because of the control overhead and energy cost involved during QoS state dissemination. Future works should consider the various networking environments and topologies while optimizing the multiconstraint routing in terms of type (throughput, delay, PDR, etc.) and level of requirements.
- Most of the previously studied QoS protocols' performance evaluation is done on metrics, such as throughput, delay, and PDR. However, the received quality of multimedia traffic is greatly affected by delay variance and peak SNR (PSNR). In the future, the QoS solutions should evaluate their effectiveness in terms of these metrics. Furthermore, in the case of multimedia traffic, providing QoS is not enough, but at the same time, QoE should also be provided. In MANETs, providing QoE is much more challenging than providing QoS because of a finite channel bandwidth. The routing protocols should find a way to manage the available resources efficiently while maintaining user satisfaction.

Researchers working toward developing suitable and efficient solutions for provisioning QoS over wireless networks should keep the above issues and problems in mind.

22.8 Summary

We have provided a specification and classification of the challenges involved in QoS provisioning over MANETs and the mechanisms proposed to address them. Because most solutions proposed in the literature are based on an underlying MAC layer protocol, we classified the existing QoS-aware solutions into two board classes: (1) those that are dependent on MAC protocols for QoS provisioning and (2) those that are independent of MAC protocols. Because of the strict constraints on available resources, frequently occurring transmission and route errors, and dynamic topology, MANETs are a challenging environment for streaming applications, such as news on demand (NOD), video conferencing, and surveillance systems. This chapter mainly concentrated on the two most important components of a system for providing QoS guarantees in MANETs: (1) the cross-layer design (CLD), which actively exploits the dependence between various protocol layers to enhance performance; and (2) the CAC design, which admits only those data sessions whose QoS requirements match with the available network resources without violating the previously admitted data sessions.

The purpose of this chapter is to provide an overview of high-quality theoretical and practical work on the QoS solutions used in MANETs. The protocols discussed in this chapter were selected in such a way as to highlight different approaches to QoS provisioning in MANETs. We summarized the basic functionality, advantages, and shortcomings of these protocols in order to highlight the variety of methods proposed and the current trends from the designers' point of view. We have also identified issues for future research work.

Key Terms and Definitions

Average throughput: Average throughput for a network is the ratio of the sum of throughput of all destination nodes to the total number of destinations.

Average end-to-end delay: Average end-to-end delay for a network is the ratio of the sum of average end-to-end delay at each destination node to the number of destination nodes in the network.

Average jitter: Average jitter is the ratio of total packet jitter for all received packets to the number of packets received.

Packet delivery ratio: Packet delivery ratio is the ratio of total packets sent from the source to total packets received at the destination.

Radio link: A radio link is a logical wireless entity between two nodes that lie within each other's transmission range.

Channel contention: A broadcast channel access method.

CS-range: The CS-range of a node consists of all the nodes belonging to its carrier sensing range.

Quality of service (QoS): QoS is a collection of constraints that a connection must guarantee to meet the requirements of an application.

Quality of experience (QoE): QoE is a subjective measure of a customer's experiences with a vendor.

Proactive protocols: In proactive routing protocols, the routes to all the destinations (or parts of the network) are determined at the start-up and maintained by using a periodic route update process.

Reactive protocols: In reactive protocols, routes are determined when they are required by the source using a route discovery process.

References

1. Chen, S., and K. Nahrstedt. "Distributed quality-of-service routing in ad hoc networks." *Selected Areas in Communications, IEEE Journal,* Aug. 17, 1999, 1488–1505.
2. Chakrabarti, S., and A. Mishra. "QoS issues in ad hoc wireless networks." *Communications Magazine, IEEE* 39, (2001): 142–148.
3. Crawley, E., R. Nair, B. Rajagopalan, and H. Sandick. A Framework for QoS-based Routing in the Internet. *RFC2386, IETF,* 1998.
4. Reddy, T. B., I. Karthigeyan, B. S. Manoj, and C. S. R. Murthy. "Quality of service provisioning in ad hoc wireless networks: A survey of issues and solutions." *Ad Hoc Networks,* Jan. 4, 2006, 83–124.
5. Saunders, S. *Antennas and Propagation for Wireless Communication Systems Concepts and Design.* Wiley, 1999.
6. Yang, Y., and R. Kravets. "Contention-aware admission control for ad hoc networks." *Mobile Computing, IEEE Transactions,* July–Aug. 4, 2005, 363–377.
7. Grossglauser, M., and D. N. C. Tse. "Mobility increases the capacity of ad hoc wireless networks." *Networking, IEEE/ACM Transactions,* Aug. 10, 2002, 477–486.
8. Neely, M. J., and E. Modiano. "Capacity and delay tradeoffs for ad hoc mobile networks." *Information Theory, IEEE Transactions* 51, (2005): 1917–1937.
9. Haenggi, M., and D. Puccinelli. "Routing in ad hoc networks: A case for long hops. *Communications Magazine, IEEE* 43, (2005): 93–101.
10. Perkins, C. E., and P. Bhagwat. "Highly Dynamic Destination-Sequenced Distance-Vector Routing (DSDV) for Mobile Computers." *Proceedings of the Conference on Communications Architectures, Protocols and Applications SIGCOMM '94,* Oct. 24, 1994, 234–244.
11. Stine, J. A., and G. de Veciana. "A paradigm for quality-of-service in wireless ad hoc networks using synchronous signaling and node states." *Selected Areas in Communications, IEEE Journal,* Sept. 22, 2004, 1301–1321.
12. Kim, D., C.-H. Min, and S. Kim. "On-demand SIR and bandwidth-guaranteed routing with transmit power assignment in ad hoc mobile networks." *Vehicular Technology, IEEE Transactions* 53, (2004): 1215–1223.
13. Perkins, C. E., and E. M. Royer. "Ad hoc On-Demand Distance Vector Routing." *Proceedings of the 2nd IEEE Workshop, Mobile Computing Systems and Applications,* New Orleans, LA, 1999, 90–100.

14. Wang, N.-C., and C. Y. Lee. "A reliable QoS aware routing protocol with slot assignment for mobile ad hoc networks. *Journal of Network and Computer Applications* 32, (2009): 1153–1166.

15. Lin, C. R., and J.-S. Liu. "QoS routing in ad hoc wireless networks." *IEEE JSAC*, Aug, 17, 1999, 1426–1438.

16. Calafate, C. T., M. P. Malumbres, J. Oliver, J. C. Cano, and P. Manzoni. "QoS support in MANETs: A modular architecture based on the IEEE 802.11e technology." *Circuits and Systems for Video Technology, IEEE Transactions*, May, 19, 2009, 678–692.

17. Van der Auwera, G., P. T. David, and M. Reisslein. "Traffic and quality characterization of single-layer video streams encoded with the H.264/MPEG-4 advanced video coding standard and scalable video coding extension." *Broadcasting, IEEE Transactions* 54, (2008): 698–718.

18. Chaparro, P. A., J. Alcober, J. Monteiro, C. T. Calafate, J. C. Cano, and P. Manzoni. "Supporting Scalable Video Transmission in MANETs through Distributed Admission Control Mechanisms." Parallel, Distributed and Network-Based Processing (PDP), 2010 18th Euromicro International Conference, Feb. 2010, 238–245.

19. Hanzo, II, L., and R. Tafazolli. "Throughput Assurances through Admission Control for Multi-hop MANETs." Personal, Indoor and Mobile Radio Communications, 2007. PIMRC 2007. IEEE 18th International Symposium, Sept. 1–5, 2007.

20. Hanzo, II, L., and R. Tafazolli. "Quality of service routing and admission control for mobile ad-hoc networks with a contention-based MAC layer." *Mobile Adhoc and Sensor Systems (MASS), 2006 IEEE International Conference*, Oct. 2006, 501–504.

21. Hanzo, L., and R. Tafazolli. "QoS-aware routing and admission control in shadow-fading environments for multirate MANETs." *Mobile Computing, IEEE Transactions*, May, 10, 2011, 622–637.

22. Pei, Y., and V. Ambetkar. "Distributed flow admission control for multimedia services over wireless ad hoc networks." *Wireless Personal Communication* 42, (2007): 23–40.

23. Johnson, D., Y. Hu, and D. Maltz. The Dynamic Source Routing Protocol (DSR). *IETF Internet Draft*, Oct. 2007.

24. De Renesse, R., V. Friderikos, and H. Aghvami. "Cross-layer cooperation for accurate admission control decisions in mobile ad hoc networks." *Communications, IET*, Aug. 1, 2007, 577–586.

25. Sarma, N., and S. Nandi. "Route stability based QoS routing in mobile ad hoc networks." *Wireless Personal Communication* 54, (2009): 203–224.

26. Tao, J., G. Bai, H. Shen, and L. Cao. ECBRP: An efficient cluster-based routing protocol for real-time multimedia streaming in MANETs." *Wireless Personal Communications*, May 2010, 1–20.

27. Jiang, M., J. Li, and Y. C. Tay. "Cluster Based Routing Protocol (CBRP)." *Internet Draft, MANET Working Group*, July, 1999.

28. Ivascu, G. I., S. Pierre, and A. Quintero. "QoS routing with traffic distribution in mobile ad hoc networks." *Computer Communication* 32, (2009): 305–316.

29. Wang, M., and G.-S. Kuo. "An Application-Aware QoS Routing Scheme with Improved Stability for Multimedia Applications in Mobile Ad Hoc Networks. *Vehicular Technology Conference, 2005. VTC-2005-Fall. 2005 IEEE*, Sept. 2005, 1901–1905.

30. Chen, Q., Z. M. Fadlullah, X. Lin, and N. Kato. "A clique-based secure admission control scheme for mobile ad hoc networks (MANETs)." *Journal of Network and Computer Applications* 34, (2011): 1827–1835.

31. Stojmenovic, I., A. Nayak, and J. Kuruvila. "Design guidelines for routing protocols in ad hoc and sensor networks with a realistic physical layer." *Communications Magazine, IEEE* 43, (2005): 101–106.

32. Karpinski, S., E. M. Belding, and K. C. Almeroth. "Wireless traffic: The failure of CBR modeling." *Broadband Communications, Networks and Systems, 2007. BROADNETS 2007*, Sept. 2007, 660–669.

Additional Reading

Boukerche, A. *Algorithms and Protocols for Wireless, Mobile Ad Hoc Networks*. Wiley-IEEE Press, November, 2008.

Fiedler, M., T. Hossfeld, and T. G. Phuoc. "A generic quantitative relationship between quality of experience and quality of service." *Network, IEEE*, March_April 24, 2010, 36–41.

Abolhasan, M., T. Wysocki, and E. Dutkiewicz. "A review of routing protocols for mobile ad hoc networks. *Ad Hoc Networks* 2, (2004): 1–22.

Tarique, M., K. E. Tepe, S. Adibi, and S. Erfani. "Survey of multipath routing protocols for mobile ad hoc networks." *Journal of Network and Computer Applications* 32, (2009): 1125–1143.

Hanzo, L., and R. Tafazolli. "Admission control schemes for 802.11-based multi-hop mobile ad hoc networks: A survey." *IEEE Communications Surveys and Tutorials*, Oct. 11, 2009, 78–108.

Chen, L., and W. B. Heinzelman. "A survey of routing protocols that support QoS in mobile ad hoc networks." *Network, IEEE*, Nov.–Dec. 21, 2007, 30–38.

Lindeberg, M., S. Kristiansen, T. Plagemann, and V. Goebel. "Challenges and techniques for video streaming over mobile ad hoc networks." *Multimedia Systems* 17, (2011): 51–82.

Reddy, T. B., I. Karthigeyan, B. S. Manoj, and C. Siva Ram Murthy. "Quality of service provisioning in ad hoc wireless networks: A survey of issues and solutions." *Ad Hoc Wireless Networks* 4, (2006): 83–124.

Hanzo, II, L., and R. Tafazolli. "A survey of QoS routing solutions for mobile ad hoc networks," *Communications Surveys and Tutorials, IEEE* 9, no. 2, (2007): 50–70.

Index

Page numbers followed by *f* and *t* indicate figures and tables, respectively.